# Endodontic Microbiology

# Endodontic Microbiology

**Edited by**

**Ashraf F. Fouad**
BDS, DDS, MS
Professor and Chairman, Department of Endodontics, Prosthodontics and Operative Dentistry
Director, Advanced Specialty Program in Endodontics
Dental School, University of Maryland
Baltimore, MD, USA

**WILEY-BLACKWELL**

A John Wiley & Sons, Ltd., Publication

Edition first published 2009
© 2009 Wiley-Blackwell

Blackwell Publishing was acquired by John Wiley & Sons in February 2007. Blackwell's publishing program has been merged with Wiley's global Scientific, Technical, and Medical business to form Wiley-Blackwell.

*Editorial Office*
2121 State Avenue, Ames, Iowa 50014-8300, USA

For details of our global editorial offices, for customer services, and for information about how to apply for permission to reuse the copyright material in this book, please see our website at www.wiley.com/wiley-blackwell.

**Disclaimer**

*Library of Congress Cataloging-in-Publication Data*

Endodontic microbiology / [edited by] Ashraf F. Fouad.
    p. ; cm.
  Includes bibliographical references and index.
  ISBN 978-0-8138-2646-2 (hardback : alk. paper)
  1. Dental pulp–Diseases–Microbiology. I. Fouad, Ashraf F.
  [DNLM: 1. Dental Pulp Diseases–drug therapy. 2. Dental Pulp Diseases–microbiology.
3. Anti-Infective Agents–therapeutic use. 4. Periapical Diseases–drug therapy. 5. Periapical Diseases–microbiology.
6. Root Canal Therapy. WU 230 E563 2009]
  RK351.E48 2009
  617.6′342071–dc22

                                                                                    2008040238

A catalog record for this book is available from the U.S. Library of Congress.

Set in 9/12 pt Dutch801BT by Aptara® Inc., New Delhi
Printed in Singapore by Markono Print Media Pte Ltd

1   2009

# Dedication

To Amal, Fikry, Lori, Amani, Anthony and Edward; thank you for providing me the opportunity, the inspiration, the motivation and the love.

*Ashraf F. Fouad*

# Contents

# Contributors

**Editor**

**Ashraf F. Fouad**, BDS, DDS, MS
Professor and Chairman
Department of Endodontics,
  Prosthodontics and Operative
  Dentistry
Director, Postgraduate
  Endodontics
Dental School, University of
  Maryland
Baltimore, MD, USA

**Authors**

**B. Güniz Baksi**, DDS, PhD
Professor
Department of Oral Diagnosis
  and Radiology
School of Dentistry
Ege University
Izmir, Turkey

**J. Craig Baumgartner**, DDS, MS,
  PhD
Chairman, Department of
  Endodontology
School of Dentistry
Oregon Health & Science
  University
Portland, OR, USA

**Domenick P. Coletti**, DDS, MD,
  FACS
Associate Professor and Division
  Chief
Department of Oral and
  Maxillofacial Surgery

University of Maryland Medical
  Center
University of Maryland School of
  Dentistry
Baltimore, MD, USA

**Gunnar Dahlén**, BSc, DDS, PhD,
  (Dr Odont)
Professor and Chairman
Department of Oral Microbiology
Institute of Odontology
Sahlgrenska Academy at
  Göteborg University
Goteborg, Sweden

**Anibal Diogenes**, DDS, MS, PhD
Department of Endodontics
UTHSCSA Dental School
San Antonio, TX, USA

**Shimon Friedman**, DMD
Professor and Head, Discipline of
  Endodontics
Director, MSc Program in
  Endodontics
University of Toronto Faculty of
  Dentistry
Toronto, Ontario, Canada

**Ken M. Hargreaves**, DDS, PhD
Professor and Chair
  Department of Endodontics
UTHSCSA Dental School
San Antonio, TX, USA

**George T.-J. Huang**, DDS, MSD,
  DSc
Associate Professor
Division of Endodontics
University of Maryland
Baltimore College of Dental
  Surgery
Baltimore, MD, USA

**Robert M. Love**, BDS, MDS,
  PhD, FRACDS
Professor, Head of Department
Department of Oral Diagnostic
  and Surgical Sciences
School of Dentistry
University of Otago
Dunedin, New Zealand

**Robert A. Ord**, DDS, MD, FRCS,
  FACS, MS
Chairman and Professor
Department of Oral and
  Maxillofacial Surgery
University of Maryland Medical
  Center
University of Maryland School of
  Dentistry
Baltimore, MD, USA

**Dag Ørstavik**, Cand Odont, Dr
  Odont
Professor
Head, Department of
  Endodontics
Institute for Clinical Dentistry
University of Oslo
Oslo, Norway

**Isabela N. Rôças**, DDS, MSc, PhD
Assistant Professor
Department of Endodontics
Faculty of Dentistry
Estácio de Sá University
Rio de Janeiro, RJ, Brazil

**Mohammad Sabeti, DDS, MA**
Diplomate, American Board of Endodontics
Associate Professor
University of Southern California
Los Angeles, CA, USA

**Christine M. Sedgley**, BDS, MDSc, MDS, FRACDS, PhD
Assistant Professor
Department of Cariology, Restorative Sciences, and Endodontics
School of Dentistry
University of Michigan
Ann Arbor, MI, USA

**Bilge Hakan Sen**, DDS, PhD
Professor, Division of Endodontology
School of Dentistry
Ege University
Izmir, Turkey

**José F. Siqueira**, Jr, DDS, MSc, PhD
Chairman and Professor
Department of Endodontics
Faculty of Dentistry
Estácio de Sá University
Rio de Janeiro, RJ, Brazil

**John R. Smith**, PhD
Professor of Pharmacology
Oregon Health & Science University
Portland, OR, USA

**Leif Tronstad**, DMD, MS, PhD
Emeritus Professor
Faculty of Dentistry
University of Oslo
Oslo, Norway

**William Wade**, BSc, PhD
Professor and Head
Infection Research Group
King's College London Dental Institute
London, UK

**Tuomas Waltimo**, DDS, PhD
Professor
Institute for Preventive Dentistry and Oral Microbiology
School of Dentistry
University of Basel
Basel, Switzerland

**Matthias Zehnder**, DMD, PhD
Department of Preventive Dentistry, Periodontology and Cariology
University of Zurich
Center of Dental Medicine
Zurich, Switzerland

# Preface

Endodontic infections are very prevalent, because they mostly represent complications of dental caries and its treatment, as well as traumatic injuries to teeth, which are all very prevalent occurrences. Collectively, they represent the majority of dental infections that present with significantly acute local and systemic signs and symptoms. This is the first textbook devoted to the study of endodontic infections, which hitherto has been limited to isolated single chapters in endodontic textbooks. This textbook is intended to provide a collection of work showing the state of the knowledge in this field. It is also intended to provide some research questions and hypotheses that, hopefully, will stimulate more efforts to understand the disease process and identify effective treatment methods.

The study of endodontic microbiology has been complicated by difficulty in epidemiological data in obtaining adequate endodontic diagnosis on large numbers of nonpatient populations. In addition, sampling is a major challenge in endodontics. Contamination from the tooth surface, caries, or saliva must first be avoided. Access to the potentially very complex root canal anatomy and disruption of biofilm on the majority of canal walls in these areas are necessary. It is almost impossible to differentiate specimens obtained from the apical and coronal portions of the root canals; thus, the effect of location of microflora within the canal is poorly understood, and can only be studied in teeth that are extracted. Finally, sampling after completion of treatment to assess effectiveness of treatment and determine the long-term outcome risk is complicated by the fact that only the areas that reached could be sampled.

The differences in sensitivity between traditional culturing and modern molecular methods are especially important in endodontic microbiology, because the endodontic specimen has so little material, and sensitivity, therefore, plays a major role in microbial identification. The description of traditional bacterial pathogens and their virulence factors represents most of the available literature today. The contributions of the not-yet-cultivated bacteria and the bacteria rendered temporarily uncultivable by traditional treatment methods have not been adequately studied. Likewise, we are just beginning to understand some of the contributions of fungi and viruses to the pathogenesis of endodontic infections.

The debate on viable versus dead microorganisms that are detected by molecular techniques must be resolved by using more accurate technologies that assess microbial counts, their viability, and their pathogenicity. Likewise, consistent and stringent methodologies, including sequencing of amplification products, are essential for assuring accurate results and enabling comparisons among studies.

Persistent endodontic pathosis may be due to persistent infection or new infection after treatment. Sampling of apical lesions during periapical surgery is complicated by the lack of sterility of the surgical field. Therefore, the microbiology of nonhealing endodontic cases is still in its infancy at this time.

It is clear that in order to determine effective treatment modalities, better sampling and identification techniques must be employed and more adequately designed outcome studies need to be performed.

Finally, the relationship between endodontic pathosis and systemic disease must be more comprehensively studied. Endodontic infections were historically thought to contribute to numerous systemic diseases. While the potential for systemic spread of an acute endodontic infection is well-known and documented, earlier studies have failed to demonstrate that chronic endodontic infections contribute to systemic diseases. However, these hypotheses must be now reexamined that we have more accurate research tools. In addition, the creation of large patient databases for longitudinal analysis of treatment outcomes and their relationships with systemic disease will be imperative in future studies that address this issue.

*Ashraf F. Fouad*

# Endodontic Microbiology

# Chapter 1
# Microbial Perspectives in the Twenty-First Century

*William Wade*

---

| | | | |
|---|---|---|---|
| 1.1 | Introduction | 1.6 | Bacterial–bacterial communication |
| 1.2 | Genomics | 1.7 | Host–bacterial interactions |
| 1.3 | Molecular microbial ecology and the | 1.8 | Complex infectious diseases |
| | study of uncultivable bacteria | 1.9 | The future |
| 1.4 | Intraspecies variation | 1.10 | References |
| 1.5 | Metagenomics | | |

---

## 1.1 Introduction

The final quarter of the nineteenth century was arguably the golden age of medical microbiology. The groundbreaking work of Pasteur, Koch, and others led to the development of broth and agar media that were able to support the growth in the laboratory of major bacterial pathogens affecting man. The ability to grow these organisms in pure culture led to the production of vaccines for many of the diseases they caused. These advances, and the subsequent discovery and development of antimicrobials, led to the mistaken belief that infectious disease had been beaten.

Of course, it is now realized that this optimistic viewpoint is not justified, not least because of the rapid emergence of bacterial resistance to antimicrobials. Indeed, the consensus view is that the battle against bacterial resistance is currently being lost, because of both the difficulty and costs associated with developing new antimicrobials and the indiscriminate use of those currently available. The predicted ultimate failure of antimicrobial strategies has led to renewed interest in elucidating the pathogenic mechanisms used by bacteria to cause disease, with the ultimate aim of devising new methods of disease prevention and treatment.

At the same time, interest in the microbial populations of the earth has been intense and new techniques have become available to characterize the bacterial communities found in every ecosystem on the planet. These have revealed the quite astonishing diversity of microbial life on earth and the extreme complexity of most bacterial communities. Furthermore, the extent of subspecific diversity is only now being fully appreciated. Bacteria readily exchange DNA and can "shuffle" their own genomes to generate diversity, with the ultimate aim of responding and adapting to environmental changes. As discussed below, bacteria in communities communicate with each other and, in the case of commensals living with plants and animals, their hosts. These interactions operate at various levels and can be remarkably sophisticated. The twenty-first century will be a period of tremendous advances in our understanding of the microbial world.

The aim of this chapter is to review recent developments in microbiology and to highlight selected areas that are likely to change our conceptual view of infectious disease as a whole and oral and endodontic infections in particular. Inevitably, a single short chapter cannot provide a comprehensive overview of an entire discipline, but the interlinked topics covered are those that will undoubtedly change our view of the microbial world and its relationship with the human host.

## 1.2 Genomics

The sequence of the human genome was published in 2001. The benefits of this outstanding achievement are now being realized with the identification of genes responsible for or causing a predisposition for, a large number of diseases (Wellcome Trust Case Control Consortium 2007). At the same time, and largely possible because of the technical advances made as part of the human genome sequencing effort, genomes of other organisms are being sequenced, including those of bacteria.

As of October 2007, 548 bacterial genomes had been sequenced and published, while over 1,400 more were in progress (for more information, see http://www.genomesonline.org/). As expected, the data obtained have revealed the enormous genetic potential contained within bacterial genomes; in each genome sequenced, around a third of the genes present have been novel and the function of a significant proportion remains unknown.

The availability of genome sequence data is allowing a far more robust bacterial classification to be constructed than previously possible. Bacterial taxonomy was once based purely on phenotypic characters and was very inexact because of the difficulties involved in obtaining and interpreting such data compared to plants and animals where differences in phenotype are far more obvious. In recent years, genetic information has been increasingly used, but on a limited scale, and typically only the sequences of the 16S rRNA and other housekeeping genes have been used. New methods are now being introduced to make use of the sequence data available for complete genomes (Konstantinidis and Tiedje 2005). In general, the results of using such methods have supported the 16S rRNA gene taxonomy at species and genus level but, in addition, have provided improved clarity of the relationships among and between the higher taxonomic ranks, where substantial overlap between ranks has been observed.

The results of the analysis of some genomic data have been extremely surprising. A Gram-positive coccus found in amoebae could not be identified by the conventional molecular analysis of 16S ribosomal RNA (rRNA) gene sequencing since no ribosomal genes could be amplified for sequencing. Genomic data explained this difficulty by revealing that the organism was actually a virus, the largest yet discovered. Now named *Mimivirus*, the large virus particles are up to 0.8 µm in diameter, the size of many bacteria. It primarily infects amoebae, but has been implicated as a cause of pneumonia on serologic grounds and has caused a laboratory-acquired pneumonia in a researcher (Raoult et al. 2007). At the other end of the bacterial scale, members of the genus *Epulopiscium*, found in the intestine of certain surgeonfish (Angert et al. 1993), have been discovered that are visible with the naked eye.

In addition to correctly identifying evolutionary oddities, genomic data have identified numerous novel biochemical pathways with the potential for exploitation. Among these are some novel antimicrobials although the range of targets within bacterial cells that has arisen by natural evolution is rather narrow. A more promising avenue to the development of novel antimicrobials is to use genomic data to identify novel targets for antimicrobial treatments (Pucci 2006). Predictions can be made from genome data as to how essential a given gene is to an organism and therefore how disrupting the gene would affect the vitality of the organism. These predictions can then be tested in an appropriate manner experimentally using a wide range of methods that have been developed in response to the availability of genomic data. These include random mutagenesis mediated by transposons or insertion of plasmids, targeted gene disruption or in vivo techniques such as signature-tagged mutagenesis and in vivo expression technology. Structural genomics, where sequence data is used to predict the structure of essential bacterial proteins, is also being used to identify potential targets for antimicrobials. Finally, comparative genomics can be used to identify common features of pathogens affecting a particular body site to custom design antimicrobials for specific purposes, for example, respiratory tract infection.

New sequencing technologies have been and are being introduced that will bring the ability to sequence bacterial genomes within the reach of individual laboratories. The challenge, though, is to accurately interpret the data, and new methods of bioinformatic analysis will be required. The information obtained thus far has been of extraordinary value in understanding the role of pathogenic bacteria in disease and is the fundamental basis of other new technologies such as transcriptomics and proteomics. The next task will be to understand how gene products interact both within a bacterial cell and in response to external stimuli from the environment and other organisms.

## 1.3 Molecular microbial ecology and the study of uncultivable bacteria

Almost without exception, oral infections are polymicrobial in nature and difficult to study because around half of the bacteria present in the oral cavity cannot be grown using conventional culture media. It has long been recognized that not all bacteria from a given habitat can be cultured on artificial media in the laboratory. Indeed, it has been estimated that less than 2% of bacteria on earth can be cultured.

Methods for the characterization of complex bacterial communities were developed as a consequence of the use of DNA sequence data for the construction of evolutionary trees. This was done by comparing the sequences of genes encoding essential functions, the so-called housekeeping genes that are found in all cellular organisms. The gene most commonly used to date has been that encoding the small subunit (16S) rRNA molecule. Ribosomes have the essential function of translating messenger RNA into amino acid chains and, because of the need to preserve function, have evolved slowly. Some of the regions of the gene have changed very little over time and are therefore virtually identical in all bacteria. These regions are very useful for the design of *universal* PCR primers that can amplify the gene from a wide range of different bacteria. Other regions are more variable and can be used to discriminate between organisms, almost to species level. Woese and colleagues used small subunit rRNA comparisons to construct a tree of life (Fig. 1.1), which showed that bacteria had evolved into two domains, the *Archaea* and *Bacteria*, while eukaryotic organisms fell into a single third domain, the Eukarya (Woese 1987). It was originally thought that organisms found in the domain *Bacteria* were those found in *normal* environments, while the *Archaea* were present in extreme environments such as the deep sea and associated with volcanoes and so on. However, these associations have since been shown not to be true and members of *Archaea* are now known to be widely distributed, and an archaeal genus, *Methanobrevibacter*, can be found in the human mouth.

One major consequence of the availability of this tree is that unknown organisms of any type can be identified simply by sequencing their rRNA gene and adding the sequence to the tree or by directly comparing the sequence with the hundreds of thousands of bacterial sequences held in the sequence databases. Complex bacterial communities can be characterized by the PCR, cloning, and sequencing of 16S rRNA.

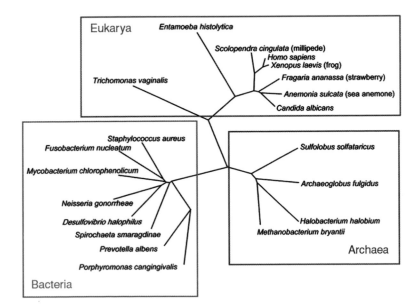

**Fig. 1.1** Phylogenetic tree showing representatives of the domains Eukarya, Archaea, and Bacteria.

Such studies have been performed with samples from the healthy and diseased human mouth and are described in more detail in Chapter 4. A common finding of every study to date has been to confirm that around half of the oral microbiota is uncultivable. Currently, around 36 divisions within the domain *Bacteria* are recognized, 13 of which have been detected in the human mouth. Some divisions have no cultivable representatives including TM7 and SR1, clones of which have been found in the mouth. There are also deep branches of the phyla Bacteroidetes and Firmicutes from which no cultivable members have yet been identified.

Major efforts are now being made to improve our understanding of currently unculturable bacteria. These include the development of new culture media that better mimic the natural environment. Very often, laboratory culture media are far richer in nutrients than the natural habitat, and the use of dilute media or filtered natural substrate has been successful in culturing previously uncultured organisms. This approach has not been applied systematically to the study of oral unculturable bacteria, but should be possible. In addition, methods to sequence the genomes of unculturable bacteria are being developed. The availability of 16S rRNA gene sequence data for these organisms makes it possible to design specific DNA probes that can be labeled with fluorochromes. The probes can then be used in conjunction with separation methods such as flow cytometry, cell sorting, or isolation with optical tweezers to yield small numbers of the target cells in pure culture. Whole genome amplification techniques are now available to obtain sufficient DNA for sequencing from a single cell. The successful partial sequencing of a representative of the TM7 division isolated from a complex community has recently been reported (Podar et al. 2007).

## 1.4 Intraspecies variation

The majority of microbiological diagnostic methods identify the target organism to species level. However, it is now recognized that individual strains within a species often vary markedly in their virulence. Within a species, some strains may be pathogenic while others are harmless. The extent of the genetic variation within species, however, has only been fully realized by the sequencing of the genomes of multiple representatives of the same species.

In one such study, three strains of *Escherichia coli* were compared: the well-known harmless laboratory strain K12, an enterohemorrhagic serotype O157 strain of the group associated with beef products, and a uropathogenic strain. It was found that they had only 39% of their genes in common, a surprisingly small number (Welch et al. 2002). These common genes encoded the functions that gave the strains their identity as members of the species *E. coli*, while the remaining genes gave them the ability to colonize particular body sites and/or damage the host by means of a specialized set of virulence factors appropriate for their natural habitat and lifestyle. Genes acquired from other organisms by horizontal gene transfer can be critical to that organism's behavior, and in the past may have been the reason a species was given a particular name. For example, if the mainly harmless environmental organism *Bacillus cereus* acquires plasmids pXO1 and pXO2, which carry genes coding for four toxins and the enzymes required to make a capsule, it becomes *Bacillus anthracis*, the causative agent of anthrax.

This work has given rise to some new genomic concepts. The core genome is that shared by all strains of the species, while the peripheral or accessory genome includes genes found in some strains but not others, but which nonetheless may be important in pathogenesis. Some bacteria go further and have two chromosomes; in this case, one normally encodes housekeeping genes and the second genes that confer fitness for competition in the environment.

The range of genes encoded by the peripheral genome can be extensive. In a study of the genome sequences of eight *Streptococcus agalactiae* strains, the authors calculated the number of strains of the species that would have to be sequenced to reveal the full genetic diversity of the species (Tettelin et al. 2005). The result was infinity. In other words, *Streptococcus galactiae* can incorporate DNA into its genome from such a wide range of sources so that all the possible genes that could be found within this species will never be known.

The implications of these findings are significant. Although much work has been invested in the development of rapid assays to detect the presence of specific organisms in clinical samples, including those collected from oral diseases, the association of a species with disease may be insufficient for diagnostic purposes. Detection of the presence (and expression) of specific virulence genes may be required to provide a meaningful microbiological diagnosis. This will clearly be difficult for diseases where the virulence determinants important in a disease are currently unknown or where multiple virulence mechanisms are operating.

## 1.5 Metagenomics

Since individual bacterial strains can vary greatly in their genetic composition and the assignation of an isolate to a species alone is likely to give a poor indication of its pathogenicity, alternative methods of analysis need to be developed to determine the role of bacterial communities in human diseases. New methods are particularly needed for complex diseases since the bacterial communities associated with the mucous membranes where these diseases primarily occur are so diverse that their routine characterization is not practicable. One novel approach does not attempt to isolate and purify all of the component species and strains, but rather it considers the whole community and all of its constituent genes as a whole. The bacterial community found at a habitat is termed the *microbiome* and the genetic material of the community members is the *metagenome* (Rondon et al. 2000).

An overview of typical metagenomic analyses is shown in Fig. 1.2. The first stage in any such analysis is to extract DNA from all of the bacteria present in the sample. This is done carefully to prevent shearing of the DNA since large fragments are required. The fragments are then incorporated into different vectors

depending on the type of analysis to be performed. If random sequencing is to be performed in order to obtain the metagenomic sequence, then conventional plasmid vectors are normally used to obtain inserts that can be easily sequenced. Larger fragments are required when the intention is to express the cloned DNA. In this case, bacterial artificial chromosomes (BACs) are normally used since these allow DNA fragments up to 100 kb in length to be stably maintained in an *E. coli* host. Alternatively, fosmid libraries can be constructed in which inserts up to around 40 kb can be successfully cloned.

Most antibiotics obtained from natural sources are synthesized by bacteria found in soil. One approach to the discovery of new antimicrobials is therefore to use metagenomic techniques. The inserts that can be cloned into BAC vectors are large enough to include complete pathways for the synthesis of novel antimicrobials. Using such techniques, a number of new antibiotics have been discovered from metagenomic analyses of soil and marine environments. For example, turbomycin A and B were discovered in this way (Gillespie et al. 2002). Interestingly, a single gene was responsible for the activity that was mediated by an interaction between indole, normally produced by *E. coli*, and the gene product. Although this single gene could have been identified by conventional small-insert cloning, the availability of large inserts greatly facilitates the screening process. The success rate in identifying novel antimicrobials in metagenomic libraries has been low; typically, several hundred

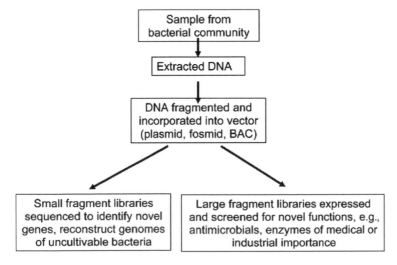

**Fig. 1.2**  Overview of metagenomic analyses.

thousands of clones have to be screened to find one new active compound.

This relative lack of success partly reflects the methodology where *E. coli* is used as the host. Successful expression may require the presence of specific promoters or other accessory molecules and factors such as the G+C content of the insert, and codon usage may adversely affect expression. Thus, in general, cloned fragments of DNA are expressed most easily in their natural host, and the more phylogenetically distant the clone host, the less likely it is that expression will be successful. To overcome this, new vector/host systems are being introduced for the expression of metagenomic libraries to enable a better match between the insert and the host in which it is being expressed. These include *Streptomyces* and *Pseudomonas*, two genera the members of which naturally produce secondary metabolites with properties of interest.

Genome sequences of bacteria are normally obtained from pure cultures. Attempts are now being made to directly sequence the genomes of bacteria, which may be unculturable, from biomass obtained from natural communities. This has been successfully performed for the relatively restricted microbiota found in a subterranean acid mine drainage biofilm where near-complete bacterial genomes were reconstructed for two previously uncultured bacteria (Tyson et al. 2004). More ambitiously, Venter et al. (2004) randomly sequenced the metagenome of water specimens collected from the Sargasso Sea. Over 1,000 billion base pairs of DNA were sequenced, which were found to be derived from 1,800 species including 148 not previously characterized. More than 1 million novel genes were found. It proved difficult, however, to reconstruct complete genomes from these data because of the number of closely related species present and the large numbers of mobile elements with high levels of sequence similarity. Although extremely high-volume DNA sequencing is now feasible, new tools are urgently needed to help with the analysis and interpretation of the data.

## 1.6 Bacterial–bacterial communication

Originally thought of as simple dumb solitary creatures, it is now known that bacteria live together in communities with a number of features in common with multicellular organisms. The basis for the cooper-

ation of individual bacterial cells within a community is communication. Communication is mediated by the production of signaling molecules often generically described as quorum-sensing molecules, after the first bacterial signaling system to be described.

In many circumstances, the total number of bacterial cells in a community is important to the overall health of the community. In the environment, the availability of nutrients and external stresses are factors that cause the community to behave in a particular way. This may increase or decrease its overall rate of growth, become more motile to move to a new habitat to obtain nutrients, and, once there, switch to a biofilm mode of growth to colonize the new environment. For pathogenic bacteria of exogenous source, bacterial–bacterial communication is particularly important. The first bacteria to colonize the host will necessarily be present in small numbers and will want to multiply without alerting the host to their presence in order to avoid the host's immune system. Since virulence factors such as protein toxins are typically highly antigenic, the pathogen will not produce them until the community is sufficiently numerous to resist the host's defenses. Once this *quorum* has been achieved, the members of the community will turn on the production of their virulence genes in order to damage the host and cause disease.

The molecular basis for quorum sensing was first elucidated for the bioluminescent marine bacterium *Vibrio fischeri*. This organism is commonly found as a symbiont in the light-producing organs of luminescent fish or squid. The signaling mechanism is a two-component system; the *luxI* gene produces autoinducer 1 (AI-1), an acyl homoserine lactone. This is produced constitutively, but when sufficient numbers of *V. fischeri* are present, the high concentration of AI-1 enables binding to the receptor, the product of *luxR*, which activates transcription of the luciferase operon and leads to production of the light-emitting compounds.

Following its discovery in *V. fischeri*, AI-1 analogs have been found in a wide range of Gram-negative bacteria. The AI-1 of a particular species is normally specific to that species so that cross talk is avoided within multispecies communities.

Gram-positive bacteria also produce signaling molecules, but those so far described have all been peptides derived from larger precursors by posttranslation modification. One important group of signaling molecules in Gram-positives is of those that induce competence. Competence is the ability of bacteria to take up DNA present in the environment. This is an important mechanism of genetic exchange and is particularly common among members of the oral microbiota such as the oral streptococci. In many streptococci, the precursor molecule is Com C that is modified as it is transported out of the cell by the ComAB transporter. The signaling molecule itself is the C-terminal end of Com C and is termed the *competence-stimulating peptide* (CSP). When the bacterial numbers reach their quorum, CSP binds to the receptor, the histidine kinase Com D, which then stimulates the response regulator Com E.

In addition to the species-specific quorum-sensing mechanisms described above, bacteria also make use of nonspecific systems that allow general communication between bacteria in communities, across species barriers. One such molecule, AI-2, is produced by, and can be detected by, a wide range of both Gram-negative and Gram-positive bacteria. AI-2 forms spontaneously from 4,5-dihydroxy-2,3-pentanedione (DPD), which is a product of the Lux S enzyme in the catabolism of S-ribosylhomocysteine. A number of oral bacteria have been shown to produce AI-2 and it has been found to play an important role in dental plaque formation. For example, *Streptococcus oralis* and *Actinomyces naeslundii* are known to co-aggregate early in the development of dental plaque biofilms and grow together in in-vitro models, forming a profuse plaque with physical interaction between cells of the two species. A *luxS* mutant of *S. oralis* that did not produce AI-2, however, did not form such biofilms with *A. naeslundii*, while the mutualistic activity was restored by *luxS* complementation (Rickard et al. 2006).

Another group of bacterial cell-signaling molecules are the family proteins related to resuscitation-promoting factors (Rpf). Originally discovered in *Micrococcus luteus* where they were able to revive *M. luteus* cells that had entered a dormant phase, they were subsequently found to be widespread among members of the phylum *Actinobacteria*, the high G+C Gram-positives (Mukamolova et al. 1998). Interestingly, the growth of *Mycobacterium tuberculosis*, which is normally extremely slow *in vitro*, is greatly stimulated by Rpf. Rpf is a protein, structurally similar to lysozyme, and can exert its effects at extremely low concentrations. It has therefore been termed a bacterial cytokine because of its resemblance to mammalian cytokines that have similar properties. The molecular basis of its action has yet to be determined, but it would ap-

pear to cleave the peptidoglycan of dormant cells and either release a second messenger or physically allow the cells to resume growth. Paralogues of Rpf have been found among the phylum *Firmicutes*, the low G+C Gram-positives, and a similar molecule has recently been reported in *Salmonella*, a Gram-negative organism.

Novel mechanisms of bacterial communication are being discovered all the time, and it is extremely likely that a network of sophisticated interactions exists among the bacterial community in dental plaque. Many of these are clearly relevant to endodontic infection and the survival of bacteria under restorations and in the treated root canal. The realization that vegetative bacterial cells can go into a dormant state, distinct from endospore production, and survive for many years may explain how bacteria survive under restorations or despite calcium hydroxide treatment in the root canal. Furthermore, a change in the environment may stimulate the production of broad-range growth stimulation factors that cause the community to undergo rapid growth, causing damage to the affected tooth and pain to the patient.

## 1.7 Host–bacterial interactions

All plants and animals are colonized by bacteria. Mammals are born sterile but extremely quickly become colonized with the microbiota characteristic for their species. The commensal microbiota associated with mammals has evolved over millions of years, and it is possible to reconstruct the evolution of the commensal microbiota in parallel with each mammalian host, the phenomenon of cospeciation. Thus, for the majority of bacteria found in the human mouth, there are versions of that organism found in other animals. For example, among the mutans group streptococci, associated with dental caries, *S. mutans* and *S. sobrinus* are found in human, *S. ferus* and *S. rattus* in rats, *S. cricetus* in hamsters, and so on.

The commensal microbiota is extremely important. First, there are a lot of it. It has been calculated that of the $10^{14}$ cells in the human body, only $10^{13}$ are human with the remainder primarily bacterial (Savage 1977). Thus, around 90% of the cells in the human body are bacterial. Our normal microbiota protects us from exogenous infection via the phenomenon of colonization resistance. All external surfaces of the body are normally covered in bacteria and thus potential binding sites for exogenous pathogens are blocked. In addition, members of the normal microbiota can produce antimicrobial substances that inhibit the growth of other organisms. However, if the commensal microbiota is disturbed, infection can result. For example, it is well-known that treatment with antibiotics can disrupt the normal microbiota to such a degree that opportunistic infection with other organisms such as coliform bacteria or the yeast *Candida albicans* can occur. Vaginal thrush and antibiotic sore tongue are examples of such conditions.

The presence of the normal microbiota is essential for the proper development of the gut. The intestinal microbiota is highly diverse with over 1,000 bacterial species present. One commonly found species, *Bacteroides thetaiotaomicron*, has profound effects on the development of the blood supply to the gut. In germfree mice, introduction of *B. thetaiotaomicron* induced intestinal angiogenesis (Stappenbeck et al. 2002). Interestingly, the genome of *B. thetaiotaomicron* includes an unusually high number of genes encoding signaling molecules of both the one- and two-component types (Xu et al. 2003). Clearly *B. thetaiotaomicron* is just one of the many species found in the large intestine, and it remains to be seen whether this species is truly unusual in this respect or whether most gut bacteria communicate in this way. What is clear however is that such communication networks are typical of multispecies bacterial communities and that given the significant associations between the composition of the gut microbiota and intestinal health, communication between the commensal microflora and the host is undoubtedly important.

## 1.8 Complex infectious diseases

"Classical" infectious diseases normally occur when a pathogen infects a susceptible host and produces a specific virulence factor that damages the host in a characteristic way, causing the signs and symptoms of the disease. For many diseases, particularly those associated with the mucous membranes, no single pathogen has been identified, but instead the disease appears to be the result of an aberrant interaction between the host and its normal resident, microbiota. These so-called complex infectious diseases include the inflammatory bowel diseases and oral infections such as chronic periodontitis and, to some extent, endodontic infections.

Host susceptibility is of primary importance in these diseases, but typically the susceptibility is conferred by multiple genes with no single genotype responsible. For oral diseases such as periodontitis, the genes responsible have yet to be discovered although there is growing evidence that increased susceptibility is due to subtle differences in the immune and inflammatory responses. For example, genetic polymorphisms associated with cytokines such as interleukin-1 have been identified, which are associated with increased cytokine secretion and severity of chronic inflammatory disease (Brett et al. 2005). Another key factor is the environment in its widest sense. Host factors such as stress are known to contribute to the severity of complex diseases presumably by adversely affecting the immune system. Diet and social factors such as smoking can also be important, particularly in the principal oral diseases such as dental caries and the periodontal diseases.

It is likely that in the investigation of oral infections and diseases, we have clung too long to the classical infectious disease model and have sought single infectious causes for them in the hope that antimicrobials could be used in a targeted way to treat them. It must, however, be remembered that these diseases are bacterial diseases and the presence of the normal microbiota is required. By mechanisms as yet unknown, it appears that the communication and cooperation between the host and its commensal microbiota breaks down, resulting in damage to the host. Much work on these diseases is therefore currently being focused on better understanding health, the question being that if the human gut is colonized by so many bacteria with the potential to cause disease, how do the majority of individuals remain healthy? Better understanding of how this healthy balance is maintained will permit insights into how disease arises when the homeostasis breaks down. In this way, we can better identify how complex diseases arise.

## 1.9 The future

We are still in the first decade of the twenty-first century, so what do we have to look forward to in terms of how microbiology will impact on our understanding of infectious disease, including oral and endodontic infections? Advances in genetics will undoubtedly give us a far better understanding of individuals' susceptibility to disease, and the challenge will be put into this context with new knowledge of the genetic potential of the commensal microbiota. A plan is in place to sequence the genomes of all the bacteria associated with man—the so-called human microbiome. The greatest challenge of all may well be the analysis and assimilation of all the new data that will become available.

## 1.10 References

Angert ER, Clements KD, and Pace NR. 1993. The largest bacterium. *Nature* 362: 239–41.

Brett PM, Zygogianni P, Griffiths GS, Tomaz M, Parkar M, D'Aiuto F, and Tonetti M. 2005. Functional gene polymorphisms in aggressive and chronic periodontitis. *J Dent Res* 84(12): 1149–53.

Gillespie DE, Brady SF, Bettermann AD, Cianciotto NP, Liles MR, Rondon MR, Clardy J, Goodman RM, and Handelsman J. 2002. Isolation of antibiotics turbomycin A and B from a metagenomic library of soil microbial DNA. *Appl Environ Microbiol* 68: 4301–6.

Konstantinidis KT and Tiedje JM. 2005. Towards a genome-based taxonomy for prokaryotes. *J Bacteriol* 187: 6258–64.

Mukamolova GV, Kaprelyants AS, Young DI, Young M, and Kell DB. 1998. A bacterial cytokine. *Proc Natl Acad Sci USA* 95: 8916–21.

Podar M, Abulencia CB, Walcher M, Hutchison D, Zengler K, Garcia JA, Holland T, Cotton D, Hauser L, and Keller M. 2007. Targeted access to the genomes of low-abundance organisms in complex microbial communities. *Appl Environ Microbiol* 73: 3205–14.

Pucci MJ. 2006. Use of genomics to select antibacterial targets. *Biochem Pharmacol* 71: 1066–72.

Raoult D, La Scola B, and Birtles R. 2007. The discovery and characterization of *Mimivirus*, the largest known virus and putative pneumonia agent. *Clin Infect Dis* 45: 95–102.

Rickard AH, Palmer RJ, Jr, Blehert DS, Campagna SR, Semmelhack MF, Egland PG, Bassler BL, and Kolenbrander PE. 2006. Autoinducer 2: a concentration-dependent signal for mutualistic bacterial biofilm growth. *Mol Microbiol* 60: 1446–56.

Rondon MR, August PR, Bettermann AD, Brady SF, Grossman TH, Liles MR, Loiacono KA, Lynch BA, MacNeil IA, Minor C, Tiong CL, Gilman M, Osburne MS, Clardy J, Handelsman J, and Goodman RM. 2000. Cloning the soil metagenome: a strategy for accessing the genetic and functional diversity of uncultured microorganisms. *Appl Environ Microbiol* 66: 2541–47.

Savage DC. 1977. Microbial ecology of the gastrointestinal tract. *Annu Rev Microbiol* 31: 107–33.

Stappenbeck TS, Hooper LV, and Gordon JI. 2002. Developmental regulation of intestinal angiogenesis by indigenous microbes via Paneth cells. *Proc Natl Acad Sci U S A* 99: 15451–55.

Tettelin H, Masignani V, Cieslewicz MJ, Donati C, Medini D, Ward NL, Angiuoli SV, Crabtree J, Jones AL, Durkin AS, Deboy RT, Davidsen TM, Mora M, Scarselli M, Margarit y Ros I, Peterson JD, Hauser CR, Sundaram JP, Nelson WC, Madupu R, Brinkac LM, Dodson RJ, Rosovitz MJ, Sullivan SA, Daugherty SC, Haft DH, Selengut J, Gwinn ML, Zhou L, Zafar N, Khouri H, Radune D, Dimitrov G, Watkins K, O'Connor KJ, Smith S, Utterback TR, White O, Rubens CE, Grandi G, Madoff LC, Kasper DL, Telford JL, Wessels MR, Rappuoli R, and Fraser CM. 2005. Genome analysis of multiple pathogenic isolates of *Streptococcus agalactiae*: Implications for the microbial "pan-genome." *Proc Natl Acad Sci USA* 102: 13950–55.

Tyson GW, Chapman J, Hugenholtz P, Allen EE, Ram RJ, Richardson PM, Solovyev VV, Rubin EM, Rokhsar DS, and Banfield JF. 2004. Community structure and metabolism through reconstruction of microbial genomes from the environment. *Nature* 428: 37–43.

Venter JC, Remington K, Heidelberg JF, Halpern AL, Rusch D, Eisen JA, Wu D, Paulsen I, Nelson KE, Nelson W, Fouts DE, Levy S, Knap AH, Lomas MW, Nealson K, White O, Peterson J, Hoffman J, Parsons R, Baden-Tillson H, Pfannkoch C, Rogers YH, and Smith HO. 2004. Environmental genome shotgun sequencing of the Sargasso Sea. *Science* 304: 66–74.

Welch RA, Burland V, Plunkett G, III, Redford P, Roesch P, Rasko D, Buckles EL, Liou SR, Boutin A, Hackett J, Stroud D, Mayhew GF, Rose DJ, Zhou S, Schwartz DC, Perna NT, Mobley HL, Donnenberg MS, and Blattner FR. 2002. Extensive mosaic structure revealed by the complete genome sequence of uropathogenic *Escherichia coli*. *Proc Natl Acad Sci USA* 99: 17020–24.

Wellcome Trust Case Control Consortium. 2007. Genome-wide association study of 14,000 cases of seven common diseases and 3,000 shared controls. *Nature* 447: 661–78.

Woese CR. 1987. Bacterial evolution. *Microbiol Rev* 51: 221–71.

Xu J, Bjursell MK, Himrod J, Deng S, Carmichael LK, Chiang HC, Hooper LV, and Gordon JI. 2003. A genomic view of the human-*Bacteroides thetaiotaomicron* symbiosis. *Science* 299: 2074–76.

# Chapter 2

# Diagnosis, Epidemiology, and Global Impact of Endodontic Infections

*Dag Ørstavik*

## 2.1 Endodontic disease: irritation, inflammation, and infection of the pulp and periapical tissues

Endodontics deals with diseases of the pulp–dentin organ and the periapical tissues. For practical purposes, these are infectious processes. Noninfectious conditions affecting the pulp or apical periodontium are rare and are seldom dealt with by specific endodontic treatment; they do, however, represent important differential diagnostic challenges.

The sources of pulpal and apical periodontal infections are numerous. Traditionally, endodontic disease has been seen as a sequel to dental caries; however, bacteria find their way to a vulnerable pulp in many other instances as well. Dental trauma is one well-known situation, so is pulp damage and infection following preparation and restoration of teeth. Low-grade irritation of pulpal nervous elements may occur following attrition and erosion, sometimes developing into pulpal necrosis and infection.

Historically, the focus has been on the inflammatory reactions of the pulp and periapical tissues, associating clinical disease with the tissue response. The usually limited extent of the inflammatory reactions has been related to not only infection, but also tissue damage during treatment and the toxic effects of medicaments and materials. It is clearly an improvement in the concept of diagnosis and treatment planning that there has been a shift toward stressing the level and extent of the infectious process, rather than wild-guessing the type and degree of the inflammatory reaction. Inflammation is a sign of infection; clinically progressing disease is hardly ever caused by trauma or materials. This concept has been productive because virtually all successful therapeutic measures are directed toward combating or preventing infection, with reduced or eliminated inflammation following as a consequence.

Moreover, the concept of endodontic diseases as infections has implications for public oral health assessment in general, and places the local infectious disease in perspective relative to systemic health issues, particularly cardiovascular disease (Caplan et al. 2006; Joshipura et al. 2006). In this context, it is important to relate epidemiological aspects of pulpal and periapical disease to endodontic microbiology.

## 2.2 Primary diagnostic criteria: subjective symptoms and radiographic changes

### 2.2.1 Pulpal involvement

Initial pulpal infection is recognized primarily by clinical symptoms or through explorative excavation of involved dentin. While conventional radiography may suggest that a resorptive or carious process is impinging on the pulp, such methods do not allow definitive assessment of pulpal involvement.

Many studies have tried to establish a correlation between the clinical and histological or bacteriologic features of pulpal inflammation, but with very limited success, if any (Cisneros-Cabello and Segura-Egea 2005). In recognition of this fact, the diagnosis of pulpal infection/inflammation has to be an operational one: based on experience and, to a degree, on the knowledge of the underlying biological processes, we categorize pulpitis as either reversible or irreversible. This scheme sidesteps the need to give a precise description of the extent and severity of inflammation in the pulp. It is assumed that in the case of reversible pulpitis, pulp vitality may be preserved with proper treatment; irreversible pulpitis implies that no treatment short of pulp extirpation and root filling can eliminate the disease. Briefly, reversible pulpitis causes clinical symptoms of short (seconds) duration and only when irritated by external stimuli, and the pulp itself is either not exposed or traumatically exposed for a short period only (<2 days). By contrast, irreversible pulpitis gives rise to symptoms of longer duration (minutes) that may also occur spontaneously, and an exposure of the pulp to the oral environment through caries, fractures, or cracks is suspected or confirmed. This concept is supported by clinical experience (Sigurdsson 2003; Iqbal et al. 2007) and by experimental studies on the effects of pulpal inflammation on nerve activity (Rodd and Boissonade 2000; Bletsa et al. 2006; Kokkas et al. 2007).

Sensitivity testing by temperature can give reasonably accurate assessment of nerve tissue activity in the pulp, but relating such recordings to the degree of pulpal inflammation is difficult considering the large variation in such measurements (Fischer et al. 1991).

Radiography is useful in special circumstances, such as for detection of internal and external cervical resorption. Pulp calcifications (diffuse and globular) and obliteration as seen radiographically may give indications of the physiological state of the pulp, but little information about pulpal infection or inflammation.

### 2.2.2 Periapical diagnosis

Periapical disease has also a significant clinical component. In comparison with symptomatic pulpitis, symptomatic apical periodontitis is typically characterized by dull, rather than sharp pain, and positive percussion and palpation tests (Iqbal et al. 2007). Total infection of the pulp with virulent organisms may give rise to acute apical abscess, a very painful and potentially harmful condition exemplifying a disease that historically defined the dental profession. Longstanding pulp infections may similarly exacerbate with symptomatic, acute apical periodontitis. Apart from distinguishing such conditions from marginal periodontal inflammation, they are seldom difficult to diagnose.

Chronic apical periodontitis is, on the other hand, largely dependent on radiographic signs for diagnosis. In its early stages and during healing, this may be very difficult, whereas a well-established, chronic apical periodontitis is a simple condition to diagnose. In teleological terms, an infected root canal of a tooth is probably perceived by the body as a risk zone for invasion by (life-threatening) microbes. A defense region is then established in which the tissue architecture is changed to prepare for the containment of invading microorganisms (Ørstavik and Pitt Ford 2008; see Chapter 11). Bone is gradually replaced by a granulomatous tissue with vascular and cellular components mobilized for host defense. These initial events produce changes in bone structure at the apex, which may be very hard to detect radiographically (Brynolf 1967), and they may occur with teeth that still has vitality or at least nervous activity in the pulp (Fig. 2.1). When periapical tissue remodeling has reached a state of complete transformation to a granuloma, the lesion is very characteristic and easily diagnosed in the radiograph

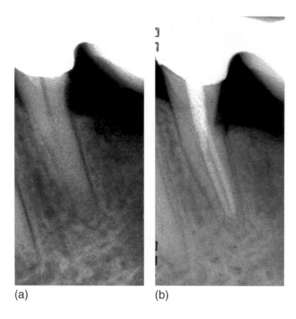

(a)                              (b)

**Fig. 2.1**   Minimal bone structural changes at the apex in conjunction with chronic pulpitis (a), necessitating endodontic treatment (b).

(Fig. 2.2). Treatment decision is then easy if the tooth does not respond to sensitivity testing. On the other hand, there may be total pulp necrosis and no infection or inflammation at the apex, such as when the pulp is devitalized by traumatic injury (Sundqvist 1976).

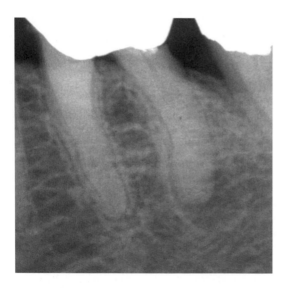

**Fig. 2.2**   Chronic apical periodontitis: incipient at mesial root, established at distal root of mandibular left first molar.

## 2.3 Pulpal inflammation versus infection: public health consequences

The clinical aspects of endodontic diseases may be serious and with some consequences for individual and public health. Symptomatic, acute pulpitis and apical periodontitis dominate as sources for acute dental pain in children and adults (Locker and Grushka 1987; Zeng et al. 1994; Lygidakis et al. 1998), which may be debilitating to the patient and lead to absence from work and involvement of costly health services. While we know that emergency dental services are in great demand in most countries, in urban as well as rural areas, there is very scant information on the actual incidence and prevalence of acute pulpal and apical periodontal disease. Therefore, one can only speculate that there is still, even in communities with well-developed dental services, a significant impact on the general well-being by acute pulpal and periodontal conditions (Sindet-Pedersen et al. 1985; Richardsson 2005).

A frequently overlooked situation is the association of pulpal and apical disease with tooth loss in the elderly. Whereas marginal periodontal disease is generally accepted as a significant cause of tooth loss, pulpal and apical diseases are important causes for extraction (Eckerbom et al. 1992), and may dominate

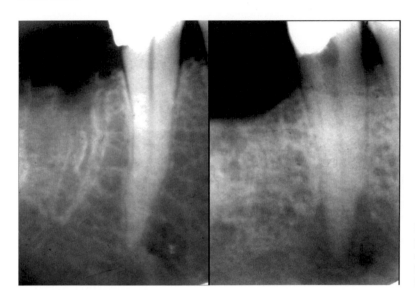

**Fig. 2.3**  Chronic apical periodontitis developing in 6 months after inadequate emergency treatment. The prognosis is reduced from >95% to <85%.

after the age of approximately 50 years (Eriksen and Bjertness 1991).

The tooth with pulpitis is obviously in danger of becoming infected and developing apical periodontitis. Correct and prompt treatment of the acute situation is therefore important not only to curb the pain and reestablish a functional tooth, but also to reduce or eliminate the risk for the insidious spreading of the infection and the emergence of a periapical lesion. It has been known for a very long time that the prognosis for treatment of apical periodontitis is much poorer than expected treatment outcome after vital pulpectomy (see Chapter 15). Early detection and root canal treatment of teeth at definitive risk of developing root canal infection are therefore essential. Failure to provide adequate treatment early will facilitate the development of an infection (Fig. 2.3), which will reduce the prognosis.

## 2.4 Epidemiology of pulpal inflammation

Information in the literature about the incidence of dental and oral pain is scarce in itself (Lipton et al. 1993; Pau et al. 2003), and the separation of the pulpal or periapical component from the inclusive diagnosis is difficult if at all possible. What little there is of targeted epidemiological data, point to a limited, but significant occurrence of acute pain of pulpal

origin (Sindet-Pedersen et al. 1985; Zeng et al. 1994; Lygidakis et al. 1998). This is an area in need of continued and extensive research. The incidence and prevalence of symptomatic pulpitis and apical periodontitis are obviously important for the targeting of dental services, and form important background knowledge for the design of dental curricula and for public health measures.

## 2.5 Criteria for assessing periapical health

Asymptomatic, chronic apical periodontitis poses a different challenge from pulpitis and symptomatic, acute apical periodontitis. The insidious nature and frequently pain-free course of this disease makes it evasive to detection outside of the dental treatment situation. The fact that asymptomatic, chronic apical periodontitis relies on radiography for detection poses limitations on the possibilities for screenings and population surveys. Moreover, when radiographic data have been made available for analysis, lack of standardization in scoring makes comparisons across studies difficult.

The radiographic technique may also influence the ability to detect with certainty the occurrence of asymptomatic, chronic apical periodontitis. For population surveys, panoramic radiography provides information at far less radiation dosage than full-mouth periapical examinations, but the possibilities of detection of apical lesions may be diminished. Comparisons of panoramic and periapical films for diagnosis of apical periodontitis suggest that there is some, but not a dramatic, reduction in the detectability of periapical lesions (Sameshima and Asgarifar 2001; Ridao-Sacie et al. 2007). Newer methods, such as tomography (Tammisalo et al. 1996), computed tomography (Huumonen et al. 2006), and cone-beam tomography (Lofthag-Hansen et al. 2007; Patel et al. 2007; Estrela et al. 2008), are more sensitive and probably more specific than periapical radiographs, but the radiation dose strongly limits their application for use in surveys.

### 2.5.1 Radiographic characteristics of chronic apical periodontitis

Verbal descriptors of the radiographic characteristics of asymptomatic, chronic apical periodontitis have included a widened periodontal space: interruptions of the lamina dura and/or presence of a radiolucent area at the site of exit of the pulp to the periodontal membrane (Ørstavik and Larheim 2008). Only when there is an overt radiolucency associated with the root tip and a concomitant finding of a necrotic pulp are the signs pathognomonic (Ørstavik and Larheim 2008). While it is possible to make assumptions from dif-

ferent studies with similar descriptions of the criteria used for detection of asymptomatic, chronic apical periodontitis, it is not possible to draw conclusions with any certainty.

### 2.5.2 The periapical index

The periapical index (PAI) was developed with the aim of improving comparisons between studies (Fig. 2.4). It makes use of an ordinal scale with five steps indicating increasing severity of apical periodontitis (Ørstavik et al. 1986). The steps are represented by radiographs that have histological verification from an extensive study on human cadavers (Brynolf 1967). This makes possible a visual reference scale that reduces the risk of personal bias otherwise associated with subjective radiographic assessments. Also, the system is used after extensive and standardized calibration of the observers, which facilitates comparisons of different studies and pooling of data. While developed for clinical follow-up studies of endodontic treatment in prospective studies, the PAI scoring system can easily be modified for use in epidemiological surveys (Eriksen 1991). A general principle in epidemiology is to avoid scoring a healthy condition wrongly as disease. This is accomplished by restricting the categorization as *diseased* (i.e., with apical periodontitis) to teeth with scores 3–5 (Fig. 2.5). In this way, some cases of asymptomatic, chronic apical periodontitis will go

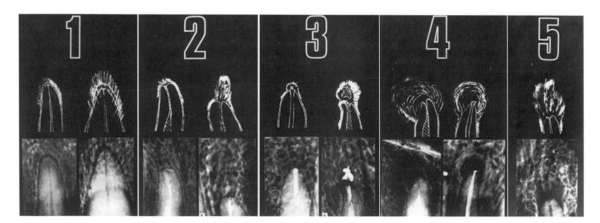

**Fig. 2.4**   The periapical index. The periapical condition is scored by comparison with a series of reference radiographs of teeth with known histology. (Reproduced with permission from Ørstavik et al. 1986.)

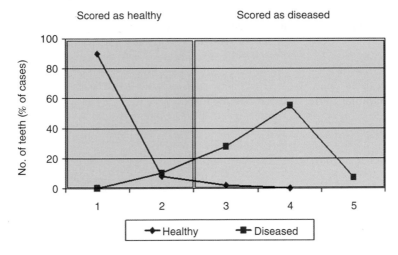

**Fig. 2.5** Dichotomization of PAI scores applied to epidemiology: teeth without apical periodontitis (blue line); teeth with apical periodontitis (red line). A minimum of false positives (healthy apical periodontium scored as diseased; blue cases in red sector) is acceptable at the expense of some false negatives (diseased teeth registered as healthy; red cases in blue sector).

undetected, but only a minimal number of healthy teeth will be scored as diseased.

Irrespective of the radiographic method of detection, it is apparent that radiographic assessments of apical periodontitis on the whole will underestimate its true incidence or prevalence (Brynolf 1967). Even with all these provisos, it may still be prudent to review and compare results from different areas and cohorts, as long as the shortcomings of the radiographic methods are kept in mind.

## 2.6 Results of epidemiological surveys of asymptomatic, chronic apical periodontitis

When periapical disease was seen only as an extension of caries, epidemiological studies paid little, if any, attention to the incidence and prevalence of apical periodontitis. On the basis of numerous institutional studies on the outcome of endodontic treatment, the notion that endodontic treatment was predictable and generally successful was accepted (Strindberg 1956; Grossman et al. 1964; Kerekes and Tronstad 1979; Ørstavik et al. 1987), and the extent and importance of apical periodontitis in the general population was largely overlooked.

In a series of studies, Eriksen and coworkers (Eriksen 1991; Eriksen and Bjertness 1991; Eriksen et al. 1995; Marques et al. 1998; Sidaravicius et al. 1999; Aleksejuniene et al. 2000; Skudutyte-Rysstad

and Eriksen 2006) examined the general prevalence of apical periodontitis and placed it in its proper perspective. The PAI scoring system was used together with simple criteria for the assessment of root-filling quality. A primary aim was to reassess the association of the quality of the root filling as seen on the radiograph with the periapical status of the teeth. Similar to what had been documented in the institutional follow-up studies, there was a clear association between poor root-filling quality and the presence of apical periodontitis, emphasizing the need for focus on high-quality technical performance during the endodontic procedures.

However, there was also an unexpectedly high prevalence of apical periodontitis in most populations and age groups. This was a source of concern and had to be considered in oral health assessments in general. Moreover, the finding that pulpal and periapical diseases were a major reason for extractions in adults, surpassing marginal periodontitis around the fifth decade of life, emphasized the impact of periapical health for retention of the dentition into old age (Eriksen and Bjertness 1991; Eckerbom et al. 1992).

These studies have later been supplemented by several others from almost all corners of the globe, and with few exceptions, the results are quite disheartening in different countries and populations, regardless of the degree and perceived quality of the dental services offered. Many studies have made use of the PAI scoring system; others rely on a simple assessment on

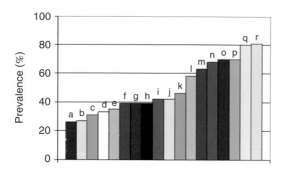

**Fig. 2.6** The prevalence of apical periodontitis in different populations: (a) Dugas et al. (2003); (b) Marques et al. (1998); (c) Frisk et al. (2003); (d) Loftus et al. (2005); (e) Buckley and Spangberg (1995); (f) De Cleen et al. (1993); (g) Eriksen and Bjertness (1991); (h) Dugas et al. (2003); (i) Kirkevang et al. (2001); (j) Frisk et al. (2003); (k) Chen et al. (2007); (l) Jiménez-Pinzón et al. (2004); (m) De Moor et al. (2000); (n) Saunders et al. (1997); (o) Sidaravicius et al. (1999); (p) Tsuneishi et al. (2005); (q) Kabak and Abbott (2005); (r) Segura-Egea et al. (2005).

the presence or absence of a radiolucent area indicating periodontitis. Fig. 2.6 shows the prevalence of apical periodontitis in people from 18 different populations in different countries. Apical periodontitis occurs with a prevalence of 30–80% in different populations, generally increasing in older age groups (Chen et al. 2007) and in populations at high risk of infectious disease (diabetes; Britto et al. 2003).

Figures produced by these kinds of surveys generally do not account for alternative ways of dealing with apical periodontitis in different environments. It is tempting to speculate that populations with low prevalence have had teeth with apical periodontitis extracted; indeed, for the Portuguese population studied by Marques et al. (1998), which showed the lowest prevalence, it was found that they had a lower mean number of remaining teeth than a comparable Norwegian population with higher prevalence of apical periodontitis (Eriksen and Bjertness 1991).

## 2.7 Quality of root canal treatment and the development and persistence of apical periodontitis

Institutional follow-up studies and epidemiological surveys have all documented that there is a very clear

correlation between presence of apical periodontitis and inadequate technical quality of the root filling as it appears in the radiograph. The association is strongest for teeth that are diagnosed with apical periodontitis at the start of treatment, and far less dominant when the root filling is placed in teeth with no lesion prior to treatment (Sjögren et al. 1990). In the latter situation, typically less than 10% of treated cases develop apical periodontitis; contrarily, teeth treated for primary apical periodontitis show persistence of lesions in 20–25% cases of institutional studies. In all likelihood, there is a poorer outcome for both preoperative diagnoses in practice compared to the institutional setting. By inference, when epidemiological surveys indicate that 30–40% of root-filled teeth have apical periodontitis, it seems fair to assume that less than 50% of teeth with apical periodontitis are cured in the average treatment setting in practice.

This should not be placed in a context to advocate more radical treatment or prophylaxis of apical periodontitis. The preservation of teeth by endodontic procedures is, after all, a clinically very successful and predictable procedure. The sequels to extractions and various prosthetic procedures, as alternative treatments, are numerous and often of greater consequence. However, these epidemiological findings clearly point to a need for improvements in the quality of endodontic care.

## 2.8 General oral health, oral health strategies, and tooth preservation as risk factors for oral infections

The infectious aspects of endodontic diseases pose the question whether the nature of the organisms and their activities in the pulp space and beyond are a source of concern for the general health of the individual. Marginal periodontitis seems to have a definitive, albeit limited, association with cardiovascular disease, and there are data emerging in the literature indicating that this is the case also for apical periodontitis (Caplan et al. 2006; Olsen et al. 2007); however, others have failed to establish such an association (Frisk et al. 2003).

The concept of endodontic diseases primarily as infections with the potential to spread and thereby to affect organs at distant sites may be important for patients' systemic health, particularly the risk of cardiovascular events (Ørstavik and Pitt Ford 2008; see also Chapter 16). On one hand, this affects the decision whether to provide antibiotic coverage prior to surgery in patients at risk of infective endocarditis or infection of vascular implants; on the other hand, the possible association of pulp and periapical infection with the risk of developing cardiovascular disease has a major impact on the rationale and case selection for endodontic treatment, and especially on prophylactic efforts to prevent pulpal infection in the first place.

The notion that granulomas may be *sterile* or caused by medicaments or materials has been abandoned. Periapical lesions are virtually all apical periodontitis, and apical periodontitis is caused by microbial infection of the root canal system. Imminent or established infections of the pulp and periapical tissues need to be contained or eliminated. Early and appropriate endodontic intervention is necessary in such cases, with emphasis on proper case selection and highly skilled technical performance of treatment.

The provision of high-quality endodontic care at all levels of dental service to the individual patient as well as to populations is therefore crucial for optimum, long-term oral health. The goals for these services are several: to prevent pulpal infection by effective caries prevention, by protection against dental trauma, and by appropriate dentin treatment under restorations; to limit pulpal pain as a source of discomfort and loss of work; and to eliminate dental infection and prevent its recurrence by root filling and surgical endodontic procedures.

## 2.9 References

Aleksejuniene J, Eriksen HM, Sidaravicius B, and Haapasalo M. 2000. Apical periodontitis and related factors in an adult Lithuanian population. *Oral Surg Oral Med Oral Pathol Oral Radiol Endod* 90(1): 95–101.

Bletsa A, Berggreen E, Fristad I, Tenstad O, and Wiig H. 2006. Cytokine signalling in rat pulp interstitial fluid and transcapillary fluid exchange during lipopolysaccharide-induced acute inflammation. *J Physiol* 573(Pt 1): 225–36. Epub Mar 9, 2006.

Britto LR, Katz J, Guelmann M, and Heft M. 2003. Periradicular radiographic assessment in diabetic and control individuals. *Oral Surg Oral Med Oral Pathol Oral Radiol Endod* 96(4): 449–52.

Brynolf L. 1967. Histological and roentgenological study of periapical region of human upper incisors. *Odontologisk Revy* 18(Suppl 11): 1–176.

Buckley M and Spangberg LS. 1995. The prevalence and technical quality of endodontic treatment in an American subpopulation. *Oral Surg Oral Med Oral Pathol Oral Radiol Endod* 79(1): 92–100.

Caplan DJ, Chasen JB, Krall EA, Cai J, Kang S, Garcia RI, Offenbacher S, and Beck JD. 2006. Lesions of endodontic origin and risk of coronary heart disease. *J Dent Res* 85(11): 996–1000.

Chen CY, Hasselgren G, Serman N, Elkind MS, Desvarieux M, and Engebretson SP. 2007. Prevalence and quality of endodontic treatment in the Northern Manhattan elderly. *J Endod* 33(3): 230–34.

Cisneros-Cabello R and Segura-Egea JJ. 2005. Relationship of patient complaints and signs to histopathologic diagnosis of pulpal condition. *Aust Endod J* 31(1): 24–27.

De Cleen MJ, Schuurs AH, Wesselink PR, and Wu MK. 1993. Periapical status and prevalence of endodontic treatment in an adult Dutch population. *Int Endod J* 26(2): 112–19.

De Moor RJ, Hommez GM, De Boever JG, Delme KI, and Martens GE. 2000. Periapical health related to the quality of root canal treatment in a Belgian population. *Int Endod J* 33(2): 113–20.

Dugas NN, Lawrence HP, Teplitsky PE, Pharoah MJ, and Friedman S. 2003. Periapical health and treatment quality assessment of root-filled teeth in two Canadian populations. *Int Endod J* 36(3): 181–92.

Eckerbom M, Magnusson T, and Martinsson T. 1992. Reasons for and incidence of tooth mortality in a Swedish population. *Endod Dent Traumatol* 8(6): 230–34.

Eriksen HM. 1991. Endodontology—epidemiologic considerations. *Endod Dent Traumatol* 7(5): 189–95.

Eriksen HM, Berset GP, Hansen BF, and Bjertness E. 1995. Changes in endodontic status 1973–1993 among 35-year-olds in Oslo, Norway. *Int Endod J* 28(3): 129–32.

Eriksen HM and Bjertness E. 1991. Prevalence of apical periodontitis and results of endodontic treatment in middle-aged adults in Norway. *Endod Dent Traumatol* 7(1):1–4.

Estrela C, Bueno MR, Leles CR, Azevedo B, and Azevedo JR. 2008. Accuracy of cone beam computed tomography and panoramic and periapical radiography for detection of apical periodontitis. *J Endod* 34(3): 273–79.

Fischer C, Wennberg A, Fischer RG, and Attström R. 1991. Clinical evaluation of pulp and dentine sensitivity after supragingival and subgingival scaling. *Endod Dent Traumatol* 7(6): 259–65.

Frisk F, Hakeberg M, Ahlqwist M, and Bengtsson C. 2003. Endodontic variables and coronary heart disease. *Acta Odontol Scand* 61(5): 257–62.

Grossman LI, Shepard LI, and Pearson LA. 1964. Roentgenologic and clinical evaluation of endodontically treated teeth. *Oral Surg Oral Med Oral Pathol Oral Radiol Endod* 17: 368–74.

Huumonen S, Kvist T, Grondahl K, and Molander A. 2006. Diagnostic value of computed tomography in re-treatment of root fillings in maxillary molars. *Int Endod J* 39(10): 827–33.

Iqbal M, Kim S, and Yoon F. 2007. An investigation into differential diagnosis of pulp and periapical pain: a PennEndo database study. *J Endod* 33(5): 548–51. Epub Mar 21, 2007.

Jiménez-Pinzón A, Segura-Egea JJ, Poyato-Ferrera M, Velasco-Ortega E, and Ríos-Santos JV. 2004. Prevalence of apical periodontitis and frequency of root-filled teeth in an adult Spanish population. *Int Endod J* 37(3): 167–73.

Joshipura KJ, Pitiphat W, Hung HC, Willett WC, Colditz GA, and Douglass CW. 2006. Pulpal inflammation and incidence of coronary heart disease. *J Endod* 32(2): 99–103.

Kabak Y and Abbott PV. 2005. Prevalence of apical periodontitis and the quality of endodontic treatment in an adult Belarusian population. *Int Endod J* 38(4): 238–45.

Kerekes K and Tronstad L. 1979. Long-term results of endodontic treatment performed with a standardized technique. *J Endod* 5(3): 83–90.

Kirkevang LL, Horsted-Bindslev P, Ørstavik D, and Wenzel A. 2001. Frequency and distribution of endodontically treated teeth and apical periodontitis in an urban Danish population. *Int Endod J* 34(3): 198–205.

Kokkas AB, Goulas A, Varsamidis K, Mirtsou V, and Tziafas D. 2007. Irreversible but not reversible pulpitis is associated with up-regulation of tumour necrosis factor-alpha gene expression in human pulp. *Int Endod J* 40(3): 198–203.

Lipton JA, Ship JA, and Larach-Robinson D. 1993. Estimated prevalence and distribution of reported orofacial pain in the United States. *J Am Dent Assoc* 124(10): 115–21.

Locker D and Grushka M. 1987. Prevalence of oral and facial pain and discomfort: preliminary results of a mail survey. *Community Dent Oral Epidemiol* 15(3): 169–72.

Lofthag-Hansen S, Huumonen S, Grondahl K, and Grondahl HG. 2007. Limited cone-beam CT and intraoral radiography for the diagnosis of periapical pathology. *Oral Surg Oral Med Oral Pathol Oral Radiol Endod* 103(1): 114–19. Epub Apr 24, 2006.

Loftus JJ, Keating AP, and McCartan BE. 2005. Periapical status and quality of endodontic treatment in an adult Irish population. *Int Endod J*. 38(2): 81–86.

Lygidakis NA, Marinou D, and Katsaris N. 1998. Analysis of dental emergencies presenting to a community paediatric dentistry centre. *Int J Paediatr Dent* 8(3): 181–90.

Marques MD, Moreira B, and Eriksen HM. 1998. Prevalence of apical periodontitis and results of endodontic treatment in an adult, Portuguese population. *Int Endod J* 31(3): 161–65.

Olsen I, Nafstad P, Schwarze P, Rønningen KS, and Håheim LL. 2007. Tooth Extractions, Bacterial Antibodies, C-reactive Protein and Myocardial Infarction Abstract No. 2291, 85th General Session and Exhibition of the IADR, New Orleans, LA.

Ørstavik D, Kerekes K, and Eriksen HM. 1986. The periapical index: a scoring system for radiographic assessment of apical periodontitis. *Endod Dent Traumatol* 2: 20–34.

Ørstavik D, Kerekes K, and Eriksen HM. 1987. Clinical performance of three endodontic sealers. *Endod Dent Traumatol* 3: 178–86.

Ørstavik D and Larheim TA. 2008. Radiology of apical periodontitis. In: Ørstavik D and Pitt Ford TR. (eds), *Essential Endodontology: Prevention and Treatment of Apical Periodontitis*, 2nd edn. Oxford: Blackwell.

Ørstavik D and Pitt Ford TR. 2008. Apical periodontitis: microbial infection and host responses. In: Ørstavik D and Pitt Ford TR. (eds), *Essential Endodontology: Prevention and Treatment of Apical Periodontitis*, 2nd edn. Oxford: Blackwell.

Patel S, Dawood A, Ford TP, and Whaites E. 2007. The potential applications of cone beam computed tomography in the management of endodontic problems. *Int Endod J* 40(10): 818–30.

Pau AK, Croucher R, and Marcenes W. 2003. Prevalence estimates and associated factors for dental pain: a review. *Oral Health Prev Dent* 1(3): 209–20.

Richardsson PS. 2005. Dental morbidity in United Kingdom Armed Forces, Iraq 2003. *Mil Med* 170(6): 536–41.

Ridao-Sacie C, Segura-Egea JJ, Fernández-Palacín A, Bullón-Fernández P, and Ríos-Santos JV. 2007. Radiological assessment of periapical status using the periapical index: comparison of periapical radiography and digital panoramic radiography. *Int Endod J*. 40(6): 433–40. Epub Apr 19, 2007.

Rodd HD and Boissonade FM. 2000. Substance P expression in human tooth pulp in relation to caries and pain experience. *Eur J Oral Sci* 108(6): 467–74.

Sameshima GT and Asgarifar KO. 2001. Assessment of root resorption and root shape: periapical vs panoramic films. *Angle Orthod* 71(3): 185–89.

Saunders WP, Saunders EM, Sadiq J, and Cruickshank E. 1997. Technical standard of root canal treatment in an adult Scottish sub-population. *Br Dent J* 182: 382–86.

Segura-Egea JJ, Jiménez-Pinzón A, Ríos-Santos JV, Velasco-Ortega E, Cisneros-Cabello R, and Poyato-Ferrera M. 2005. High prevalence of apical periodontitis amongst type 2 diabetic patients. *Int Endod J* 38(8): 564–69.

Sidaravicius B, Aleksejuniene J, and Eriksen HM. 1999. Endodontic treatment and prevalence of apical periodontitis in an adult population of Vilnius, Lithuania. *Endod Dent Traumatol* 15(5): 210–15.

Sigurdsson A. 2003. Pulpal diagnosis. *Endodontic Topics* 5: 12–23.

Sindet-Pedersen S, Petersen JK, and Götzsche PC. 1985. Incidence of pain conditions in dental practice in a Danish county. *Community Dent Oral Epidemiol* 13(4): 244–46.

Sjögren U, Hägglund B, Sundqvist G, and Wing K. 1990. Factors affecting the long-term results of endodontic treatment. *J Endod*. 16(10): 498–504.

Skudutyte-Rysstad R and Eriksen HM. 2006. Endodontic status amongst 35-year-old Oslo citizens and changes over a 30-year period. *Int Endod J* 39(8): 637–42.

Strindberg LZ. 1956. The dependence of the results of pulp therapy on certain factors: an analytic study based on radiographic and clinical follow-up examination. *Acta Odontol Scand* 14(Suppl 21): 1–175.

Sundqvist G. 1976. Bacteriological Studies of Necrotic Dental Pulps. Umeå University Odontological Dissertations, Umeå, Sweden.

Tammisalo T, Luostarinen T, Vahatalo K, and Neva M. 1996. Detailed tomography of periapical and periodontal lesions. Diagnostic accuracy compared with periapical radiography. *Dentomaxillofac Radiol* 25(2): 89–96.

Tsuneishi M, Yamamoto T, Yamanaka R, Tamaki N, Sakamoto T, Tsuji K, and Watanabe T. 2005. Radiographic evaluation of periapical status and prevalence of endodontic treatment in an adult Japanese population. *Oral Surg Oral Med Oral Pathol Oral Radiol Endod* 100(5): 631–35.

Zeng Y, Sheller B, and Milgrom P. 1994. Epidemiology of dental emergency visits to an urban children's hospital. *Pediatr Dent* 16(6): 419–23.

# Chapter 3

# Microbiology of Caries and Dentinal Tubule Infection

*Robert M. Love*

## 3.1 Introduction

It is well established that bacteria are the prime etiologic factor in the development and progression of dental caries and pulp and periapical disease. In the late nineteenth and early twentieth centuries, W.D. Miller demonstrated bacterial invasion of dentinal tubules of both carious and noncarious dentin and reported that the tubule microflora consisted of cocci and rods (Miller 1890). However, during the first half of the twentieth century, the role of bacteria in pulp and periapical disease was overshadowed by the hollow tube theory (Rickert and Dixon 1931; Dixon and Rickert 1933) and it was not until the 1960s that sound experimental evidence clearly established the fundamental role of bacteria in dental disease. Keyes (1960) demonstrated that dental caries did not develop in gnotobiotic (germfree) animals fed a cariogenic diet, while Kakehashi et al. (1965) showed that pulp and periapical disease occurred in surgically exposed rat molar pulp only when bacteria were present in the oral cavity. Indeed, in germfree rats exposed pulps remained healthy and were able to initiate repair by way of dentin bridging of the exposure, demonstrating the innate regenerative capacity of the dental pulp.

The pulp and dentin share a common embryonic origin and form a functional complex (the pulp–dentin complex) that is primarily involved in the production of dentin and in tooth sensibility. Dentin is a porous structure due to the development, by odontoblast processes, of dentinal tubules (Fig. 3.1) that run from the outer surface of the dentin (dentinoenamel or dentinocemental junction) to the inner (pulpal) surface of the dentin.

The tubules provide a diffusion or invasion channel for exogenous substances or bacteria in the oral cavity; however, the overlying enamel and cementum protect the pulp–dentin complex, and a breach in the protective layer may result in disease. The pulp–dentin complex possesses a number of defensive processes, such as an outward flushing flow of dentinal fluid, buffering capacity by dentin matrix and tubule fluid, dentinal

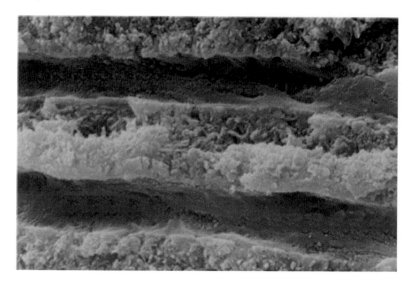

**Fig. 3.1** Scanning electron micrograph of a section of fractured dentin showing two dentinal tubules of approximately 2 µm in diameter. It can be seen that the intratubular dentin forming the tubule wall has a more homogenous appearance than the intertubular dentin between the tubules. This is due to the more highly mineralized nature if the intratubular dentin compared to intertubular dentin.

tubule sclerosis, reparative dentin formation, and a pulpal inflammatory response, which react to bacterial invasion of dentinal tubules and exposed pulp in an attempt to eliminate them. If, however, the route of infection is not eradicated by these natural processes, or by operative procedures, then the invading bacteria overcome the defenses and cause pulpitis, pulp necrosis and infection of the pulp chamber and root canal system, and progressing to inflammatory periapical disease (Fig. 3.2).

Invasion of dentinal tubules by bacteria from supra- or subgingival plaque occurs whenever dentin is exposed in the oral cavity. This can be through carious lesions, restorative or periodontal procedures, tooth wear, enamel or dentin cracks, or dental trauma (Tronstad and Langeland 1971; Pashley 1990; Peters et al. 1995; Love 1996a). Bacteria invading coronal dentinal tubules (Fig. 3.2a) may cause pulpal disease (Brännström and Nyborg 1971) and subsequently take part in infection of the root canal system (Fig. 3.2b). Species such as streptococci and *Actinomyces* are major bacterial species of dental plaque (Jenkinson and Lamont 1997) and may initiate tubule and pulpal infection; however, the dynamics of the root canal flora are such that obligate anaerobic bacteria commonly dominate the microflora of the infected root canal.

As infection of the pulp space progresses, bacteria invade the radicular dentinal tubules (Figs 3.2b–c). If these bacteria are not removed or killed during endodontic treatment, the presence of vital bacteria within radicular dentin may be responsible for continued root canal infection (Haapasalo and Ørstavik 1987) and persistent apical periodontitis (Fig. 3.2e).

The study and practice of endodontics is aimed at preventing and treating pulp and periapical disease, and an understanding of the mechanisms of bacterial invasion of dentinal tubules is central to these aims.

## 3.2 Colonization of dentinal tubules

It is generally accepted that before bacteria can invade the human body, they must first colonize the host by undergoing a series of interactive events. To begin the process of colonization, bacteria must first adhere to the host tissue. Initially, this is by a loose physical association of the organism to the surface of a tissue that allows stronger and more permanent bonds to be established through binding of microbial cell surface adhesins and/or fimbriae to complementary host surface receptors or to enamel and dentin. Once the microbial cells are bound, they must be able to utilize available nutrients, compete or cooperate with other species in the immediate environment, and contend with host defense mechanisms before accumulation of microorganisms can occur by cell division and growth. Once these processes occur at multiple sites, the host becomes colonized. As such, bacterial invasion of dentinal tubules must follow these principles and recent evidence demonstrates these principles.

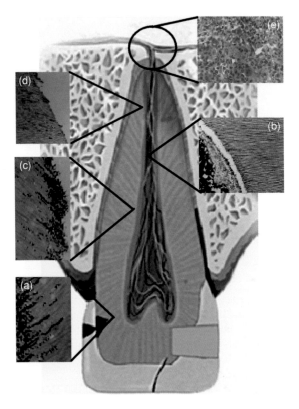

tissue proteins or glycoproteins. Dentinal tubules contain unmineralized collagen (Dai et al. 1991) and oral streptococci bind to collagen type I when adsorbed onto hydroxyapatite surfaces (Liu and Gibbons 1990; Liu et al. 1990) to unmineralized collagen and to root dentin (Switalski et al. 1993). Binding to collagen was shown to be mediated by expression of oral streptococci antigen I/II polypeptide adhesins on the surface of the bacteria (Love et al. 1997), and it has also been shown that these polypeptides are necessary for bacterial invasion of dentin (Love et al. 1997). Similarly, experiments using *Enterococcus faecalis* mutants deficient in serine protease and the collagen-binding protein (Ace) demonstrated that these molecules contribute to cell adhesion to radicular dentin (Hubble et al. 2003). Thus, evidence suggests that bacterial cell recognition of collagen may facilitate bacterial adhesion to exposed dentin or cementum (Fig. 3.3), upregulate production of antigen I/II polypeptide, and induce a morphological growth response, manifested by long chaining of streptococcal cells, facilitating tubule invasion (Love et al. 1997) (Fig. 3.4).

Nutritional supply within a dentinal tubule may influence the depth of bacterial penetration. This is partly dependent on the patency of the tubule as diffusion of substances into dentinal tubules from the oral cavity, pulpal fluid, or periradicular tissues is proportional to tubule diameter. This may account for the higher numbers of cariogenic bacteria present within superficial dentin (Edwardsson 1987), where

**Fig. 3.2** Diagram showing potential routes of infection of coronal and radicular dentin. Bacterial invasion of coronal dentinal tubules toward the pulp space (a) occurs as a result of a breach in the integrity of the enamel from dental caries, enamel cracks/fractures, or restorative procedures. Invasion of tubules toward the pulp also occurs when the cementum is breached as a consequence of periodontal disease or procedures. If unchecked, bacteria within dentinal tubules will enter and infect the pulp chamber and root canal space, and bacterial biofilms (b) will develop. Subsequently, bacterial invasion of radicular dentin occurs from the pulpal surface toward the dentinocemental junction. Invasion in cervical and mid-root radicular dentin readily occurs (heavy invasion shown in (c)), while the amount and depth of invasion in apical dentin is low (d). Inflammatory periradicular disease (e) results from the bacterial infection. (Diagram reproduced from Love 2004.)

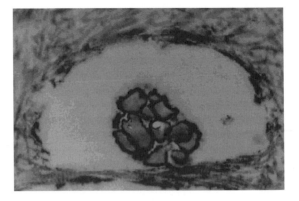

**Fig. 3.3** Transmission electron micrograph demonstrating a colony of bacterial cells invading a radicular dentinal tubule. Note the close approximation of peripheral cells of the colony with the wall of the tubule indicative of cell attachment to tubule structure, an essential step in colonization. (Reproduced from Love 2004.)

Oral bacteria can express multiple cell surface adhesins that allow adhesion to structures in the oral cavity; however, attachment does not readily occur to the mineral component, and adhesion to dentin usually requires the cell to attach to some proteinaceous portion of the dentin matrix such as deposited salivary or

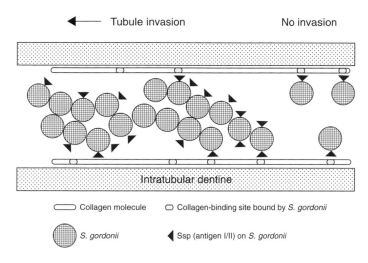

◄——— Tubule invasion          No invasion

Intratubular dentine

⊂⊃ Collagen molecule     ◯ Collagen-binding site bound by *S. gordonii*

◉ *S. gordonii*          ◀ Ssp (antigen I/II) on *S. gordonii*

**Fig. 3.4** Diagram demonstrating a model for tubule invasion by primary colonizing streptococci. Cell surface adhesins on *S. gordonii* attach the cells to unmineralized collagen within a tubule and the cells undergo a series of reactions resulting in chaining growth and invasion of the tubule. (Reproduced from Love and Jenkinson 2002.)

the presence of fermentable carbohydrates and oxygen from the oral cavity is likely to be higher than in deeper dentin. Similarly, the anaerobic environment and possible presence of tissue components, for example, hemin, within coronal or radicular dentin close to the pulp space is likely to favor growth and survival of fastidious organisms such as *Prevotella intermedia* and *Peptostreptococcus micros* (now *Parvimonas micra*) (Love 2007).

Studies demonstrate that bacteria may compete for invasion of dentinal tubules and also that they may cooperate in invasion. This may be due to a number of factors such as bacterial-induced alteration of the tubule environment, for example, oxygen tension facilitating or inhibiting growth of other organisms, bacterial production of bacteriocins that will inhibit growth of other bacteria, or cooperative bacterial co-aggregation reactions. Nagaoka et al. (1995) showed that invasion of dentin by *Streptococcus sobrinus* or *Actinomyces naeslundii* (*viscosus*) was inhibited in the presence of *Lactobacillus casei*, while invasion by *L. casei* was enhanced. Similarly, it has been shown that dentinal tubule invasion by *Porphyromonas gingivalis* was promoted when cocultivated with *Streptococcus gordonii*, but not in the presence of *Streptococcus mutans*. This invasion pattern was related to a specific co-aggregation interaction between *S. gordonii* and *P. gingivalis* cells mediated by the streptococcal antigen I/II polypeptides promoting tubule invasion by *P. gingivalis* (Fig. 3.5), while the antigen I/II polypeptide SpaP of *S. mutans* did not have the same binding capacity or invasive effect on *P. gingivalis*, indicating a

species-specific function of antigen I/II polypeptides (Love et al. 2000).

The pulp–dentin complex defensive processes can effectively inhibit bacterial invasion of dentin. These are particularly active when a vital pulp is present; however, as the pulp looses structure and function as a result of the bacterial insult, such as from the effect of bacterial products like lipopolysaccharide, bacteria are able to invade more readily. The contents of a dentinal tubule play an important role in defense functions and colonization of a tubule. Although the composition of dentinal tubule fluid in vital dentin is not fully known, it resembles serum with proteins such as albumin and immunoglobulin G (IgG) being present (Knutsson et al. 1994); additionally other blood proteins, such as fibrinogen, may be found in dentinal tubules after cavity preparation (Knutsson et al. 1994; Izumi et al. 1998). Similar molecules, derived from fluid originating from alveolar bone and periodontal ligament and saliva, are present in nonvital radicular and coronal dentinal tubules. It has been shown that albumin, fibrinogen, and IgG present within dentinal tubules decrease fluid flow through dentin in vitro (Pashley et al. 1982; Hahn and Overton 1997) and inhibit bacterial invasion of radicular dentinal tubules (Love 2002) by interacting with bacterial cells or physically occluding tubules and reducing dentin permeability. Bacterial cells within dentinal tubules also reduce dentin permeability (Michelich et al. 1980; Love et al. 1996) and this would suggest that bacteria themselves would inhibit subsequent invasion; however, reduced fluid flow might promote disease pathogenesis

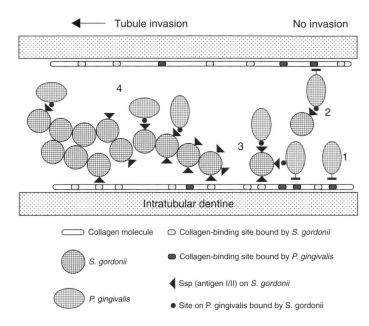

**Fig. 3.5** Diagram demonstrating a model for tubule invasion by secondary colonizers. Secondary colonizers (*P. gingivalis*) may be able to attach to components of the dentin matrix, but this does not allow them to invade a tubule (1, 2). Rather the cells attach to primary colonizing bacteria (*S. gordonii*), which allows them to invade a tubule (3, 4). (Reproduced from Love and Jenkinson 2002.)

by allowing increased diffusion of destructive or toxic bacterial products toward the pulp and/or periradicular tissues (Pashley 1996).

These studies demonstrate that tubule invasion and development of the intertubular bacterial flora follow that seen in colonization of tooth surfaces and the formation of plaque biofilms, that is, initial attachment and colonization by primary streptococcal colonizers, which then allow colonization by late colonizers such as *P. intermedia* and *P. gingivalis*. It is highly likely that other bacterial interactions between host proteins and other bacteria influence tubule invasion.

The relationship between enamel and cementum to dentin and the structure of the dentinal tubules also influence the process of bacterial invasion of dentinal tubules. The number of dentinal tubules per square mm varies from 15,000 at the dentinoenamel junction to 45,000 at the pulp (Garberoglio and Brännström 1976). Deposition of intratubular dentin within the tubule (Fig. 3.1) results in narrowing of the tubule (Nalbandian et al. 1960; Linde and Goldberg 1993) and is more advanced in superficial older dentin compared with dentin closer to the pulp. This results in a tapered tubule with the largest dimension at the pulp (~2.5 μm in diameter) and the smallest dimension at the dentinoenamel or dentinocemental junction (~0.9 μm in diameter). Therefore, a tubule is normally larger in diameter than an average oral streptococcal

cell (0.5–0.7 μm) and will not physically impede invasion. Tertiary, reactionary, or reparative dentin, which is laid down by newly differentiated odontoblasts as a consequence of loss of primarily differentiated odontoblasts, although permeable does not have a regular tubular form and may provide a more physical barrier to cell invasion.

The degree of permeability varies between different areas of a tooth, the number of patent dentinal tubules present, and by tubule contents (Pashley 1990), while the diameter of dentinal tubules influences the depth of bacterial invasion since this determines the rate of solute diffusion (Pashley 1992). The collagen matrix of intratubular dentin is highly mineralized (~95 vol % mineral phase) compared with the less mineralized (about 30 vol % mineral phase) intertubular dentin (Marshall 1993) (Fig. 3.1), and becomes more mineralized with increasing age. This results in a decrease in size, and ultimately obliteration, of the dentinal tubules with about 40% decrease in the overall numbers between the ages of 20 and 80 years (Nalbandian et al. 1960; Tronstad 1973; Carrigan et al. 1984). The mean numbers of tubules at any given age within coronal, cervical, and mid-root dentin are similar (Carrigan et al. 1984). However, significantly fewer dentinal tubules are found in apical dentin (Nalbandian et al. 1960; Carrigan et al. 1984), indicating that the formation of intratubular dentin occurs more rapidly in the

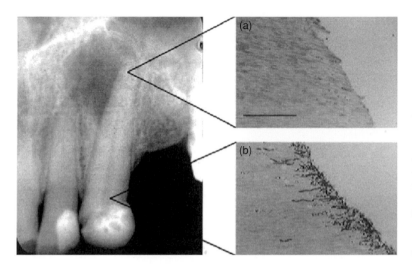

**Fig. 3.6**   Diagram demonstrating regional variation in bacterial invasion of radicular dentin of an upper left canine with infection of the root canal system and resulting periapical inflammatory disease. Bacterial colonization of the root canal from a coronal to apical direction and more advanced dentinal tubule sclerosis in the apical radicular dentin result in low and superficial invasion of apical dentin (a) and heavy and deep invasion in mid-root and cervical radicular dentin (b).

apical region of the root and progresses toward the crown as a tooth matures (Nalbandian et al. 1960). Sclerotic or obliterated tubules will physically impede bacterial invasion and reduce the diffusion of bacterial and tissue products through dentin. The pattern of dentinal tubule sclerosis results in regional differences of invasion between different areas of a tooth. Invasion of cervical and mid-root radicular dentinal tubules occurs readily, while the extent and depth of invasion are significantly less in apical tubules (Love 1996b) (Fig. 3.6). As discussed, this is because of the lower number of patent tubules in this region due to dentinal sclerosis, which is always more advanced in the apical region compared with cervical and mid-root radicular dentin at any age (Nalbandian et al. 1960). This pattern of invasion of radicular dentin supports the concept that chemomechanical root canal preparation with coronal flaring to remove the heavier mid-root and cervical dentin infection and a moderate apical enlargement can effectively reduce the bulk of infected radicular dentin (Love 1996b).

The presence of a dentinal smear layer, subsequent to tooth or canal instrumentation such as filling a root canal, obliterates the superficial portion of tubules (Fig. 3.7) and prevents penetration of coronal or radicular dentinal tubules by bacteria (Vojinovic et al. 1973; Olgart et al. 1974; Michelich et al. 1980; Love et al. 1996). As such the production of a smear layer on the root canal wall may help inhibit subsequent recolonization of the root canal space during the canal medication stage by vital bacteria remaining in radicu-

lar dentinal tubules; however, the smear layer will also inhibit diffusion of the intracanal medicament into the radicular tubules, thus reducing its efficacy.

Once pathosis, trauma, restorative, or periodontal procedures breach the integrity of enamel or cementum, the underlying dentinal tubules provide diffusion channels to the pulp or periradicular tissues. Intact cementum is crucial to limiting bacterial invasion of radicular dentinal tubules from the pulpal surface of an infected root canal system. Bacterial penetration is enhanced when the overlying cementum is resorbed (Valderhaug 1974; Haapasalo and Ørstavik 1987; Love 1996b), a common occurrence in the presence

**Fig. 3.7**   A scanning electron micrograph of a dentin smear layer produced by instrumenting a root canal wall with an endodontic hand file; note the presence of bacteria on the smear layer. The smear layer has occluded the underlying dentinal tubules and will inhibit the penetration of antimicrobial medicaments into dentinal tubules.

**Fig. 3.8**  Photomicrograph showing resorption of cementum and dentin and exposure of dentinal tubules to the periradicular tissues. As a consequence, the tubules are patent on both the pulpal (inner) and external dentin surfaces, and this results in enhanced permeability of the dentin and deeper bacterial invasion from the pulpal to the external root surface (hemotoxin and eosin stain, 200× magnification).

of inflammatory periapical disease (Fig. 3.8) and after traumatic injuries that damage the periradicular tissues. Thus, dentin infection will be heavier and deeper in areas where the cementum has been resorbed and diffusion of bacterial products toward the periradicular tissues will be enhanced, resulting in an increased periradicular inflammatory response.

## 3.3 Microflora of dentinal tubule infection

Several hundred bacterial species make up the oral microflora; however, only relatively few species appear to be able to invade coronal dentin, infect the root canal system through intact dentin, and subsequently invade radicular dentinal tubules (Kantz and Henry 1974; Sundqvist 1976; Dahlén and Bergenholtz 1980). This suggests that many species of oral bacteria do not have the necessary properties that allow invasion of tubules and their survival within the intratubular environment. The bacterial flora associated with invasion of radicular dentinal tubules is determined by the species of bacteria that invade coronal carious and noncarious dentin and infect the root canal.

### 3.3.1 Coronal dentinal tubule infection

The cariogenic microflora present on the surface of teeth consists mainly of streptococci, lactobacilli, and *Actinomyces* spp. Members of the mutans group streptococci, in particular *S. mutans* and *S. sobrinus*, are

likely to be the main bacteria initiating the induction of coronal and root caries (Bowden 1990; van Houte 1994; van Houte et al. 1994). Different regions of carious dentin contain different proportions of bacteria in their microflora with greater numbers of bacteria present in superficial infected dentin compared with deeper dentin (Hoshino 1985). Carious dentin from the outer surfaces of teeth contains *Streptococcus* spp., *Lactobacillus* spp., *Actinomyces* spp., and other Gram-positive rods (Loesche and Syed 1973). In contrast, the pulpal side (deeper dentin) contains larger numbers of Gram-positive anaerobic rods of *Eubacterium*, *Propionibacterium*, and *Bifidobacterium* species, with *Actinomyces* and *Lactobacillus* being the most prevalent facultative bacteria isolated, with streptococci comprising only a minor group of the total isolates (Edwardsson 1974).

Mutans group streptococci have been shown to be the predominant bacteria within dentin from fissure and smooth surface coronal caries, with higher numbers in the shallow and middle layers of dentin compared with deep dentin (Ozaki et al. 1994). Other bacteria identified as being dominant members of the microflora of carious human dentin, such as *Lactobacillus* spp., *Eubacterium alactolyticum*, and *Fusobacterium nucleatum* (Edwardsson 1974; Hoshino 1985; Byun et al. 2004), are frequently detected though their relative proportions are low (Table 3.1). The environment of deep carious dentin promotes the survival of obligately anaerobic bacteria, as such species of *Propionibacterium*, *Eubacterium*, and *Bifidobacterium*

**Table 3.1**   Bacterial species identified in superficial or deep carious coronal dentin.

| Bacterial genus or species | Isolation frequency | | Bacterial genus or species | Isolation frequency | |
|---|---|---|---|---|---|
| | Superficial | Deep | | Superficial | Deep |
| *Streptococcus* | High | Low/moderate | *Propionibacterium* | Moderate/high | High |
| S. mutans | | | P. acnes | | |
| S. sobrinus | | | P. avidum | | |
| S. intermedius | | | P. lymphophilum | | |
| S. morbillorum | | | P. propionicum | Low | Moderate |
| S. sanguinis | | | | | |
| *Peptostreptococcus* | Low | Low | *Lactobacillus* | High | High |
| P. anaerobius | | | L. casei | | |
| P. parvulus | | | L. plantarum | | |
| Parvimonas micra | | | L. minutus | | |
| *Actinomyces* | High | Moderate | *Fusobacterium nucleatum* | Low | Low |
| A. israelii | | | | | |
| A. naeslundii | | | | | |
| A. odontolyticus | | | | | |
| *Eubacterium* | High | High | *Peptococcus* spp. | Low | Low |
| E. alactolticum | | | | | |
| E. aerofaciens | | | | | |
| E. saburreum | | | | | |
| *Veillonella* spp. | Moderate | Low | *Clostridium* spp. | Low | Low |
| *Bifidobacterium* spp. | High | High | *Porphyromonas* spp. | Low | Low |
| | | | *Prevotella* spp. | Low | Low |

dominate the microflora of deep carious dentin, with *Actinomyces, Lactobacillus,* and some streptococci being present, but rarely *S. mutans.* However, Gramnegative obligate anaerobes, for example, *Fusobacterium,* are recovered in only very low numbers (Hoshino 1985; Edwardsson 1987; Ozaki et al. 1994; Munson et al. 2004). These observations demonstrate that the environment (e.g., available nutrients and oxygen tension) within superficial carious dentin favors the growth of facultative anaerobes that are associated with the carious process, for example, mutans streptococci, while obligate anaerobic organisms that are associated with soft tissue infection dominate the microflora deep within the dentin.

In contrast to the microflora of fissure and smooth surface carious dentin, *A. naeslundii* is the major species associated with dentin invasion in root surface caries. *Actinomyces* species are found in shallow, middle, and deep dentin, with higher numbers of cells in deeper dentin. Mutans streptococci are frequently detected at all levels of carious root dentin, though they are mainly located in the shallow layer and do not make up a high proportion of the microflora. On the other hand, lactobacilli and Gram-negative organ-

isms are found in low numbers, or not at all (Syed et al. 1975; Hill et al. 1977; Hoshino 1985; Edwardsson 1987; Ozaki et al. 1994) (Table 3.2). It can be seen that the composition of the microflora associated with carious dentin differs between coronal and root caries.

The microflora of carious and cavitated dentin of teeth with pulpitis is similar to that reported for intact carious dentin (Hahn et al. 1990). Gram-positive organisms predominate, especially *Lactobacillus* spp. and streptococci. Gram-negative bacteria, for example, *P. intermedia,* are found in lower numbers in superficial dentin, but are more prevalent within dentin at the pulpal wall, and a positive correlation between the presence of *P. intermedia* and *P. melaninogenica* and extensive pulp inflammation has been demonstrated (Massey et al. 1993).

In vivo studies show that bacteria are able to penetrate into the tubules of noncarious coronal dentin exposed to the oral environment. Invasion of tubules occurs readily and is evident within a week of exposure (Lundy and Stanley 1969; Olgart et al. 1974). With time, the number of tubules infected and the depth of infection increases (Lundy and Stanley 1969). The pattern of invasion is characterized by variable numbers

**Table 3.2**   Bacterial species identified in carious root dentin.

| Bacterial species | Isolation frequency | Bacterial species | Isolation frequency |
|---|---|---|---|
| *Streptococcus* | Low/High | *Propionibacterium* | Low/moderate |
| S. sanguinis | | P. acnes | |
| S. mitis | | *Lactobacillus* | Low |
| S. mutans | | L. casei | |
| S. sobrinus | | L. plantarum | |
| *Actinomyces* | High | *Parvimonas micra* | Low |
| A. naeslundii | | *Fusobacterium nucleatum* | Low |
| A. odontolyticus | | *Porphyromonas endodontalis* | Low |
| A. viscosus | | *Veillonella* spp. | Low |
| *Eubacterium* | Low/moderate | | |
| E. alactolyticum | | | |

of tubules penetrated and variable depths of penetration between different areas of dentin (Tronstad and Langeland 1971; Olgart et al. 1974). Inflammatory changes within the pulp are commonly observed and can be seen within a week of exposure (Olgart et al. 1974). The composition of the microflora invading exposed noncarious dentin has not been fully elucidated but is dominated by Gram-positive cells (Lundy and Stanley 1969; Brännström and Nyborg 1971; Tronstad and Langeland 1971; Vojinovic et al. 1973; Olgart et al. 1974) and probably resembles the composition of the biofilm infiltrating the tooth–restoration interface (Edwardsson 1987). This biofilm resembles mature plaque and is composed mainly of streptococci and *Actinomyces* spp. Anaerobic Gram-positive cocci, for example, *P. micros* (now *P. micra*), and Gram-negative organisms tend to be present in only low numbers (Mejàre et al. 1979, 1987).

### 3.3.2 Microflora of the infected root canal

An understanding of the dynamics of root canal infection is necessary to appreciate radicular tubule invasion and this is summarized (the reader is referred to Chapters 4 and 5 for an extensive discussion of this topic). Almost all bacteria recovered from primary infections of the root canal systems of intact teeth, that is, teeth with canals that are not physically exposed to the oral cavity, belong to a select group of the oral microflora (Wittgow and Sabiston 1975; Sundqvist 1976; Le Goff et al. 1997). Various factors such as nutritional supply, oxygen tension, and bacterial interactions influence the development of the root canal flora (Sundqvist 1992a, b). Most of the oxygen-sensitive members of the root canal microflora are not readily cultivable without strict application of anaerobic methods (Carlsson et al. 1977), and this may explain why, in early studies, many teeth with apical periodontitis did not appear to harbor bacteria in the root canal. In fact, many of the species that were recently identified with molecular techniques have not been cultivated yet. Utilizing strictly anaerobic sampling techniques, it has been shown that in addition to streptococci, lactobacilli, and *Actinomyces*, obligately anaerobic species of *Fusobacterium*, *Peptostreptococcus*, *Eubacterium*, *Propionibacterium*, *Veillonella*, *Wolinella*, *Prevotella*, and *Porphyromonas* dominate the established root canal microflora (Wittgow and Sabiston 1975; Sundqvist 1976, 1992a; Möller et al. 1981; Le Goff et al. 1997; Gomes et al. 2004) (Table 3.3). Other bacteria such as spirochetes, for example, *Treponema* spp. (Jung et al. 2000; Rôças et al. 2001), and other microorganisms such as yeasts, for example, *Candida* and *Saccharomyces* (Lana et al. 2001), have also been recovered from an infected root canal. Recently, the use of molecular techniques to identify root canal bacteria have confirmed the presence of the dominant species (Table 3.3) as well as the presence of previously unidentified or not-yet-cultivated bacteria such as *Dialister pneumosintes* (Rolph et al. 2001; Siqueira et al. 2002).

Persistent periapical pathosis subsequent to endodontic treatment is due to either intraradicular infection (Nair et al. 1990a), extraradicular infection (Sjögren et al. 1988), or other pathosis, for example, a true cyst (Nair et al. 1990b; Nair 1998). Intraradicular infection is the most common cause of persistent pathology, and the reasons for infection are numerous

**Table 3.3**  Bacterial species commonly found in infected root canals.

| Gram-positive cocci | Gram-positive rods | Gram-negative cocci | Gram-negative rods |
|---|---|---|---|
| *Streptococcus* | *Actinomyces* | *Capnocytophaga* | *Fusobacterium nucleatum* |
| S. anginosus | A. israeli | C. ochracea | *Prevotella* |
| S. sanguinis | A. naeslundii | C. sputigena | P. intermedia |
| S. mitis | *Eubacterium* | *Veillonella parvula* | P. melaninogenica |
| S. mutans | E. alactolyticum | *Campylobacter* | P. denticola |
| *Enterococcus faecalis* | E. lentum | C. rectus | P. buccae |
| *Peptostreptococcus* | E. nodatum | C. curvus | P. buccalis |
| P. anaerobius | E. timidum | | P. oralis |
| *Parvimonas* | *Propionibacterium* | | *Porphyromonas* |
| P. micra | P. propionicum | | P. gingivalis |
| | P. granulosum | | P. endodontalis |
| | *Lactobacillus* spp. | | *Bacteroides gracilis* |

including lack of coronal seal, missed canals, insufficient debridement and disinfection of the root canal system, and the presence of therapy-resistant bacteria. Unlike primary root canal infections, which are typically mixed consisting of between two and eight bacterial species with obligate anaerobic bacteria dominating the microflora and streptococci making up a significant proportion of the facultative species, the root canal flora from failed cases is primarily Gram-positive facultative anaerobes and consists of one to two species per canal. The most frequently cultivable microorganisms include bacterial species from *Enterococcus*, *Streptococcus*, *Peptostreptococcus*, and *Actinomyces* (Molander et al. 1998; Sundqvist et al. 1998; Gomes et al. 2004). While molecular analysis of refractory endodontic cases has demonstrated the ability to detect the presence of not-yet-cultivated bacteria such as *Dialister* (Rolph et al. 2001; Siqueira and Rocas 2004). Additionally yeasts, notably *Candida albicans*, have been isolated from cases of endodontic failure (Nair et al. 1990a; Molander et al. 1998; Sundqvist et al. 1998; Peciuliene et al. 2001). *E. faecalis*, which makes up a small percentage of the flora in primary root canal infection, is the bacterial species most frequently recovered in root-filled teeth with apical periodontitis, and often as a pure culture (Sundqvist et al. 1998; Peciuliene et al. 2001).

### 3.3.3 Bacterial invasion of radicular dentin from the root canal

When bacteria gain access into the root canal system, they invade radicular dentinal tubules (Shovelton 1964) (Fig. 3.2c) and may be responsible for persistent root canal infection (Haapasalo and Ørstavik 1987; Ørstavik and Haapasalo 1990). The numbers of radicular tubules containing bacteria and the depth of penetration are variable from tooth to tooth and among sections of an individual tooth (Shovelton 1964; Love 1996b). It was noted that the presence of bacteria within radicular tubules was related to the clinical history of the tooth such that long-standing infections had more bacterial invasion and that tubule invasion did not occur immediately after the bacteria appeared in the root canal (Shovelton, 1964). These observations are similar to studies on invasion of noncarious coronal dentin (Lundy and Stanley 1969; Brännström and Nyborg 1971; Tronstad and Langeland 1971; Vojinovic et al. 1973; Olgart et al. 1974).

The microflora within radicular dentinal tubules of teeth with infected root canals resembles that of deep layers of carious coronal dentin (Edwardsson 1974; Hoshino 1985) (Table 3.1). Lactobacilli, streptococci, and *Propionibacterium* spp. are predominant members of the flora, with other bacteria such as Gram-positive anaerobic cocci, *Eubacterium* spp., and *Veillonella* spp. being present in lower numbers. Obligate anaerobic Gram-negative bacteria such as *F. nucleatum*, *P. gingivalis*, and *P. intermedia* are recovered in variable numbers (Edwardsson 1974; Hoshino 1985; Ando and Hoshino 1990; Peters et al. 2001). Additionally, yeasts have been reported to invade radicular dentinal tubules in teeth with root canals heavily colonized by yeasts (Sen et al. 1995). The inability to detect fastidious anaerobes within invaded coronal or radicular dentin may be due to difficulties in cultivating these bacteria. Other techniques, such as immunohistochemical staining using specific antisera, demonstrate the presence of bacteria that are difficult to cultivate such as *Porphyromonas endodontalis* (Ozaki et al. 1994). These observations show that bacteria that are able to invade coronal dentinal tubules maintain the ability, under different environmental conditions in the root canal, to invade radicular dentin.

It is noteworthy that *E. faecalis* and other bacterial species identified in refractory cases such as *S. gordonii/sanguinis*, *Streptococcus mitis*, *P. micra*, *A. naeslaundi*, and *P. intermedia* possess the ability to invade dentinal tubules as mono-cultures in ex vivo experiments (Love 2001, 2004). For bacteria to be involved in the pathogenesis and maintenance of persistent apical periodontitis, they must be able to survive in the inhospitable environment of the filled root canal where the nutrient supply is limited. Investigators have focused on *E. faecalis* in an attempt to elicit mechanisms involved in pathogenesis. Studies have shown that *E. faecalis* is able to withstand a high alkaline environment such as produced in radicular dentin by calcium hydroxide (Haapasalo and Ørstavik 1987), which appears to be related to a cell proton pump that is necessary for its survival at high pH (Evans et al. 2002), and can form biofilms in calcium hydroxide-medicated canals (Distel et al. 2001). In addition, under starved conditions, it shows resistance to sodium hypochlorite (LaPlace et al. 1997), heat, hydrogen peroxide, acid, and ethanol (Giard et al. 1996). *E. faecalis* can also survive extended periods of starvation in water (Hartke et al. 1998) and within water-filled dentinal tubules (Ørstavik and Haapasalo 1990) and human serum (Love 2001), which likely reflects the nutritional supply within nonvital radicular dentinal tubules. The upregulation of stress-induced proteins has been shown to be important for cell survival in a stressed environment (Hartke et al. 1998). It is highly likely that bacterial cells within dentinal tubules would be in a state of starvation and some of these mechanisms may come into play. It is suggested that following root canal therapy this ability may allow residual bacterial cells in radicular dentinal tubules to recolonize the obturated root canal and participate in chronic failure of endodontically treated teeth.

### 3.3.4 Bacterial invasion of radicular dentin from a periodontal pocket

Bacterial invasion of radicular dentin from the periodontal pocket of periodontally diseased teeth has been demonstrated by light microscopy (Kopczyk and Conroy 1968; Langeland et al. 1974; Adriaens et al. 1987b) and by microbiological studies (Adriaens et al. 1987a; Giuliana et al. 1997). The majority of species recovered from radicular dentin are Gram-positive bacteria (*P. micra*, *Streptococcus intermedius*, *A. naeslundii*), with lower numbers of Gram-negative organisms (*P. gingivalis*, *P. intermedia*, *Tannerella forsythia*, *F. nucleatum*, *Veillonella parvula*) (Giuliana et al. 1997).

While it is clear that bacteria are able to invade radicular dentin from the periodontal pocket, it is not clear whether bacteria invade healthy cementum prior to dentin penetration or if bacteria gain assess to dentin only via breaches in the cementum layer. A number of studies have described invasion of the cementum of periodontally diseased teeth (Daly et al. 1982; Adriaens et al. 1987a, b; Giuliana et al. 1997). However, it was not evident from any of these studies if the invaded cementum was intact, healthy, or diseased. Exposed cementum is a thin, often discontinuous layer (Moskow 1969), and commonly shows surface defects, for example, at sites where Sharpey's fibers attach to the cementum matrix (Adriaens et al. 1987b). Exposure of cementum to crevicular fluid, bacterial enzymes, or acidic metabolites may induce physicochemical and structural alterations, such as localized resorptive lacunae or demineralization (Daly et al. 1982; Eide et al. 1984; Adriaens et al. 1987b). It seems likely therefore that bacterial invasion of exposed cementum associated with periodontal disease occurs after the cementum has been altered by physiological, bacterial, or environmental factors. Similarly, removal of cementum as a consequence of periodontal treatment, for example, scaling will enhance bacterial invasion of radicular dentin.

## 3.4 Clinical aspects of dentinal tubule infection

### 3.4.1 Invasion of coronal dentin: Influence on the progression and management of pulp disease

Bacterial invasion through coronal dentin is the primary route of infection of the pulp and root canal system. Bacteria may enter the root canal system directly via carious lesions or via pulp exposure following trauma. However, many infections of the pulp occur as a result of supragingival or subgingival bacteria penetrating exposed dentin, enamel-dentin cracks, and around restorations (Pashley 1990; Love 1996a; Peters et al. 2001) and then invading dentinal tubules. Microleakage is defined as the clinically undetectable passage of bacteria, fluids, molecules, or ions between a cavity wall and the restorative material applied to it (Kidd 1976), and studies have demonstrated that microleakage of oral bacteria around restorations allows bacterial invasion of exposed dentinal tubules at the base of the cavity (Brännström and Nyborg 1971; Vojinovic et al. 1973), resulting in pulpal inflammation (Vojinovic et al. 1973) or periapical disease (Ray and Trope 1995). Likewise, microleakage through enamel cracks and fractures as a result of trauma may lead to bacterial invasion of the pulp–dentin complex and act as a cause of endodontic disease (Love 1996a). Hence, sealing of dentin from exogenous substances and bacteria in the oral cavity, in both vital and nonvital teeth, is a critical step in tooth restoration.

### 3.4.2 Invasion of radicular dentin: influence on the progression and management of periapical disease

The principles of complete debridement of vital and necrotic pulp tissue, removal of microorganisms, and affected dentin and disinfection of the root canal system and radicular dentin are cornerstones in successful root canal therapy. The ability to achieve these aims is partly dependent on the choice of root canal preparation technique and disinfection regimen (for a more extensive discussion of endodontic disinfection procedures, see Chapter 14). Here, the basic principles of canal preparation and disinfection are outlined.

A preparation technique that results in a flared root canal with minimal canal transportation is favored (Fig. 3.9) as this results in the most effective debridement of the intraradicular space (Walton 1976) and radicular dentin, and produces a root canal shape that is conducive to filling and sealing the root canal (Ayar and Love 2004). In addition, the regional differences in the pattern of tubule invasion in the root suggests that a technique that results in moderate apical preparation, because of mild and limited depth of dentin infection in this region, but that flares coronally debriding heavily infected mid-root and cervical dentin (Fig. 3.6), should be employed (Love 1996b).

The complex anatomy of the root canal system is not readily cleaned by mechanical means alone. The concept of chemomechanical preparation, utilizing a

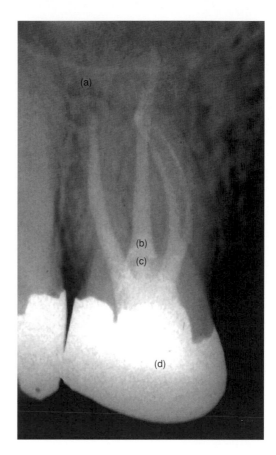

**Fig. 3.9** Radiograph of an endodontically treated upper right first molar with four root canals. Successful disinfection of the root canal and radicular dentin is dependant on chemomechanical instrumentation of all of the root canal system to produce a well-centered canal flaring from the apical stop 1 mm from the radiographic apex (a) to the canal orifice (b) at the pulp chamber. The root-filling materials should three-dimensionally fill the root canal space and form an apical seal at the apical stop and extend coronally to fill the root canal at or 1–2 mm short of the canal orifice; however, the root filling should not extend into the pulp chamber. A coronal seal (c) is formed over the coronal extent of the root filling using an appropriate permanent restorative material and a well-sealed crown (d) that returns the tooth to function and form is placed.

chemical irrigant as an adjunct to mechanical instrumentation, maximizes soft tissue and microorganism removal. Traditionally, sodium hypochlorite (NaOCl) has been used to accomplish both functions (Byström and Sundqvist 1983) and is commonly used in solutions ranging from 0.5% to 5.25%. Its soft tissue dissolving and antimicrobial properties may be enhanced when the solution is warmed; similarly, its antimicrobial effectiveness may also be improved in conjunction with ultrasonic energy.

The chemicals should also aid penetration of disinfection agents into the tubules. Whenever dentin is cut or abraded, a smear layer of debris packs into and occludes the dentinal tubules. This layer is burnished onto the surface of the dentin and cannot be readily removed (Fig. 3.7). Dentinal smear layers decrease the area available for diffusion and markedly reduce the permeability of dentin. This inhibits the ability of intracanal medicaments to diffuse into infected dentinal tubules and eliminate microorganisms. To facilitate penetration of irrigants and intracanal medicaments into radicular dentin, the smear layer should be removed during chemomechanical preparation (Fig. 3.10).

The most common method is to use ethylenediaminetetraacetic acid (EDTA 17% w/v), a chelation agent that dissolves the hard tissue components of the smear layer, in conjunction with sodium hypochlorite irrigation. Several other irrigants are commercially available to the clinician and have been developed to enhance penetration into radicular dentin, for example, by the addition of acids to remove smear layer and detergents to reduce surface tension and to be more effective against resistant bacteria such as *E. faecalis*.

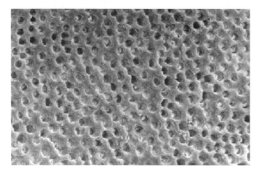

**Fig. 3.10** A scanning electron micrograph of an instrumented root canal wall that was treated with an ethylenediaminetetraacetic acid/sodium hypochlorite regime to remove the smear layer. The patent dentinal tubules allow maximum diffusion of antimicrobial medicaments into tubules to eradicate invading bacteria. Additionally, root canal filling materials can better form a seal with the root canal wall and occlude the tubules when the smear layer is removed.

Chemomechanical preparation greatly reduces the number of bacteria in the root canal. However, to date no preparation technique or irrigation material can predictably render a root canal and radicular dentin sterile. As a consequence, bacteria can survive and multiply within the root canal system and dentinal tubules necessitating the use of intracanal medicaments between endodontic appointments. Intracanal medicaments have been recommended for a number of different reasons, for example, to reduce periapical inflammation, induce healing of calcified tissue, eliminate apical exudate, and neutralize tissue debris. However, the main reason is to eliminate residual bacteria within the root canal and dentinal tubules after chemomechanical preparation. Clinicians should choose an intracanal medicament that is capable of disinfecting deeply infected dentin so as to effectively eradicate potentially heavy and deep infection of cervical and mid-root dentinal tubules and deeply penetrating bacteria at sites associated with external resorption of the cementum. A number of intracanal medicaments are available; however, calcium hydroxide is a popular medicament that has been shown to eliminate bacteria within dentinal tubules (Ørstavik and Haapasalo 1990), and effectively disinfect the prepared root canal system after 1 week (Sjögren et al. 1991).

Theoretically, a root-filling material ought to be placed into a sterile root canal system that can be sealed from the oral cavity to maintain sterility. However, clinically it is not possible to predictably render a root canal system and radicular dentin sterile, and it would take only a few viable bacteria to survive within dentinal tubules to subsequently reinfect the root canal system. As such, a root filling is placed within the root canal system (Fig. 3.9) to control reinfection of the root canal and it does this primarily by producing an environment in the root canal and radicular dentin that is inhospitable for bacterial growth; for example, it denies bacteria nutrition by inhibiting microleakage into the root canal and tubules of tissue fluid that would sustain growth, it denies space for bacteria to multiply in (an essential part of colonization), and it alters the redox conditions within the root canal. There are numerous root-filling techniques and materials, and they all conform to the same principles and attempt to produce a three-dimensional fill of the root canal that is closely associated with the radicular dentin so as to form a fluid-tight seal along the whole length of the root canal from the apical preparation (apical seal) to the canal orifice (coronal seal) (Fig. 3.9). As such, a root-fillings function to (i) entomb residual viable bacteria within dentinal tubules, and as long as the root-filling produces an environment (primarily by an apical seal) that inhibits their growth, the bacteria will not reinfect the root canal or cause inflammatory periapical disease and (ii) prevent reinfection of the root canal system from the oral cavity by producing a coronal seal at the canal orifice.

The provision of a coronal seal is mandatory to prevent reinfection of the intraradicular space from oral bacteria both between endodontic appointments and after filling of the canal. A number of materials can be used as a coronal seal during endodontic therapy; however, the material must not only provide a microbial seal but also be able to withstand forces of mastication in order not to compromise the seal. A functional restoration placed immediately after root canal filling is the most effective coronal seal. The provision of a temporary restoration during an endodontic success/failure assessment period is not warranted as the coronal seal may break down over prolonged periods and allow reinfection of the intraradicular space.

Persistent (refractory) periapical pathology subsequent to endodontic treatment is due to three main reasons: (i) intraradicular infection as a consequence of reinfection of the root canal either by residual bacteria or by coronal microleakage from the oral cavity as discussed, (ii) extraradicular infection where certain bacteria, for example, *Actinomyces* spp. and *Proprionibacterium* spp., can colonize the external root surface, or (iii) other pathological condition such as an apical cyst. The most common cause of persistent periapical pathosis is intraradicular infection, and this may be due to a number of factors such as the presence of therapy-resistant bacteria, for example, *E. faecalis*, ineffective chemomechanical instrumentation, for example, failure to locate and instrument canals leaving viable bacteria within the root canal system and/or radicular dentinal tubules, or an ineffective apical or coronal seal as discussed. To effectively treat the root canal and radicular dentin infection, conventional endodontic retreatment should be performed, as this will be the most effective means of removing the cause. If it is not possible to adequately clean/debride and shape the full extent of the root canal system and radicular dentin, for example, a blocked canal prevents instrumentation of a canal to length or removal of a post may compromise remaining tooth structure then endodontic surgery may be indicated. However,

it should be borne in mind that since the cause of the treatment failure is due to intraradicular infection, endodontic surgery may not eradicate the cause and failure may result. Treatment of an extraradicular infection or a true apical cyst requires a surgical approach; however, conventional retreatment is commonly completed prior to surgery so as to exclude intraradicular infection as the cause of the refractory disease.

### 3.4.3 Invasion of radicular dentin: influence on the progression and management of periodontal disease

The effect of periodontal disease and associated infected radicular dentin on the development of pulp and periapical disease is not clear. The effects on the pulp are probably degenerative and inflammatory (Langeland et al. 1974); however, studies and clinical practice suggest that if pulpal pathosis does develop as a consequence of periodontal disease, it occurs as a result of exposure of a lateral/accessory canal (Rubach and Mitchell 1965) or develops late in the periodontal disease progression, for example, when it involves the apex of the tooth (Czarnecki and Schilder 1979). However, the effect of infected radicular dentin on progression of periodontal disease may be more significant, it has been suggested that the dentinal tubule microflora associated with a periodontal pocket could act as a reservoir for recolonization of the pocket after debridement (Adriaens et al. 1987a, b; Giuliana et al. 1997). The presence of an infected root canal system and associated radicular dentinal tubules in treated periodontitis-prone patients was shown to result in increased pocket depth and radiographic evidence of loss of attachment (Jansson et al. 1993a, b), and impaired healing of pockets (Ehnevid et al. 1993). Similarly, the presence of a root canal infection in patients with periodontitis was associated with higher levels of attachment loss as assessed by probing depths and radiographs. It was suggested that a root canal infection in periodontitis-involved teeth may potentiate progression of the periodontal disease by spreading of endodontic pathogens through patent accessory canals and dentinal tubules (Jansson et al. 1995; Jansson and Ehnevid 1998). As a consequence, asymptomatic endodontic pathosis should be promptly treated in patients with periodontal disease, while the presence of an infected root canal system should be investigated in teeth where the response to periodontal treatment is less than expected or in teeth with isolated loss of periodontal attachment. Similarly, if advanced periodontal techniques are proposed, for example, tissue regeneration then the presence of root canal infection in associated teeth should be assessed and treated as necessary prior to the periodontal treatment.

## 3.5 Conclusion

Bacterial invasion of dentinal tubules and the clinical consequences have been recognized for over a century. However, while many components of the infected dentinal tubule microflora have been identified, it is likely that there are etiologic agents involved in endodontic infections that have not yet been recognized. Bacterial invasion of coronal dentinal tubules occurs in caries, when the dentin is exposed to the oral environment and of radicular dentinal tubules subsequent to infection of the root canal system or as a consequence of periodontal disease. The content and architecture of a dentinal tubule can influence bacterial invasion with tubule patency being important. This can account for regional variations in bacterial invasion and is particularly seen with dentinal sclerosis where more advanced sclerotic changes in apical radicular tubules, especially in elderly individuals, limits bacterial invasion in this area.

While several hundred bacterial species are known to inhabit the oral cavity, a relatively small and select group of bacteria are involved in invasion of dentinal tubules. Gram-positive organisms dominate the tubule microflora in both carious and noncarious dentin, while the relatively high numbers of obligate anaerobes present, such as *Eubacterium* spp., *Propionibacterium* spp., *Bifidobacterium* spp., *P. micra*, and *Veillonella* spp., suggest that the environment favors growth of these bacteria. Gram-negative obligate anaerobic rods, for example, *Porphyromonas* spp., are less frequently recovered; however, with time, fastidious obligately anaerobic bacteria become established as principal components of the microflora and can be found within the deep dentin layers. Although clinical techniques can manage infected coronal and radicular dentin, the complexity of the microbial interactions and bacterial resistance to treatment will require the development of materials and techniques to both prevent and treat bacterial invasion of dentinal tubules.

## 3.6 References

Adriaens PA, De Boever JA, and Loesche WJ. 1987a. Bacterial invasion in root cementum and radicular dentin of periodontally diseased teeth in humans. *J Periodontol* 59: 222–30.

Adriaens PA, Edwards CA, De Boever JA, and Loesche WJ. 1987b. Ultrastructural observations on bacterial invasion in cementum and radicular dentin of periodontally diseased human teeth. *J Periodontol* 59: 493–503.

Ando N and Hoshino E. 1990. Predominant obligate anaerobes invading the deep layers of root canal dentine. *Int Endodon J* 23: 20–27.

Ayar LR and Love RM. 2004. Shaping ability of ProFile and K3 rotary Ni–Ti instruments when used in a variable tip sequence in simulated curved root canals. *Int Endod J* 37: 593–601.

Bowden GHW. 1990. Microbiology of root surface caries in humans. *J Dent Res* 69: 1205–10.

Brännström M and Nyborg H. 1971. The presence of bacteria in cavities filled with silicate cement and composite resin materials. *Swed Dent J* 64: 149–55.

Byström A and Sundqvist G. 1983. Bacteriological evaluation of the effect of 0.5 percent sodium hypochlorite in endodontic therapy. *Oral Surg Oral Med Oral Pathol* 55: 1307–12.

Byun R, Nadkarni MA, Chhour KL, Martin FE, Jacques NA, and Hunter NJ. 2004. Quantitative analysis of diverse lactobacillus species present in advanced dental caries. *Clin Microbiol* 42: 3128–36.

Carlsson J, Frolander F, and Sundqvist G. 1977. Oxygen tolerance of anaerobic bacteria isolated from necrotic dental pulps. *Acta Odontol Scand* 35: 139–45.

Carrigan P, Morse JDR, Furst ML, and Sinai IH. 1984. A scanning electron microscope evaluation of human dentinal tubules according to age and location. *J Endodon* 10: 359–63.

Czarnecki RT and Schilder H. 1979. A histological evaluation of the human pulp in teeth with varying degrees of periodontal disease. *J Endodon* 5: 242–53.

Dahlén G and Bergenholtz G. 1980. Endotoxic activity in teeth with necrotic pulps. *J Dent Res* 59: 1033–40.

Dai X-F, Ten Cate AR, and Limeback H. 1991. The extent and distribution of intratubular collagen fibrils in human dentine. *Arch Oral Biol* 36: 775–78.

Daly C, Seymour G, Kieser J, and Corbet E. 1982. Histological assessment of periodontally involved cementum. *J Clin Periodontol* 9: 266–74.

Distel J, Hatton J, and Gillespie MJ. 2001. *Enterococcus faecalis* colonization and biofilm formation in medicated root canals. *J Endodon* 28: 689–93.

Dixon C and Rickert U. 1933. Tissue tolerance to foreign materials. *J Am Dent Assoc* 20: 1458–72.

Edwardsson S. 1974. Bacteriological studies on deep areas of carious dentine. *Odontol Revy* 25(Suppl 32): 1–143.

Edwardsson S. 1987. Bacteriology of dentin caries. In: Thylstrup A, Leach SA, and Qvist V (eds), *Dentine and Dentine Reactions in the Oral Cavity*. Oxford: IRL Press Ltd, 95–102.

Ehnevid H, Jansson L, Lindskog S, and Blomlof L. 1993. Periodontal healing in teeth with periapical lesions. A clinical retrospective study. *J Clin Periodontol* 20(4): 254–58.

Eide B, Lie T, and Selvig KA. 1984. Surface coatings on dental cementum incident to periodontal disease. II. Scanning electron microscope confirmation of a mineralized cuticle. *J Clin Periodontol* 11: 565–75.

Evans M, Davies JK, Sundqvist G, and Figdor D. 2002. Mechanisms involved in the resistance of *Enterococcus faecalis* to calcium hydroxide. *Int Endod J* 35: 221–28.

Garberoglio R and Brännström M. 1976. Scanning electron microscopic investigation of human dentinal tubules. *Arch Oral Bio* 21: 355–62.

Giard JC, Hartke A, Flahaut S, Benachour A, Boutibonnes P, and Auffray Y. 1996. Starvation-induced multiresistance in *Enterococcus faecalis* JH2-2. *Curr Microbiol* 32: 264–71.

Giuliana G, Ammatuna P, Pizzo G, Capone F, and D'Angelo M. 1997. Occurrence of invading bacteria in radicular dentin of periodontally diseased teeth: microbiological findings. *J Clin Periodontol* 24: 478–85.

Gomes BPFA, Pinheiro ET, Gadê-Neto CR, Sousa ELR, Ferraz CCR, Zaia AA, Teixeira FB, and Souza-Filho FJ. 2004. Microbiological examination of infected root canals. *Oral Micobiol Immunol* 19: 71–76.

Haapasalo M and Ørstavik D. 1987. *In vitro* infection and disinfection of dentinal tubules. *J Dent Res* 66: 1375–79.

Hahn C-L, Falkler WA, Jr, and Minah GE. 1990. Microbiological studies of carious dentine from human teeth with irreversible pulpitis. *Archs Oral Biol* 36: 147–53.

Hahn C-L and Overton B. 1997. The effects of immunoglobulins on the convective permeability of human dentine *in vitro*. *Arch Oral Biol* 42: 835–43.

Hartke A, Giard JC, LaPlace JM, and Auffray Y. 1998. Survival of *Enterococcus faecalis* in an oligotrophic microcosm: Changes in morphology, development of general stress resistance, and analysis of protein synthesis. *Appl Environ Microbiol* 64: 4238–45.

Hill PE, Knox KW, Schamschula RG, and Tabua J. 1977. The identification and enumeration of actinomyces from plaque of New Guinea indigenes. *Caries Res* 11: 327–35.

Hoshino E. 1985. Predominant obligate anaerobes in human carious dentin. *J Dent Res* 64: 1195–98.

Hubble TS, Hatton JF, Nallapareddy SR, Murray BE, and Gillespie MJ. 2003. Influence of *Enterococcus faecalis* proteases and the collagen-binding protein, Ace, on adhesion to dentin. *Oral Microbiol Immunol* 18: 121–26.

Izumi T, Yamada K, Inoue H, Watanabe K, and Nishigawa Y. 1998. Fibrinogen/fibrin and fibronectin in the dentin-pulp complex after cavity preparation in rat molars. *Oral Surg Oral Med Oral Pathol Oral Radiol Endod* 86: 587–91.

Jansson LE and Ehnevid H. 1998. The influence of endodontic infection on periodontal status in mandibular molars. *J Periodontol* 69(12): 1392–96.

Jansson LE, Ehnevid H, Lindskog SF, and Blomlof LB. 1993a. Relationship between periapical and periodontal status. A clinical retrospective study. *J Clin Periodontol* 20(2): 117–23.

Jansson LE, Ehnevid H, Lindskog SF, and Blomlof LB. 1993b. Radiographic attachment in periodontitis-prone teeth with endodontic infection. *J Periodontol* 64(10): 947–53.

Jansson LE, Ehnevid H, Lindskog SF, and Blomlof LB. 1995. The influence of endodontic infection on progression of marginal bone loss in periodontitis. *J Clin Periodontol* 22(10): 729–34.

Jenkinson HF and Lamont RJ. 1997. Streptococcal adhesion and colonization. *Crit Rev Oral Biol Med* 8: 175–200.

Jung I-Y, Choi B-K, Kum K-Y, Roh B-D, Lee S-J, Lee C-Y, and Park DS. 2000. Molecular epidemiology and association of putative pathogens in root canal infection. *J Endodon* 26: 599–604.

Kakehashi S, Stanley HR, and Fitzgerald RJ. 1965. The effects of surgical exposures of dental pulps in germ-free and conventional laboratory rats. *Oral Surg Oral Med Oral Path* 20: 340–49.

Kantz WE and Henry CA. 1974. Isolation and classification of anaerobic bacteria from intact pulp chambers of nonvital teeth in man. *Arch Oral Biol* 19: 91–96.

Keyes PH. 1960. The infectious and transmissible nature of experimental caries. Findings and implications. *Arch Oral Biol* 1: 304–20.

Kidd EA. 1976. Microleakage: a review. *J Dent* 4(5): 199–206.

Knutsson G, Jontell M, and Bergenholtz G. 1994. Determination of plasma proteins in dentinal fluid from cavities prepared in healthy young human teeth. *Archs Oral Biol* 39: 185–90.

Kopczyk R and Conroy C. 1968. The attachment of calculus to root planed surfaces. *Periodontics* 6: 78–83.

Lana MA, Ribeiro-Sobrinho AP, Stehling R, Garcia GD, Silva BKC, Hamdan JS, Nicoli JR, Carvalho MA, and Farias Lde M. 2001. Microorganisms isolated from root canals presenting necrotic pulp and their drug susceptibility *in vitro*. *Oral Microbiol Immunol* 16: 100–105.

Langeland K, Rodrigues H, and Dowden W. 1974. Periodontal disease, bacteria, and pulpal histopathology. *Oral Surg Oral Med Oral Pathol* 37: 257–70.

LaPlace J-M, Thuault M, Hartke A, Boutibonnes P, and Auffray Y. 1997. Sodium hypochlorite stress in *Enterococcus faecalis:* influence of antecedent growth conditions and induced proteins. *Curr Microbiol* 34: 284–89.

Le Goff A, Bunetel L, Mouton C, and Bonnaure-Mallet M. 1997. Evaluation of root canal bacteria and their antimicrobial susceptibility in teeth with necrotic pulp. *Oral Microbiol Immunol* 12: 318–22.

Linde A and Goldberg M. Dentinogenesis. 1993. *Crit Rev Oral Biol Med* 45: 679–728.

Liu T and Gibbons RJ. 1990. Binding of streptococci of the mutans group to type 1 collagen associated with apatitic surfaces. *Oral Microbiol Immunol* 5: 131–36.

Liu T, Gibbons RJ, and Hay DI. 1990. *Streptococcus cricetus* and *Streptococcus rattus* bind different segments of collagen molecules. *Oral Microbiol Immunol* 5: 143–48.

Loesche WJ and Syed SA. 1973. The predominant cultivable flora of carious plaque and carious dentin. *Caries Res* 7: 201–16.

Love RM. 1996a. Bacterial penetration of the root canal of intact incisor teeth after a simulated traumatic injury. *Endod Dent Traumatol* 12: 289–93.

Love RM. 1996b. Regional variation in root dentinal tubule infection by *Streptococcus gordonii*. *J Endodon* 22: 290–93.

Love RM. 2001. *Enterococcus faecalis*—a mechanism for its role in endodontic failure. *Int Endodon J* 34: 399–405.

Love RM. 2002. The effect of tissue molecules on bacterial invasion of dentine. *Oral Microbiol Immunol* 17: 32–37.

Love RM. 2004. Invasion of radicular dentinal tubules by root canal bacteria. *Endod Topics* 9: 52–65.

Love RM. 2007. Hemin nutritional stress inhibits bacterial invasion of radicular dentine by two endodontic anaerobes. *Int Endodon J* 40: 94–99.

Love RM, Chandler NP, and Jenkinson HF. 1996. Penetration of smeared or nonsmeared dentine by *Streptococcus gordonii*. *Int Endodon J* 29: 2–12.

Love RM and Jenkinson HF. 2002. Invasion of dentinal tubules by oral bacteria. *Crit Rev Oral Biol Med* 13: 171–83.

Love RM, McMillan MD, and Jenkinson HF. 1997. Invasion of dentinal tubules by oral streptococci is associated with collagen recognition mediated by the antigen I/II family of polypeptides. *Infect Immun* 65: 5157–64.

Love RM, McMillan MD, Park Y, and Jenkinson HF. 2000. Coinvasion of dentinal tubules by *Porphyromonas gingivalis* and *Streptococcus gordonii* depends upon binding specificity of streptococcal antigen I/II adhesin. *Infect Immun* 68: 1359–65.

Lundy T and Stanley HR. 1969. Correlation of pulpal histopathology and clinical symptoms in human teeth subjected to experimental irritation. *Oral Surg* 27: 187–201.

Marshall GW. 1993. Dentin microstructure and characterization. *Quintessence Int* 24: 606–17.

Massey WLK, Romberg DM, Hunter N, and Hume WR. 1993. The association of carious dentin microflora with tissue changes in human pulpitis. *Oral Microbiol Immunol* 8: 30–35.

Mejàre B, Mejàre I, and Edwardsson S. 1979. Bacteria beneath composite restorations—a culturing and histobacteriological study. *Acta Odont Scand* 37: 267–75.

Mejàre B, Mejàre I, and Edwardsson S. 1987. Acid etching and composite resin restorations. A culturing and histologic study on bacterial penetration. *Endod Dent Traumatol* 3: 1–5.

Michelich VJ, Schuster GS, and Pashley DH. 1980. Bacterial penetration of human dentin *in vitro*. *J Dent Res* 59: 1398–403.

Miller WD. 1890. *The Micro-Organisms of the Human Mouth*. Philadelphia: The S.S. White Dental Manufacturing Co.

Molander A, Reit C, Dahlen G, and Kvist T. 1998. Microbiological status of root-filled teeth with apical periodontitis. *Int Endod J* 31: 1–7.

Möller ÅJR, Fabricius L, Dahlén G, Öhman AE, and Heyden G. 1981. Influence on periapical tissues of indigenous oral bacteria and necrotic pulp tissue in monkeys. *Scand J Dent Res* 89: 475–84.

Moskow B. 1969. Calculus attachment in cemental separations. *J Periodontol* 40: 125–30.

Munson MA, Banerjee A, Watson TF, and Wade WG. 2004. Molecular analysis of the microflora associated with dental caries. *J Clin Microbiol* 42: 3023–29.

Nagaoka S, Liu H-J, Minemoto K, and Kawagoe M. 1995. Microbial induction of dentinal caries in human teeth *in vitro*. *J Endodon* 21: 546–51.

Nair PNR. 1998. New perspectives on radicular cysts: Do they heal? *Int Endod J* 31: 155–60.

Nair PNR, Sjögren U, Krey G, Kahnberg KE, and Sundqvist G. 1990a. Intraradicular bacteria and fungi in root-filled, asymptomatic human teeth with therapy-resistant periapi-cal lesions: a long-term light and electron microscopic follow-up study. *J Endodon* 16: 580–88.

Nair PNR, Sjögren U, Krey G, and Sundqvist G. 1990b. Therapy-resistant foreign body giant cell granuloma at the periapex of a root-filled human tooth. *J Endodon* 16: 589–95.

Nalbandian J, Gonzales F, and Sognnaes RF. 1960. Sclerotic age changes in root dentin of human teeth as observed by optical, electron, and x-ray microscopy. *J Dent Res* 39: 598–607.

Olgart L, Brännström M, and Johnson G. 1974. Invasion of bacteria into dentinal tubules. Experiments *in vivo* and *in vitro*. *Acta Odontol Scand* 32: 61–70.

Ørstavik D and Haapasalo M. 1990. Disinfection by endodontic irrigants and dressings of experimentally infected dentinal tubules. *Endodon Dent Traumatol* 23: 20–27.

Ozaki K, Matsua T, Nakae H, Noiri Y, Yoshiyama M, and Ebisu S. 1994. A quantative comparison of selected bacteria in human carious dentine by microscopic counts. *Caries Res* 28: 137–45.

Pashley DH. 1990. Clinical considerations in microleakage. *J Endodon* 16: 70–77.

Pashley DH. 1992. Dentin permeability and dentin sensitivity. *Proc Finn Dent Soc* 88(Suppl 1): 215–24.

Pashley DH. 1996. Dynamics of the pulpo–dentin complex. *Crit Rev Oral Biol Med* 7: 104–33.

Pashley DH, Nelson R, and Kepler E. 1982. The effects of plasma and salivary constituents on dentin permeability. *J Dent Res* 61: 978–81.

Peciuliene V, Reynaud AH, Balciuniene I, and Haapasalo M. 2001. Isolation of yeasts and enteric bacteria in root-filled teeth with chronic apical periodontitis. *Int Endod J* 34: 429–34.

Peters LB, Wesselink PR, Buijs JF, and van Winkelhoff AJ. 2001. Viable bacteria in root dentinal tubules of teeth with apical periodontitis. *J Endodon* 27: 76–81.

Peters LB, Wesselink PR, and Moorer WR. 1995. The fate and role of bacteria left in root dentinal tubules. *Int Endodon J* 28: 95–99.

Ray HA and Trope M. 1995. Periapical status of endodontically treated teeth in relation to the technical quality of the root filling and the coronal restoration. *Int Endodon J* 28: 12–18.

Rickert U and Dixon C. 1931. The controlling of root surgery. *Proceeding of the 8th International Dental Congress, Paris*, Section IIIa: 15–22.

Rôças IN, Siqueira JF, Jr, Santos KRN, and Coelho AMA. 2001. "Red complex" (*Bacteroides forsythus, Porphyromonas gingivalis*, and *Treponema denticola*) in endodontic infections: a molecular approach. *Oral Surg Oral Med Oral Pathol Oral Radiol Endod* 91: 468–71.

Rolph HJ, Lennon A, Riggio MP, Saunders WP, MacKenzie D, Coldero L, and Bagg J. 2001. Molecular identification of microorganisms from endodontic infections. *J Clin Microbiol* 39: 3282–89.

Rubach WC and Mitchell DF. 1965. Periodontal disease, accessory canals and pulp pathosis. *J Periodontol* 36: 34–38.

Sen BH, Piskin B, and Demirci D. 1995. Observation of bacteria and fungi in infected root canals and dentinal tubules by SEM. *Endod Dent Traumatol* 11: 6–9.

Shovelton DH. 1964. The presence and distribution of micro-organisms within non-vital teeth. *Brit Dent J* 117: 101–7.

Siqueira JF, Jr, and Rôças IN. 2004. Polymerase chain reaction-based analysis of microorganisms associated with failed endodontic treatment. *Oral Surg Oral Med Oral Pathol Oral Radiol Endod* 97: 85–94.

Siqueira JF, Jr, Rôças IN, Souto R, de Uzeda M, and Colombo A. 2002. *Actinomyces* species, *Streptococci*, and *Enterococcus faecalis* in primary root canal infections. *J Endodon* 28: 168–72.

Sjögren U, Figdor D, and Sundqvist G. 1991. The antimicrobial effect of calcium hydroxide as a short-term intracanal dressing. *Int Endodon J* 24: 119–25.

Sjögren U, Happonen RP, Kahnberg K-E, and Sundqvist G. 1988. Survival of *Arachnia propionica* in periapical tissue. *Int Endod J* 22: 277–82.

Sundqvist G. 1976. Bacteriological studies of necrotic dental pulps. *Odontotological Dissertations*, no. 7. Umeå, Sweden: University of Umeå.

Sundqvist G. 1992a. Associations between microbial species in dental root canal infections. *Oral Microbiol Immunol* 7: 257–62.

Sundqvist G. 1992b. Ecology of the root canal flora. *J Endodon* 18: 427–30.

Sundqvist G, Fidgor D, and Sjogren U. 1998. Microbiology analyses of teeth with endodontic treatment and the outcome of conservative retreatment. *Oral Surg Oral Med Oral Pathol* 85: 86–93.

Switalski LM, Butcher WG, Caufield PC, and Lantz MS. 1993. Collagen mediates adhesion of *Streptococcus mutans* to human dentin. *Infect Immun* 61: 4119–25.

Syed SA, Loesche WJ, Pape HL, and Grenier E. 1975. Predominant cultivable flora isolated from human root surface caries plaque. *Infect Immun* 11: 727–31.

Tronstad L. 1973. Ultrastructural observations on human coronal dentine. *Scand J Dent Res* 81: 101–11.

Tronstad L and Langeland K. 1971. Effect of attrition on subjacent dentin and pulp. *J Dent Res* 50: 17–30.

Valderhaug J. 1974. A histologic study of experimentally induced periapical inflammation in primary teeth in monkeys. *Int J Oral Surg* 3: 111–23.

van Houte J. 1994. Role of microorganisms in caries etiology. *J Dent Res* 73: 672–81.

van Houte J, Lopman J, and Kent R. 1994. The predominant cultivable flora of sound and carious human root surfaces. *J Dent Res* 73: 1727–34.

Vojinovic O, Nyborg H, and Brännström M. 1973. Acid treatment of cavities under resin fillings: bacterial growth in dentinal tubules and pulpal reactions. *J Dent Res* 52: 1189–93.

Walton R. 1976. Histological evaluation of enlarging the pulp canal space. *J Endodon* 2: 304–11.

Wittgow WC and Sabiston CB. 1975. Microorganisms from pulpal chambers of intact teeth with necrotic pulps. *J Endodon* 1: 168–71.

# Chapter 4
# Culture-Based Analysis of Endodontic Infections

*Gunnar Dahlén*

## 4.1 Introduction

Most of our knowledge of the endodontic microflora is based on culture studies. This is simply because we had no real alternatives in the past. Microscopic studies have been used; however, they have severe limitations when it comes to deeper identification, to evaluate the composition, to characterize various microorganisms, and to do further experimental studies on isolated species. Microscopy of smears from the root canal is limited to main morphotypes. Microbial staining of histological sections has the advantage of localizing the microbes in situ (Fig. 4.1). Electron-microscopic pictures (transmission or scanning) have been valuable to distinguish main morphotypes in various locations (Nair 1987). During the last decade, numerous studies using various types of molecular biology techniques have been used to characterize more closely the microbial composition of the root canal flora (see Chapter 5). These methods have definitely showed that the root canal flora is much more complex than previously thought. This has made the clinical interpretations, diagnosis, and treatment strategies more difficult. Still, culture is a "gold standard" for dentists in the clinic to identify specific targets for treatment and to evaluate treatment strategies due to its easier accessibility than the new techniques. This chapter is aimed to describe the knowledge we have gained by culture studies for the microbial composition in various phases of the endodontic infection. Culture is also used in experimental models, which have disclosed the dynamics of the infection and the nature of the

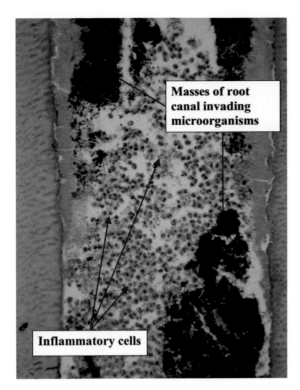

Fig. 4.1 Histological section showing a stained (blue) microbial invasion into the root canal. (Courtesy of Dr Dominico Riccuci.)

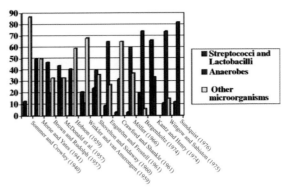

**Fig. 4.2** Microorganisms (percent of total isolates) isolated by different investigators in initial samples from root canals with necrotic pulps.

microorganisms. This is outlined, with the insight that we still do not have the complete picture of the microflora and that new techniques will complement or even change our future opinions on the natural history and complexity of the root canal flora.

## 4.2 Historical perspectives

Miller in the late nineteenth century clearly described that the open pulp chamber was filled with various bacteria (Miller 1894). Endodontic problems were early at focus in the discussion on focal infections and difficulties to adequately diagnose and treat endodontically involved teeth frequently led to extraction. Culture studies at this point recovered predominantly aerobic and facultatively anaerobic microorganisms. With the historical perspective in mind, we can today conclude that the studies in the first half of the twentieth century were hampered by insufficient culture techniques, poor sampling methods, neglected

antiseptic measures, and efforts to avoid contaminations (Fig. 4.2). The microbial findings from infected root canals were predominated by facultatives and by species and groups that we today regard as contaminants from the oral cavity or from the surroundings by careless handling in the dental office or laboratory. Anaerobes were seldom isolated from the root canals during the first half of the century. It can also be noted from this figure (Fig. 4.2) the low recovery or lack of streptococci/lactobacilli in some studies (Bergenholtz 1974; Sundqvist 1976) that have used traumatized teeth to study rather than teeth endodontically involved due to caries. Even if all these problems still in part exist in practice, a number of important milestone studies have been carried out through the years, throwing light on the complexity of endodontic microbiology. The importance of adequate antiseptic measures, sampling procedures, transportation, and laboratory techniques (media, anaerobiosis, identification, and interpretation) were shown in the classical work by Möller (1966). He described how to avoid contaminations, how to take a relevant sample, the invention of the VMG III transport medium (today designated VMGA III after some later modifications; Dahlén et al. 1993), and the evaluation of various culture media. His recommendations are still in use today.

Anaerobic techniques were introduced by several researchers in the 1960s and 1970s (Möller 1966; Bergenholtz 1974; Kantz and Henry 1974; Wittgow and Sabiston 1975; Sundqvist 1976). The study by Sundqvist (1976) illustrates well the complexity of the root canal microbiology, as it utilized anaerobic techniques combined with a thorough phenotypical

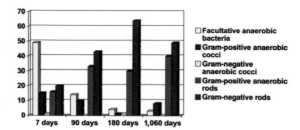

**Fig. 4.3**  Mean percent of anaerobic and facultatives after different times of experimental infection (Fabricius et al. 1982a).

identification of the bacterial isolates. In this study, a number of bacterial species were discovered that were not previously detected in the root canal environment.

Kakehashi et al. (1965) are often mentioned as having performed the key study on germfree animals to prove that microbes were necessary for the development of apical periodontitis. Möller et al. (1981) also showed in monkeys that infected teeth developed periapical lesions while noninfected did not. The importance of anaerobes in the pathogenesis of root canal infection was further emphasized in experimental studies (Fig. 4.3) on monkeys by Fabricius et al. (1982a, b). Table 4.1 shows the microbiological outcome in some later studies on primary endodontic infections. The heterogeneity within and between the studies is quite substantial. However, the general trend is that 1–12 predominating species in each specimen are involved and the number of recovered cells may vary from $<10^2$ to over $10^8$ (Sundqvist 1992a). Anaerobic streptococcal species, Gram-positive anaerobic rods, and species of *Prevotella*, *Fusobacterium*, and *Campylobacter* are usually present. *Treponema* species are also likely to be present; however, they are regularly missed in culture analysis. It has to be clear that the microflora varies between the studies due to a number of factors, for example, the diagnosis, type of teeth, sampling procedures, and laboratory measures. Conclusively, the primary endodontic infection is a polymicrobial, predominantly anaerobic infection with little microbial specificity.

## 4.3 Culture-based analysis in clinical practice

### 4.3.1 Sampling

A most important measure for a correct sampling and avoiding false-positive samples is the sterilization and control of the operation field (Fig. 4.4). After rubber dam application, prewashing with hydrogen peroxide (30%) should be carried out in order to make the following sterilization procedure more efficient. Leakage between the rubber dam and the tooth must be carefully controlled. Disinfection of the operation field is performed by flushing with 10% iodine tincture. The surface layer of the temporary filling can preferentially be removed and the procedure be repeated to eliminate microorganisms at the borderline between the filling and the tooth. Likewise, if caries or defective restorations are present, these should be removed and the disinfection protocol repeated. Möller (1966) recommended a control sample from the operation field in order to check the sterility. This is especially recommended for the unexperienced dentists to learn how to avoid contaminations and to work aseptically. If a control sample is taken, an inactivation procedure of the iodine with 5% thiosulfate solution should be carried out to avoid false-negative samples, for example, viable but not culturable bacteria that may be bacteriostatically affected by the iodine.

The root canal sample is taken after removing the temporary filling (Fig. 4.5). Interappointment dressings should be removed by irrigation using saline or sampling solution (VMG I; Möller 1966). Execute pumping movements with a file is recommended in order to get a suspension with bacteria from the root canal wall, dentine, and apex delta (Fig. 4.6), that is, to disrupt the biofilm in these areas. Sampling is then performed using charcoaled paper points that are transferred to a transport medium, for example, VMGA III (Möller 1966; Dahlén et al. 1993). The sampling procedure is repeated with additional points until all liquid is absorbed. It should be noticed that the last point is the most important because it will absorb the liquid from the most peripheral areas of the apical region.

### 4.3.2 Transportation

The purpose of a transport medium VMGA III (Möller 1966; Dahlén et al. 1993) is to keep the viability of the microorganisms during transportation and being bacteriostatic in the sense that no multiplication takes place. The VMGA III medium has also a general inactivating ability, which in addition to the charcoaled paper points will inactivate medicals and antiseptic substances used in the root canal, which otherwise

**Table 4.1**   Frequency (percent of total number of isolated strains) of microorganisms in root canal samples from teeth with necrotic pulps in some culture studies.

| Microorganisms | Wasfy et al. (1992) | Sundqvist (1992a) | Le Goff et al. (1997) | Lana et al. (2001) | Peters et al. (2002) | Gomes et al. (2004) | Chu et al. (2005)[a] |
|---|---|---|---|---|---|---|---|
| *S. aureus* | – | – | – | – | – | 0.6 | – |
| Other *Staphylococcus* spp. | 2 | – | 2 | 0.7 | 2 | 4 | 0.5 |
| *Streptococcus* spp. (PSP)[b] | 6 | 4 | 1 | 4 | – | **11** | 2 |
| *Streptococcus* spp. (non-PSP)[c] | **17** | 4 | 3 | 9 | 5 | **11** | 5 |
| *E. faecalis* | – | 2 | – | 0.7 | – | 1 | 0.5 |
| *P. micra*[d] | 2 | 6 | 3 | 1 | **13** | **27** | 5 |
| Other anaerobic streptococci[e] | 0.4 | 9 | – | 7 | 2 | – | **16** |
| *Neisseria* spp. | – | – | – | – | – | 0.6 | 5 |
| *Veillonella* spp. | 5 | 2 | 2 | 3 | 3 | 4 | 4 |
| *Bacillus* spp. | – | – | – | – | – | – | – |
| *Clostridium* spp. | – | – | – | 8 | – | 2 | 1 |
| *Corynebacterium* spp. | – | – | 6 | – | – | – | 2 |
| *Lactobacillus* spp.[f] | – | 7 | 3 | **14** | 3 | 2 | 6 |
| *Propionibacterium* spp. | 3 | 2 | **13** | 0.7 | 8 | 1 | 3 |
| *Actinomyces* spp. | 8 | 4 | 3 | 4 | **15** | 5 | **12** |
| *Eubacterium* spp.[g] | **20** | **17** | 7 | 2 | 7 | 4 | 4 |
| Enteric rods | – | 0.3 | – | – | – | – | – |
| *Capnocytophaga* spp. | – | 2 | 7 | 0.7 | 7 | 0.6 | 5 |
| *Campylobacter* spp.[h] | – | 5 | **12** *C. gracilis* | – | – | 2 *C. gracilis* | 8 |
| *Eikenella* spp. | – | 0.3 | – | – | – | – | 0.5 |
| *Porphyromonas* spp. | 2 | 9 | 2 | – | – | 2 | 3 |
| *Prevotella intermedia/ nigrescens* | **12** | 6 | 2 | 6 | **15** | 5 | 4 |
| Other *Prevotella* spp.[i] | **14** | 6 | **14** | 9 | **11** | 8 | **18** |
| *Fusobacterium* spp. | 6 | **14** | 7 | 8 | 8 | **11** | 6 |
| Spirochetes | – | – | – | – | – | – | – |
| *Candida* spp. | – | – | – | 2 | – | – | – |
| Total number of isolated strains | 259 | 353 | 84 | 138 | 131 | 171 | 395 |
| Number of teeth | 85 | 65 | 26 | 31 | 58 | 41 | 88 |

Species >10% in bold.
[a]Chu et al. (2005) includes both samples from root canals "exposed" and "nonexposed" to the oral cavity.
[b]Polysaccharide-producing (PSP) streptococci including *S. sanguis, S. salivarius, S. mutans, S. oralis,* and *S. mitis.*
[c]Non-polysaccharide-producing (non-PSP) streptococci including *S. anginosus, S. constellatus, S. intermedius,* and *Gemella morbillorum.*
[d]*Parvimonas micra* formerly *Micromonas micros* earlier *Peptostreptococcus micros.*
[e]Other anaerobic streptococci including peptostreptococci, peptococci, and *Finegoldia* spp.
[f]*Lactobacillus* includes both anaerobic and facultative species, *Olsenella uli* and *Bifidobacterium* spp.
[g]*Eubacterium* also including *Colinsella aerofaciens, Eggerthella lenta,* and *Filifactor alocis.*
[h]*Campylobacter* includes formerly designated *Wolinella* spp. and *Bacteroides gracilis.*
[i]Other *Prevotella* includes both pigmented and nonpigmented species and *Bacteroides* species such as *B. capillosus, B. uniformis,* and more.

would prevent bacterial cells from growing in the laboratory media. Furthermore, the VMGA III medium contains reducing substances (cysteine) to keep the medium from being oxygenized and a redox indicator showing that this does not happen. VMGA III cannot be obtained commercially; however, a full description of composition and procedures is published (Dahlén et al. 1993).

Prewashing with
$H_2O_2$ and taking
away the surface
of the sealing cement

Sterilization of the
operation field

Iodine tincture

Inactivation of
the iodine

Thiosulfate solution

Control of the sterility

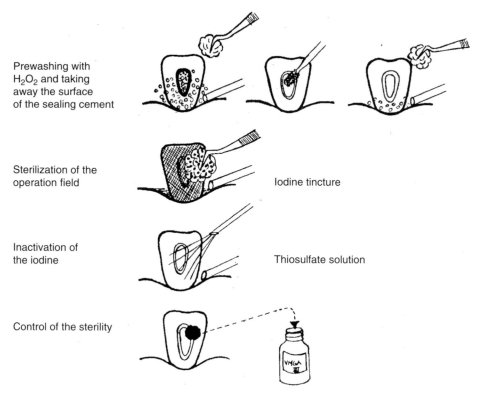

**Fig. 4.4**    Drawing showing the main steps in the preparation and sterilization of the operation field before entering the root canal. (Drawing by Mrs Gunilla Hjort.)

### 4.3.3 Laboratory considerations

The goals of the laboratory measures are as follows:

- To decide whether or not viable bacteria are present in the sample
- To identify microorganisms in the sample according to species or genus levels that are of clinical importance
- To obtain a semi-quantitative measure of bacterial load in the sample

One fairly simple methodology that could be used by most laboratories is shown in the Fig. 4.7. This is to serve the clinician with the important information for the diagnosis of infection indicated by growth/no growth and for treatment decision by the number and type of present microorganisms.

Liquid media are primarily recommended since they most often allow fastidious and/or dormant bacterial cells to grow. For practical purposes, media such as thioglycolate, trypticase broth, or brain heart infusion broth could be used. However, it is clear that more sophisticated media such as HCMG SuIa (Möller 1966) give a higher frequency of samples with growth. The liquid media also have the advantage that no extra equipment is necessary for the anaerobic incubation if the tubes are flushed with oxygen-free gas (e.g., nitrogen) when inoculated. If the tubes are prepared under anaerobic conditions and the sample is inoculated in the bottom layer of the tube and the tubes are capped tightly (rubber stopper), the medium itself will ascertain anaerobic conditions even for the most oxygen-sensitive bacterial species. Another important factor for the growth of a sample with few bacterial cells, sometimes in a "bad" condition, is to allow them sufficient incubation time. In the Laboratory of Oral Microbiology at Göteborg University, we give a preliminary reply to the dentist after 5–6 days on growth/no growth and then continue the incubation for 14 days. If growth appears after the preliminary

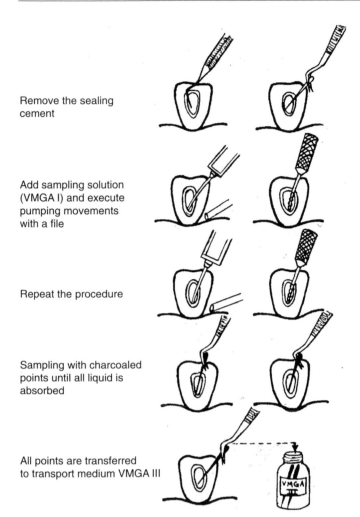

Remove the sealing cement

Add sampling solution (VMGA I) and execute pumping movements with a file

Repeat the procedure

Sampling with charcoaled points until all liquid is absorbed

All points are transferred to transport medium VMGA III

**Fig. 4.5** Microbiological sampling from the root canal. (Drawing by Mrs Gunilla Hjort.)

reply has been delivered, the dentist is informed by a phone call.

Solid media (e.g., Brucella blood agar supplemented with hemolytic blood and vitamins such as hemin and menadione), one for aerobic incubation and one for incubation in anaerobic jars, are used to complement the liquid medium in order to disclose the diversity and presence of various microbial species and to give a semi-quantification of each colony morphology type. Specification to genus or species level is based on Gram stain, selective media, and simple biochemical tests. In practice, a detailed specification to species level on all present bacteria is not necessary since this will have little impact on the diagnosis and the choice of treatment procedure. The final reply

after 14 days also includes the finding recovered from the solid media including semi-quantification.

### 4.3.4 Sensitivity and specificity

The sensitivity of cultural methods is fairly good and acceptable for practical purpose as long as it is carried out appropriately. Molecular biology methods have disclosed more microbial species. Many of these are representing not-yet-cultured species, but in other cases they may also represent nonviable (dead) cells or even remaining DNA. We do not know the truth due to lack of an absolute gold standard, and all sensitivity calculations of each method will be hampered by its limitations. For the more easily cultured species

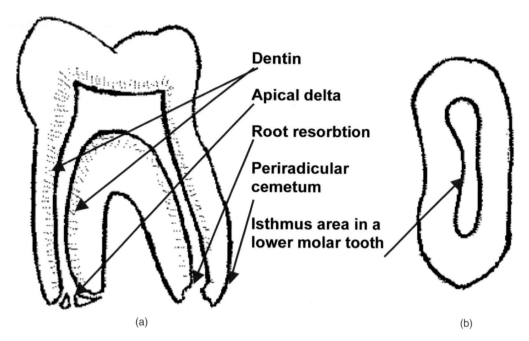

(a)                                          (b)

**Fig. 4.6**   Schematic drawing of a molar tooth indicating the location where special attention has to be taken for reaching bacteria at sampling: (a) longitudinal section; (b) horizontal or cross-sectional section of a lower molar tooth.

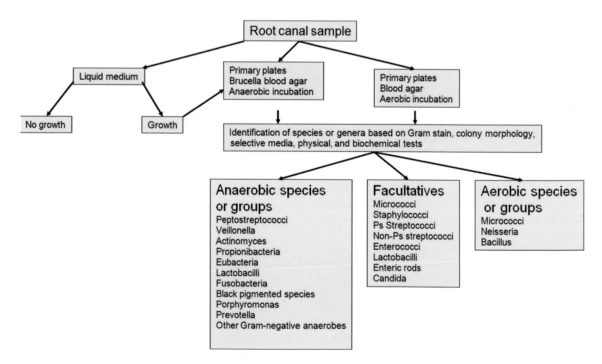

**Fig. 4.7**   The laboratory procedures of root canal samples in the Laboratory of Oral Microbiology at Göteborg University.

such as facultatives, theoretically only one viable cell is needed for growth in the liquid medium or to form a colony on the agar plate. In that sense, the sensitivity of the culture analysis should be regarded as high. It should be noted, on the other hand, that in samples with a high number of bacterial cells such as in primary infected and untreated teeth, bacteria in low numbers will not be detected due to dilution or overgrowth of predominant species. An important factor for the sensitivity, irrespective of the detection method used, is the sampling method and how well the present microorganisms can be reached and sampled (Fig. 4.6). The sensitivity of the complete procedure, including both sampling and analyses, is complex and cannot be fully evaluated.

A sample showing no growth (negative sample) is indicating that the root canal is free from viable microorganisms and the treatment goal is achieved. To avoid false negatives, an adequate and representative sample must be taken, which might be somewhat demanding. During treatment, microorganisms are most easily eliminated from the main root canal, while it is a lot more difficult to eradicate them from dentinal tubules, lateral canals, and apical deltas (Fig. 4.6). All possible precautions should be taken for transportation and culture to give viable and culturable bacteria the best chance to grow. Still, there is the problem with uncultivable/difficult to culture microbial cells that give an unknown number of false negatives. Future studies should focus on the problem whether remaining uncultivable microorganisms are of significant importance for failing outcome of the endodontic treatment.

False-negative samples are especially difficult to avoid when taking samples at revision of a previously root-filled tooth. Even if the gutta-percha or sealer is removed mechanically, the remaining bacteria may hide in peripheral parts of the root canal system and they may be in a stressed situation, which does not allow them to grow instantly in the laboratory. This reason for false-negative samples is rather likely when using dressings and antiseptic irrigations that do not kill the microorganisms but leave them in a *dormant phase* or in a biofilm (Fig. 4.8). This can be avoided by the use of specific or unspecific inhibitors. Such inhibitors are sodium thiosulfate (5%) for halogen-containing antiseptics (iodine and chlorine), L-alpha lecithin in Tween 80 for chlorhexidine (Zamany and Spangberg 2002), or more unspecific inactivating

agents such as charcoal or VMGA III. Another possibility is to leave the canal free from dressings between appointments. This was shown by Reit and Dahlén (1988) that a second sample (*true sample*) did show more culture positives than a first sample (*indicator sample*), indicating a significant risk of false negatives if the sample is taken immediately after removing the $Ca(OH)_2$ paste. A third possibility for a false-negative sample is when the bacteria are retained on the external root surface, in the apical root cementum or in resorptions around the apical orifice of the root canal (Fig 4.9), these bacteria cannot be reached by sampling through the root canal but only by surgical access to the root tip (see Chapter 5). This can only be performed in specific nonhealing cases, and, it is not a recommended procedure in the general dental practice.

Specificity in endodontic sampling is high because the number of false positives can be significantly reduced and controlled. A false-positive test means that the samples show growth by contaminating microorganisms of various kinds. The most common reason is probably due to a not adequately sterilized operation field or leakage, despite rubber dam application. It was recommended by Möller (1966) to take a separate sample from the operation field as a control of the used antiseptic technique. Bacteria present in the saliva and plaque may appear in the operation field samples and if they occur concomitantly in the root canal sample a contamination may be suspected. Facultative anaerobic species such as polysaccharide-producing streptococci (*S. mutans*, *S. sanguis*, *S. oralis*, and *S. salivarius*), *Corynebacterium* spp., *Neisseria* spp., and *Haemophilus* spp. are oral bacteria that we empirically know that they do not establish themselves in the anaerobic and non-saccharolytic environment in the root canal, and thereby strongly indicate a leakage. Also, micrococci, coagulase-negative staphylococci, spore-forming bacteria (e.g., *Bacillus* spp.), and enteric rods are most likely contaminants from the surroundings by careless handling of the samples in the office or laboratory (Table 4.2). It should be noted, however, that enteric rods and *Staphylococcus aureus*, sometimes although rarely, may usually appear in root canals as monoinfections. Such infections are important to disclose since serious complications (e.g., osteomyelitis and other acute dentoalveolar infections) may follow and special treatment strategies have to be considered.

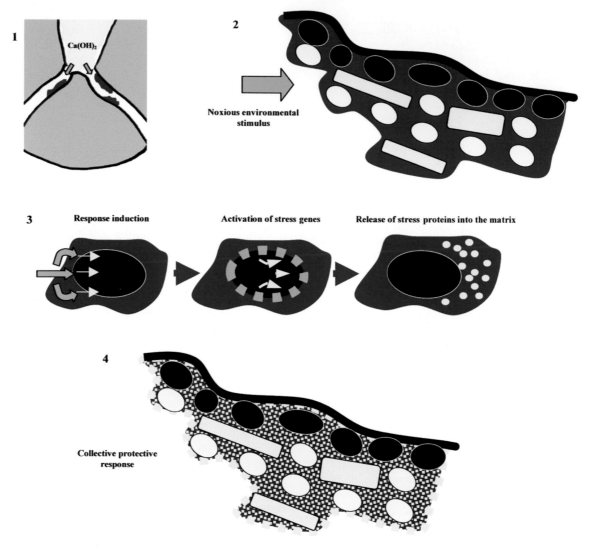

**Fig. 4.8**   Hypothetical outline of a stress protein response in biofilm communities of root canal bacteria. The stress induces production of stress proteins, which are released into the biofilm matrix. These proteins will provide beneficial effects for the community. (Figure from Chavez de Paz et al. 2004.)

## 4.4 Clinical interpretations

### 4.4.1 How to interpret the primary infection

The reason for taking a sample from a primary infected tooth is not always rational. An untreated root canal with a necrotic pulp and an apical lesion, with or without symptoms, is always infected. We know that in most of these cases the microflora is polymicrobial, predominantly anaerobic, and treatment procedure using mechanical debridement and antiseptic irrigation is the first choice. In case of an acute infection with general symptoms and risk of spreading, the administration of systemic antibiotics must be done instantly; however, there is generally little benefit of a microbiological sample at this stage.

**Fig. 4.9** Histological section of the apical region of a root showing resorption, where present bacteria will be difficult to sample. (Figure from Chavez de Paz, 2004.)

### 4.4.2 Interpretation of growth during treatment

A microbiological sample taken during treatment as a control for bacterial growth may be important. Some studies have shown that if there is absence of viable bacteria prior to filling, the prognosis of root canal treatment (Sjögren et al. 1997) or retreatment (Sundqvist et al. 1998) is improved. If cultivable bacteria persist in the canal, the long-term outcome may depend on the quality of the root canal filling (Möller et al. 2004; Fabricius et al. 2006) and the healing of lesions may be delayed (Waltimo et al. 2005). The use of intracanal medicaments between appointments improves the chances of bacterial elimination, but does not guarantee it (see Chapter 14). If the bacterial sample shows a low number of persisting bacteria, additional antiseptic procedures and interappointment dressings should be considered. If a high number of bacteria of polymicrobial and anaerobic nature are still present, a leakage (through rubber dam, fractures,

remaining fillings, and crowns) should be suspected. In rare cases, more specific infections, for example *S. aureus*, enteric rods, and *Candida* spp., could be disclosed, which need special considerations and treatment procedures.

### 4.4.3 Use of specific antiseptics and antibiotics

The choice between various antiseptics may, to a certain degree, be based on the type of remaining bacteria. Anaerobes are usually very sensitive for most antiseptics used for irrigation and as interappointment dressings, and no specific considerations are necessary. It should be remembered that irrigation is usually of short duration, and the effect is limited compared to an interappointment dressing that can extend its effect over days. A very important basis for the antiseptic effect is that as much organic materials (necrotic pulp material and microorganisms) as possible should

**Table 4.2**  Clinical significance of various microorganisms occurring in endodontic samples.

| Microorganism[a] | Frequency | Pathogen[b] | Persistence to treatment[c] | Risk to contamination | Main appearance |
|---|---|---|---|---|---|
| Micrococci | + | − | − | +++ | Careless handling in clinic or in laboratory |
| *Staphylococcus aureus* | + | +++ | +++ | ++ | |
| *Staphylococcus epidermidis* | + | − | Not known | ++ | |
| Streptococci (polysaccharide producing) | + | + | ++ | +++ | Primary sample, through leakage or saliva contamination |
|   *S. mutans* | | | | | |
|   *S. salivarius* | | | | | |
|   *S. sanguis* | | | | | |
| Streptococci (other) | +++ | ++ | ++ | + | Primary samples, treatment persistence, and root-filled teeth with apical periodontitis |
|   *S. anginosus (S. milleri)* | | | | | |
|   *S. oralis (S. mitior)* | | | | | |
|   *S. mitis* | | | | | |
|   *S. intermedius* | | | | | |
| Enterococci | ++ | + | +++ | + | Treatment persistence and at root-filled teeth with apical periodontitis |
|   *E. faecalis* | | | | | |
| Peptostreptococci (*P. micra*)[d] | ++ | ++ | + | − | Primary samples and abscesses |
| Other anaerobic streptococci | ++ | ++ | + | − | |
| Gram-negative cocci | | | | | |
|   *Neisseria* spp. | (+) | − | − | +++ | Saliva contamination |
|   *Veillonella* spp. | + | + | − | ++ | Primary samples or through leakage |
| Sporeformers | | | | | |
|   *Bacillus* spp. | (+) | − | Not known | +++ | Careless handling in clinic or in laboratory |
| Gram-positive rods | | | | | |
|   *Corynebacterium* spp. | + | − | + | +++ | Primary samples, persistence to treatment and at root-filled teeth with apical periodontitis |
|   *Actinomyces* spp. | ++ | ++ | ++ | + | |
|   *Lactobacillus* spp. | ++ | ++ | ++ | + | |
|   *Propionibacterium* spp. | ++ | + | + | + | |
| Other Gram-positive anaerobic rods | ++ | + | + | + | Primary samples, saliva contamination |
|   *Eubacterium* spp. | ++ | ++ | − | − | Primary samples |
| Enterobacteriaceae (enteric rods) | (+) | ++ | ++ | +++ | Careless handling in clinic or in laboratory |
| Gliders and corroding rods | + | ++ | − | − | Primary samples and abscesses |
|   *Capnocytophaga* spp. | | | | | |
|   *Campylobacter* spp. | | | | | |
|   *Eikenella* spp. | | | | | |
| *Prevotella/Porphyromonas* | | | | | |
|   *P. gingivalis/P. endodontalis* | ++ | +++ | − | − | Primary samples and abscesses |
|   *P. intermedia*[e] | ++ | +++ | − | − | Primary samples and abscesses |
|   Other *Prevotella* spp. | ++ | ++ | + | − | |
|   *Fusobacterium* spp. | ++ | ++ | + | − | Primary samples and abscesses |
| Spirochetes | + | ++ | − | − | Microscopy |
| Fungi (yeasts) | | | | | Seldom found in culture analysis |
| *Candida* spp. | + | + | + | ++ | Through leakage, persistence to treatment, and careless handling in clinic or in laboratory |

[a]The microorganisms are grouped as species, genus, or other groups according to its clinical relevance.
[b]The pathogenicity is based on test in animal experiments and their presence in clinical samples.
[c]Resistance against adequate mechanical and chemical treatment.
[d]*Parvimonas micra* formerly *Micromonas micros* earlier *Peptostreptococcus micros*.
[e]Formerly *Bacteroides intermedius*.

be removed. Otherwise, this will rapidly inactivate any antimicrobials. Calcium hydroxide paste has become useful as an interappointment dressing due to its ability to fill up the root canal lumen and prevent the remaining bacteria to grow. However, calcium hydroxide has a rather weak (bacteriostatic) antimicrobial effect and many bacteria may survive, especially Gram-positive facultative species. If such bacteria remain in the root canal, other antiseptics such as iodine or chlorhexidine should be considered.

There is no antibiotic that is efficient for all types of microorganisms occurring in the infected root canal and local antibiotics in the canal is, therefore, not recommended at this time. In primary acute infections with pus formation, antibiotics may be used systemically in order to prevent spreading of the infection. Since the main character of this infection is polymicrobial predominantly anaerobic, penicillin or amoxicillin with or without metronidazole remains the drugs of choice (see Chapter 13).

## 4.5 Route of infection in vital and necrotic pulp

### 4.5.1 Root canals with vital pulps

The bacteria gain access to the pulp and root canal through different routes: (1) through the dentinal tubules, for example, the caries process or by deep mechanical preparations; (2) through the exposed pulp by trauma or through fractures; (3) through the apical foramen either through the periodontal ligament (deep pockets or trauma) or hematogenously (Bergenholtz 1977, 1981). Caries is probably the most common cause for bacterial access to the pulp. Bacteria invading the dentine cause inflammation in the pulp tissue that by time leads to necrosis. If the pulp loses its vitality prior to becoming infected (by trauma and physical and chemical injuries), the bacteria are attracted to the necrotic tissue through the dentinal tubules, fractures, and other routes and subsequently grow and infect the root canal.

As long as the pulp is vital, there is a host response mechanism in function that prevents the bacteria from invading deeper into the pulp tissues. There is no infection in strict sense, but bacteria are present in the carious dentine adjacent to the exposed pulp (Bergenholtz 1977, 1981). Bacteria that could be isolated here

are predominantly species that are associated with the caries process, for example, various *Lactobacillus* and *Bifidobacterium* species and species of *Streptococcus*, *Propionibacterium*, *Actinomyces*, *Corynebacteria*, and *Eubacterium* but very few Gram-negative anaerobic species (Edwardsson 1974; Hoshino 1985). Conclusively, vital pulps including those exposed to caries and those exposed by trauma may have accidentally bacteria on the surface, however, usually in low numbers and without penetration into the vital pulp tissues.

### 4.5.2 Root canals with necrotic pulps

The pulp can easily go into necrosis due to its limited and fragile blood supply at the apical foramen. Younger teeth with open apices often withstand injuries better than older ones where the apical foramen is narrow. The necrosis is the terminal end of the inflammatory process which, when it becomes large enough, causes the tissue to collapse due to the heavy bacterial load. Bacteria easily invade the necrotic pulp tissue since there is no host defense. The invading bacteria seem to go through some selection mechanism by the route of infection and through the ecological pressure in the root canal system. A random process occurs only if the pulp chamber is left open to the oral cavity. We should therefore not expect the numbers of participating species in the closed root canal to be as rich and diverse as those root canals that have been left open to the oral cavity.

The intensity of the infection process is related to the bacterial activity and presence of certain growth and virulence factors that favor some, but not all, bacteria in their root canal environment (Fig. 4.10). The bacterial growth is due to an anaerobic, proteolytic bacterial metabolism in the root canal system and necrotic protein containing pulp. Lack of carbohydrates and especially sugars disfavor saccharolytic bacterial species and the low oxygen level does not give the facultatives any advantage over the strict anaerobes. The local and systemic host defense system in the root canal is destroyed, and cannot act until the infection frontline reaches the vital tissues in the apical region. This is the environment in which the root canal flora of teeth with necrotic pulps develops (Table 4.1). An inflammatory reaction is formed in the periapical tissues (apical periodontitis) that can either be acute or chronic.

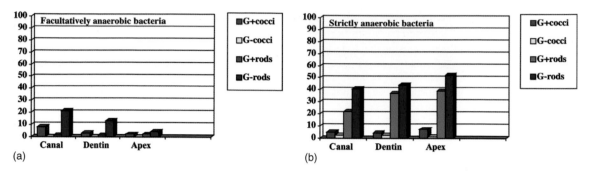

**Fig. 4.10**  The predominant bacteria ((a) facultative anaerobic and (b) anaerobic bacteria) in the main canal, dentine, and apical region of root canals of monkey teeth left open to the cavity for 7 days and then sealed for 6 months. Note the high number of enterics (G-facultative rods) typical for monkeys. (From Fabricius et al. 1982a.)

## 4.6 The apical periodontitis

### 4.6.1 The acute infection

The acute infection is characterized by an increased bacterial metabolism and nonregulated multiplication. The body response is rapid and mainly unspecific and the neutrophilic granulocytes (polymorphonuclear leukocytes or PMNs) are predominating in the periapical area. The battle between the growing bacteria and the phagocytic cells (PMN cells) can be very dramatic and the tissues are destroyed (collapses). At this stage, the body has one main goal and that is to prevent the infection from spreading. A fibrotic capsule can be formed in order to build a barrier more difficult for the bacteria to penetrate, however, at the cost of a total destruction of the tissues within the barrier. This results in an abscess with pus. This is a common situation in the clinic because the patients usually have symptoms, sometimes severe and seek the dentist for immediate treatment. Table 4.3 summarizes the microbial composition in acute endodontic infections with periapical abscess and presence of pus. Species of *Peptostreptococcus* (including *Parvimonas micra* formerly *Peptostreptococcus micros*), *Prevotella* (formerly *Bacteroides*), and *Fusobacterium* are prevailing together with various Gram-positive anaerobic/microaerophilic rods and cocci.

### 4.6.2 Bacteria and symptoms

The acute infection is also a common reason to investigate the microbial composition in order to find out if there is some kind of specificity and if this has some treatment implications. Table 4.4 shows the results of some studies that evaluated the association between specific bacteria and symptoms such as pain, swelling, tenderness, abscess formation, and sinus tract. Conclusively, anaerobes, predominantly Gram-negatives rods, are most commonly present; however, the specificity is low. Correlations with some anaerobes (*Peptostreptococcus* spp., *Eubacterium* spp., *Prevotella* spp., and *Fusobacterium* spp.) have been claimed. Especially, the BPB (black-pigmented bacteria) have gained much attention. Notably, *Porphyromonas* spp. (*P. gingivalis* and *P. endodontalis*), which are considered more virulent than *Prevotella* species and major pathogens in periodontitis, are less frequent in acute endodontic infections and are outnumbered of black-pigmented *Prevotella* species, including *P. intermedia* (including *P. nigrescens*) in particular. The importance of anaerobes in acute infections has been confirmed in numerous experimental animal studies (for review, see Dahlén 2002). In one study, 8 different bacterial species isolated from the same infected root canal in a monkey were inoculated into 12 experimentally devitalized teeth of monkeys in the same proportions (Fabricius et al. 1982b). Fig. 4.11 shows the proportion of those 8 species in the 12 teeth after being followed for 6 months. Note that the four anaerobic species (*F. nucleatum*, *F. necrophorum*, *P. anaerobius*, and *Prevotella oralis* formely *Bacteroides oralis*) are predominating both in the original infection as well as in the 12 experimentally infected teeth. These 8 strains were also inoculated into steel net wound chambers implanted in the back of rabbits and the dynamics were followed for >30 days (Dahlén et al. 1987). It was clear that the pus and abscess were

**Table 4.3** Frequency (percent of total number of isolated strains) of microorganisms in root canal samples from teeth with abscess (pus) in some culture studies.

| Microorganisms | Brook et al. (1981) | Oguntebi et al. (1982) | Williams et al. (1983) | Lewis et al. (1986) | Brook et al. (1991) | Sakamoto et al. (1998) | Khemaleelakul et al. (2002) |
|---|---|---|---|---|---|---|---|
| *S. aureus* | – | – | – | – | 1 | – | – |
| Other *Staphylococcus* spp. | – | 4 | 5 | – | – | 3 | 9 |
| *Streptococcus* spp. (PSP)[a] | **10** | **24** (*S. mitis*) | – | 2 | **18** | 7 | **14** |
| *Streptococcus* spp. (non-PSP)[b] | 8 | 4 | 2 | **17** | 4 | **16** | 9 |
| *E. faecalis* | – | **12** | – | – | 4 | – | – |
| *P. micra*[c] | **20** | 4 | **12** | NS[d] | **23** | 3 | 6 |
| Other anaerobic streptococci | – | 8 | 7 | **28** | – | **10** | – |
| *Neisseria* spp. | – | – | – | – | – | – | – |
| *Veillonella* spp. | **12** | – | – | 2 | 2 | 6 | 2 |
| *Bacillus* spp. | – | – | – | – | – | – | – |
| *Clostridium* spp. | – | – | – | – | – | – | 3 |
| *Corynebacterium* spp. | – | – | – | – | – | – | **11** |
| *Lactobacillus* spp. | 6 | – | 5 | 2 | – | 1 | 3 |
| *Propionibacterium* spp. | – | – | – | 1 | 1 | 1 | 3 |
| *Actinomyces* spp. | 6 | **12** | 5 | 1 | – | 1 | 4 |
| *Eubacterium* spp. | 2 | – | – | – | 2 | – | 4 |
| Enteric rods | – | – | – | – | – | – | – |
| *Capnocytophaga* spp. | – | – | – | – | – | – | – |
| *Campylobacter* spp. | – | – | – | – | – | 3 | – |
| *Eikenella* spp. | – | – | – | – | – | – | 1 |
| *Porphyromonas* spp. | 4 | – | – | 8 | **10** | 4 | 2 |
| *Prevotella intermedia/nigrescens* | 2 | 8 | – | 3 | 3 | 4 | 5 |
| Other *Prevotella* spp.[e] | **29** | – | **48** | **23** | **17** | **21** | **19** |
| *Fusobacterium* spp. | 8 | **28** | **15** | 4 | **12** | **11** | 4 |
| Spirochetes | – | – | – | – | – | – | – |
| *Candida* spp. | – | – | – | – | – | – | – |
| Total number of isolated strains | 59 | 25 | 40 | 168 | 78 | 112 | 118 |
| Number of teeth | 12 | 10 | 10 | 50 | 32 | 23 | 17 |

Species >10% in bold.
[a]PSP, polysaccharide-producing streptococci.
[b]Non-PSP, non-polysaccharide-producing streptococci.
[c]*Parvimonas micra* formerly *Micromonas micros* earlier *Peptostreptococcus micros*.
[d]NS, not specified.
[e]Including isolates designated as *Bacteroides* spp.

formed when 3 of the anaerobic species started to grow and multiply (Fig. 4.12). On the other hand, it was necessary to include facultative anaerobes in the bacterial collection in order to let the anaerobes survive the initial phase of the infection, supposedly by reducing the redox potential by consuming available oxygen. This appearance has been noticed and confirmed in numerous other experimental animal studies using subcutaneous injections (for review, see Dahlén 2002). Sundqvist et al. (1979) did a transmission study where they found that a *Bacteroides* strain (later iden-

tified as *P. endodontalis*) was essential for the transfer of an infection between animals by bacterial combinations originating from infected root canals. No infections developed using this strain alone. It seems that the specificity in these anaerobic infections is low and numerous combinations of normally low virulent oral bacterial species have the capacity to induce an acute infection in the root canal and periapical tissues. The low virulence is compensated by the increase in numbers by the growth and multiplication and by the polymicrobial nature of the primary endodontic

**Table 4.4**    Studies indicating microorganisms associated to symptoms.

| Study | Number of teeth | Method of detection | Microorganisms associated with symptoms | Frequency in teeth with symptoms (%) |
|---|---|---|---|---|
| Griffee et al. (1980) | 12 | Culture | *B. melaninogenicus*[a] | 92% |
| Van Winkelhoff et al. (1985) | 17 | Culture | *P. endodontalis* | 53% |
| Haapasalo (1986) | 35 | Culture | *B. buccae*[b] | 37% |
| Haapasalo et al. (1986) | 31 | Culture | Black-pigmented | 54% |
| | | | *Bacteroides* | 32% |
| | | | *B. intermedius*[b] | 19% |
| | | | *B. gingivalis*[c] | |
| Yoshida et al. (1987) | 11 | Culture | *P. magnus*[d] | 55% |
| Sundqvist et al. (1989) | 72 | Culture | Black-pigmented bacteria | 73% |
| | | | *B. intermedius*[b] | |
| Hashioka et al. (1992) | 25 | Culture | *Eubacterium* spp. | 15–35% |
| | | | *Peptococcus* spp. | 7–24% |
| | | | *Peptostreptococcus* spp. | 11–18% |
| | | | *Porphyromonas* spp. | 9–24% |
| Gomes et al. (1994) | 30 | Culture | *P. micra*[e] | 17% (of isolates) |
| | | | *F. nucleatum* | 7% (of isolates) |
| | | | Black-pigmented bacteria | 13% (of isolates) |
| | | | *S. milleri* group[f] | <23% (of isolates) |
| Baumgartner et al. (1999) | 40 | PCR | Black-pigmented bacteria | 55% |
| | | | *P. nigrescens* | 50% |
| | | | *P. intermedia* | 36% |
| Chavez de Paz (2002) | 28 | Culture | *F. nucleatum* | 36% |
| Rocas et al. (2002) | 20 | PCR | *T. denticola* | 50% |
| | | | *T. forsythia* | 40% |
| | | | *P. endodontalis* | 40% |
| | | | *P. gingivalis* | 30% |
| Fouad et al. (2002) | 24 | PCR | *Streptococcus* spp. | Odds ratio 13 |
| | | | *F. nucleatum* | Odds ratio 3.2 |
| Foschi et al. (2005) | 62 | PCR | *T. denticola* | 56% |

[a]Now divided up in 6–9 different black-pigmented bacteria.
[b]Now *Prevotella* spp. (*P. buccae, P. intermedia, P. nigrescens*).
[c]Now *Porphyromonas* spp. (*P. gingivalis, P. endodontalis*).
[d]Now *Fingoldia magna.*
[e]*Parvimonas micra* formerly *Micromonas micros* earlier *Peptostreptococcus micros.*
[f]Now including *S. anginosus, S. intermedia,* and *S. constellatus.*

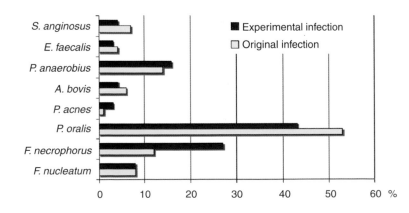

**Fig. 4.11**  Mean viable counts of 8 strains in percent of total counts in samples from 12 teeth after 6 months (experimental infection). The 8 strains were originally isolated from a monkey tooth (original infection), pure cultured in the laboratory and inoculated in equal numbers experimentally in 12 monkey teeth. (From Fabricius et al. 1982b.)

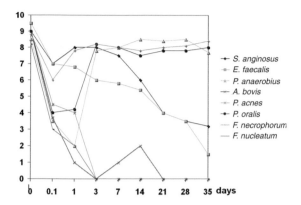

**Fig. 4.12** Total viable counts (10th logarithm) at various time points of each strain of the eight strain collection in wound chambers in rabbits followed for 35 days. (From Dahlén et al. 1987.)

infection. The character of anaerobic infections, in general, is that they develop when the local and general defense is hampered (Finegold 1977). When the blood supply is inhibited or strangled, the decrease in oxygen level disfavors the PMNs' oxygen-dependent killing mechanisms and the growth of anaerobic bacteria is concomitantly favored. This is very much true in the root canal with a necrotic pulp, where the

bacteria can grow extensively without being reached by the defense system (PMNs, antibodies, complement factors) and by their production of toxic metabolites, proteolytic enzymes, and so on, which increase the challenge of the host defense system (Olsen and Dahlén 2004). The concomitant outgrowth of bacteria through apical foramen into the external periapical tissues cannot be prevented since the bacteria are in an active growing phase, sometimes even stimulated by host factors such as blood components and serum. The fate of the periapical acute infection/abscess is probably much dependent on the communication through the apical foramen (Sundqvist 1992a). If that communication is wide, for example, as in younger teeth, this will probably favor the bacteria due to a better nutrient supply and the infection route may be more dramatic. Even if the root canal infection is polymicrobial and unspecific, it does not mean that specific features do not exist. Some bacterial species are more common in these infections than others and some bacteria do produce unique virulence factors (capsule, leukotoxins, complement resistance, and immunoglobulin-degrading enzymes), which make them more adapted to survive and grow in the lesion than others and to invade the tissues and actively participate in the pathological destruction (Table 4.5) (see also Chapter 7).

**Table 4.5** Commonly isolated bacterial species in teeth with acute periapical lesions and some of their virulence factors.

| Bacterial species | Virulence factors | Toxins |
| --- | --- | --- |
| *P. anaerobius/P. micra* | Weak proteolytic activity | Lipoteichoic acid<br>Metabolic acids |
| *F. nucleatum/F. necrophorum* | Capsule polysaccharides (not known)<br>Proteolytic activity | Endotoxin<br>Wide spectrum of metabolic acids<br>Sulfur products<br>Leukotoxin (*F. necrophorum*) |
| *P. intermedia/P. nigrescens* | Thin polysaccharide capsule<br>Proteolytic activity | Endotoxin<br>Metabolic acids |
| *P. endodontalis* | Capsule polysaccharides (not known)<br>Proteolytic enzymes (gingipains) | Endotoxin<br>Wide spectrum of metabolic acids |
| *P. gingivalis* | Thick polysaccharide capsule<br>Strong proteolytic activity (gingipain R and K, and collagenase)<br>Some strains highly invasive | Endotoxin<br>Wide spectrum of metabolic acids |
| *Treponema* spp./*T. denticola* | Strong proteolytic activity (gingipains) | Endotoxins<br>Wide spectrum of metabolic acids<br>Sulfur products<br>Ammonia |

From Olsen and Dahlén (2004).

### 4.6.3 Abscess and fistula formation

When bacteria grow in the necrotic root canal system, the periapical tissues are involved and the host defense system against infections is activated and attracted to the area. It is important to emphasize that this reaction is general for all acute infections of the body. If the bacteria maintain a high metabolic activity and growth, the body's main defense goal is to prevent the infection from spreading. The infection process may be encapsulated by a fibrotic barrier (abscess). This reaction is time dependent and it is sometimes too late to prevent bacteria and bacterial products already from spreading through the tissues. Such bacteria will be taken care of by the lymphatic drainage and the local lymph nodes that become swollen and painful. This is a stage where systemic antibiotic treatment is indicated with the purpose to inhibit the bacterial growth and spread of the infection. In the spreading periapical abscess (if no antibiotics is given), the bacteria may still grow leading to an expansion of the abscess through the tissues. The nature of this expansion follows the route of the least resistance and in most cases ends up with drainage into the oral cavity through either the periodontal pocket or the mucosal membrane. The latter condition is called a sinus tract and is frequently seen in the clinic. The microflora in sinus tract is also mixed anaerobic (Haapasalo et al. 1987). Fortunately, less frequently the infection is spread to other compartments of the head and neck region, where serious complications can follow (see Chapter 10). The sinus tract is usually the termination of the acute phase of the infection, the symptoms decline, and the whole process becomes chronic. However, as long as the primary root canal infection is not subjected to intervention, the bacteria still remain in the tooth and maintain the process and the sinus tract can remain for a long period. Periapical infections are further considered in Chapters 5 and 10.

### 4.6.4 The chronic infection

The chronic infection is characterized by remaining/persisting bacteria that are in a low metabolic stage with no or little multiplication. The body defense reaction is also changing into a chronic inflammation, predominating by lymphocytes and antibody-producing plasma cells. The tissue is reorganized into a granulomatous tissue that has its main purpose to keep the infection and bacteria localized and to prevent the bacteria from spreading. The formation of granulomatous tissue is favored by components of the immune system, for example, antibodies. This process is usually quiescent, with no or little symptoms and the risk of spreading is small. This was shown in monkeys where immunization was performed with those species that were later experimentally introduced in the root canal (Dahlén et al. 1982a). The periapical lesions in immunized monkeys that developed at infected teeth were clearly visible on radiographs due to a sharp demarcation and sometimes even a sclerotic zone in the bone surrounding the lesion. Histologically, the sharp demarcation was confirmed and the inflammatory infiltration was only seen adjacent to the apical foramen and surrounded by a thick fibrotic capsule (Fig. 4.13). However, in the nonimmunized control monkeys this capsule was not formed, the inflammatory cells were spread deeper in the tissues including the bone (osteitis), and the radiographical lesions were more diffuse and sometimes not detectable. In patients, the same type of lesion as seen in the immunized animals is frequently seen among adults and elderly. Longtime exposure for the antigens present in the tooth may stimulate the immune response and antibody formation. In monkeys, it was shown that antigens, for example, lipopolysaccharides present in the root canal, can induce a specific antibody response (Dahlén et al. 1982b). The patients are usually not aware of this type of chronic lesions that are only detected on radiographs. These lesions should be treated because the infection may sooner or later exacerbate.

### 4.6.5 The microflora of the root canal versus the deep periodontal pocket

The bacterial flora of the untreated root canal (Table 4.1, Fig. 4.14) is characterized predominantly by anaerobic bacteria. They are mainly Gram-negative anaerobic rods and the whole flora resembles that of the deep periodontal pocket (Haffajee and Socransky 1994; Socransky et al. 1998; Marsh 2004). Thus, the main metabolic activities at both sites are anaerobic and proteolytic and the access to oxygen and sugars/carbohydrates are limited. On the other hand, there are different selection mechanisms and prevailing ecological pressure that lead to striking differences (Sundqvist and Fidgor 2003). The access to nutrients, for example, blood and serum, is much higher in the periodontal pocket by the excessive exudate

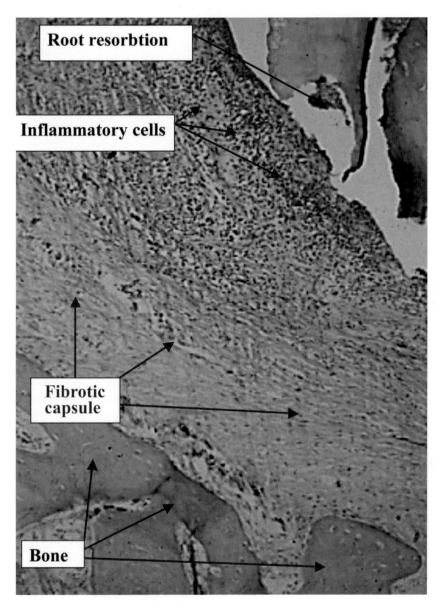

**Fig. 4.13**  Histological picture of the periapical area of a tooth in an immunized monkey. Note root resorption, cell infiltrate adjacent to the root, and the thick fibrotic capsule. (From Dahlén et al. 1982a.)

flow due to inflammation. This leads to a higher bacterial activity and much higher numbers of bacterial cells (Table 4.6). In addition, there is an open communication between the periodontal pocket and the oral cavity, which results in higher number of species and a more complex flora with hundreds of different microorganisms. While some are more virulent and have etiologic relation to periodontitis, others are innocent bystanders. The endodontic microflora shows some distinct features. First, there is a fewer number of species in endodontic infections due to limited communication with the oral cavity, unless the root canal

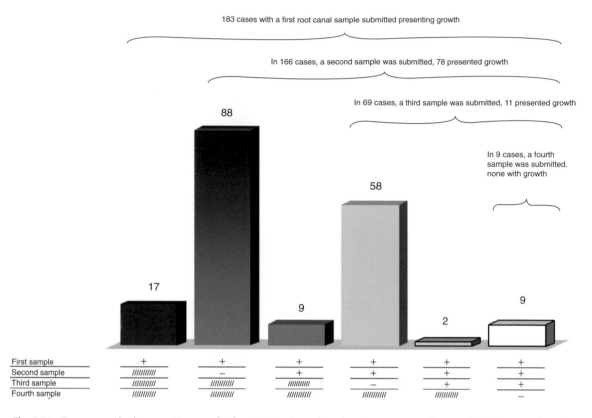

**Fig. 4.14** Frequency of culture positive samples from 183 root canals undergoing treatment (first sample), 166 cases after the second appointment (second sample), 69 cases after the third appointment (third sample), and 9 cases after the fourth appointment (fourth sample). (Courtesy of Dr Chavez de Paz.)

has been left open. This is an important reason for not leaving the pulp chamber open to avoid a more heavily invasion of bacteria. An open communication can only favor the infection and is of no benefit to the patient. Second, there is a selective ecological mechanism that allows some species to be more common in the root canal flora as detected by culture studies than that of the periodontal pocket. Table 4.6 shows microbiological characteristics of endodontic and periodontal infections. Species such as *Aggregatibacter* (formerly *Actinobacillus*) *actinomycetemcomitans*, *Haemophilus* spp., *Neisseria* spp., some *Streptococcus* species (especially polysaccharide-producing *S. salivarius*, *S. sanguis*, *S. oralis*, *S. mutans*) can regularly be found in subgingival samples but seldom in root canals. The exemplified bacteria are facultative and mainly saccharolytic. They cannot apparently compete in the anaerobic environment in the root canal with no ac-

cess to sugars, and they have no advantage of their strong adherent capacity both to epithelial cells (mucosa) and to enamel (salivary glycoproteins), which makes them predominant on the oral mucosa and teeth. On the contrary, *Lactobacillus* spp., *Bifidobacterium* spp., *Propionibacterium* spp., and enterococci have low adhering capacity to the enamel surface, while they apparently have easier to be established in the root canal system, especially under more hash conditions. Some species/genera, for example, *Tannerella forsythia* (formerly *Bacteroides forsythus*), *Treponema* species (spirochetes), *Eubacterium* spp., *Selenomonas* spp., and *Camplyobacter* spp. are underscored in culture studies because they are difficult to grow. Studies to disclose more specific but less frequent bacteria by culture have shown that species may be detected in the root canal but in lower frequency than in the deep periodontal pocket. By molecular biology methods, a

**Table 4.6**  Comparison between the root canal flora versus the flora in the deep periodontal pocket.

| Similarities and differences | Root canal flora at primary endodontic infection | Deep periodontal pocket at periodontitis |
|---|---|---|
| Similarities | Polymicrobial predominantly anaerobic infection | Polymicrobial predominantly anaerobic infection |
| Differences | <12 predominant species | > hundreds of species |
| General characteristics | $10^5$–$10^6$ bacterial cells in a paperpoint sample | $10^7$–$10^8$ bacterial cells in a paperpoint sample |
|  | Low access to nutrients | High access to nutrients |
| Differences | Highly frequent species or groups | Highly frequent pathogens |
| Specific species or groups | *Prevotella* spp. | The red complex |
|  | *Fusobacterium* spp. | *P. gingivalis* |
|  | *Prevotella intermedia* | *T. forsythia* |
|  | *Eubacterium* spp. | *T. denticola* |
|  | *Propionibacterium* spp. | *A. actinomycetemcomitans* |
|  | *Actinomyces* spp. | Frequent pathogens |
|  | Non-polysaccharide-producing streptococci | *P. intermedia/nigrescens* |
|  | *P. micra*[a] | *C. rectus* |
|  | Other anaerobic streptococci | *P. micra*[a] |
|  | *Treponema* spp. (underscored in culture studies) | *P. tannerae* |
|  | Frequent species or groups | *F. alocis* |
|  | Polysaccharide-producing streptococci | *P. endodontalis* |
|  | *Lactobacillus/Bifidobacterium* | Frequent less pathogenic species |
|  | *Porphyromonas* spp. | *Prevotella* spp. |
|  |  | *F. nucleatum* |
|  |  | *E. corrodens* |
|  |  | *S. intermedia* and more |

[a]*Parvimonas micra* formerly *Micromonas micros* earlier *Peptostreptococcus micros.*

number of new bacterial species, not yet cultured or difficult to culture, have been discovered both in the periodontal pocket and in the primary infected root canals. Thus, more quantitative and qualitative differences might be disclosed between the two types of infections in the future.

The root canal represents a special environment in which selective pressures result in the establishment of a restricted number of microorganisms (Sundqvist 1992b; Sundqvist and Fidgor 2003). In the primary infection, the microflora commonly consists of 10–12 predominantly anaerobic bacteria. Bacterial interactions and access to nutrients are key factors in determining the outcome of the infection. Endodontic treatment should not only eliminate bacteria but also disrupt the balance within the microbial community established in the root canal. This balance is stabilized by the fact that persisting bacteria remains in biofilm communities in various parts of the root canal system (Fig. 4.8). The root canal biofilm may not be as complex as the dental plaque biofilm; however, it

gives the microorganisms a number of advantages and support (Table 4.7). Dental biofilms are considered difficult therapeutic targets (Socransky and Haffajee 2002) and similar aspects can be made for biofilms in the root canal system. The increased resistance for antimicrobials should be specifically emphasized because it explains the difficulties in completely eliminating the remaining microorganisms from the root canal (Chavez et al. 2007).

## 4.7 Treatment aspects

The endodontic treatment has two major aims. First, the infections and the microorganisms have to be eliminated. A root canal free from microorganisms is a primary goal. Second, the root canal should be mechanically debrided and enlarged in a way that it can be adequately obturated with a permanent root filling. These two goals go well together since it is quite clear that a well-prepared and filled root canal system also has

**Table 4.7**   General properties of a biofilm.

| General properties |
| --- |
| Protection from host defense |
| Protection from dehydration |
| Protection from antimicrobial agents (antiseptics and antibiotics) |
|    Surface-associated phenotype[a] |
|    Slow growth rate |
|    Poor penetration |
|    Inactivation and neutralization |
| Novel gene expression and phenotype[a] |
| Persistence in a flowing system |
| Spatial and environmental heterogeneity |
| Metabolic interaction and food web |
| Elevated concentrations of nutrients |

Adapted from Marsh and Martin (1999).
[a]Increased resistance to antimicrobial agents may be due to altered gene expression.

the best chance to be free of microorganisms. These two goals are currently being pursued through a combination of mechanical debridement, irrigation, and interappointment dressings. These goals are not contradictory and a thorough debridement and enlargement of the root canal lumen is highly favoring the chance to eliminate the bacteria from the root canal. The use of irrigation and interappointment dressing is to further make it possible to kill/reduce/eliminate the bacteria from the root canal. There is an ongoing debate on whether this is necessary/possible and if this should/could be performed in one visit or if two steps or multi-appointments and repeatable treatments are necessary, or at least gives better results. The success rate in practice is measured by the rate of future failures. However, what is a failure? Is it restricted to future exacerbations and new acute infections, remaining symptoms, persistent lesions observed radiographically, and/or those lesions above a certain size? All these questions are related to the attitudes of the dentist, the possibility to perform the procedures aseptically, whether the tooth could be restored and the patient's wish, together with practical and economical factors. A series of publications by Chavez de Paz et al. (2003, 2004, 2005, 2007) illustrate not only the possibility to render the root canal bacteria free but also the problem with the strategy to eliminate persistent bacteria (Fig. 4.15) as analyzed by culture. Taken together, the root canal displayed a negative culture sample after one to four repeated treatments in 155

of 183 cases. The remaining 28 cases were considered dropouts due to a final sample was not obtained. A negative pre-obturation sample will probably improve the prognosis of the tooth; however, very few longitudinal studies over many years have been performed with this aspect as a purpose (Sjögren 1997; Sundqvist et al. 1998; Waltimo 2005).

### 4.7.1 Why is it so difficult to eliminate root canal microorganisms?

There are several reasons why it is difficult in practice to eliminate the microorganisms from the root canal. Some even believe that it is impossible or doubt its importance because bacteria may be entombed and die in the filled root canal. Even if it is difficult to eliminate bacteria, this argument cannot be accepted as a reason for not trying hard enough, for doing a number of shortcuts or even for neglecting the importance of the antimicrobial efforts. First, the importance of careful antiseptic measures in all treatment steps cannot be overestimated. It seems obvious that the tooth should be isolated by a rubber dam and a careful disinfection of the operative field is a must (as discussed previously) to avoid reinfection. Second, the anatomical condition and variations can be extremely problematic (Fig. 4.6) especially in molars with three to four canals, and with apical resorptions, isthmi, fins, and accessory canals. Bacteria may penetrate, especially in long-standing infections, into the dentinal tubules that make it difficult to reach them by mechanical and chemical procedures. Cross-sectionally, it is obvious that the canals are very seldom circular and especially the isthmus area in roots with two canals (lower molars) is difficult to reach mechanically (Fig. 4.6). It is even more difficult in teeth of older people with a lot of hard tissue formation (calcification and secondary dentine) on the root canal walls. In the apical region, this hard tissue may give rise to a very complicated apex delta that will not be possible to reach mechanically. Bacteria could also penetrate through apex and be present externally of the tooth and, in particular, in the acute and abscess situation. Bacteria may thus be present on the periapical root surface that may be difficult to reach by the intracanal treatment procedures. This problem is aggravated in case of root surface resorptions (Fig. 4.9). Third, mechanically unreachable bacteria are remaining in niches of the root canal system, forming the biofilms (Nair et al. 2005). Biofilms are of the benefit for the bacteria and offer

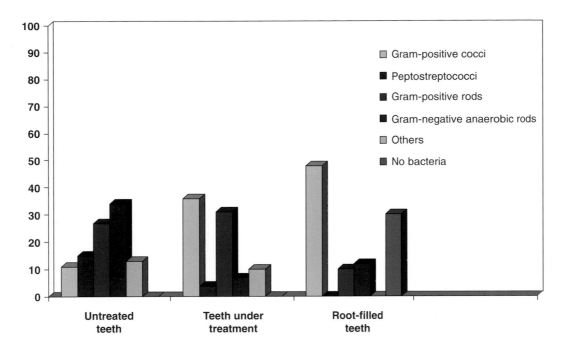

**Fig. 4.15**  Microbial composition (%) between untreated teeth (Sundqvist 1992a), teeth under treatment (Chavez de Paz et al. 2003), and root-filled teeth (Molander et al. 1998).

them a number of favors (Fig. 4.8). Notably, the increase of resistance against all types of antimicrobial agents should be considered. Finally, a number of species are true "persisters" by the fact that they have a natural ability to survive more harsh environments and stressed conditions (Chavez de Paz 2007). Gram-positive bacteria generally survive better than Gram-negatives when it comes to dehydration, lack of nutrients, changed ion strength/osmotic pressure, and presence of antiseptics. This has been appreciated specifically for enterococci and the use of Ca(OH)$_2$ paste as an interappointment dressing (Chavez de Paz et al. 2007). However, this is also true for other Gram-positives, for example, streptococci, lactobacilli, actinomyces, propionibacteria, and yeasts (Waltimo et al. 1997; Chavez de Paz et al. 2003, 2004, 2005). Conclusively, there is a striking difference between the primary infected teeth and teeth undergoing treatment (Fig. 4.16). Particularly, the increase of Gram-positive facultative cocci and the reduction of Gram-negative rods should be noted (Chavez de Paz 2004).

## 4.8 Persisting infections at root-filled teeth

Apical periodontitis associated with root-filled teeth is very common among endodontic patients. It seems to be an increasing problem associated with the increasing number of root canal treatments and fillings that are performed. The process is not noticed by the patient since they are chronic infections in nature with little or no symptoms. The remaining microorganisms have a low metabolic activity, and the apical lesion formed is characteristically a granulomatous tissue with predominance of lymphocytes and plasma cells. Apical periodontitis lesions are usually well recognized on radiographs due to the concomitant loss of bone. The size can differ greatly from small and hardly discernable lesions to lesions 10 mm or more in diameter. Their progression is usually slow and makes the dentist unsure whether to do revision/retreatment or not. They are often just followed and therapy is postponed until a period of 2–4 years after initial therapy, when a definitive decision is usually made.

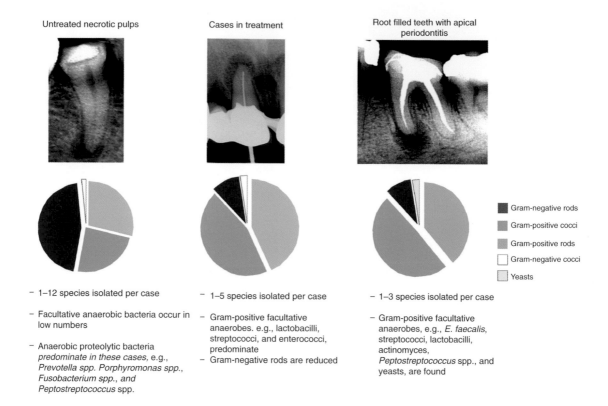

**Fig. 4.16**    Pie charts showing the proportions of organisms isolated in studies of untreated necrotic pulps, cases undergoing treatment, and root-filled teeth with apical periodontitis. (Courtesy of Dr Chavez de Paz.)

It is an ongoing discussion whether and when these lesions should be retreated. A number of culture studies have been conducted to see to what extent these lesions show presence of microorganisms and of what kind. This is not easy since it is very likely that the present microorganisms are unreachable for sampling or they are eliminated when the root-filling material is removed. Therefore, the risk of false-negative samples is high when culture is used. The frequency of positive samples by culture or molecular methods has increased in recent studies, when the difficulties have been more carefully considered (Table 4.8). Only one to three species are generally isolated and in low numbers. They are predominately Gram-positive cocci and rods, for example, *Enterococcus faecalis*, streptococci, lactobacilli, actinomyces, peptostreptococci, and yeast (Fig. 4.17). In Fig. 4.17, the change in microbiological character from the primary untreated tooth, cases in treatment, and the root-filled tooth with apical pe-

riodontitis is also illustrated. Molander et al. (1998) reached a frequency of 68% positive samples and it was suggested that the prevalence is, in fact, 100%. Later studies have confirmed an even higher prevalence of positive bacterial cultures from root-filled teeth (Table 4.8). With the microbial pattern of the persisting flora during treatment in mind, it is likely that the bacteria observed in the root-filled teeth persist from the initial treatment (Fig. 4.16). A secondary invasion by coronal leakage or other routes is possible, but unlikely in the adequately treated and restored case. This has been experimentally proved in monkeys (Möller et al. 2004; Fabricius et al. 2006), where 175 teeth were infected using a collection of four to five strains previously isolated from an infected root canal of a monkey (Fabricius et al. 1982a). After establishing an infection in all teeth, they were subjected to treatment with mechanical debridement and irrigation with NaOCl but no interappointment dressing. After

**Table 4.8**  Microflora (percent of isolates) in root-filled teeth with apical periodontitis as indicated in some recent studies.

| Microorganisms | Sirén et al. (1997) | Molander et al. (1998) | Sundqvist et al. (1998) | Pinheiro et al. (2003) | Gomes et al. (2004) | Adib et al. (2004) |
|---|---|---|---|---|---|---|
| *S. aureus* | – | – | – | – | – | 1 |
| Other *Staphylococcus* spp. | 7 | 6 | 5 | 2 | 3 | 21 |
| *Streptococcus* spp. (PSP) | **26** | 6 | 6 | 9 | **14** | **13** |
| *Streptococcus* spp. (non-PSP) | NS[a] | 6 | 12 | 6 | **11** | **19** |
| *E. faecalis* | **16** | **27** | **29** | **25** | **17** | **11** |
| *P. micra*[b] | 5 | 1 | 6 | 5 | 8 | – |
| Other anaerobic streptococci | – | – | – | 7 | 8 | 7 |
| *Neisseria* spp. | 2 | – | – | – | – | – |
| *Veillonella* spp. | – | 1 | – | 4 | – | – |
| *Bacillus* spp. | – | 1 | – | – | – | – |
| *Clostridium* spp. | – | – | – | 1 | – | – |
| *Corynebacterium* spp. | – | – | – | – | – | – |
| *Lactobacillus* spp. | 1 | **12** | 3 | 4 | 3 | 2 |
| *Propionibacterium* spp. | – | 3 | **10** | – | 6 | 4 |
| *Actinomyces* spp. | 2 | 2 | 6 | 4 | 6 | 9 |
| *Eubacterium* spp. | – | 1 | 6 | – | 4 | 6 |
| Enteric rods | 7 | **13** | – | – | – | 9 |
| *Capnocytophaga* spp. | – | – | – | 1 | – | 2 |
| *Campylobacter* spp. | 1 | 1 | **10** (*C. gracilis*) | – | – | – |
| *Eikenella* spp. | – | – | – | – | – | – |
| *Porphyromonas* spp. | 2 | – | – | – | 3 | – |
| *Prevotella intermedia/nigrescens* | 5 | – | – | 3 | 3 | – |
| Other *Prevotella* spp. | – | 4 | – | 6 | **11** | 1 |
| *Fusobacterium* spp. | **12** | 4 | 3 | 3 | – | 2 |
| Spirochetes | – | – | – | – | – | – |
| *Candida* spp. | 3 | 3 | 6 | 2 | – | 3 |
| Total number of isolated strains | 147 | 117 | 31 | 108 | 36 | 90 |
| Number of teeth | 40 | 100 | 40 | 60 | 19 | 8 |
| Teeth with detected microorganisms | 100% (selected cases) | 68% | 44% (root canals) | 85% | No data | 100% (selected cases) |

Species frequency >10% in bold.
[a]NS, not specified.
[b]*Parvimonas micra* formerly *Micromonas micros* earlier *Peptostreptococcus micros*.

two appointments, all teeth were permanently root-filled and followed for 2 years. Eighty teeth were radiographically and histologically healed, while 95 did not heal (Fig. 4.17). Only 19 (21%) of the teeth that contained bacteria healed, while 61 (72%) teeth with a negative sample at the root-filling occasion healed. This gives an odds ratio of 3.2 for healing in the absence of cultured bacteria. In the nonhealed teeth, 71 (79%) contained bacteria compared to 24 (28%) that did not show growth. This gives an odds ratio of 3.0 for a nonhealed periapical lesion in case of remaining

bacteria. It was also found that no other bacteria occurred than those primarily inoculated into the root canals, indicating low risk for coronal leakage if adequately sealed. Furthermore, the root-filling quality was of less importance in the sense that in root-filled teeth without bacteria the lesions healed irrespective if the root filling was extended beyond or short of apex. On the other hand, lesions remained at many teeth with bacteria even if the root filling was adequate. Conclusively, remaining bacteria at permanent root filling is a stronger risk factor for a nonhealed

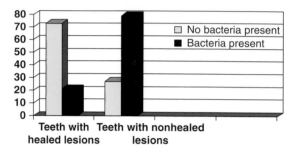

**Fig. 4.17**  Healed and nonhealed periapical lesions (%) in relation to persisting bacterial infection after 2–2.5 years of root filling of 175 teeth (90 with persisting infection and 85 without bacteria) in 8 monkeys. (From Fabricius et al. 2006.)

periapical lesion than the technical quality of the root filling. These studies clearly show that the root canal should be bacteria free according to culture before root filling and that careful antiseptic measures at the appointments are necessary.

## 4.9 Culture versus molecular biology methods

It is clear that the endodontic microbiology has developed tremendously through the years by using the culture analysis. Much of the current knowledge is based on samples and experiments where culture analysis has been used. We have also learned about the characteristics of the isolated bacteria and under what conditions they participate in various stages of the root canal infection. We could also do experimental infections to prove their pathogenic nature. We must also admit that the culture analysis has only disclosed part of the true microbial condition of the infected root canal. The problems of avoiding false-negative samples due to sampling problems are highlighted as well as to get dormant (viable but uncultivable/difficult to grow) bacteria to grow simply because we do not have the right medium or conditions. By molecular biology techniques, a significant number of microorganisms, designated as *not-yet-cultured* have been detected (see Chapter 5). It is not clear whether all these new findings, in fact, correspond to viable infectious cells or to what degree they are remains of dead cells. Conclusively, culture will for a number of years be the gold standard for most clinicians, while the new techniques are further developed, disclosing new knowledge and complementing the present knowledge in endodontic microbiology.

## 4.10 Summary

Endodontic infections are major problems in the dental clinic and treatment of the root canal constitutes a serious challenge for the dentist. The root canal constitutes an excellent environment for microorganisms because it offers a protected compartment with no or weak host defense system and little chance for the host to eliminate them. The common situation involves necrosis of the pulp, which leaves a nutrient source for the bacteria to grow and multiply. The infection reaches the root apex, penetrates through the apical foramen, and forms a periapical lesion that may develop into an acute infection with symptoms or into a chronic infection where there are little symptoms. In both situations, the microflora is a polymicrobial and predominantly anaerobic, harboring 10–12 species according to culture-based analysis. Species of anaerobic streptococci, *Eubacterium*, *Campylobacter*, *Prevotella*, and *Fusobacterium* usually prevail and the flora thus resembles the flora of the periodontal pocket. The treatment includes mechanical debridement, irrigation, and interappointment dressings performed under strict aseptic measures. The combination of the high risk of reinfection during the treatment and the difficulties to reach all the bacteria in the root canal system leads to an unacceptable situation with root canals often permanently filled with remaining microorganisms. Culture analysis of samples from root canals undergoing treatment shows that the persisting microorganisms are mainly Gram-positive and facultative such as streptococci, enterococci, lactobacilli, propionibacteria, and *Actinomyces*. These microorganisms are resistant against most antimicrobial agents and form biofilms in locations in the root canal system, particularly where it is especially difficult to reach them. They can survive for years after the permanent root filling is placed and are the predominating flora in teeth that are subjected to revision of treatment.

## 4.11 References

Adib V, Spratt S, Ng Y-L, and Gulabivala K. 2004. Cultivable microflora associated with persistent periapical disease and coronal leakage after root canal treatment: a preliminary study. *Int Endod J* 37: 542–51.

Baumgartner JC, Watkins BJ, Bae KS, and Xia T. 1999. Association of black-pigmented bacteria with endodontic infections. *J Endod* 6: 413–15.

Bergenholtz G. 1974. Microorganisms from necrotic pulps of traumatized teeth. *Odontol Revy* 25: 347–58.

Bergenholtz G. 1977. Effect of bacterial products on inflammatory reactions in the dental pulp. *Scand J Dent Res* 85: 122–29.

Bergenholtz G. 1981. Inflammatory response of the dental pulp to bacterial irritation. *J Endod* 7: 100–104.

Brook I, Frazier EH, and Gher ME. 1991. Aerobic and anaerobic microbiology of periapical abscess. *Oral Microbiol Immunol* 6: 123–25.

Brook I, Grimm S, and Kielich RB. 1981. Bacteriology of acute periapical abscess in children. *J Endod* 7: 378–80.

Brown LR, Jr, and Rudolph CE, Jr. 1957. Isolation and identification of microorganisms from unexposed canals of pulp-involved teeth. *Oral Surg Oral Med Oral Pathol* 10: 1094–99.

Chavez de Paz LE. 2002. Fusobacterium nucleatum in endodontic flare ups. *Oral Surg Oral Med Oral Pathol Oral Radiol Endod* 93: 179–83.

Chavez de Paz LE. 2004. Gram-positive organisms in endodontic infections. *Endod Top* 9: 79–96.

Chavez de Paz LE. 2007. Redefining the persistent infection in root canal: possible role of biofilm communities. *J Endod* 33: 652–62.

Chavez de Paz LE, Bergenholtz G, Dahlén G, and Svensäter G. 2007. Response to alkaline stress by root canal bacteria in biofilms. *Int Endod J* 48: 344–55.

Chavez de Paz LE, Molander A, and Dahlén G. 2004. Gram-positive rods prevailing in teeth with apical periodontitis undergoing root canal treatment. *Int Endod J* 37: 579–87.

Chavez de Paz LE, Molander A, Dahlén G, Möller ÅJR, and Bergenholtz G. 2003. Bacteria recovered from teeth with apical periodontitis and antimicrobial endodontic treatment. *Int Endod J* 36: 500–508.

Chavez de Paz LE, Svensäter G, Dahlén G, and Bergenholtz G. 2005. Streptococci from root canals in teeth with apical periodontitis receiving endodontic treatment. *Oral Surg Oral Med Oral Pathol Oral Radiol Endod* 100: 232–41.

Chu FCS, Tsang CS, Chow TW, and Samaranayake LP. 2005. Identification of cultivable microorganisms from primary endodontic infections with exposed and unexposed pulp space. *J Endod* 31: 424–29.

Crawford JJ and Shankle RJ. 1961. Application of newer methods to study the importance of root canal and oral microbiota in endodontics. *Oral Surg Oral Med Oral Pathol* 14: 1109–23.

Dahlén G. 2002. Microbiology and treatment of dental abscesses and periodontal-endodontic lesions. *Periodontology 2000* 28: 206–39.

Dahlén G, Fabricius L, Heyden G, Holm SE, and Möller ÅJR. 1982a. Apical periodontitis induced by selected bacterial strains in root canals of immunized and nonimmunized monkeys. *Scand J Dent Res* 90: 207–16.

Dahlén G, Fabricius L, Holm SE, and Möller ÅJR. 1982b. Circulating antibodies after experimental chronic infection in the root canal of teeth in monkeys. *Scand J Dent Res* 90: 338–44.

Dahlén G, Fabricius L, Holm SE, and Möller ÅJR. 1987. Interactions within a collection of eight bacterial strains from a monkey dental root canal. *Oral Microbiol Immunol* 2: 164–70.

Dahlén G, Pipattanagovit P, Rosling B, and Möller ÅJR. 1993. A comparison between two transport media for saliva and subgingival samples. *Oral Microbiol Immunol* 8: 375–82.

Edwardsson S. 1974. Bacteriological studies on deep areas of carious dentin. *Odontol Revy* 25: 1–143.

Engström B and Frostell G. 1961. Bacteriological studies of the non-vital pulp in cases with intact pulp cavities. *Acta Odontol Scand* 19: 23–39.

Fabricius L, Dahlén G, Holm SE, and Möller ÅJR. 1982a. Influence of combinations of oral bacteria on periapical tissues of monkeys. *Scand J Dent Res* 90: 200–206.

Fabricius L, Dahlén G, Öhman AE, and Möller ÅJR. 1982b. Predominant indigenous oral bacteria isolated from infected root canals after varied time of closure. *Scand J Dent Res* 90: 134–44.

Fabricius L, Dahlén G, Sundqvist G, Happonen R-P, and Möller ÅJR. 2006. Influence of residual bacteria on periapical tissue healing after chemomechanical treatment and root filling of experimentally infected monkey teeth. *Eur J Oral Sci* 114: 278–85.

Finegold S. 1977. *Anaerobic Bacteria in Human Disease*. New York: Academic Press.

Foschi F, Cavrini F, Montebugnoli L, Stashenko P, Sambri V, and Prati C. 2005. Detection of bacteria in endodontic samples by polymerase chain reaction assays and association with defined clinical signs in Italian patients. *Oral Microbiol Immunol* 20: 289–95.

Fouad AF, Barry J, Caimano M, Clawson M, Zhu Q, Carver R, Hazlett K, and Radolf JD. 2002. PCR-based identification of bacteria associated with endodontic infections. *J Clin Microbiol* 40: 3223–31.

Gomes BPFA, Drucker DB, and Lilley JD. 1994. Association of specific bacteria with some endodontic signs and symptoms. *Int Endod J* 27: 291–98.

Gomes BPFA, Pinheira ET, Gade-Neto CR, Sousa ELR, Ferrez CCR, Zaia AA, Teixeira FB, and Souza-Filho FJ. 2004. Microbiological examination of infected dental root canals. *Oral Microbiol Immunol* 19: 71–76.

Griffee MB, Patterson SS, Miller CH, Kafrawy AH, and Newton CW. 1980. The relationship of Bacteroides melaninogenicus to symptoms associated with pulpal necrosis. *Oral Surg Oral Med Oral Pathol* 50: 457–61.

Haapasalo M. 1986. *Bacteroides buccae* and related taxa in necrotic root canal infection. *J Clin Microbiol* 24: 940–44.

Haapasalo M, Ranta H, Ranta K, and Shah H. 1986. Black-pigmented *Bacteroides* spp. in human apical periodontitis. *Infect Immun* 53: 149–53.

Haapasalo M, Ranta K, and Ranta H. 1987. Mixed anaerobic periapical infection with sinus tract. *Endod Dent Traumatol* 3: 83–85.

Haffajee AD and Socransky SS 1994. Microbial etiological agents of destructive periodontal disease. *Periodontology 2000* 5: 78–111.

Hashioka K, Yamasaki M, Nakane A, Horiba N, and Nakamura H. 1992. The relationship between clinical symptoms and anaerobic bacteria from infected root canals. *J Endod* 18: 558–61.

Hobson P. 1959. An investigation into the bacteriological control of infected root canals. *Br Dent J* 106: 63–70.

Hoshino E. 1985. Predominant obligate anaerobes in human carious dentine. *J Dent Res* 64: 1195–98.

Kakehashi S, Stanley H, and Fitzgerald R. 1965. The effect of surgical procedures of dental pulps in germ-free and conventional laboratory rats. *Oral Surg Oral Med Oral Pathol* 20: 340–49.

Kantz WE and Henry CA. 1974. Isolation and classification of anaerobic bacteria from intact pulp chambers of nonvital teeth in man. *Arch Oral Biol* 19: 91–96.

Khemaleelakul S, Baumgartner JC, and Pruksakorn S. 2002. Identification of bacteria in acute endodontic infections and their antimicrobial susceptibility. *Oral Surg Oral Med Oral Pathol Oral Radiol Endod* 94: 746–55.

Lana MA, Riberio-Sobrinho AP, Stehling R, Garcia GD, Silva BKC, Hamadan JS, Nicoli JR, Carvalho MAR de M, and Farias L. 2001. Microorganisms isolated from root canals presenting necrotic pulp and their drug susceptibility in vitro. *Oral Microbiol Immunol* 16: 100–105.

Le Goff, Bunetel L, Mouton C, and Bonnaure-Mallet M. 1997. Evaluation of root canal bacteria and their antimicrobial susceptibility in teeth with necrotic pulp. *Oral Microbiol Immunol* 12: 318–22.

Lewis MAO, MacFarlane TW, and McGowan DA. 1986. Quantitative bacteriology of acute dentoalveolar abscess. J Med Microbiol 27: 101–4.

Marsh P. 2004. Dental plaque as a microbial biofilm. *Caries Res* 38: 204–11.

Marsh P and Martin MV. (eds) 1999. *Oral Microbiology*, 4th edn. Oxford: Wright, 35–57.

McDonald JB, Hare GC, and Wood AWS. 1957. The bacteriologic status of the pulp chambers in intact teeth found to be non-vital following trauma. *Oral Surg Oral Med Oral Pathol* 10: 318–22.

Miller WD. 1894. An introduction in the study of the bacteriopathology of the dental pulp. *Dental Cosmos* 36: 505–28.

Molander A, Reit C, Dahlén G, and Kvist T. 1998. Microbiological status of root-filled teeth with apical periodontitis. *Int Endod J* 31: 1–7.

Möller ÅJR. 1966. Microbiological examination of root canals and periapical tissues of human teeth. Methodological studies. *Odontol Tidskr* 74: 1–380.

Möller ÅJR, Fabricius L, Dahlén G, Öhman AE, and Heyden G. 1981. Influence on periapical tissues of indigenous oral bacteria and necrotic pulp tissue in monkeys. *Scand J Dent Res* 89: 475–84.

Möller ÅJR, Fabricius L, Dahlén G, Sundqvist G, and Happonen R-P. 2004. Apical periodontitis development and bacterial response to endodontic treatment. Experimental root canal infections in monkeys with selected bacterial strains. *Eur J Oral Sci* 112: 207–15.

Morse FW, Jr, and Yates MF. 1941. Follow-up studies of root-filled teeth in relation to bacteriologic findings. *J Am Dent Assoc* 28: 956–71.

Nair PN, Henry S, Cano V, and Vera J. 2005. Microbial status of apical root canal system of human mandibular first molars with primary apical periodontitis after "one-visit"

endodontic treatment. *Oral Surg Oral Med Oral Pathol Oral Radiol Endod* 99: 231–52.

Nair R. 1987. Light and electronmicroscopic studies of root canal flora and periapical lesions. *J Endod* 13: 29–39.

Oguntebi B, Slee AM, Tanzer JM, and Langeland K. 1982. Predominant microflora associated with human dental periradicular abscesses. *J Clin Microbiol* 15: 964–66.

Olsen I and Dahlén G. 2004. Salient virulence factors in anaerobic bacteria, with emphasis on *Porphyromonas*, *Prevotella*, *Fusobacterium*, and *Peptostreptococcus* spp. *Endod Top* 9: 15–26.

Peters LB, Wesselink PR, and Van Winkelhoff AJ. 2002. Combinations of bacterial species in endodontic infections. *Int Endod J* 35: 698–702.

Pinheiro ET, Gomes BPFA, Ferraz CCR, Sousa ELR, Teixeira FB, and Souza-Filho FJ. 2003. Microorganisms from canals of root-filled teeth with periapical lesions. *Int Endod J* 36: 1–11.

Reit C and Dahlén G. 1988. Decision making analysis of endodontic treatment strategies in teeth with apical periodontitis. *Int Endod J* 21: 291–99.

Rocas IN, Siquiera JF, Jr, Andrade AFB, and de Uzeda M. 2002. Identification of selected putative oral pathogens in primary root canal infections associated with symptoms. *Anaerobe* 8: 200–208.

Sakamoto H, Hato H, Sato T, and Sasaki J. 1998. Semiquatitative bacteriology of closed odontogenic abscesses. *Bull Tokyo Dent Coll* 39: 103–7.

Shovelton DS and Sidaway DA. 1960. Infections in root canals. *Br Dent J* 108: 115–18.

Sirén EK, Haapasalo MPP, Ranta K, Salmi P, and Kerouso ENJ. 1997. Microbiological findings and clinical treatment procedures in endodontic cases for microbiological investigation. *Int Endod J* 30: 91–95.

Sjögren U, Figdor D, Persson S, and Sundqvist G. 1997. Influence of infection at the time of root filling on the outcome of endodontic treatment of teeth with apical periodontitis. *Int Endod J* 30: 297–306.

Socransky SS and Haffajee AD. 2002. Dental biofilms: Difficult therapeutic targets. *Periodontology 2000* 28: 12–55.

Socransky SS, Haffajee AD, Cugini MA, Smith C, and Kent RL, Jr. 1998. Microbial complexes in subgingival plaque. *J Clin Periodontol* 25: 134–44.

Sommer RF and Crowley MC. 1940. Bacteriologic verification of roentgenographic findings in pulp-involved teeth. *J Am Dent Assoc* 27: 723–34.

Sundqvist G. 1976. *Bacteriological Studies of Necrotic Dental Pulps*. Umeå, Sweden: Umeå University Odontological Dissertation.

Sundqvist G. 1992a. Associations between microbial species in dental root canal infections. *Oral Microbiol Immunol* 7: 257–62.

Sundqvist G. 1992b. Ecology of the root canal flora. *J Endod* 16: 427–30.

Sundqvist G and Fidgor D. 2003. Life as an endodontic pathogen. Ecological differences between the untreated and root filled root canal. *Endod Top* 6: 3–28.

Sundqvist G, Figdor D, Persson S, and Sjögren U. 1998. Microbiologic analysis of teeth with failed endodontic treatment and outcome of conservative re-treatment.

*Oral Surg Oral Med Oral Pathol Oral Radiol Endod* 85: 86–93.

Sundqvist G, Johansson E, and Sjögren U. 1989. Prevalence of black-pigmented bacteroides species in root canal infections. *J Endod* 15: 13–19.

Sundqvist GK, Eckerbom MI, Larsson AP, and Sjögren UF. 1979. Capacity of anaerobic bacteria from necrotic dental pulps to induce purulent infections. *Infect Immun* 25: 685–93.

Van Winkelhoff AJ, Carlee AW, and deGraaff J. 1985. Bacteriodes endodontalis and other black-pigmented Bacteriodes species in odontogenic abscesses. *Infect Immun* 49: 494–97.

Waltimo T, Sirén E, Torkko H, Olsen I, and Haapasalo M. 1997. Fungi in therapy resistant apical periodontitis. *Int Endod J* 30: 96–101.

Waltimo T, Trope M, Haapasalo M, and Örstavik D. 2005. Clinical efficacy of treatment procedures in endodontic infection control and one year follow-up of periapical healing. *J Endod* 31: 863–68.

Wasfy MG, McMahgon KT, Minah CE, and Falkler WA, Jr. 1992. Microbiological evaluation of periapical infections in Egypt. *Oral Microbiol Immunol* 7: 100–105.

Williams BL, McCann GF, and Schoenknecht FD. 1983. Bacteriology of dental abscesses of endodontic origin. *J Clin Microbiol* 18: 770–74.

Winkler KC and van Amerongen J. 1959. Bacteriologic results from 4000 root canal cultures. *Oral Surg Oral Med Oral Pathol* 12: 857–75.

Wittgow WC and Sabiston CB. 1975. Microorganisms from pulpal chambers of intact teeth with necrotic pulp. *J Endod* 1: 168–71.

Yoshida M, Fukushima H, Yamamoto K, Ogawa K, Toda T, and Sagawa H. 1987. Correlation between symptoms and microorganisms isolated from root canals of teeth with apical pathosis. *Int Endod J* 13: 24–28.

Zamany A and Spangberg L. 2002. An effective method of inactivating chlorhexidine. *Oral Surg Oral Med Oral Pathol Oral Radiol Endod* 93: 617–20.

# Chapter 5
# Molecular Analysis of Endodontic Infections

*José F. Siqueira, Jr, and Isabela N. Rôças*

## 5.1 Introduction

Research in endodontic microbiology is moving at a rapid pace and, as a consequence, our understanding of the etiology and pathogenesis of endodontic diseases has intensified and continues to evolve. Specifically, the past decade has witnessed an over-whelming volume of new information about diverse aspects of endodontic infections. Much of the substantial progress in this area has been a result of improvement in laboratory techniques, particularly the introduction and further widespread use of culture-independent molecular biology techniques. This chapter reviews the molecular biology techniques that have been or have the potential to be used in endodontic microbiology research, their advantages and limitations as well as the contribution they have made to the field of endodontic microbiology.

## 5.2 Limitations of culture methods

Traditionally, microbiological culture has been the preferred means for examination of the endodontic microbiota, as discussed in Chapter 4. Culture is the process of propagating microorganisms in the laboratory by providing them with proper environmental conditions. Ingredients necessary for microbial pathogens can be supplied by living systems (e.g., growth in an animal host or in cell culture) or artificial systems (by gathering the required nutrients and conditions for growth). Artificial systems have been widely used for microbiological diagnosis of most bacterial and fungal infections that affect humans. In order for microorganisms to multiply on or in artificial media, they must have available the required nutrients and proper physicochemical conditions, including temperature, moisture, atmosphere, salt concentration, and pH (Slots 1986).

Essentially, culture analyses involve the following steps: sample collection and transport, dispersion, dilution, cultivation, isolation, and identification. Oral samples are collected and transported to the laboratory in a viability-preserving, nonsupportive, anaerobic medium. They are then dispersed by sonication or by vortex mixing, diluted, distributed onto various types of agar media and cultivated under aerobic or anaerobic conditions. After a suitable period of incubation, individual colonies are subcultivated and identified on the basis of multiple phenotype-based aspects, including colony and cellular morphology, Gram-staining pattern, oxygen tolerance, comprehensive biochemical characterization, and metabolic end product analysis by gas-liquid chromatography. The outer cellular membrane protein profile as examined by gel electrophoresis, fluorescence under ultraviolet light, and susceptibility tests to selected antibiotics can be needed for identification of some species (Engelkirk et al. 1992). Marketed packaged kits that test for preformed enzymes have also been used for rapid identification of several species.

Culture analyses of endodontic infections have provided a substantial body of information about the etiology of apical periodontitis, composition of the endodontic microbiota in different clinical conditions, effects of treatment procedures in microbial elimination, susceptibilities of endodontic microorganisms to antibiotics, and so on (for review, see Chapter 4). Culture has its advantages and limitations, which are listed in Table 5.1. As one can tell, some important

**Table 5.1**   Advantages and limitations of culture methods.

| Advantages | Limitations |
|---|---|
| 1. Broad-range nature, identification of unexpected species | 1. Impossibility of culturing a large number of extant microbial species |
| 2. Allow quantification of all major viable microorganisms in the samples | 2. Not all viable microorganisms can be recovered |
| 3. Allow determination of antimicrobial susceptibilities of the isolates | 3. Once isolated, microorganisms require identification using a number of techniques |
| 4. Physiological studies are possible | 4. Misidentification of strains with ambiguous phenotypic behavior |
| 5. Pathogenicity studies are possible | 5. Low sensitivity |
| 6. Widely available | 6. Strict dependence on the mode of sample transport |
| | 7. Samples require immediate processing |
| | 8. Costly, time consuming and laborious |
| | 9. Specificity is dependent on the composition of media and experience of the microbiologist |
| | 10. Extensive expertise and specialized equipment are needed to isolate strict anaerobes |
| | 11. It takes several days to weeks to identify most anaerobic bacteria |

limitations of culture methods make a comprehensive analysis of the endodontic microbiota difficult to achieve. The difficulties in culturing or in identifying many microbial species are of special relevance and deserve more discussion.

### 5.2.1 Difficulties in culturing—the huge as-yet-uncultivated majority

Microorganisms survive and reproduce in their natural environments, where their nutritional and physiological needs are met. Successful cultivation of these microorganisms relies on our ability to determine and reproduce their growth requirements in the laboratory. Unfortunately, not all microorganisms can be cultivated under artificial conditions, and this is simply because the nutritional and physiological needs of most microorganisms are still unknown. There are several instances of microbial ecosystems that were thought to be well-characterized by culture-dependent approaches, but which proved to be far different when assessed by culture-independent techniques (Hugenholtz and Pace 1996). Investigations of many aquatic and terrestrial environments using culture-independent methods have revealed that the cultivable members of these systems represent less than 1% of the total extant population (Ward et al. 1990; Amann et al. 1995). These figures are usually calculated by comparing the number and types of bacteria directly observed by microscopy to the number and types of bacteria that are cultivated from the same sample. The discrepancy between the directly observed and the cultivable bacteria has been referred to as the *great plate count anomaly* (Handelsman 2004). This problem assumes prominence when one considers that there is a strong bias toward the cultivable minority—it is estimated that more than 99.9% of the microbiologists work on the 1% of cultivable microorganisms (Lewis 2007).

Culture-independent molecular biology methods that involve amplification of the 16S rRNA gene followed by cloning and sequencing (discussed below) have been recently used to determine the bacterial diversity in diverse environments. Perhaps not surprisingly, the number of recognized bacterial phyla has exploded from the original estimate of 11 in 1987 to near 53, of which one-half has only as-yet-uncultivated representatives (Rappe and Giovannoni 2003; Keller and Zengler 2004). Of the phyla containing cultivable

members, a few possess many cultivable taxa while the great majority contain too few cultivable representatives to represent the full spectrum of diversity in the phylum (Hugenholtz 2002; Riesenfeld et al. 2004).

Several terms, such as uncultivable, as-yet-uncultivated, noncultured, nonculturable, not-yet-cultured, and so forth, have been used to refer to bacteria that are known only through culture-independent approaches. It has been suggested that terms like *not-yet-cultured* or *as-yet-uncultivated bacteria* be used instead of nonculturable or uncultivable bacteria, since conceptually all bacteria are able to grow under the proper nutritional and physicochemical conditions (Clarridge 2004). In this chapter, we refer to these bacteria as as-yet-uncultivated. Also, the term *phylotype* is used for those as-yet-uncultivated species that are known only by a 16S rRNA gene sequence.

It is not difficult to understand that many microorganisms are not capable of adapting to the unfamiliar, artificial, and rather restrictive conditions of laboratory cultures (Rodriguez-Valera 2002). There are many possible reasons for the fact that several bacteria have not been cultivated yet. These include (a) lack of essential nutrients or growth factors in the artificial culture medium; (b) overfeeding conditions; (c) toxicity of the culture medium itself, which can inhibit bacterial growth; (d) production of substances inhibitory to certain bacteria by other species present in a mixed consortium; (e) metabolic dependence on other species for growth; (f) disruption of bacterial *quorum-sensing* systems induced by separation of bacteria on solid culture media; and (g) cells in a "viable but noncultivable" state, that is, a state of low metabolic activity or dormancy, in which cells are unable to divide or form colonies onto agar plates (Kell and Young 2000; Wade 2002; Siqueira and Rôças 2005a).

Obviously, if bacteria cannot be cultivated, they cannot be identified by phenotype-based methods. While we stay relatively unaware of the requirements of many bacteria to grow, identification methods that are not based on cultivability are required. This may avoid that many pathogens pass unnoticed when one is microbiologically surveying clinical samples.

It is worth pointing out that the fact that a given species has not been cultivated does not imply that this species will remain indefinitely impossible to cultivate. For instance, a myriad of obligate anaerobic bacteria were unable to be cultivated 100 years ago, but further developments in anaerobic culturing techniques have to a large extent helped to solve this problem. It must be assumed that no single method or culture medium is suitable for isolating the vast diversity of microorganisms present in most environments (Green and Keller 2006). There is a growing trend to develop specific approaches and culture media that allow cultivation of previously uncultivated bacteria, many of which can be clinically important. Strategies may rely on application of culturing conditions that are as close as possible to the natural environment from which samples were taken. Recent efforts to accomplish this objective have met with some success by including the following: the use of agar media with little or no added nutrients (traditional culturing procedures usually supply excessive nutrients to a system, resulting in the overgrowth of less nutrient-demanding bacteria); relatively lengthy periods of incubation (more than 30 days); and inclusion of substances that are typical of the natural environment in the artificial growth media (Breznak 2002; Stevenson et al. 2004).

A still more sophisticated strategy to devise specific culture media to as-yet-uncultivated bacteria will soon become available. Procedures are being developed that allow sequencing of the complete genome of individual bacterial cells, even those belonging to as-yet-uncultivated species, directly from the environment (Zhang et al. 2006; Dethlefsen and Relman 2007). The availability of sequenced genomes of as-yet-uncultivated bacteria furnishes opportunities to define culture media for growth of these bacteria with computer modeling of metabolic networks. For instance, by analyzing the sequenced genome of bacteria recalcitrant to culture, one can identify missing genes and consequent metabolic deficiencies and use this information to design a culture medium containing substances that complement such deficient metabolic pathways (Renesto et al. 2003).

### 5.2.2 Difficulties in identification—species with uncommon phenotypes

One should be mindful that in some circumstances even the successful cultivation of a given microorganism does not necessarily mean that this microorganism can be successfully identified. Culture-dependent identification is based on phenotypic traits observed in reference strains, with predictable biochemical and physical properties under optimal growth conditions. However, many phenotype-related factors can lead to difficulties in identification and even to misidentifica-

tion. These factors include (a) not all strains within a given species may exhibit a common phenotype, with some strains showing a divergent behavior (Beighton et al. 1991; Tanner et al. 1992); (b) strains of different species may show a similar phenotype, characterizing a convergent behavior (Tanner et al. 1992; Siqueira and Rôças 2005a); (c) the phenotype is not static and can change under some conditions, such as stress (Ochman et al. 2005; Petti et al. 2005); (d) the same strain may show different results after repeated tests (Tardif et al. 1989); (e) databases do not usually include newly named species and, obviously, as-yet-uncharacterized species; (f) test results are sensitive to even small alterations in the assay, with consequent false results (Bosshard et al. 2004); and (g) test results rely on individual interpretation and expertise (Bosshard et al. 2004). As a consequence of all these factors, phenotype-based identification does not always allow an unequivocal identification.

The 16S rRNA gene sequencing approach has become the reference method for bacterial identification and taxonomy (Patel 2001; Clarridge 2004). In addition to being widely used to identify both cultivable and as-yet-uncultivated bacteria without the need for cultivation, the 16S rRNA gene sequencing approach can also be used for identification of bacterial isolates. By this method, an isolate can be identified after obtaining its 16S rRNA gene sequence and comparing it to sequences deposited in public databases. This molecular technique can provide a more precise and reliable identification of bacteria that are difficult to identify or that cannot be precisely identified by available phenotypic tests (Tang et al. 1998; Drancourt et al. 2000; Bosshard et al. 2003; Song et al. 2003; Petti et al. 2005; Siqueira and Rôças 2005a). Also, the 16S rRNA gene sequencing approach has the advantage of being able to accurately identify rare isolates, poorly described bacteria, as-yet-uncultivated and uncharacterized bacteria, and newly named species.

It should be assumed that some of the as-yet-uncultivated bacteria revealed by molecular studies are indeed cultivable but as-yet-uncharacterized species that can grow in ordinary culture media. If previously cultivated, they may have been misidentified or incompletely identified by phenotype-based approaches. Misidentification may be related to the fact that these not-yet-cultivated bacteria still remain to be phenotypically characterized, and there are no described biochemical and physical attributes of a reference strain that can be used as parameters for precise identification. It is possible that many of these species were cultivated in previous studies but may have been identified only at the genus level or may have been assigned to another species based on incomplete data. Identification of isolates by 16S rRNA gene sequencing may help solve this problem by offering a more precise identification of these bacteria.

## 5.3 Molecular biology techniques

It has now been recognized that culture-dependent techniques can strongly underestimate the diversity of microbial populations. The recognition that the as-yet-uncultivated microbial world far outsizes the cultivable world has caused a great revolution in microbiology. Fortunately, tools and procedures have become available and have been substantially improved to achieve a more realistic description of this unseen world.

A significant contribution of molecular biology methods to medical microbiology relates to the identification of previously unknown human pathogens (Fredricks and Relman 1999; Relman 1999). Furthermore, molecular studies have revealed that over 50% of the bacterial species in the oral cavity (Paster et al. 2001; Aas et al. 2005b; Kumar et al. 2005) and about 80% of the species in the gut and colon (Suau et al. 1999; Eckburg et al. 2005) represent unknown and as-yet uncultivated bacteria. As a consequence, it is fair to realize that there can exist a number of uncharacterized pathogens in this uncultivated proportion of the human microbiota.

There are a plethora of molecular biology methods for the study of microorganisms and the choice of a particular approach depends on the questions being addressed. This chapter restricts discussion to the most commonly used approaches applied to the research of the endodontic microbiota and some with potential to be used with this intent.

## 5.4 Gene targets for microbial identification

Each living organism carries sequences within certain genes that are uniquely and specifically present only in its own species. Indeed, each particular individual within a species has also its signature DNA sequences. These unique sequences bring important genomic information that makes it possible to identify each species and even each individual within a species by using molecular biology methods.

Molecular approaches for microbial identification rely on certain genes that contain revealing information about the microbial identity. Ideally, a gene to be used as a target for microbial identification should contain regions that are unique to each species. Genes encoding housekeeping functions are preferable to infer phylogenetic classification since they are usually ubiquitous and tend to exhibit functional constancy, evolving slowly with time (Woese 2000; Wade 2004).

Several genes have been chosen as targets for bacterial identification. Some of these genes are shared by a vast majority, if not all, of bacterial species. Genes proposed for bacterial identification include the 16S rRNA and 23S rRNA genes, the 16S-23S rRNA gene internal transcribed sequences, the *rpoB* gene encoding the β-subunit of RNA polymerase, the *groEL* gene encoding the heat-shock protein, the *gyrB* gene encoding the β-subunit of DNA gyrase, the *tuf* gene, and homologous recombination-encoding *recA* (Ke et al. 1999; Drancourt and Raoult 2005). Of these, the gene encoding the 16S rRNA has been widely accepted and used for bacterial identification.

Following the pioneer studies by Woese (Woese 1987), the genes encoding rRNA molecules, which are present in all cellular forms of life, namely, the domains Bacteria, Archaea, and Eucarya, have been extensively used for comprehensive identification of virtually all living organisms and inference of their natural relationships. The rRNA is the central component of the highly complex translation apparatus of the cell, and because fidelity and maintenance of this translation function are critical, some regions of the rRNA are so highly conserved that they can be used to align genes from different organisms (Woese 2000). Other regions less critical to translation of the code are under less selective pressure and show enough variation so that each species has a unique sequence. The advantages of using the small subunit rRNA genes for microbial identification is that it is found in all organisms, is long enough to be highly informative and short enough to be easily sequenced (particularly with the advent of automated DNA sequencers), and affords reliability for inferring phylogenetic relationships (Woese 1987). Thus, the 16S rRNA gene (or 16S rDNA) of bacteria and archaea and the 18S rRNA gene (or 18S rDNA) of fungi and other eukaryotes have been extensively examined and used for identification

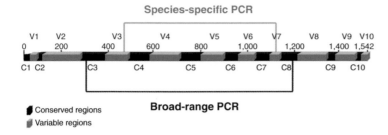

**Fig. 5.1** Schematic drawing of the 16S rRNA gene (rDNA). Areas in yellow correspond to variable regions, which contain information about the genus and the species. Primers designed on these regions are used in species-specific assays. Red areas correspond to conserved regions of the gene. Primers designed on these areas are used in broad-range assays.

and phylogenetic studies. The conserved regions of these genes are virtually identical in all representatives of each domain, while the variable regions contain unique signatures of the genus and species of the organism.

Data from small subunit rRNA gene sequences can be used for accurate and rapid identification of known and unknown bacterial species, using techniques that do not require cultivation. For instance, the 16S RNA gene of virtually all bacterial species in an environment, including as-yet-uncultivated and uncharacterized bacteria, can be amplified by polymerase chain reaction (PCR) using broad-range (or universal) primers that are complementary to conserved regions of this gene. Sequencing of the variable regions flanked by the broad-range primers will provide information for accurate bacterial identification. Primers or probes that are complementary to variable regions can also be designed to detect specific target species directly in clinical samples (Fig. 5.1). The analytical sensitivity of most species-specific PCR assays is usually higher when compared to broad-range PCR assays (Maiwald 2004).

## 5.5 PCR and its derivatives

The PCR process was conceived by Kary Mullis in 1983 and ever since has revolutionized the field of molecular biology by enabling the amplification of as few as one copy of a gene into millions to billions of copies of that gene in just a matter of minutes to a few hours (Mullis et al. 1994). The impact of PCR on biological and medical research has been remarkable. Nowadays, it is possible to isolate essentially any gene from any organism using PCR, which makes this technique a cornerstone of genome sequencing projects (Lee and Tirnady 2003). Since its introduction, PCR has spawned an increasing number of associated tech-

nologies for diverse applications. Perhaps the most widespread advance in clinical diagnostic technology has come from the application of PCR for detection of microbial pathogens (Whelen and Persing 1996; Tang et al. 1997; Louie et al. 2000).

The PCR method is based on the in vitro replication of DNA through repetitive cycles of denaturation, primer annealing, and extension steps. Briefly, the method consists of three steps that are repeated in several cycles of amplification:

1. The target DNA serving as template is denatured (melted) at temperatures high enough to break the hydrogen bonds holding the strands together, thus liberating single strands of DNA.
2. Two short oligonucleotides (primers) anneal to complementary sequences on opposite strands of the target DNA. Primers define the two ends of the amplified stretch of DNA.
3. A complementary second strand of new DNA is synthesized through the extension of each annealed primer by a thermostable DNA polymerase in the presence of excess deoxyribonucleoside triphosphates. All previously synthesized products act as templates for new primer extension reactions in each ensuing cycle. The result is the exponential amplification of new products.

PCR has unrivaled sensitivity. While it can detect as few as 10 bacterial cells in a sample, other methods of identification show too higher detection limits. For instance, culture using nonselective media can detect $10^4$–$10^5$ cultivable cells in a sample (Zambon and Haraszthy 1995). It is easy to understand this low sensitivity if we do some calculations. After 10-fold serial dilutions of the sample for cultivation, bacterial counts are usually performed on plates containing 30–300 colonies. As a consequence, a single colony would represent 0.3–3% of the total cultivable population. A negative result for a target species usually means that it was absent or at least present at lower amount that is less than 0.3% of the population. Based on the calculation that we can detect one colony of a given species among 300 colonies grown on the surface of a blood agar plate (0.3% of the population), then in a root canal sample containing a total number of $10^8$ cells, roughly $3 \times 10^5$ cells of this species must be present so that it can be detected by culture. When selective media are used, the sensitivity of culture method can increase to $10^3$ cells (Zambon and Haraszthy 1995). Immunologic methods have a detection limit ranging from $10^3$ to $10^4$ cells. DNA–DNA hybridization assays can detect $10^3$–$10^4$ cells in a sample. Thus, PCR methodology is at least 10- to 100-fold more sensitive than the other more sensitive identification method (Siqueira and Rôças 2003d).

There are several methods to check if the intended PCR product was generated. The most commonly used method for detecting PCR products is electrophoresis in an agarose gel. Aliquots of the PCR reaction are loaded into the gel and an electrical gradient is applied through a buffer solution. The products migrate through the gel according to size, with larger products running a shorter distance in the gel, because they experience more resistance in the gel matrix. DNA ladder digests represent DNA fragments of known size and are run in the same gel to serve as molecular size standard. This allows the size of the PCR products to be estimated. The gel is usually visualized using ethidium bromide staining and ultraviolet transillumination. Designed primers are expected to generate a PCR product of a given size and observation of a band of the predicted size in the electrophoretic gel is consistent with a positive PCR result. Identity of PCR products should be confirmed by one of the following methods: sequencing of the PCR product; hybridization of a specific oligonucleotide probe to a region of the PCR product that is internal to the priming sites; or restriction enzyme cleavage of the PCR product using an enzyme that is known to cut a specific sequence within the product (restriction fragment length polymorphism—RFLP).

Numerous derivatives in PCR technology have been developed since its inception. The most used PCR-derived assays are described in the following sections.

### 5.5.1 Species-specific PCR

One of the simplest approaches to detect a target species in a sample is to use a species-specific PCR assay. By this method, primers designed to anneal to signature genomic sequences of a given species are used to detect this species directly in clinical samples even in the presence of a background of nontargeted species and without the need for cultivation. Most assays use the variable regions of the 16S rRNA gene to design primers specific for bacterial species. Using public databases that contain the 16S rRNA gene sequences from a vast number of oral bacteria, primers can be designed to specifically detect virtually every cultivable and as-yet-uncultivated oral species. The presence of a species-specific PCR product of predicted size is usually determined by agarose gel electrophoresis and represents a positive result for the occurrence of the target species in the sample. The best way to check the specificity of the assay is by sequencing the PCR product. This approach can be used not only in single PCR assays, but also in nested PCR and multiplex PCR, furnishing qualitative results (presence or absence) about one (the two former techniques) or more (the latter technique) target species. Species-specific detection can also be performed using a quantitative real-time PCR assay, which detects and monitors the appearance of the amplification product throughout the reaction.

Specificity of the PCR assay can be increased by using a touchdown procedure. By this approach, the annealing temperature in the initial PCR cycle is set several degrees above the calculated melting temperature ($T_m$) of the primers. In subsequent cycles, the annealing temperature is decreased in steps of 0.5–2°C per cycle until a temperature is reached that is equal to, or 2–5°C below, the $T_m$ of the primers. Touchdown techniques have been considered useful to avoid the amplification of spurious DNA fragments (non-rRNA gene fragments and/or fragments with improper sizes) (Don et al. 1991).

### 5.5.2 Multiplex PCR

Most PCR assays have concentrated on the detection of a single species by means of individual reactions. Multiplex PCR is a process where multiple primer pairs are used to simultaneously amplify several sequences in a single reaction (Chamberlain et al. 1988). Since more than one unique target sequence in a clinical specimen can be amplified at the same time, multiplex PCR assays permit the concomitant detection of different species. Multiplex PCR assays have been used to minimize the time and expenditure needed for detection approaches. Primers used in multiplex assays must be designed carefully to have similar annealing temperatures and avoid complementarity among them (Dieffenbach and Dveksler 1995; Hayden 2004).

### 5.5.3 Nested PCR

Nested PCR (nPCR) is a conventional PCR method that amplifies a target region of DNA with an outer primer pair in an initial reaction, followed by a second amplification using an internal primer pair (Haqqi et al. 1988). The first PCR products are used as template in the second round of amplification with a separate primer set, which anneals internally to the first products and generates a shorter amplified fragment. This approach shows increased sensitivity when compared to single PCR. Increased sensitivity is due to the large total number of cycles. In addition, target DNA is amplified in the first round of amplification, with subsequent reduction of nontargeted DNA and inhibitors present in the sample. The set of primers used in the second round of PCR results in additional specificity. The second reaction is performed with reduced background of eukaryotic DNA and other regions of the bacterial DNA (Siqueira and Rôças 2003d). Even if nonspecific DNA amplification occurs in the first round of amplification, the nonspecific PCR product does not serve as template in the second reaction, since it is highly unlikely to possess regions of DNA complementary to the second set of specific primers (McPherson and Moller 2000).

### 5.5.4 Reverse transcriptase-PCR

Reverse transcriptase-PCR (RT-PCR) was developed to amplify RNA targets and exploits the use of the enzyme reverse transcriptase, which can synthesize a strand of complementary DNA (cDNA) from an RNA template. Most RT-PCR assays employ a two-step approach. In the first step, reverse transcriptase converts RNA into single-stranded cDNA. In the second step, PCR primers, DNA polymerase, and nucleotides are added to create the second strand of cDNA. Once the double-stranded DNA is formed, it can be used as template for amplification as in conventional PCR (Sambrook and Russell 2001). The RT-PCR process may be modified into a one-step approach by using it directly with RNA as the template. In this approach, an enzyme with both reverse transcriptase and DNA

polymerase activities is used, such as that from the bacteria *Thermus thermophilus* (*Tth*).

### 5.5.5 Quantitative PCR

Conventional PCR assays are qualitative or can be adjusted to be semi-quantitative. Some PCR technologies have been introduced that provide quantitative results. Basically, quantitative PCR (qPCR) can be performed using three distinct assays: most probable number (MPN)-PCR, competitive PCR, and real-time PCR (Sharma et al. 2007). Of these, the real-time PCR method has been the most widely used approach, mostly because it is a high-throughput technique that is more accurate and precise than the other qPCR assays, and requires no post-PCR manipulation steps, reducing the risks of contamination (Raoult et al. 2004; Sharma et al. 2007). Real-time PCR assays allow the quantification of individual target species as well as total bacteria in clinical samples.

Basically, by adding a fluorescent dye to the assay and monitoring the appearance of fluorescence during the reaction, the amount of synthesis of new DNA can be measured. The fluorescent signal is proportional to the amount of DNA being synthesized and is measured in a closed tube format by a fluorimeter combined to the PCR thermocycler. There are several different real-time PCR approaches. The three most used real-time PCR chemistries are SYBR Green (Higuchi et al. 1992), TaqMan (Heid et al. 1996), and molecular beacon (Tyagi and Kramer 1996). SYBR Green is the simplest and most affordable method, and consists of a fluorescent dye that binds to double-stranded DNA. During extension, increasing amounts of dye bind to the increasing amount of newly formed double-stranded DNA. Fluorescence is measured at the end of the extension step of every PCR cycle to monitor the increasing amount of amplified DNA. Dye that remains unbound exhibits little fluorescence in solution. The SYBR Green assay is very sensitive but has diminished specificity, as the dye binds to all double-stranded DNA present, and primer dimers may result in a false reading (Bustin 2000). However, a study found no significant difference between the TaqMan and SYBR Green chemistry with regard to specificity, quantitativity, and sensitivity for detection of oral bacteria (Maeda et al. 2003). The advantage of the SYBR Green assay over the TaqMan assay is that the protocol is usually simpler and less expensive (Maeda et al. 2003).

The TaqMan method can be more specific than the SYBR Green assay. Increased specificity of the TaqMan assay results from the utilization of a specific labeled oligonucleotide probe along with the primers (Holland et al. 1991; Heid et al. 1996). The TaqMan probe is a 20- to 30-base-long oligonucleotide sequence that specifically anneals to a sequence flanked by the two primers. The TaqMan probe contains a reporter fluorescent dye at the 5′ end and a quencher dye at the 3′ end that quenches the emission spectrum of the reporter dye. As long as the probe remains unbound, it is intact and no signal is generated. During the extension step of real-time PCR, the *Thermus aquaticus* (*Taq*) DNA polymerase enzyme cleaves the TaqMan probe, resulting in separation of the reporter from the quencher. This results in increased fluorescence emission.

Another real-time PCR assay uses molecular beacons, which are single-stranded nucleic acid molecules with a stem-and-loop structure (Mackay 2004). The loop portion is complementary to a sequence in the target DNA. The stem is formed by the annealing of complementary arm sequences located on either side of the probe. A fluorescent marker is attached to the end of one arm, and a quencher is attached to the end of the other arm. Free molecular beacons acquire a hairpin structure when in solution, and the stem keeps the arm in close proximity. This results in efficient quenching of the fluorescent dye. When hybridizing to their complementary target, molecular beacons are forced to undergo a conformational change, with consequent formation of probe that hybridizes to the template. The conformational change forces the fluorescent dye and the quencher apart, generating fluorescence (Mhlanga and Malmberg 2001).

### 5.5.6 PCR-based microbial typing

PCR technology can also be used for clonal analysis of microorganisms. An example of the PCR techniques used for this purpose includes the arbitrarily primed PCR (AP-PCR), also referred to as random amplified polymorphic DNA (RAPD) (Welsh and McClelland 1990; Power 1996). AP-PCR is a relatively rapid tool to determine whether two isolates of the same species are related. This method is based on the use of a single 10- to 20-base-long random sequence primer that anneals to unspecified DNA target sites under conditions that allow for mismatched base pairing. The use of a random sequence primer at low stringency allows for priming at sites with imperfect matches.

Amplification will only occur when two primers anneal close enough to one another, in the proper *forward* and *reverse* directions necessary for the product to be formed. Genetic variations between two DNA templates result in discriminative DNA fingerprints because of the differences in the priming sites. The amplicons generated form a strain-specific pattern of about 5–15 bands per species in the electrophoretic gel (Spiegelman et al. 2005). The advantage of AP-PCR is its ability to furnish highly specific DNA profiles with no prerequisite for knowing the DNA sequences. Primers may also be designed to target known genetic elements, such as enterobacterial repetitive intergenic consensus sequences (ERIC-PCR) (de Bruijn 1992; Arora et al. 1996) and repetitive extragenic palindromic sequences (REP-PCR) (Higgins et al. 1982). Clonal analysis may help elucidate whether certain strains of a given species are more associated with signs or symptoms of a given disease. Clonal analysis may also help to track the origin of microorganisms infecting a given site. For instance, by comparing bacterial strains isolated from the root canal and the gingival sulcus or other oral sites, one can have information as to where bacteria present in the root canal system came from. Clonal analysis can also track the origin of the microorganisms present in a suspected focal disease by comparing the isolates found in the secondarily infected site with others present in the suspected original focus of infection.

### 5.5.7 Broad-range PCR and clone library analysis

Broad-range PCR has been extensively used to investigate the whole microbial diversity in diverse environments. In broad-range PCR, primers are designed to be complementary to conserved regions of a particular gene that is shared by a group of microorganisms. For instance, primers that are complementary to conserved regions of the 16S rRNA gene have been used with the intention of exploiting the variable internal regions of the amplified sequence for sequencing and further bacterial identification (Göbel 1995). The strength of broad-range PCR lies in the relative absence of selectivity, so that, in principle, any kind of bacteria present in a sample can be detected and identified. This aspect is in analogy to cultivation and in contrast to species-specific molecular approaches (Maiwald 2004). Thus, broad-range PCR can detect the unexpected and in this regard it is far more effective and accurate than culture.

Broad-range PCR has allowed the identification of several novel fastidious or as-yet-uncultivated bacterial pathogens directly from diverse human sites (Relman 1997; Pitt and Saunders 2000; Wade 2002). By using broad-range PCR and 16S rRNA gene clone library construction, one can identify virtually every bacterial species present in a sample, regardless of whether it can be cultivated or has been previously unknown. The most used protocol is as follows. Initially, bulk DNA is extracted directly from samples. Afterward, the 16S rRNA gene is isolated from the bulk DNA via PCR with primers specific for conserved regions of the gene (broad-range or universal primers). Longer distances between primer pairs generally result in less sensitivity, but this provides more variable sequence information for accurate identification and may reduce the risks of amplifying DNA from contaminants in PCR reagents, which appear to be fragmented into smaller sizes (Maiwald 2004). Amplification with broad-range primers results in a mixture of the 16S rRNA gene amplified from nearly all bacteria in the sample. In mixed infections, direct sequencing of the PCR products cannot be performed because there are mixed products from the different species composing the consortium. PCR products are then cloned into a plasmid vector, which is used to transform *Escherichia coli* cells, establishing a library of 16S rRNA genes from the sample. The cloning procedure is used to separate the sequences so that they can be characterized individually by sequencing. After the cloned genes are individually sequenced, the obtained sequences are submitted for identification to databases, usually via the World Wide Web, such as the National Center for Biotechnology Information (http://www.ncbi.nlm.nih.gov/) or the Ribosomal Database Project II (http://rdp.cme.msu.edu/). Preliminary identification can be performed by using similarity searches in public databases. A 98.5–99% identity in 16S rRNA gene sequence has been the most accepted criterion used to identify a bacterium to the species level (Drancourt et al. 2000, 2004; Paster et al. 2006). If a sequence exhibits low similarity scores (<98.5–99%) to the other sequences from defined species or clones in public databases, it potentially represents a new species (Drancourt et al. 2000, 2004; Paster et al. 2006). These new species are usually considered as uncultivated and hitherto unknown bacterial taxa and an unofficial name is assigned. Generally, the genus of their closest neighbor in the phylogenetic tree is used followed by a letter and number code, for example, *Synergistes*

oral clone BA121, *Dialister* oral clone BS095, and *Peptostreptococcus oral* clone CK035.

In addition to the similarity search, phylogenetic analysis should be accomplished since it provides a much more accurate assessment (Lepp and Relman 2004; Maiwald 2004). By this approach, the 16S rRNA gene sequences from different species are aligned based on conserved regions, and computations based on the number of dissimilarities present in variable regions are then used to construct a phylogenetic tree (Leys et al. 2006). The relationships among various bacterial phylotypes (or taxa) are analyzed by special bioinformatic software and the results are shown in dendrograms, or phylogenetic trees, such as that depicted in Fig. 5.2. The 16S rRNA gene can be used to establish phylogenetic relationships among bacteria even when sequences are derived from previously uncultivated and uncharacterized bacteria.

Since broad-range primers are used, there is a high risk for amplification of contaminant DNA from sources other than the site sampled (Millar et al. 2002). Contaminating DNA can be introduced by inadvertent tube-to-tube contamination or by the use of contaminated reagents. In PCR experiments, in general, and especially in broad-range reactions, a number of precautions are necessary to avoid contamination. These include separate room for pre- and post-PCR work, ultraviolet decontamination of surface areas, use of high-quality reagents, and adequate sampling techniques and vials for clinical specimens (Dragon 1993; McPherson and Moller 2000; Millar et al. 2002; Maiwald 2004).

Information brought about by 16S rRNA gene clone library analysis includes sequences from both cultivable and as-yet-uncultivated bacteria. The number of bacterial species that have been discovered in the oral cavity after the advent of this technology has more than doubled when compared to species revealed by culturing approaches (Leys et al. 2006). DNA probes or primers based on the 16S rRNA gene sequences can be designed to specifically detect any of these species and then be used in clinical studies with large numbers of samples to associate species or phylotypes with a particular disease.

## 5.6 Denaturing gradient gel electrophoresis

Techniques for genetic fingerprinting of microbial communities can be used to determine the diversity of

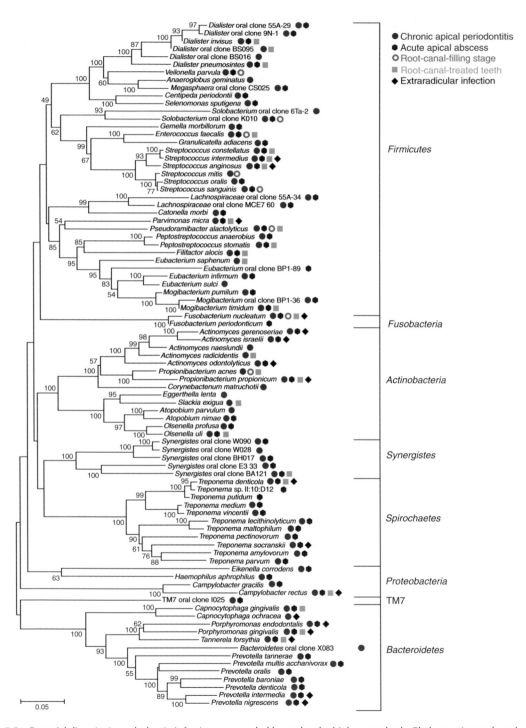

**Fig. 5.2** Bacterial diversity in endodontic infections as revealed by molecular biology methods. Phylogenetic tree based on 16S rRNA gene comparisons showing several candidate endodontic pathogens, their respective phyla, and the clinical conditions they have been associated with. Scale bar shows number of nucleotide substitutions per site.

different microorganisms living in diverse ecosystems and to monitor microbial community behavior over time. A commonly used strategy for genetic finger-printing of complex microbial communities encompasses DNA extraction, amplification of the 16S rRNA genes using broad-range primers, and then the analysis of PCR products by denaturing gradient gel electrophoresis (DGGE).

In DGGE, DNA fragments of the same length but with different sequences can be separated (Myers et al. 1985; Muyzer et al. 1993). The DGGE technique is based on electrophoresis of PCR-amplified 16S rRNA gene (or other genes) fragments in poly-acrylamide gels containing a linearly increasing gradient of DNA denaturants (a mixture of urea and for-mamide). As the PCR product migrates in the gel, it encounters increasing concentrations of denaturants and, at some position in the gel, it will become partially or fully denatured. Partial denaturation causes a significant decrease in the electrophoretic mobility of the DNA molecule. Molecules with different sequences may have a different melting behavior and will therefore stop migrating at different positions in the gel. The position in the gel at which the DNA melts is determined by its nucleotide sequence and composition (Gasser 1998). A GC-rich sequence (GC-clamp) is added to the 5′-end of one of the primers used in the PCR reaction and makes the DNA unable to denature completely in the gel (Muyzer and Smalla 1998). DNA bands in DGGE can be visualized using ethidium bromide, SYBR Green, or silver staining.

In DGGE, multiple samples can be analyzed concurrently, making it possible to compare the structure of the microbial community of different samples and to follow changes in microbial populations over time, including after antimicrobial treatment (Siqueira et al. 2005d). If identification is desired, specific bands can be excised from the gels, re-amplified and sequenced (Machado de Oliveira et al. 2006).

## 5.7 Terminal restriction fragment length polymorphism

The terminal restriction fragment length polymorphism (T-RFLP) approach can be used to explore the microbial diversity in the environment and provide insight into the structure of microbial communities over time and after treatment (Marsh 1999). T-RFLP analysis measures the size polymorphism of terminal restriction fragments from a PCR-amplified marker. When T-RFLP is used to analyze bacterial communities, PCR is first carried out to amplify the 16S rRNA gene from different species in the sample. One of the PCR primers is labeled with a fluorescent dye (Clement et al. 1998). PCR amplicons are then digested with restriction enzymes, generating fluorescently labeled fragments of different lengths (the terminal fragments). These fragments are separated on high-resolution sequencing gels in an automated DNA sequencer, which is used to read both the size and the intensity of terminally labeled restriction fragments (T-RF), creating a typical profile. In such a profile, size is represented on the horizontal axis and intensity (relative to the abundance of a given fragment size) is represented on the vertical axis (Spiegelman et al. 2005). In theory, each T-RF represents a single species. Extensive databases exist for 16S rRNA gene sequences and can be used to identify all T-RFs predicted from known sequences, considering a given set of primers and restriction enzymes (Matsumoto et al. 2005). T-RF lengths are predicted by finding the restriction site closest to the site where the labeled primer will anneal and counting the number of nucleotides in between. Multiple restriction enzymes (four or five) are usually necessary to provide reliable identification since distinct species may generate the same T-RF when only one enzyme is used (Sakamoto et al. 2005).

Through application of automated DNA sequencer technology, T-RFLP has considerably greater resolution than gel-based community fingerprinting techniques, such as DGGE (Clement et al. 1998; Marsh 1999). The digital output of the method also eliminates an element of human error from the analysis process (Spiegelman et al. 2005).

## 5.8 DNA–DNA hybridization assays

DNA–DNA hybridization methodology is the process of annealing the complementary bases of two single-stranded DNA molecules. It employs labeled single-stranded DNA probes that can locate and bind to a target sequence, forming a new duplex molecule. The labeled duplex can then be detected (Li and Hanna 2004). Probes can be constructed from either whole genomic DNA or oligonucleotides. Whole genomic probes are more likely to cross-react with nontargeted microorganisms due to the presence of homologous sequences between different species. Oligonucleotide probes based on signature sequences of specific genes (such as the 16S rRNA gene) may display higher specificity, since the known probe sequence allows calculation of stringent hybridization temperatures and mismatches are not tolerated due to the considerable reduction of bond strength between the short probe and the target (Theron and Cloete 2000; Juretschko et al. 2004). Also, oligonucleotide probes can differentiate between closely related species or even subspecies and can be designed to detect as-yet-uncultivated bacteria.

Some DNA–DNA hybridization techniques are suitable for large-scale clinical studies. They include the checkerboard DNA–DNA hybridization and DNA microarray techniques.

### 5.8.1 Checkerboard DNA–DNA hybridization

This technique was introduced by Socransky et al. (1994) for hybridizing large numbers of DNA samples against large numbers of digoxigenin-labeled whole genomic DNA on a single support membrane. Briefly, denatured DNA from clinical samples is loaded in lanes on a nylon membrane using a Minislot apparatus. After fixation of the samples to the membrane, the membrane is placed in a Miniblotter 45 apparatus with the lanes of samples at $90°$ to the lanes of the device. Digoxigenin-labeled whole genomic DNA probes are then loaded in individual lanes of the Miniblotter. After hybridization, the membranes are washed at high stringency and the DNA probes detected using antibody to digoxigenin conjugated with alkaline phosphatase and chemifluorescence or chemiluminescence detection. The presence of a spot on the membrane in the crossing lanes means that hybridization of one probe to one sample occurred. The intensity of the spot is proportional to the amount of DNA from the target species in the sample. The checkerboard method permits the simultaneous determination of the presence of a multitude of bacterial species in single or multiple clinical samples.

A modification of the checkerboard method was proposed by Paster et al. (1998) and consists of a PCR-based, reverse-capture checkerboard hybridization methodology. In this assay, the probe rather than the sample is fixed first to the membrane—so the name *reverse-capture checkerboard*. The procedure circumvents the need for in vitro bacterial culture, necessary for preparation of whole genomic probes in the original checkerboard approach. Up to 30 reverse-

capture oligonucleotide probes that target regions of the 16S rRNA gene are deposited on a nylon membrane in separate horizontal lanes, using a Minislot apparatus. Probes are synthesized with a polythymidine tail, which are cross-linked to the membrane via ultraviolet irradiation or heat, leaving the probe available for hybridization. The 16S rRNA gene from clinical samples is PCR amplified using one of the primers labeled with digoxigenin. Hybridizations are performed in vertical channels in a Miniblotter apparatus with digoxigenin-labeled PCR amplicons from up to 45 samples. Hybridization signals are detected using chemifluorescence or chemiluminescence procedures. The reverse-capture checkerboard has the potential to be more specific than the original method, mostly because of the use of oligonucleotide probes. Furthermore, oligonucleotide probes can be designed to detect both cultivable and as-yet-uncultivated bacteria, while in the original method using whole genomic probes only cultivable species are targeted.

### 5.8.2 DNA microarrays

In essence, DNA microarrays can be regarded as a miniaturization of the checkerboard approach (Leys et al. 2006). DNA microarrays were first described in 1995 (Schena et al. 1995) and consist of a high-density matrix of DNA probes, which are printed or synthesized on a glass or silicon slide (Mothershed and Whitney 2006). Targets incorporate either a fluorescent label or some other moiety that permits subsequent detection with a secondary label. Targets are applied to the array and those that hybridize to complementary probes are detected using some type of reporter molecule. Following hybridization, arrays are imaged using a high-resolution laser scanner and analyzed by sophisticated computer software programs.

DNA microarrays can be used to enhance PCR product detection and identification. When broad-range PCR is used to amplify DNA from samples of mixed infections, microarrays can then be used to identify the PCR products by hybridization to an array that is composed of hundreds to thousands of species-specific probes (Palmer et al. 2006). A DNA microarray approach to detect oral species has been developed—the Human Oral Microbe Identification Microarray (Boches et al. 2006). In this assay, 16S rRNA gene-based oligonucleotide probes targeting 200 oral species/phylotypes are covalently attached to a slide. Broad-range PCR is used to amplify the

16S rRNA genes from clinical samples. PCR products are labeled with a fluorescent dye during the reaction. When labeled bacterial DNA hybridizes to a species-specific probe on the slide, it provides a fluorescent signal that can be read with a microarray laser scanner. By using this method, a single hybridization approach can provide results as to the presence and levels of 200 target species/phylotypes at a time.

## 5.9 Fluorescence in situ hybridization

This method uses fluorescently labeled rRNA-directed probes and fluorescence microscopy to detect intact microbial cells directly in clinical specimens, in situ (Moter and Gobel 2000). One of the greatest advantages of using fluorescence in situ hybridization (FISH) is that this technique allows identification while providing information about morphology, number, community architecture, and spatial relationships of microorganisms (Amann et al. 2001). Because oligonucleotide probes can be designed for use, FISH allows not only the detection of cultivable species, but also the detection of as-yet-uncultivated phylotypes (Moter et al. 1998a, b). In FISH, microbial cells are fixed and then hybridized with rRNA-directed probes on a glass slide. Probes are short (15- to 25-base long) labeled covalently at one end with a fluorescent dye. After stringent washing, cells are visualized by using a conventional epifluorescence microscope or a confocal laser scanning microscope. Multiple species-specific probes, each labeled with different colored fluorescent dyes, can be used at the same time.

## 5.10 Metagenomics

The two fundamental questions in microbial ecology are *who is there* and *what are they doing*. Molecular biology methods have provided a great deal of information about the species composition in diverse environments. Now, the important question to be answered refers to the role of different species in the consortium, that is, what they are doing there.

Data from 16S rRNA gene clone libraries have opened a window into a world of bacterial diversity that is astonishing in its breadth. The challenge now is to develop methods to move beyond cataloging 16S rRNA gene sequences toward an understanding of the physiology and functional roles of bacteria in dif-

ferent environments. The 16S rRNA gene represents only about 0.05% on the average of the genome of a bacterial cell and has very little value to predict the physiology and functional roles of such cell. For instance, assigning physiologies and functions to as-yet-uncultivated bacteria based on 16S rRNA gene sequences is complicated in many cases by the lack of characterized close relatives or even by the diversity of phenotypes among close relatives (Janssen 2006). There are many examples of strains that are related by more than 98% similarity at the 16S rRNA gene level but behave very differently physiologically and ecologically (Rodriguez-Valera 2002). While the 16S rRNA gene often provides accurate identification, the other 99.95% of the genome provides the blueprint for the vast array of metabolic, structural, and virulence abilities of each bacterium. Because as-yet-uncultivated bacteria make up a large proportion of most environments, studies of the physiological and functional roles of the community members should also rely on culture-independent approaches.

Metagenomics is the culture-independent analysis of the collective microbial genomes (termed the *metagenome*) in an environmental community, using an approach based either on expression or on sequencing (Handelsman 2004). Metagenomics treats the genomes of all microorganisms present in a specific habitat as an entity. Theoretically, a metagenomic library will contain DNA sequences for all the genes in the microbial community. Metagenomic libraries permit analyses of species diversity based on a PCR-independent approach as well as a comprehensive description of functionalities of the whole ecosystem.

The metagenomic approach typically begins with the construction of a clone library from DNA retrieved from environmental or clinical samples. Extracted DNA is cloned into large-insert cloning vectors, such as fosmids or bacterial artificial chromosomes (BACs). This results in access to large contiguous portions of microbial genomes (35–40 kb for fosmids and up to 200 kb for BACs). BACs have the advantage that they can be used to maintain and express the insert genes in the host, harboring the vector (Xu 2006). *E. coli* is the preferred host for the cloning and expression of metagenome-derived genes. Clones are then selected for screening using either functional or sequence-based approaches. In the functional approach, genes retrieved from the environment are heterologously expressed in a host, such as *E. coli*, and sophisticated functional screens are employed to detect clones expressing functions of interest. In the sequence-based approach, clones are selected for sequencing based on the presence of either phylogenetically informative genes, such as the 16S rRNA gene, or other genes of interest (Hugenholtz 2002). Recently, facilitated by the increasing capacity of sequencing centers, whole-genome shotgun sequencing of the entire clone library has emerged as a third approach to metagenomics. Unlike previous approaches, which typically study a single gene or individual genomes, this approach offers a more global view of the community, allowing the better assessment of levels of phylogenetic diversity and intraspecies polymorphism, study of the metabolic pathways in the community, and in some cases, reconstruction of the near-complete genome sequences (Chen and Pachter 2005). Shotgun sequencing also has the potential to disclose new genes that are too different from known genes to be amplified with PCR or heterologously expressed in common hosts (Hugenholtz 2002; Chen and Pachter 2005).

In the near future, metagenomic analysis of the oral microbiome will provide invaluable information about the physiological and functional roles of the oral microbiota, including bacteria that have not yet been cultivated.

## 5.11 Advantages and limitations of molecular methods

Most of the advantages of molecular methods were already pointed out in the previous sections. Like other methods, molecular biology techniques have also their own limitations. Advantages and limitations are depicted in Table 5.2. The issues related to the ability of PCR to detect either an extremely low number of cells or dead cells are of special interest when one interprets the results of PCR identification procedures in endodontic microbiology research. Therefore, these issues deserve a separate discussion.

### 5.11.1 The too-high sensitivity issue

The high detection rate of PCR may be a reason of concern, particularly when nonquantitative assays are employed. It has been claimed that because PCR can detect a very low number of cells of a given target species, the results obtained by this method may have no significance with regard to disease causation. Nonetheless, the method's high sensitivity can represent a great advantage for microbiological diagnosis in endodontics.

**Table 5.2**   Advantages and limitations of molecular biology methods.

| Advantages | Limitations |
|---|---|
| 1. Detect both cultivable and as-yet-uncultivated species or strains | 1. Most assays are qualitative or semi-quantitative (exceptions: real-time PCR, DNA microarrays) |
| 2. High specificity and accurate identification of strains with ambiguous phenotypic behavior | 2. Most assays only detect one species or a few different species at a time (exceptions: broad-range PCR, DGGE, T-RFLP, checkerboard, DNA microarrays, metagenomics) |
| 3. Detect species directly in clinical samples | 3. Most assays detect only the target species and fail to detect unexpected species (exceptions: broad-range PCR, DGGE, T-RFLP, metagenomics) |
| 4. High sensitivity | 4. Some assays can be laborious and costly (e.g., broad-range PCR, T-RFLP, and metagenomics) |
| 5. Rapid—most assays take no more than minutes to a few hours to identify a microbial species | 5. Biases in broad-range PCR introduced by homogenization procedures, preferential DNA amplification, and differential DNA extraction |
| 6. Do not require carefully controlled anaerobic conditions during sampling, transportation, and handling | 6. Hybridization assays using whole genome probes detect only cultivable species |
| 7. Can be used during antimicrobial treatment | 7. Detect dead microorganisms |
| 8. Samples can be stored frozen for later analysis | |
| 9. DNA can be transported easily between laboratories | |
| 10. Detect dead microorganisms | |

When taking samples from endodontic infections, difficulties posed by the physical constraints of the root canal system and by the limitations of the sampling techniques can make difficult the attainment of a representative sample from the main canal (Siqueira and Rôças 2005a). If cells of a given species are sampled in a number below of the detection rate of the diagnostic test, species prevalence will be underestimated.

It is also important to take into consideration the analytical sensitivity needed for the specific clinical sample. For example, a sensitivity of no more than $10^4$ microbial cells/mL is required for urine, while a sensitivity of one cell may be of extreme relevance for samples from blood or cerebrospinal fluid (Boissinot and Bergeron 2002). There is no clear evidence as to the microbial load necessary for apical periodontitis to be induced. Endodontic infections are characterized by a mixed community, and individual species can play different roles in the consortium or dominate various stages of the infection. At least theoretically, all bacterial species established in the infected root canal have the potential to be considered endodontic pathogens (Sundqvist and Figdor 2003). Also, from an ecological point of view, one cannot dismiss the ecological role of the less abundant species in the consortium, exclusively on the basis of numbers, particularly if we consider that nothing is known about their functional role in the ecosystem. In other words, even less abundant species may exert a big difference in the ecosystem. For instance, at 5% of the population, chemolithoautotrophs or secondary metabolite producers do make a big difference to an ecosystem (Strous 2007). Also, nothing is known whether the abundance of these species changes over time and what are the consequences of such changes. Based on this, it would be glaringly prudent to use the method with the highest sensitivity to detect all species colonizing the root canal.

PCR detection of very low numbers of cells in clinical samples may not be as common as anticipated. There are numerous factors that can influence PCR reactions, sometimes dramatically reducing sensitivity for direct microbial detection in clinical samples. Therefore, the analytical sensitivity of the method does not always correspond to its "clinical" sensitivity. It is well-known that the effects of inhibitors are magnified in samples with low number of target DNA and therefore can significantly decrease the sensitivity of the method (Hayden 2004). Another impediment refers to the aliquots of the whole sample used in

PCR reactions. Detection of 1–10 cells of a species in a sample would be possible only if the entire sample was used in the PCR assay. And this is not the case. In fact, aliquots that usually represent 1–5% of the entire sample are used in each amplification reaction. Thus, for the PCR assay to detect 1–10 cells in the aliquot used in each PCR tube, 20–100 (detection limit of 1 cell) to 200–1,000 (detection limit of 10 cells) cells have to be present in the original sample, not considering the effects of inhibitors (Siqueira and Rôças 2003d). These numbers are still lower than the detection limits of other methods, but can represent more significance with regard to pathogenicity.

Therefore, the use of highly sensitive techniques is welcome in the study of endodontic infections, decreasing the risks for potentially important species to pass unnoticed during sample analysis. Although qualitative results do not lack significance, the use of quantitative molecular assays, like the real-time PCR, can allow inference of the role of a given species in the infectious process while maintaining high sensitivity and the ability to detect fastidious or as-yet-uncultivated bacteria.

### 5.11.2 The dead-cell issue

Detection of dead cells by a given identification method can be at the same time an advantage and a limitation. On the one hand, this allows detection of hitherto uncultivated or fastidious bacteria that can die during sampling, transportation, or isolation procedures (Wang et al. 1996; Rantakokko-Jalava et al. 2000; Siqueira and Rôças 2005a). On the other hand, if the bacteria were already dead in the infected site, they may also be detected and this might give rise to a false assumption of their role in the infectious process (Josephson et al. 1993; Keer and Birch 2003).

Several studies show that bacterial DNA is rapidly cleared from the host sites after bacterial death and that DNA from different species may differ as to the elimination kinetics at different body sites (Malawista et al. 1994; Post et al. 1996; Aul et al. 1998; Wicher et al. 1998). It remains to be clarified how long bacterial DNA from dead cells can remain detectable in the infected root canal system.

It is true that detection of microbial DNA sequences in clinical samples does not indicate viability of the microorganism. However, the fact that some microorganisms die during the course of an infectious process does not necessarily implicate that in a determined moment these microorganisms have not participated in the pathogenesis of the disease. In addition, the fate of DNA from microorganisms that have entered and not survived in root canals is unknown. DNA from dead cells might be adsorbed to dentine due to affinity of hydroxyapatite to this molecule (Bernardi 1965). However, it remains to be shown if DNA from dead microbial cells can really be adsorbed on dentinal walls and, if even, it can be retrieved during sampling with paper points. In fact, it is highly unlikely that free microbial DNA can remain intact in an environment colonized by living microorganisms. The half-life of the DNA released in the environment is considered to be very short owing to the presence of DNases in a complex microbial background like that occurring in the infected root canal. DNases released by some living species as well as at cell death can degrade free DNA in the environment. It has been reported the presence of DNase activity on whole bacterial cells and vesicles thereof that can degrade DNA (Leduc et al. 1995). Bacteria displaying DNase activity include common putative endodontic pathogens, such as *Porphyromonas endodontalis*, *Porphyromonas gingivalis*, *Tannerella forsythia*, *Fusobacterium* species, *Prevotella intermedia*, and *Prevotella nigrescens*. Thus, the free DNA molecule faces an onslaught of bacteria that can degrade macromolecules (Paabo et al. 2004). Indeed, DNases are of concern during sample storage, since they can be carried along with the sample and cause DNA degradation, with consequent false-negative results after PCR amplification.

Under rare circumstances, such as when the tissue becomes rapidly desiccated after host death or the DNA becomes adsorbed to a mineral matrix, like bone or teeth, DNA may escape enzymatic and microbial degradation. Even so, slower but still relentless chemical processes start affecting the DNA. Many of these processes are similar to those that affect the DNA in the living cell, with the difference that, after cell death, these processes are not counterbalanced by cellular repair processes. Thus, damage accumulates progressively until the DNA loses its integrity and decomposes, with an irreversible loss of nucleotide sequence information (Paabo et al. 2004).

DNA is not a stable molecule, and chemical processes, such as oxidation and hydrolysis, induce damage over time. As a result, the DNA becomes fragmented and difficult or even impossible to be detected and/or analyzed. In paleomicrobiology, certain strategies have to be developed for successful detection of ancient DNA. One of the most important strategies consists of using primers that will amplify a small DNA target size, preferably below 200 bp. Even so, the sample has to be well preserved, usually frozen or mummified. It has been stated that it is not the age of the DNA but the environmental conditions that are critical in preservation (Donoghue et al. 2004). Thus, any comparisons between the use of molecular methods in paleomicrobiology and endodontics can be considered inappropriate at best.

Based on the discussion above, although there is a possibility of detecting DNA from dead cells in endodontic infections, this possibility is conceivably low. In the event DNA from dead cells is detected, the results by no means lack significance with regard to participation in disease causation. Nonetheless, the ability to detect DNA from dead cells poses a major problem when one is investigating the immediate effectiveness of antimicrobial intracanal treatment, as DNA released from cells that have recently died can be detected. To circumvent or at least minimize this problem, one can use some adjustments in the PCR assay or take advantage of PCR technology derivatives, such as RT-PCR. Because smaller fragments of DNA may persist for a longer time after cell death than larger sequences, designing primers to generate large amplicons may reduce the risks of positive results due to DNA from dead cells (McCarty and Atlas 1993). Moreover, assays directed toward the detection of RNA through RT-PCR can be more reliable for detection of living cells. Since RNAs are more labile and have a shorter half-life when compared to DNA, they can be rapidly degraded after cell death (Keer and Birch 2003).

Another approach has been proposed to distinguish between the DNA from viable and dead cells in a sample. A viable/dead stain, propidium monoazide (PMA) or ethidium monoazide (EMA), is used in combination with real-time PCR to inhibit amplification of DNA from dead cells that have taken up PMA or EMA (Nocker and Camper 2006; Nocker et al. 2006). These stains are selective in penetrating only into dead bacterial cells with compromised membrane/cell wall integrity but not into live cells with intact membranes/cell walls. The stain is added to the test sample, penetrates the dead cells (if present), and binds to the DNA. Exposure to bright light for 1 min leads to covalent binding and inactivation of the free stain. This process renders the DNA insoluble and results in its loss during subsequent DNA extraction.

DNA from viable cells is unstained, while the DNA from the dead cells covalently bound to the stain is selectively removed and not amplified by PCR. Thus, only DNA from viable cells is available for detection (Nocker and Camper 2006; Nocker et al. 2006). PMA shows superiority over EMA, since the latter can also penetrate live cells of some bacterial species (Nocker et al. 2006).

## 5.12 Bacterial diversity in the oral cavity

The oral microbiota is composed of diverse groups of species, each one possessing its specific nutritional and physicochemical requirements. The successful cultivation of these bacteria in the laboratory will depend on how culturing conditions are adjusted to suit these varied requirements (Leys et al. 2006). A high diversity of bacterial species has been revealed in the oral cavity by culture, but early microscopy studies had already suggested that roughly 50% of the oral microbiota cannot be cultivated (Socransky et al. 1963). Application of molecular biology methods to the analysis of the bacterial diversity has not only confirmed this picture revealed by microscopic studies, but also disclosed a still broader and more diverse spectrum of extant oral bacteria. Data from culture-dependent and more recent culture-independent molecular studies have identified over 700 bacterial species in the human oral cavity (Paster et al. 2006). Any particular individual can harbor about 100–200 of these 700 species of oral bacteria, indicating that there is substantial diversity among different people (Paster et al. 2006). While some species can be present in most individuals and common to all sites, the majority of species are selective for a particular oral site (Aas et al. 2005b).

Taken as a whole, bacteria detected from the oral cavity fall into 13 separate phyla. The majority of oral species/phylotypes belong to the phyla Firmicutes, Fusobacteria, Bacteroidetes, Actinobacteria, Proteobacteria, Spirochaetes, Synergistes, and TM7, while representatives of the phyla SR1, Chloroflexi, Cyanobacteria, Deinococcus, and Acidobacteria have been sporadically reported (Paster et al. 2001, 2002; Kazor et al. 2003; Lillo et al. 2006; Aas et al. 2007). Of these, more than 700 species, over 50%, remain to be cultivated and fully characterized. This raises the possibility that as-yet-uncultivated and uncharacterized species that have remained invisible to studies using

traditional identification methods, but actually make up a large fraction of the living oral microbiota, can play an important ecological role as well as participate in the etiology of oral diseases.

### 5.12.1 Caries

It is well-known that species of *Streptococcus*, *Lactobacillus*, and *Actinomyces* are closely associated with the etiopathogenesis of different forms and stages of caries (Marsh and Martin 1999; Bowden 2000). Recent molecular biology approaches have demonstrated that the diversity and complexity of the microbiota associated with caries are far greater than anticipated. Overall, about 40–60% of the microbiota occurring in carious lesions consists of as-yet-uncultivated species (Aas et al. 2003, 2005a; Munson et al. 2004). In addition to confirming the association of named species of *Streptococcus*, *Lactobacillus*, and *Actinomyces* with caries, as-yet-uncultivated phylotypes of *Bifidobacterium* and *Atopobium* have also been suggested to participate in the pathogenesis of this disease (Becker et al. 2002; Aas et al. 2003).

As the caries lesion advances deep into dentin, the composition of the involved microbiota shifts from a predominance of facultative and saccharolytic Gram-positive bacteria in shallow lesions to a predominance of lactobacilli and/or proteolytic anaerobic bacteria in deep dentinal lesions (Martin et al. 2002; Nadkarni et al. 2004; Chhour et al. 2005). Other taxa present in a number of dentinal caries lesions or occurring in abundance include species/phylotypes of the genera *Selenomonas*, *Dialister*, *Fusobacterium*, *Eubacterium*, *Olsenella*, *Bifidobacterium*, members of the Lachnospiraceae family, and *Pseudoramibacter alactolyticus* (Chhour et al. 2005). Most of these taxa have also been detected in infected root canals, clearly suggesting that, in addition to being involved with pulpal damage, these dentinal lesions might well be the primary source of bacteria for endodontic infections.

### 5.12.2 Halitosis

Bacteria colonizing the dorsum of the tongue have been implicated as a major source of oral malodor in subjects with halitosis (Loesche and Kazor 2002). Many oral bacteria can produce volatile sulfur compounds, short-chain fatty acids, and polyamines, which may be the major contributing substances to malodor.

A molecular study revealed that about 60% of the bacteria detected on the tongue dorsum remain uncultivated (Kazor et al. 2003). While *Streptococcus salivarius* is by far the predominant species in healthy subjects, this species is typically absent from subjects with halitosis. Species/phylotypes most associated with halitosis include *Atopobium parvulum*, *Dialister* clone BS095, *Eubacterium sulci*, TM7 clone DR034, *Solobacterium moorei*, and *Streptococcus* clone BW009 (Kazor et al. 2003).

### 5.12.3 Periodontal diseases

Periodontal diseases result from the subgingival presence of complex bacterial biofilms, although specific etiologic agents have not been unequivocally identified (Moore and Moore 1994). Important advances in understanding the infectious agents of periodontal diseases have occurred over the past two decades. The application of molecular identification approaches to the study of the microbiota associated with different forms of periodontal diseases has confirmed results from culture studies with regard to the participation of some species but has also enabled identification of new bacterial species or phylotypes possibly implicated in the etiology of these diseases (Harper-Owen et al. 1999; Paster et al. 2001; Sakamoto et al. 2002; Brinig et al. 2003; Griffen et al. 2003; Kumar et al. 2003, 2005).

Analysis of the subgingival bacterial communities in large numbers of plaque samples by using the original checkerboard approach has revealed the occurrence of five major bacterial complexes (Socransky et al. 1998). The most pathogenic complex (termed the *red complex*) comprised the cultivated species *T. forsythia*, *P. gingivalis*, and *Treponema denticola* and was strikingly related to the severity of periodontal disease.

Comprehensive studies using broad-range PCR and 16S rRNA gene clone library analysis have determined the bacterial diversity in subgingival plaque samples and reported that about 50–60% of the microbiota is made up of as-yet-uncultivated phylotypes (Paster et al. 2001; Kumar et al. 2005). Many of these phylotypes have been found in association with disease. They include *Synergistes* clones D084, BH017, and W090, *Peptostreptococcus* clone BS044, *Bacteroidetes* clone AU126, *Megasphaera* clone MCE3_141, *Desulfobulbus* clones CH031 and R004, and *Selenomonas* clones D0042 and EY047 (Paster et al. 2001; Kumar et al. 2003, 2005).

Named and cultivable species other than the members of the red complex that have also been recently suggested to be candidate periodontal pathogens on the basis of molecular studies include *Dialister pneumosintes*, *Dialister invisus*, *Filifactor alocis*, *Prevotella denticola*, *Cryptobacterium curtum*, *Treponema medium*, *Treponema socranskii*, *Eubacterium saphenum*, *Catonella morbi*, and *Selenomonas sputigena* (Doan et al. 2000; Paster et al. 2001; Kumar et al. 2003, 2005). Interestingly, *P. endodontalis*, a putative endodontic pathogen, has also been often detected in periodontally diseased subjects and rarely in healthy subjects (Paster et al. 2001; Kumar et al. 2003). In addition, the newly renamed *Aggregatibacter* (formerly *Actinobacillus*) *actinomycetemcomitans* has been confirmed by molecular studies to be an important pathogen in periodontal disease.

Knowledge of the infectious etiologic agents of periodontal diseases keeps expanding as microorganisms other than bacteria, namely archaea (Lepp et al. 2004) and herpesviruses (Slots 2005), have also been found in association with periodontal diseases.

As one can tell, a significant revolution in the knowledge of the oral microbiota in health and disease has taken place over the very recent years after the advent of new technology for microbial identification. In this context, endodontic infections are no exceptions. Bacterial diversity in the different types of endodontic infections is discussed in the next sections of this chapter.

## 5.13 Unraveling the endodontic microbiota with molecular biology methods

As aforementioned, the limitations imposed by the ability of a microorganism to be cultivated can be sidestepped by the use of molecular identification methods. Several molecular methods have been or have the potential to be applied to the study of endodontic infections. The choice of a particular approach depends on the questions to be answered. The majority of molecular studies have addressed the issue of species composition in different types of endodontic infections. Nevertheless, at this time, there is no available technique that can provide a comprehensive list of all bacterial species in a sample. As a consequence, a variety of techniques have been used to offer a better picture of the endodontic bacterial communities. If the purpose is to investigate the breadth of bacterial diversity in the endodontic environment, the broad-range PCR followed by 16S rRNA gene clone library analysis can be the method of choice, at least while the metagenomics approach does not become economically and procedurally feasible to many laboratories. Analysis of bacterial community structures and identification of most of the community members can be performed via fingerprinting techniques, such as DGGE and T-RFLP. Fluorescence in situ hybridization (FISH) can identify, measure abundance, and provide information on spatial distribution of particular species in tissues. If the purpose is to detect se-

lected target species, checkerboard assays, DNA microarrays, single PCR, nested PCR, multiplex PCR, and real-time PCR can be used to survey a large number of clinical samples, each one with its own advantages and limitations. Fig. 5.3 exemplifies the use of different molecular biology methods for the study of endodontic infections.

Traditionally, endodontic infections have been studied by means of culture-dependent approaches (see Chapter 4). Such studies have resulted in the establishment of a set of species thought to play an important role in the pathogenesis of apical

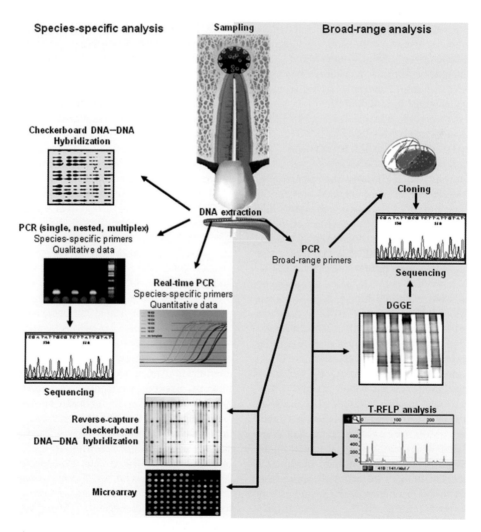

**Fig. 5.3**  Molecular biology methods used in the study of endodontic infections. The choice for a particular technique will depend on the type of analysis to be performed.

periodontitis. More recently, not only have findings from culture-based methods been confirmed but they have also been significantly supplemented with those from molecular diagnostic techniques. Better put, molecular methods have confirmed and strengthened the association of many cultivable bacterial species with apical periodontitis and have also revealed new suspected endodontic pathogens. Detection of cultivable named species in higher prevalence by molecular methods can be explained on the basis of the higher sensitivity of the methods when compared to culture, the fastidious nature of the microorganisms, or even the fact that not all strains within a species can be cultivated or accurately identified. Moreover, the list of candidate pathogens has expanded to include fastidious cultivable species or even as-yet-uncultivated bacteria that have never been previously found in endodontic infections by culturing procedures. As a consequence, the endodontic microbiota has been refined and redefined by molecular methods.

## 5.14 Primary intraradicular infections

Molecular methods have revealed that primary intraradicular infections are characterized by a mixed consortium composed of 10–30 species per canal (Siqueira and Rôças 2005b). The number of bacterial cells in an infected canal varies from $10^3$ to $10^8$ (Vianna et al. 2006b; Sakamoto et al. 2007). Primary infections display a large interindividual variability, that is, each individual shows a unique endodontic microbiota in terms of composition and species dominance as revealed by DGGE and T-RFLP approaches (Siqueira et al. 2004b; Sakamoto et al. 2006; Machado de Oliveira et al. 2007). Comparison of the bacterial community profiles between symptomatic (abscesses) and asymptomatic primary infections by molecular fingerprinting techniques has found marked differences between these conditions (Siqueira et al. 2004b; Sakamoto et al. 2006). This means that there is a significant difference in the species composition and abundance associated with symptomatic and asymptomatic infections. Symptomatic infections also harbored a significantly higher number of bacterial species. The mean number of species in symptomatic and asymptomatic cases was, respectively, 12 and 7 (DGGE), 13 and 11 (T-RFLP), and 18 and 12 (clone library analysis) (Siqueira et al. 2004b; Sakamoto et al. 2006).

Bacterial species/phylotypes detected in primary infections fall into 9 of the 13 phyla that have oral representatives, namely Firmicutes, Bacteroidetes, Fusobacteria, Actinobacteria, Proteobacteria, Spirochaetes, Synergistes, TM7, and SR1 (Munson et al. 2002; Siqueira and Rôças 2005c; Saito et al. 2006; Sakamoto et al. 2006, 2007; Rôças and Siqueira 2008). Members of the four latter phyla are usually difficult or impossible to culture and have been found in endodontic infections only after the introduction of molecular methods. Culture-independent molecular studies have also disclosed several as-yet-uncultivated phylotypes from the five other phyla that have cultivable representatives in the endodontic microbiota.

An important periodontal pathogen, *T. forsythia* (formerly *Bacteroides forsythus*), a Gram-negative obligate anaerobe that had never been reported to occur in infected root canals by culture, was for the first time detected in primary endodontic infections in a study using species-specific single PCR (Conrads et al. 1997). Subsequent studies using different PCR assays, the checkerboard approach and DNA microarray have confirmed that *T. forsythia* is a common member of the microbiota associated with different forms of apical periodontitis, including abscesses (Conrads et al. 1997; Jung et al. 2000; Siqueira et al. 2000b, 2001c, 2002b; Rôças et al. 2001; Fouad et al. 2002; Siqueira and Rôças 2003a; Foschi et al. 2005; Vianna et al. 2005; Gomes et al. 2006). Depending on the molecular identification approach, prevalence values for *T. forsythia* have ranged from 5% to 52% of the cases investigated (Fig. 5.4).

Spirochetes are highly motile spiral-shaped bacteria that have been frequently observed in samples taken from endodontic infections by microscopy, but had never been identified to the species level. The application of molecular diagnostic methods to the identification of spirochetes has demonstrated that their occurrence in infections of endodontic origin has been overlooked by technical hurdles of culture techniques. Thus far, 10 oral *Treponema* species have been cultivated and validly named. They can be classified into two groups according to the fermentation of carbohydrates: the saccharolytic species include *T. pectinovorum*, *T. socranskii*, *T. amylovorum*, *T. lecithinolyticum*, *T. maltophilum*, and *T. parvum*, and the asaccharolytic species include *T. denticola*, *T. medium*, *T. putidum*, and *T. vincentii*. Using species-specific PCR, we detected *T. denticola*, a recognized periodontal pathogen, for the first time in infected root

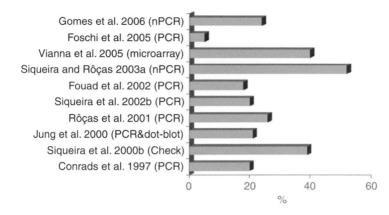

**Fig. 5.4** Prevalence of *Tannerella forsythia* in primary endodontic infections as revealed by molecular studies using different methodologies (check, checkerboard).

canals (Siqueira et al. 2000a). Several other molecular studies confirmed that this treponeme and all the other cultivable species can take part in the microbiota of primary endodontic infections, including abscesses (Siqueira et al. 2000b, 2001a; Jung et al. 2001; Fouad et al. 2002; Baumgartner et al. 2003; Rôças et al. 2003; Siqueira and Rôças 2003c, 2004c; Foschi et al. 2005; Rôças and Siqueira 2005b; Vianna et al. 2005; Gomes et al. 2006) (Figs 5.5 and 5.6). The most frequent treponemes in endodontic infections are *T. denticola* and *T. socranskii* (Siqueira et al. 2000a; Baumgartner et al.

2003; Rôças et al. 2003; Siqueira and Rôças 2004c). The species *T. parvum*, *T. maltophilum* and *T. lecithinolyticum* have been moderately prevalent (Jung et al. 2001; Baumgartner et al. 2003; Siqueira and Rôças 2003c, 2004c; Rôças and Siqueira 2005b).

*Dialister* species are asaccharolytic, obligately anaerobic Gram-negative coccobacilli that represent another example of bacteria that have been consistently detected in endodontic infections only by molecular biology techniques. *D. pneumosintes* and the recently described *D. invisus* have been frequently

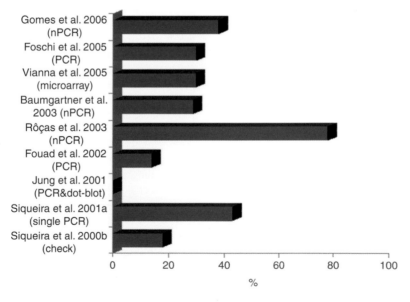

**Fig. 5.5** Prevalence of *Treponema denticola* in primary endodontic infections as revealed by molecular studies using different methodologies (check, checkerboard).

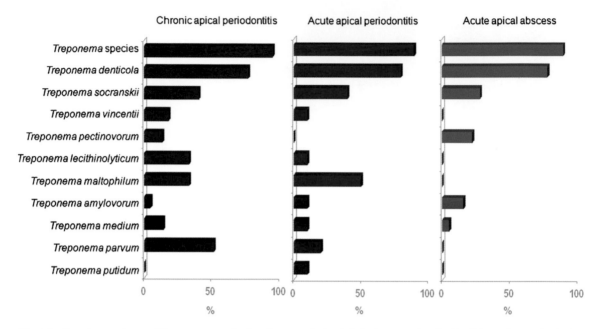

**Fig. 5.6** Prevalence of named *Treponema* species in primary endodontic infections associated with different forms of apical periodontitis. Findings from the authors' laboratory using group- and species-specific nested PCR.

detected in the microbiota associated with asymptomatic and symptomatic primary endodontic infections (Munson et al. 2002; Rôças and Siqueira 2002, 2005a; Siqueira and Rôças 2002, 2003e, 2005c; Saito et al. 2006; Sakamoto et al. 2006).

Black-pigmented Gram-negative anaerobic rods of the genera *Prevotella* and *Porphyromonas* have been commonly found in endodontic infections by culture (Haapasalo et al. 1986; Baumgartner et al. 1999), with some species suggested to be associated with symptoms (van Winkelhoff et al. 1985; Sundqvist et al. 1989). Molecular studies have shown even higher prevalence of black-pigmented bacteria in primary infections (Siqueira et al. 2001c; Baumgartner et al. 2004; Gomes et al. 2005; Seol et al. 2006), but have failed to disclose association with symptoms (Jung et al. 2000; Siqueira et al. 2001b; Fouad et al. 2002). The most prevalent species include *P. endodontalis*, *P. gingivalis*, *P. intermedia*, and *P. nigrescens* (Fig. 5.7). Some newly named *Prevotella* species, such as *Prevotella tannerae*, *Prevotella multissacharivorax*, and *Prevotella baroniae*, have been detected in endodontic infections only after the advent of molecular methods (Xia et al. 2000; Baumgartner et al. 2004; Sakamoto et al. 2006). The subgingival plaque can be a source of

black-pigmented bacteria for endodontic infections, as suggested by a study using AP-PCR to compare periodontal and endodontic isolates of two black-pigmented species (Goncalves et al. 1999).

*Fusobacterium nucleatum* is one of the most commonly encountered Gram-negative species in primary endodontic infections by culturing studies (Sundqvist 1992; Debelian et al. 1995). Molecular studies have confirmed these findings and have shown even higher prevalence values for this species in primarily infected root canals, including cases of abscesses (Jung et al. 2000; Fouad et al. 2002; Baumgartner et al. 2004). Different clonal types of *F. nucleatum* can be found in the same infected canal as revealed by PCR-based bacterial typing techniques (ERIC-PCR and AP-PCR) (Moraes et al. 2002).

Culture-dependent studies have shown that anaerobic Gram-negative bacteria are the most common microorganisms in primary endodontic infections, but some Gram-positive bacteria may also be frequent members of the endodontic microbial consortium. Culture-independent analyses of primary endodontic infections not only have supported these findings, but also have shown higher prevalence for some species and included new Gram-positive species in the set

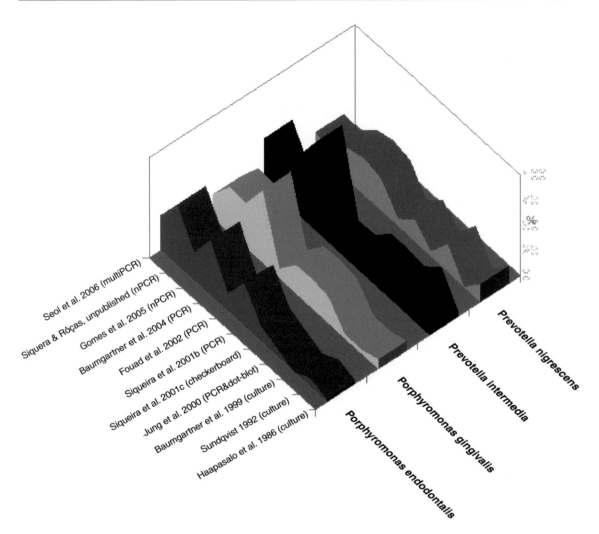

**Fig. 5.7**  Prevalence of four black-pigmented bacterial species in primary endodontic infections as revealed by culture and molecular studies.

of candidate pathogens. Gram-positive species found in similar or higher prevalence by molecular techniques when compared to culture include streptococci (Siqueira et al. 2002b, c; Fouad et al. 2003), *P. alactolyticus* (Siqueira and Rôças 2003f; Siqueira et al. 2004a), *Propionibacterium propionicum* (Siqueira and Rôças 2003g), and *Parvimonas micra* (Siqueira et al. 2003). Species isolated only sporadically or never found by culture that have been disclosed by molecular methods in relatively high frequencies include *F. alocis* (Siqueira and Rôças 2003b; Gomes et al. 2006), *Slackia exigua* (Hashimura et al. 2001), *Eubacterium*

*infirmum* (Fouad et al. 2003), and *Olsenella* species (Fouad et al. 2002; Munson et al. 2002; Rôças and Siqueira 2005c).

Other examples of named cultivable species that have been found in infected canals only by molecular methods include *C. morbi*, a saccharolytic obligately anaerobic Gram-negative rod, *Centipeda periodontii*, an obligately anaerobic Gram-negative serpentine rod, and *Granulicatella adiacens*, a facultative Gram-positive coccus (Siqueira and Rôças 2004a, 2006).

*A. actinomycetemcomitans*, which is implicated in the etiology of periodontal diseases, particularly the localized aggressive periodontitis (previously local-ized juvenile periodontitis) (Tatakis and Kumar 2005), has been a rare finding in root canal infections, even when highly sensitive molecular approaches have been used (Siqueira et al. 2002a). This suggests that this capnophilic species is not favored in the root canal environment and thus does not participate in the pathogenesis of apical periodontitis.

Fig. 5.8 shows data from the authors' laboratory as to the most frequently detected species/phylotypes

in primary endodontic infections associated with symptomatic (abscesses) and asymptomatic apical periodontitis, as revealed by a species-specific nested PCR approach.

### 5.14.1 Uncultivated bacteria

Investigations of the diversity of the endodontic microbiota by broad-range PCR and 16S rRNA gene clone library analysis have demonstrated that it is

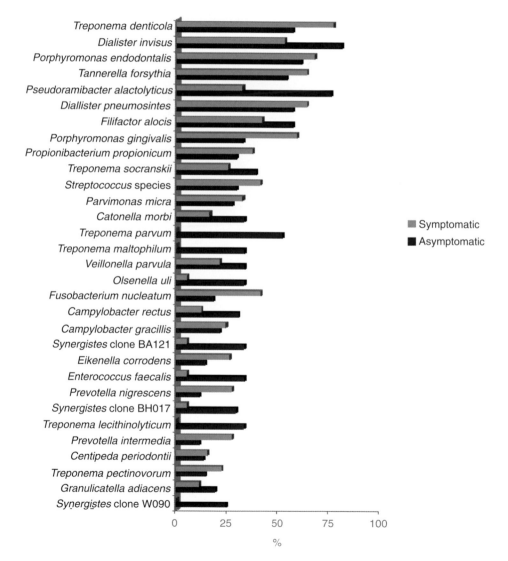

**Fig. 5.8**  Most prevalent bacterial species/phylotypes in asymptomatic and symptomatic primary endodontic infections. Findings from the authors' laboratory using species-specific nested PCR.

far more complex than previously reported by culturing studies. Noteworthy is the common occurrence of as-yet-uncultivated bacteria—about 40–55% of the endodontic microbiota is composed of as-yet-uncultivated phylotypes (Munson et al. 2002; Sakamoto et al. 2006; Sakamoto et al. 2007; Vickerman et al. 2007). Several of these phylotypes can be candidate endodontic pathogens. For instance, oral *Synergistes* clones BA121, E3_33, BH017, and W090, which had been originally assigned to the *Flexistipes* or *Deferribacteres* groups (Paster et al. 2001; Godon et al. 2005), have been commonly detected in samples from asymptomatic and symptomatic endodontic infections (Rôças and Siqueira 2005a; Siqueira and Rôças 2005c; Siqueira et al. 2005c). The great majority of *Synergistes* bacteria remain uncultivated and this can be the primary reason for the fact that their presence in endodontic infections has been overlooked by culturing studies. Similarly, clone I025 from the TM7 candidate phylum, which so far has no cultivable representative, has also been detected in primary endodontic infections (Rôças and Siqueira 2005a; Siqueira and Rôças 2005c). Results from 16S rRNA gene clone libraries have shown the occurrence of uncultivated phylotypes related to the genera *Dialister*, *Prevotella*, *Peptostreptococcus*, *Solobacterium*, *Olsenella*, *Selenomonas*, *Eubacterium*, *Megasphaera*, *Veillonella*, and *Cytophaga* as well as phylotypes related to the family Lachnospiraceae (Rolph et al. 2001; Munson et al. 2002; Saito et al. 2006; Sakamoto et al. 2006; Vickerman et al. 2007). One study (Sakamoto et al. 2006) found some uncultivated phylotypes among the most prevalent bacteria in primary intraradicular infections, including *Lachnospiraceae* oral clone 55 A-34, *Megasphaera* oral clone CS025, and *Veillonella* oral clone BP1-85. Two phylotypes—*Bacteroidetes* oral clone X083 and *Dialister* oral clone BS016—were detected only in asymptomatic teeth, while *Prevotella* oral clone PUS9.180, *Eubacterium* oral clone BP1-89, and *Lachnospiraceae* oral clone MCE7_60 were exclusively detected in symptomatic samples (Sakamoto et al. 2006). Detection of uncultivated phylotypes in samples from endodontic infections suggests that they can be previously unrecognized bacteria that play a role in the pathogenesis of different forms of apical periodontitis.

## 5.14.2 Geographical influence

Data from epidemiological microbiology studies carried out in some geographic regions have been intuitively considered as applicable to other distinct locations, with the clear idea that *everything is everywhere and playing the same role*. Nevertheless, studies have recently fostered the assumption that oral microbial communities can differ significantly according to the geographic location (Ide et al. 2000; Haffajee et al. 2004, 2005).

Findings from laboratories in different countries are often quite different regarding the prevalence of the species involved in endodontic infections. Although these differences may be attributed to variations in the identification methodologies, a geographical influence in the composition of the root canal microbiota has been suspected. Molecular approaches are the most appropriate methods to compare microbiological findings from distinct geographic locations. Because samples from different countries should ideally be analyzed in the same laboratory, the time elapsed from collection to delivery to a distant laboratory may make samples improper for culturing analysis. Molecular methods detect DNA, which can remain relatively unaltered and thereby detectable for a long time when stored under proper conditions. Thus, samples can be submitted to a distant laboratory all at once and resist longtime transportation.

Data from molecular studies that have directly compared the endodontic microbiota of patients residing in different geographic locations suggest that significant differences in the prevalence of some important species can actually exist. Comparisons of acute apical abscess samples taken from U.S. and Brazilian patients have shown that the prevalence of *P. intermedia*, *P. nigrescens*, *P. tannerae*, *F. nucleatum*, *P. gingivalis*, *T. denticola*, and *T. forsythia* markedly differed regarding the two locations (Baumgartner et al. 2004; Rôças et al. 2006) (Fig. 5.9). Analysis of samples from primary infections from Brazilian and South Korean patients revealed that the frequencies of *P. endodontalis*, *D. pneumosintes*, *F. alocis*, *T. denticola*, and *T. forsythia* were significantly different between the two geographic locations (Siqueira et al. 2005a) (Fig. 5.10). A study using the DGGE approach compared the bacterial community profiles of the microbiota associated with acute apical abscesses from Brazilian and U.S. patients (Machado de Oliveira et al. 2007). Results displayed a great variability among samples. This indicates that bacterial communities of abscesses are unique for each individual in terms of diversity. The composition of the microbiota in many samples showed a geography-related pattern. Several species were exclusive for each location and others

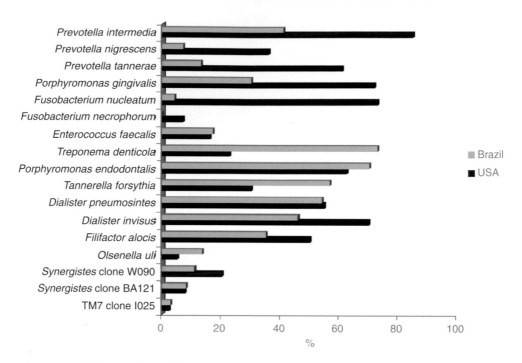

**Fig. 5.9**   Comparison of the prevalence of different bacterial species/phylotypes in acute apical abscess samples taken from two geographic locations.

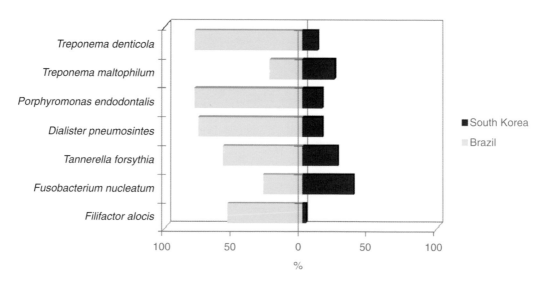

**Fig. 5.10**   Comparison of the prevalence of different bacterial species in primary endodontic infections of patients from two geographic locations.

shared by the two locations showed great differences in prevalence.

The factors that can lead to differences in the composition of the endodontic microbiota and the impact of these differences on therapy, particularly in abscessed cases requiring systemic antibiotic therapy, remain to be illuminated.

## 5.15 Persistent/secondary intraradicular infections

Persistent or secondary intraradicular infections are the major causes of endodontic treatment failures. This statement is supported by two strong evidence-based arguments. First, most (if not all) root canal–treated teeth evincing persistent apical periodontitis lesions have been demonstrated to harbor an intraradicular infection (Lin et al. 1991, 1992; Sundqvist et al. 1998; Pinheiro et al. 2003; Rôças et al. 2004a; Siqueira and Rôças 2004b). Second, it has been demonstrated that there is an increased risk of adverse treatment outcome when microorganisms are present in the canal at the time of filling (Sjögren et al. 1997; Waltimo et al. 2005; Fabricius et al. 2006).

Molecular methods have been recently applied to the study of the microorganisms found at the root canal–filling stage, which have the potential to put the treatment outcome at risk, or in root canal–treated teeth with apical periodontitis, which can be participating in the already established treatment failure.

### 5.15.1 Bacteria at the root canal–filling stage

The impact of bacterial persistence on treatment outcome has been demonstrated and studies have intended to identity species resisting root canal procedures. Most culturing studies have revealed an overall higher occurrence of Gram-positive bacteria in both post-instrumentation and post-medication samples (Sjögren et al. 1997; Chávez de Paz 2005; Chu et al. 2006). With the recent findings showing as-yet-uncultivated bacteria as constituents of a significant proportion of the endodontic microbiota, studies on the effects of intracanal antimicrobial procedures should also focus these bacteria. A study using broad-range PCR and 16S rRNA gene clone library investigated the bacteria persisting after chemome-

chanical preparation using 2.5% NaOCl as irrigant and intracanal medication with a calcium hydroxide paste (Sakamoto et al. 2007). Fifty-six percent of the taxa found in initial samples consisted of as-yet-uncultivated bacteria. A mean of 11 taxa was detected in initial (S1) samples, 4 taxa in post-instrumentation (S2) samples, and 5 taxa in post-medication (S3) samples. *Streptococcus* species were detected in all post-treatment samples and were also the most dominant taxa in these samples, except for an S2 sample in which *Solobacterium* clone K010 corresponded to 56% of the clones sequenced. Forty-two percent of the taxa found in posttreatment samples were as-yet-uncultivated bacteria. These findings suggest that previously uncharacterized bacteria may also participate in persistent endodontic infections.

A couple of studies have used quantitative real-time PCR assays with broad-range primers to analyze the reduction in bacterial numbers after endodontic treatment procedures. One study compared the bacterial reduction in infected canals after chemomechanical preparation using either 2.5% NaOCl or 2% chlorhexidine (CHX) as irrigant (Vianna et al. 2006b). Bacterial quantification was performed by real-time PCR using the SYBR Green and the TaqMan chemistries, and compared with culturing results. Using real-time PCR, the bacterial load before and after chemomechanical preparation was greater when compared with evaluation using culture. In the NaOCl group, the median of the bacterial numbers in initial samples was $2.8 \times 10^6$ (SYBR Green) and $7.6 \times 10^6$ (TaqMan). In the CHX group, the median of bacterial numbers was $2.3 \times 10^6$ (SYBR Green) and $3.0 \times 10^6$ (TaqMan). After chemomechanical preparation of the infected root canals, only one case (from the NaOCl group) showed negative results. The median of the bacterial numbers in the NaOCl group decreased to $2 \times 10^2$ (SYBR Green) and $1.6 \times 10^4$ (TaqMan). Correspondent figures for the CHX group were $6.2 \times 10^4$ (SYBR Green) and $5.3 \times 10^4$ (TaqMan). Bacterial reduction in the NaOCl group (99.99% SYBR Green and 99.63% TaqMan) was significantly greater than in the CHX group (96.62% SYBR Green and 96.6% TaqMan).

Another study used real-time PCR with a SYBR Green format to assess bacterial elimination after preparation with 2.5% NaOCl as irrigant and intracanal medication with a calcium hydroxide paste (Sakamoto et al. 2007). All initial samples were positive for the presence of bacteria as shown by PCR.

Five out of 15 samples showed negative PCR results for both S2 and S3 samples. In the other 10 samples showing bacteria in S2 and S3, the median number of bacteria in S1 samples was $9.64 \times 10^6$. After instrumentation (S2), the median value of the number of bacteria decreased to $2.44 \times 10^4$, while after intracanal medication (S3), the median number was $2.42 \times 10^4$. S2 and S3 samples showed a mean reduction of 99.67% and 99.85%, respectively, in the number of bacteria when compared to S1 samples. Although S3 samples showed an overall percentage decrease of 56% in the number of bacteria when compared to S2, there was no significant difference in bacterial counts between S2 and S3.

Because molecular methods are more sensitive and specific than culture and can detect as-yet-uncultivated bacteria, they can provide a more realistic and detailed insight into the effects of antimicrobial treatment protocols. However, as discussed early in this chapter, molecular technologies have some limitations that may affect this kind of analysis. Of particular interest, the ability to detect DNA from dead cells poses a major problem when one is investigating the immediate effectiveness of antibacterial treatment, since DNA from cells that have recently died can still be detected. The results of molecular studies that some canals are negative for the presence of bacteria and that most of the other cases show over 99% reduction in the number of bacteria might argue otherwise (Sakamoto et al. 2007). The possibility exists that DNA from dead cells can also be destroyed by the effects of substances used during treatment. NaOCl is known to kill bacteria and degrade DNA, with resultant fragments being undetectable by PCR (McCarty and Atlas 1993; Fouad and Barry 2005). Hydroxyl ions from calcium hydroxide also exert oxidative damaging effects on DNA (Siqueira and Lopes 1999), and may contribute to degradation of free DNA from dead cells.

### 5.15.2 Microbiota in root canal–treated teeth

Culture studies have demonstrated that the microbiota of root canal–treated teeth with apical periodontitis mainly comprises one to two species, which are predominantly Gram-positive bacteria with *Enterococcus faecalis* as the most prevalent one (Engström 1964; Möller 1966; Molander et al. 1998; Sundqvist et al. 1998; Peciuliene et al. 2000; Hancock et al. 2001; Pinheiro et al. 2003). Poorly filled root canals

have been shown to contain a greater number of species than canals apparently well treated (Sundqvist et al. 1998; Pinheiro et al. 2003). A study using species-specific PCR revealed that the mean number of species in adequately treated cases was 3 (range 1–5), while cases poorly treated yielded a mean of 5 species (range 2–11) (Siqueira and Rôças 2004b). This difference was statistically significant. Gram-positive bacteria were present in all cases, and at least one of the following species was detected—*E. faecalis*, *P. alactolyticus*, and *P. propionicum*—which occurred isolatedly, in pairs or in threes. PCR-DGGE analysis of the bacterial communities in root canal–treated teeth revealed an average of about 6 species per canal (range 1–26) (Rôças et al. 2004b). The structure of the bacterial communities varied from individual to individual, suggesting that distinct bacterial combinations can play a role in treatment failure.

Several molecular studies have confirmed *E. faecalis* as frequent species in root canal–treated teeth evincing apical periodontitis lesions, with prevalence values reaching up to 90% of the cases (Rôças et al. 2004a, c; Siqueira and Rôças 2004b; Foschi et al. 2005; Fouad et al. 2005; Sedgley et al. 2006; Williams et al. 2006) (Fig. 5.11). Quantitative real-time PCR analysis has revealed that *E. faecalis* may constitute a median 0.98% (range 0.14–100%) of the overall bacterial load in root canal–treated teeth (Sedgley et al. 2006). Root canal–treated teeth are about nine times more likely to harbor *E. faecalis* than cases of primary infections (Rôças et al. 2004c). This suggests that this species can be inhibited by other members of a mixed bacterial consortium commonly present in primary infections and that the bleak environmental conditions within filled root canals do not prevent its survival.

The fact that *E. faecalis* is the most commonly encountered species in treated teeth and the attributes of this species that make it to survive in treated canals have prompted many authors to nominate *E. faecalis* as the main pathogen involved in treatment failures. The consequence of this was an avalanche of in vitro papers focusing on *E. faecalis* (Spangberg 2006). However, findings from recent molecular studies carried out in independent laboratories have somewhat questioned the role of *E. faecalis* as the main causative agent of endodontic failures. Some studies have not succeeded in detecting enterococci in root canal–treated teeth with lesions (Rolph et al. 2001) or have demonstrated that *E. faecalis* is not the dominant species in most retreatment cases (Rôças et al. 2004b;

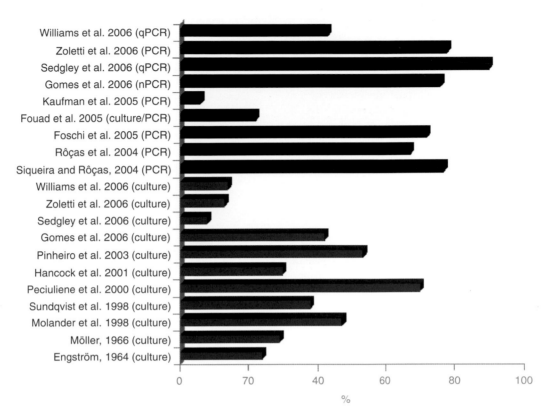

**Fig. 5.11** Prevalence of *Enterococcus faecalis* in samples from root canal–treated teeth with apical periodontitis. Data from culture and molecular studies.

Rôças et al. 2008; Sakamoto et al. 2008). Reports in the literature have demonstrated that *E. faecalis* can also be found in root canal–treated teeth with no lesions. A study detected enterococci in 6% of the root canal–treated teeth with apical periodontitis and in 23% of the treated teeth with no lesions (Kaufman et al. 2005). In another study (Zoletti et al. 2006), *E. faecalis* was found in 81.5% of the root canal–treated teeth with no apical periodontitis and in 78% of the treated teeth with disease. Although these findings apparently put into question the status of *E. faecalis* as the main species causing treatment failure, other related factors still need to be clarified before this assumption turns into certainty. Further studies addressing this issue should include a precise diagnosis of apical periodontitis (with no radiographic biases), provide quantitative results, test association of specific clonal types with disease, and check the different patterns of host response to infection (Zoletti et al. 2006).

In addition to finding *E. faecalis* in high prevalence, molecular studies have also detected streptococci and some fastidious anaerobic species—*P. alactolyticus, P. propionicum, F. alocis,* and *D. pneumosintes*—in several cases of root canal–treated teeth with apical periodontitis (Siqueira and Rôças 2004b). Taxa related to the genera *Capnocytophaga, Cytophaga, Dialister, Eubacterium, Fusobacterium, Gemella, Mogibacterium, Peptostreptococcus, Prevotella, Propionibacterium, Selenomonas, Solobacterium, Streptococcus,* and *Veillonella* and two as-yet-uncultivated bacteria have also been reported to occur in root canal–treated teeth by a study using broad-range PCR and clone library analysis (Rolph et al. 2001). All these findings and those from the structure of bacterial communities (Rôças et al. 2004b) strongly suggest that the microbiota of root canal–treated teeth with apical periodontitis is a little more complex than previously anticipated by culture studies.

As-yet-uncultivated phylotypes may correspond to 55% of the taxa detected in treated canals. Collectively, they can also be in high proportions, representing about one-half of the 16S rRNA gene sequences retrieved in clone libraries (Sakamoto et al. 2008). Some as-yet-uncultivated phylotypes, such as *Bacteroidetes* clone X083 and *Synergistes* clone BA121, have been found amongst the most prevalent taxa in treated canals (Rôças et al. 2008; Sakamoto et al. 2008). As-yet-uncultivated bacteria have been reported to dominate the microbial community in several individual cases (Sakamoto et al. 2008), helping explain why culture studies failed to detect bacteria in some treated root canals.

Molecular methods have also strengthened the association of persistent/secondary intraradicular infections with treatment failures—bacteria have been detected in virtually all treated cases with apical periodontitis (Rôças et al. 2004a, b; Siqueira and Rôças 2004b). On the other hand, previous culture and microscopic studies have failed to detect microorganisms in some cases of root canal–treated teeth with persistent disease (Molander et al. 1998; Sundqvist et al. 1998; Nair et al. 1999; Pinheiro et al. 2003). This discrepancy is better explained by the low sensitivities of culture and microscopic methods and the occurrence of as-yet-uncultivated phylotypes and strains. In addition, many microbial cells can be lost or pass unnoticed as a result of some steps during culturing procedures (e.g., sample transportation, dilutions, and short incubation time), or during sample processing for microscopic examinations, including decalcification, washings, and serial sectioning. The detection of bacteria by staining of demineralized tissue sections is only reliable when large numbers of microorganisms are present in the region under examination (Watts and Paterson 1990). For instance, it has been postulated that for every microorganism detected in histological sections, 25,000 microorganisms have to be actually present (Stanley 1977). Where smaller numbers of bacteria are expected, bacteria may only be detected if serial sections are examined under high magnification (400× or above), but the reliability of this has yet to be demonstrated. The microbiologically negative data derived from culture (Molander et al. 1998; Sundqvist et al. 1998; Pinheiro et al. 2003) and microscopic studies (Nair et al. 1990, 1999) require very careful interpretation in the light of the techniques' numerous limitations, so as to avoid reaching overestimating conclusions about the role of nonmicrobial factors in endodontic treatment failures.

## 5.16 Extraradicular infections

Studies using molecular techniques, specifically the checkerboard assay and FISH, have confirmed previous findings from some culture studies as to the occurrence of extraradicular bacteria in persistent chronic apical periodontitis lesions (Gatti et al. 2000; Sunde et al. 2000, 2003). The issue of extraradicular infections is discussed in more detail in Chapters 6 and 10.

## 5.17 Other microorganisms in endodontic infections

### 5.17.1 Fungi

Fungi are eukaryotic microorganisms that have been found only in 2% of primary root canal infections by broad-range PCR using universal fungal primers (Siqueira et al. 2002b). However, a species-specific PCR-based study has reported the occurrence of *Candida albicans* in 21% of the samples from primary root canal infections (Baumgartner et al. 2000). Another study using species-specific PCR has found *C. albicans* in 9% of the root canal–treated teeth with apical periodontitis (Siqueira and Rôças 2004b).

### 5.17.2 Archaea

Archaea comprise a highly diverse group of prokaryotes, distinct from bacteria in terms of genetic, biochemical, and structural features. For instance, they possess unique flagellins and ether-linked lipids, and lack peptidoglycan in their cell walls. Members of this domain have been traditionally recognized as extremophiles, but recently some of these microorganisms have also been found to flourish in nonextreme environments, including the human body. To date, no member of the Archaea domain has been described as a human pathogen. Although one study failed to detect archaea in necrotic root canals and abscess cases (Siqueira et al. 2005b), another study detected methanogenic archaea in 25% of the canals of untreated teeth with chronic apical periodontitis (Vianna et al. 2006a). Archaeal diversity was limited to a *Methanobrevibacter oralis*–like phylotype, and the size of the archaeal population accounted for up to 2.5% of the total prokaryotic community (i.e., bacteria and archaea), as evaluated by real-time PCR. Another study using clone library analysis confirmed that an *M. oralis*–like phylotype can be found in primary endodontic infections (Vickerman et al. 2007).

### 5.17.3 Virus

Viruses are obligate intracellular parasites in the sense that they are totally dependent on living cells to replicate. Therefore, the necrotic pulp is not a place where viruses can prosper. On the other hand, recent molecular studies using RT-PCR have detected herpesviruses, specifically human cytomegalovirus (HCMV) and Epstein–Barr virus (EBV), in apical periodontitis lesions, where living host cells abound. HCMV and EBV transcripts have been found in high frequencies in symptomatic apical periodontitis lesions (Sabeti et al. 2003a, b), in lesions exhibiting elevated occurrence of anaerobic bacteria (Sabeti and Slots 2004), and in cases of large periradicular bone destruction (Sabeti et al. 2003b; Sabeti and Slots 2004) (see Chapter 8 for more detail).

## 5.18 Concluding remarks

The impact of molecular methods in the medical microbiology scenario is remarkable and endodontics as a health care discipline that deals with infectious disorders is starting to take advantage of the power of these new technologies.

As the breadth of bacterial diversity in endodontic infections has been deciphered by molecular biology methods, the list of candidate endodontic pathogens has expanded to include several cultivable and as-yet-uncultivated species that had been underrated by culture-dependent methods. Bacteria are by far the most frequent and most diverse group of microorganisms involved with these infections. Fig. 5.2 depicts several named bacterial species and as-yet-uncultivated phylotypes that have been found by different molecular methods in samples taken from distinct clinical conditions. The occasional presence of fungi in endodontic infections has been confirmed by molecular methods. Also, other microorganisms never found previously in association with apical periodontitis, namely archaea and herpesviruses, have been recently detected by molecular approaches. A role for

archaea and viruses in the etiology of apical periodontitis remains to be clarified.

Undoubtedly, the well-directed use of molecular methods will provide additional valuable information regarding the identification and understanding of the causative factors of apical periodontitis. Molecular biology techniques hold the hope of making the knowledge of endodontic infectious processes still more accurate. Additionally, molecular methods have the potential to make diagnosis more rapid and directed evidence-based antimicrobial therapy a reality. In fact, the future looks bright for those involved with endodontic microbiology research and for those who will benefit from the refined knowledge, that is, clinicians and patients. Paraphrasing the great Dutch microbiologist Martinus Beijerink—"Happy are those who are starting now."

## 5.19 References

Aas JA, Barbuto SM, Alpagot T, Olsen I, Dewhirst FE, and Paster BJ. 2007. Subgingival plaque microbiota in HIV positive patients. *J Clin Periodontol* 34(3): 189–95.

Aas JA, Dardis SR, Griffen AL, Stokes LN, Lee AM, Olsen I, Dewhirst FE, Leys EJ, and Paster BJ. 2005a. Most of the microbiota in caries has not yet been cultivated. *J Dent Res* 84(Spec. Issue A): IADR Abstract No 2805. http://www.dentalresearch.org.

Aas JA, Dardis SR, Griffen AL, Stokes LN, Lee AMP, Olsen NAI, Dewhirst FE, Leys EJ, and Paster BJ. 2003. Molecular analysis of bacteria associated with caries in permanent teeth. *J Dent Res* 82(Spec Iss A): IADR Abstract No. 1025. http://www.dentalresearch.org.

Aas JA, Paster BJ, Stokes LN, Olsen I, and Dewhirst FE. 2005b. Defining the normal bacterial flora of the oral cavity. *J Clin Microbiol* 43(11): 5721–32.

Amann R, Fuchs BM, and Behrens S. 2001. The identification of microorganisms by fluorescence in situ hybridisation. *Curr Opin Biotechnol* 12(3): 231–36.

Amann RI, Ludwig W, and Schleifer KH. 1995. Phylogenetic identification and in situ detection of individual microbial cells without cultivation. *Microbiol Rev* 59(1): 143–69.

Arora DK, Hirsch PR, and Kerry BR. 1996. PCR-based molecular discrimination of *Verticillium chlamydosporium* isolates. *Mycol Res* 7: 801–9.

Aul JJ, Anderson KW, Wadowsky RM, Doyle WJ, Kingsley LA, Post JC, and Ehrlich GD. 1998. Comparative evaluation of culture and PCR for the detection and determination of persistence of bacterial strains and DNAs in the *Chinchilla laniger* model of otitis media. *Ann Otol Rhinol Laryngol* 107(6): 508–13.

Baumgartner JC, Khemaleelakul SU, and Xia T. 2003. Identification of spirochetes (treponemes) in endodontic infections. *J Endod* 29(12): 794–97.

Baumgartner JC, Siqueira JF, Jr, Xia T, and Rôças IN. 2004. Geographical differences in bacteria detected in endodontic infections using polymerase chain reaction. *J Endod* 30(3): 141–44.

Baumgartner JC, Watkins BJ, Bae KS, and Xia T. 1999. Association of black-pigmented bacteria with endodontic infections. *J Endod* 25(6): 413–15.

Baumgartner JC, Watts CM, and Xia T. 2000. Occurrence of *Candida albicans* in infections of endodontic origin. *J Endod* 26(12): 695–98.

Becker MR, Paster BJ, Leys EJ, Moeschberger ML, Kenyon SG, Galvin JL, Boches SK, Dewhirst FE, and Griffen AL. 2002. Molecular analysis of bacterial species associated with childhood caries. *J Clin Microbiol* 40(3): 1001–9.

Beighton D, Hardie JM, and Whiley RA. 1991. A scheme for the identification of viridans streptococci. *J Med Microbiol* 35: 367–72.

Bernardi G. 1965. Chromatography of nucleic acids on hydroxyapatite. *Nature* 206(986): 779–83.

Boches SK, Paster BJ, and Dewhirst FE. 2006. Development of a human oral microbe identification microarray. *J Dent Res* 85(Spec Issue A): IADR Abstract No 838. http://www.dentalresearch.org.

Boissinot M and Bergeron MG. 2002. Toward rapid real-time molecular diagnostic to guide smart use of antimicrobials. *Curr Opin Microbiol* 5(5): 478–82.

Bosshard PP, Abels S, Altwegg M, Bottger EC, and Zbinden R. 2004. Comparison of conventional and molecular methods for identification of aerobic catalase-negative Gram-positive cocci in the clinical laboratory. *J Clin Microbiol* 42: 2065–73.

Bosshard PP, Abels S, Zbinden R, Bottger EC, and Altwegg M. 2003. Ribosomal DNA sequencing for identification of aerobic Gram-positive rods in the clinical laboratory (an 18-month evaluation). *J Clin Microbiol* 41(9): 4134–40.

Bowden GH. 2000. The microbial ecology of dental caries. *Microb Ecol Health Dis* 12: 138–48.

Breznak JA. 2002. A need to retrieve the not-yet-cultured majority. *Environ Microbiol* 4(1): 4–5.

Brinig MM, Lepp PW, Ouverney CC, Armitage GC, and Relman DA. 2003. Prevalence of bacteria of division TM7 in human subgingival plaque and their association with disease. *Appl Environ Microbiol* 69(3): 1687–94.

Bustin SA. 2000. Absolute quantification of mRNA using real-time reverse transcription polymerase chain reaction assays. *J Mol Endocrinol* 25(2): 169–93.

Chamberlain JS, Gibbs RA, Ranier JE, Nguyen PN, and Caskey CT. 1988. Deletion screening of the Duchenne muscular dystrophy locus via multiplex DNA amplification. *Nucleic Acids Res* 16(23): 11141–56.

Chávez de Paz LE. 2005. On bacteria persisting root canal treatment. Identification and potential mechanisms of resistance to antimicrobial measures. PhD thesis. Göteborg, Sweden: Göteborg University.

Chen K and Pachter L. 2005. Bioinformatics for whole-genome shotgun sequencing of microbial communities. *PLoS Comput Biol* 1(2): 106–12.

Chhour KL, Nadkarni MA, Byun R, Martin FE, Jacques NA, and Hunter N. 2005. Molecular analysis of microbial diversity in advanced caries. *J Clin Microbiol* 43(2): 843–49.

Chu FC, Leung WK, Tsang PC, Chow TW, and Samaranayake LP. 2006. Identification of cultivable microorganisms from root canals with apical periodontitis following two-visit endodontic treatment with antibiotics/steroid or calcium hydroxide dressings. *J Endod* 32(1): 17–23.

Clarridge JE III. 2004. Impact of 16S rRNA gene sequence analysis for identification of bacteria on clinical microbiology and infectious diseases. *Clin Microbiol Rev* 17(4): 840–62.

Clement BG, Kehl LE, De Bord KL, and Kitts CL. 1998. Terminal restriction fragment patterns (TRFPs), a rapid, PCR-based method for the comparison of complex bacterial communities. *J Microbiol Methods* 31: 135–42.

Conrads G, Gharbia SE, Gulabivala K, Lampert F, and Shah HN. 1997. The use of a 16s rDNA directed PCR for the detection of endodontopathogenic bacteria. *J Endod* 23(7): 433–38.

de Bruijn FJ. 1992. Use of repetitive (repetitive extragenic palindromic and enterobacterial repetitive intergeneric consensus) sequences and the polymerase chain reaction to fingerprint the genomes of *Rhizobium meliloti* isolates and other soil bacteria. *Appl Environ Microbiol* 58(7): 2180–87.

Debelian GJ, Olsen I, and Tronstad L. 1995. Bacteremia in conjunction with endodontic therapy. *Endod Dent Traumatol* 11(3): 142–49.

Dethlefsen L and Relman DA. 2007. The importance of individuals and scale: moving towards single cell microbiology. *Environ Microbiol* 9(1): 8–10.

Dieffenbach CW and Dveksler GS. 1995. *PCR Primer. A Laboratory Manual.* Plain View, NY: Cold Spring Harbor Laboratory Press.

Doan N, Contreras A, Flynn J, Slots J, and Chen C. 2000. Molecular identification of *Dialister pneumosintes* in subgingival plaque of humans. *J Clin Microbiol* 38(8): 3043–47.

Don RH, Cox PT, Wainwright BJ, Baker K, and Mattick JS. 1991. "Touchdown" PCR to circumvent spurious priming during gene amplification. *Nucleic Acids Res* 19: 4008.

Donoghue HD, Spigelman M, Greenblatt CL, Lev-Maor G, Bar-Gal G, Matheson C, Vernon K, Nerlich AG, and Zink AR. 2004. Tuberculosis: from prehistory to Robert Koch, as revealed by ancient DNA. *Lancet Infect Dis* 4: 584–92.

Dragon EA. 1993. Handling reagents in the PCR laboratory. *PCR Methods Appl* 3(2): S8–9.

Drancourt M, Berger P, and Raoult D. 2004. Systematic 16S rRNA gene sequencing of atypical clinical isolates identified 27 new bacterial species associated with humans. *J Clin Microbiol* 42(5): 2197–202.

Drancourt M, Bollet C, Carlioz A, Martelin R, Gayral JP, and Raoult D. 2000. 16S ribosomal DNA sequence analysis of a large collection of environmental and clinical unidentifiable bacterial isolates. *J Clin Microbiol* 38(10): 3623–30.

Drancourt M and Raoult D. 2005. Sequence-based identification of new bacteria: a proposition for creation of an orphan bacterium repository. *J Clin Microbiol* 43: 4311–15.

Eckburg PB, Bik EM, Bernstein CN, Purdom E, Dethlefsen L, Sargent M, Gill SR, Nelson KE, and Relman DA. 2005. Diversity of the human intestinal microbial flora. *Science* 308(5728): 1635–38.

Engelkirk PG, Duben-Engelkirk J and Dowell VR, Jr. 1992. *Principles and Practice of Clinical Anaerobic Bacteriology.* Belmont, CA: Star Publishing Company.

Engström B. 1964. The significance of enterococci in root canal treatment. *Odontol Revy* 15: 87–106.

Fabricius L, Dahlén G, Sundqvist G, Happonen RP, and Möller AJR. 2006. Influence of residual bacteria on periapical tissue healing after chemomechanical treatment and root filling of experimentally infected monkey teeth. *Eur J Oral Sci* 114: 278–85.

Foschi F, Cavrini F, Montebugnoli L, Stashenko P, Sambri V, and Prati C. 2005. Detection of bacteria in endodontic samples by polymerase chain reaction assays and association with defined clinical signs in Italian patients. *Oral Microbiol Immunol* 20(5): 289–95.

Fouad AF and Barry J. 2005. The effect of antibiotics and endodontic antimicrobials on the polymerase chain reaction. *J Endod* 31(7): 510–13.

Fouad AF, Barry J, Caimano M, Clawson M, Zhu Q, Carver R, Hazlett K, and Radolf JD. 2002. PCR-based identification of bacteria associated with endodontic infections. *J Clin Microbiol* 40(9): 3223–31.

Fouad AF, Kum KY, Clawson ML, Barry J, Abenoja C, Zhu Q, Caimano M, and Radolf JD. 2003. Molecular characterization of the presence of *Eubacterium* spp and *Streptococcus* spp in endodontic infections. *Oral Microbiol Immunol* 18(4): 249–55.

Fouad AF, Zerella J, Barry J, and Spangberg LS. 2005. Molecular detection of *Enterococcus* species in root canals of therapy-resistant endodontic infections. *Oral Surg Oral Med Oral Pathol Oral Radiol Endod* 99(1): 112–18.

Fredricks DN and Relman DA. 1999. Application of polymerase chain reaction to the diagnosis of infectious diseases. *Clin Infect Dis* 29(3):475–86; quiz 487–88.

Gasser RB. 1998. What's in that band? *Int J Parasitol* 28: 989–96.

Gatti JJ, Dobeck JM, Smith C, White RR, Socransky SS, and Skobe Z. 2000. Bacteria of asymptomatic periradicular endodontic lesions identified by DNA–DNA hybridization. *Endod Dent Traumatol* 16(5): 197–204.

Göbel UB. 1995. Phylogenetic amplification for the detection of uncultured bacteria and the analysis of complex microbiota. *J Microbiol Methods* 23: 117–28.

Godon J-J, Morinière J, Moletta M, Gaillac M, Bru V, and Delgènes J-P. 2005. Rarity associated with specific ecological niches in the bacterial world: the "Synergistes" example. *Environ Microbiol* 7: 213–24.

Gomes BP, Jacinto RC, Pinheiro ET, Sousa EL, Zaia AA, Ferraz CC, and Souza-Filho FJ. 2005. *Porphyromonas gingivalis, Porphyromonas endodontalis, Prevotella intermedia* and *Prevotella nigrescens* in endodontic lesions detected by culture, and by PCR. *Oral Microbiol Immunol* 20(4): 211–15.

Gomes BP, Jacinto RC, Pinheiro ET, Sousa EL, Zaia AA, Ferraz CC, and Souza-Filho FJ. 2006. Molecular analysis of *Filifactor alocis, Tannerella forsythia,* and *Treponema denticola* associated with primary endodontic infections and failed endodontic treatment. *J Endod* 32(10): 937–40.

Goncalves RB, Robitaille M, and Mouton C. 1999. Identical clonal types of *Porphyromonas gingivalis* or *Prevotella nigrescens* recovered from infected root canals and subgingival plaque. *Oral Microbiol Immunol* 14(3): 197–200.

Green BD and Keller M. 2006. Capturing the uncultivated majority. *Curr Opin Biotechnol* 17(3): 236–40.

Griffen AL, Kumar PS, and Leys EJ. 2003. *A Quantitative, molecular view of Oral Biofilm Communities in Health and Disease Suggests a Role for Uncultivated Species.* Polymicrobial diseases, American Society for Microbiology conferences, Lake Tahoe, Nevada, 13.

Haapasalo M, Ranta H, Ranta K, and Shah H. 1986. Black-pigmented *Bacteroides* spp. in human apical periodontitis. *Infect Immun* 53(1): 149–53.

Haffajee AD, Bogren A, Hasturk H, Feres M, Lopez NJ, and Socransky SS. 2004. Subgingival microbiota of chronic periodontitis subjects from different geographic locations. *J Clin Periodontol* 31(11): 996–1002.

Haffajee AD, Japlit M, Bogren A, Kent RL, Jr, Goodson JM, and Socransky SS. 2005. Differences in the subgingival microbiota of Swedish and USA subjects who were periodontally healthy or exhibited minimal periodontal disease. *J Clin Periodontol* 32(1): 33–39.

Hancock HH III, Sigurdsson A, Trope M, and Moiseiwitsch J. 2001. Bacteria isolated after unsuccessful endodontic treatment in a North American population. *Oral Surg Oral Med Oral Pathol Oral Radiol Endod* 91(5): 579–86.

Handelsman J. 2004. Metagenomics: application of genomics to uncultured microorganisms. *Microbiol Mol Biol Rev* 68(4): 669–85.

Haqqi TM, Sarkar G, David CS, and Sommer SS. 1988. Specific amplification with PCR of a refractory segment of genomic DNA. *Nucleic Acids Res* 16(24): 11844.

Harper-Owen R, Dymock D, Booth V, Weightman AJ, and Wade WG. 1999. Detection of unculturable bacteria in periodontal health and disease by PCR. *J Clin Microbiol* 37(5): 1469–73.

Hashimura T, Sato M, and Hoshino E. 2001. Detection of *Slackia exigua, Mogibacterium timidum* and *Eubacterium saphenum* from pulpal and periradicular samples using the polymerase chain reaction (PCR) method. *Int Endod J* 34(6): 463–70.

Hayden RT. 2004. In vitro nucleic acid amplification techniques. In: Persing DH, Tenover FC, Versalovic J, Tang Y-W, Unger ER, Relman DA, and White TJ (eds), *Molecular Microbiology. Diagnostic Principles and Practice.* Washington, DC: ASM Press, 43–69.

Heid CA, Stevens J, Livak KJ, and Williams PM. 1996. Real time quantitative PCR. *Genome Res* 6(10): 986–94.

Higgins CF, Ames GF, Barnes WM, Clement JM, and Hofnung M. 1982. A novel intercistronic regulatory element of prokaryotic operons. *Nature* 298(5876): 760–62.

Higuchi R, Dollinger G, Walsh PS, and Griffith R. 1992. Simultaneous amplification and detection of specific DNA sequences. *Biotechnology* 10(4): 413–17.

Holland PM, Abramson RD, Watson R, and Gelfand DH. 1991. Detection of specific polymerase chain reaction product by utilizing the 5′–3′ exonuclease activity of *Thermus aquaticus* DNA polymerase. *Proc Natl Acad Sci USA* 88(16): 7276–80.

Hugenholtz P. 2002. Exploring prokaryotic diversity in the genomic era. *Genome Biol* 3(2): REVIEWS0003.

Hugenholtz P and Pace NR. 1996. Identifying microbial diversity in the natural environment: A molecular phylogenetic approach. *Trends Biotechnol* 14(6): 190–97.

Ide L, Lotufo RFM, Contreras A, Bergamashi O, and Slots J. 2000. Occurrence of seven putative periodontal pathogens in the subgingival plaque of two native populations in the Xingu Indian Park. *Anaerobe* 6: 135–37.

Janssen PH. 2006. Identifying the dominant soil bacterial taxa in libraries of 16S rRNA and 16S rRNA genes. *Appl Environ Microbiol* 72(3): 1719–28.

Josephson KL, Gerba CP, and Pepper IL. 1993. Polymerase chain reaction detection of nonviable bacterial pathogens. *Appl Environ Microbiol* 59(10): 3513–15.

Jung IY, Choi BK, Kum KY, Roh BD, Lee SJ, Lee CY, and Park DS. 2000. Molecular epidemiology and association of putative pathogens in root canal infection. *J Endod* 26(10): 599–604.

Jung IY, Choi B, Kum KY, Yoo YJ, Yoon TC, Lee SJ, and Lee CY. 2001. Identification of oral spirochetes at the species level and their association with other bacteria in endodontic infections. *Oral Surg Oral Med Oral Pathol Oral Radiol Endod* 92(3): 329–34.

Juretschko S, Buccat AM, and Fritsche TR. 2004. Applications of fluorescence in situ hybridization in diagnostic microbiology. In: Persing DH, Tenover FC, Versalovic J, Tang Y-W, Unger ER, Relman D, and White TJ (eds), *Molecular Microbiology. Diagnostic Principles and Practice.* Washington, DC: ASM Press, 3–18.

Kaufman B, Spangberg L, Barry J, and Fouad AF. 2005. *Enterococcus* spp. in endodontically treated teeth with and without periradicular lesions. *J Endod* 31(12): 851–56.

Kazor CE, Mitchell PM, Lee AM, Stokes LN, Loesche WJ, Dewhirst FE, and Paster BJ. 2003. Diversity of bacterial populations on the tongue dorsa of patients with halitosis and healthy patients. *J Clin Microbiol* 41(2): 558–63.

Ke D, Picard FJ, Martineau F, Menard C, Roy PH, Ouellette M, and Bergeron MG. 1999. Development of a PCR assay for rapid detection of enterococci. *J Clin Microbiol* 37(11): 3497–503.

Keer JT and Birch L. 2003. Molecular methods for the assessment of bacterial viability. *J Microbiol Methods* 53(2): 175–83.

Kell DB and Young M. 2000. Bacterial dormancy and culturability: the role of autocrine growth factors. *Curr Opin Microbiol* 3(3): 238–43.

Keller M and Zengler K. 2004. Tapping into microbial diversity. *Nat Rev Microbiol* 2: 141–50.

Kumar PS, Griffen AL, Barton JA, Paster BJ, Moeschberger ML, and Leys EJ. 2003. New bacterial species associated with chronic periodontitis. *J Dent Res* 82(5): 338–44.

Kumar PS, Griffen AL, Moeschberger ML, and Leys EJ. 2005. Identification of candidate periodontal pathogens and beneficial species by quantitative 16S clonal analysis. *J Clin Microbiol* 43(8): 3944–55.

Leduc A, Grenier D, and Mayrand D. 1995. Outer membrane-associated deoxyribonuclease activity of *Porphyromonas gingivalis*. *Anaerobe* 1: 129–34.

Lee HC and Tirnady F. 2003. *Blood Evidence. How DNA Is Revolutionizing the Way We Solve Crimes.* Cambridge, MA: Perseus Publishing.

Lepp PW, Brinig MM, Ouverney CC, Palm K, Armitage GC, and Relman DA. 2004. Methanogenic Archaea and human periodontal disease. *Proc Natl Acad Sci USA* 101(16): 6176–81.

Lepp PW and Relman DA. 2004. Molecular phylogenetic analysis. In: Persing DH, Tenover FC, Versalovic J, Tang Y-W, Unger ER, Relman D, and White TJ (eds), *Molecular Microbiology. Diagnostic Principles and Practice.* Washington, DC: ASM Press, 161–80.

Lewis K. 2007. Persister cells, dormancy and infectious disease. *Nat Rev Microbiol* 5(1): 48–56.

Leys EJ, Griffen AL, Kumar PS, and Maiden MF. 2006. Isolation, classification, and identification of oral microorganisms. In: Lamont RJ, Burne RA, Lantz MS, and Leblanc DJ (eds), *Oral Microbiology and Immunology.* Washington DC: ASM Press, 73–88.

Li J and Hanna BA. 2004. DNA probes for culture confirmation and direct detection of bacterial infections: a review of technology. In: Persing DH, Tenover FC, Versalovic J, Tang Y-W, Unger ER, Relman DA, and White TJ (eds), *Molecular Microbiology. Diagnostic Principles and Practice.* Washington, DC: ASM Press, 19–26.

Lillo A, Ashley FP, Palmer RM, Munson MA, Kyriacou L, Weightman AJ, and Wade WG. 2006. Novel subgingival bacterial phylotypes detected using multiple universal polymerase chain reaction primer sets. *Oral Microbiol Immunol* 21(1): 61–68.

Lin LM, Pascon EA, Skribner J, Gangler P, and Langeland K. 1991. Clinical, radiographic, and histologic study of endodontic treatment failures. *Oral Surg Oral Med Oral Pathol* 71(5): 603–11.

Lin LM, Skribner JE, and Gaengler P. 1992. Factors associated with endodontic treatment failures. *J Endod* 18(12): 625–27.

Loesche WJ and Kazor C. 2002. Microbiology and treatment of halitosis. *Periodontol 2000* 28: 256–79.

Louie M, Louie L, and Simor AE. 2000. The role of DNA amplification technology in the diagnosis of infectious diseases. *CMAJ* 163(3): 301–9.

Machado de Oliveira JC, Gama TG, Siqueira JF, Jr, Rocas IN, Peixoto RS, and Rosado AS. 2006. On the use of denaturing gradient gel electrophoresis approach for bacterial identification in endodontic infections. *Clin Oral Investig* 11(2):127–32

Machado de Oliveira JC, Siqueira JF, Jr, Rôças IN, Baumgartner JC, Xia T, Peixoto RS, and Rosado AS. 2007. Bacterial community profiles of endodontic abscesses from Brazilian and USA subjects as compared by denaturing gradient gel electrophoresis analysis. *Oral Microbiol Immunol* 22(1): 14–18.

Mackay IM. 2004. Real-time PCR in the microbiology laboratory. *Clin Microbiol Infect* 10: 190–212.

Maeda H, Fujimoto C, Haruki Y, Maeda T, Kokeguchi S, Petelin M, Arai H, Tanimoto I, Nishimura F, and Takashiba S. 2003. Quantitative real-time PCR using TaqMan and SYBR Green for *Actinobacillus actinomycetemcomitans*, *Porphyromonas gingivalis*, *Prevotella intermedia*, *tetQ* gene and total bacteria. *FEMS Immunol Med Microbiol* 39(1): 81–86.

Maiwald M. 2004. Broad-range PCR for detection and identification of bacteria. In: Persing DH, Tenover FC, Versalovic J, Tang Y-W, Relman D, and White TJ (eds), *Molecular Microbiology. Diagnostic Principles and Practice.* Washington, DC: ASM Press, 379–90.

Malawista SE, Barthold SW, and Persing DH. 1994. Fate of *Borrelia burgdorferi* DNA in tissues of infected mice after antibiotic treatment. *J Infect Dis* 170: 1312–16.

Marsh P and Martin MV. 1999. *Oral Microbiology*. Oxford: Wright.

Marsh TL. 1999. Terminal restriction fragment length polymorphism (T-RFLP): an emerging method for characterizing diversity among homologous populations of amplification products. *Curr Opin Microbiol* 2(3): 323–27.

Martin FE, Nadkarni MA, Jacques NA, and Hunter N. 2002. Quantitative microbiological study of human carious dentine by culture and real-time PCR: association of anaerobes with histopathological changes in chronic pulpitis. *J Clin Microbiol* 40(5): 1698–704.

Matsumoto M, Sakamoto M, Hayashi H, and Benno Y. 2005. Novel phylogenetic assignment database for terminal-restriction fragment length polymorphism analysis of human colonic microbiota. *J Microbiol Methods* 61(3): 305–19.

McCarty SC and Atlas RM. 1993. Effect of amplicon size on PCR detection of bacteria exposed to chlorine. *PCR Methods Appl* 3(3): 181–85.

McPherson MJ and Moller SG. 2000. *PCR*. Oxford, UK: BIOS Scientific Publishers Ltd.

Mhlanga MM and Malmberg L. 2001. Using molecular beacons to detect single-nucleotide polymorphisms with real-time PCR. *Methods* 25(4): 463–71.

Millar BC, Xu J, and Moore JE. 2002. Risk assessment models and contamination management: implications for broad-range ribosomal DNA PCR as a diagnostic tool in medical bacteriology. *J Clin Microbiol* 40(5): 1575–80.

Molander A, Reit C, Dahlen G, and Kvist T. 1998. Microbiological status of root-filled teeth with apical periodontitis. *Int Endod J* 31(1): 1–7.

Möller AJR. 1966. Microbial examination of root canals and periapical tissues of human teeth. *Odontol Tidskr* 74(Suppl): 1–380.

Moore WEC and Moore LVH. 1994. The bacteria of periodontal diseases. *Periodontol 2000* 5: 66–77.

Moraes SR, Siqueira JF, Jr, Rôças IN, Ferreira MC, and Domingues RM. 2002. Clonality of *Fusobacterium nucleatum* in root canal infections. *Oral Microbiol Immunol* 17(6): 394–96.

Moter A and Gobel UB. 2000. Fluorescence in situ hybridization (FISH) for direct visualization of microorganisms. *J Microbiol Methods* 41(2): 85–112.

Moter A, Hoenig C, Choi BK, Riep B, and Gobel UB. 1998a. Molecular epidemiology of oral treponemes associated with periodontal disease. *J Clin Microbiol* 36(5): 1399–403.

Moter A, Leist G, Rudolph R, Schrank K, Choi BK, Wagner M, and Gobel UB. 1998b. Fluorescence in situ hybridization shows spatial distribution of as yet uncultured treponemes in biopsies from digital dermatitis lesions. *Microbiology* 144(Pt 9): 2459–67.

Mothershed EA and Whitney AM. 2006. Nucleic acid-based methods for the detection of bacterial pathogens: Present and future considerations for the clinical laboratory. *Clin Chim Acta* 363(1–2): 206–20.

Mullis KB, Ferré F, and Gibbs RA. 1994. *The Polymerase Chain Reaction*. Boston, MA: Birkhäuser.

Munson MA, Banerjee A, Watson TF, and Wade WG. 2004. Molecular analysis of the microflora associated with dental caries. *J Clin Microbiol* 42(7): 3023–29.

Munson MA, Pitt-Ford T, Chong B, Weightman A, and Wade WG. 2002. Molecular and cultural analysis of the microflora associated with endodontic infections. *J Dent Res* 81(11): 761–66.

Muyzer G, de Waal EC, and Uitterlinden AG. 1993. Profiling of complex microbial populations by denaturing gradient gel electrophoresis analysis of polymerase chain reaction-amplified genes coding for 16S rRNA. *Appl Environ Microbiol* 59(3): 695–700.

Muyzer G and Smalla K. 1998. Application of denaturing gradient gel electrophoresis (DGGE) and temperature gradient gel electrophoresis (TGGE) in microbial ecology. *Antonie Van Leeuwenhoek* 73(1): 127–41.

Myers RM, Fischer SG, Lerman LS, and Maniatis T. 1985. Nearly all single base substitutions in DNA fragments joined to a GC-clamp can be detected by denaturing gradient gel electrophoresis. *Nucleic Acids Res* 13(9): 3131–45.

Nadkarni MA, Caldon CE, Chhour K-L, Fisher IP, Martin FE, Jacques NA, and Hunter N. 2004. Carious dentine provides a habitat for a complex array of novel *Prevotella*-like bacteria. *J Clin Microbiol* 42: 5238–44.

Nair PN, Sjogren U, Figdor D, and Sundqvist G. 1999. Persistent periapical radiolucencies of root-filled human teeth, failed endodontic treatments, and periapical scars. *Oral Surg Oral Med Oral Pathol Oral Radiol Endod* 87(5): 617–27.

Nair PN, Sjogren U, Krey G, and Sundqvist G. 1990. Therapy-resistant foreign body giant cell granuloma at the periapex of a root-filled human tooth. *J Endod* 16(12): 589–95.

Nocker A and Camper AK. 2006. Selective removal of DNA from dead cells of mixed bacterial communities by use of ethidium monoazide. *Appl Environ Microbiol* 72(3): 1997–2004.

Nocker A, Cheung CY, and Camper AK. 2006. Comparison of propidium monoazide with ethidium monoazide for differentiation of live vs. dead bacteria by selective removal of DNA from dead cells. *J Microbiol Methods* 67(2): 310–20.

Ochman H, Lerat E, and Daubin V. 2005. Examining bacterial species under the specter of gene transfer and exchange. *Proc Natl Acad Sci USA* 102: 6595–99.

Paabo S, Poinar H, Serre D, Jaenicke-Despres V, Hebler J, Rohland N, Kuch M, Krause J, Vigilant L, and Hofreiter M. 2004. Genetic analyses from ancient DNA. *Annu Rev Genet* 38: 645–79.

Palmer C, Bik EM, Eisen MB, Eckburg PB, Sana TR, Wolber PK, Relman DA, and Brown PO. 2006. Rapid quantitative profiling of complex microbial populations. *Nucleic Acids Res* 34(1): e5.

Paster BJ, Bartoszyk IM, and Dewhirst FE. 1998. Identification of oral streptococci using PCR-based, reverse-capture, checkerboard hybridization. *Meth Cell Sci* 20: 223–31.

Paster BJ, Boches SK, Galvin JL, Ericson RE, Lau CN, Levanos VA, Sahasrabudhe A, and Dewhirst FE. 2001. Bacterial diversity in human subgingival plaque. *J Bacteriol* 183(12): 3770–83.

Paster BJ, Falkler WA, Jr, Enwonwu CO, Jr, Idigbe EO, Savage KO, Levanos VA, Tamer MA, Ericson RL, Lau CN, and Dewhirst FE. 2002. Prevalent bacterial species and novel phylotypes in advanced noma lesions. *J Clin Microbiol* 40(6): 2187–91.

Paster BJ, Olsen I, Aas JA, and Dewhirst FE. 2006. The breadth of bacterial diversity in the human periodontal pocket and other oral sites. *Periodontology 2000* 42(1): 80–87.

Patel JB. 2001. 16S rRNA gene sequencing for bacterial pathogen identification in the clinical laboratory. *Mol Diagn* 6(4): 313–21.

Peciuliene V, Balciuniene I, Eriksen HM, and Haapasalo M. 2000. Isolation of *Enterococcus faecalis* in previously root-filled canals in a Lithuanian population. *J Endod* 26(10): 593–95.

Petti CA, Polage CR, and Schreckenberger P. 2005. The role of 16S rRNA gene sequencing in identification of microorganisms misidentified by conventional methods. *J Clin Microbiol* 43: 6123–25.

Pinheiro ET, Gomes BP, Ferraz CC, Sousa EL, Teixeira FB, and Souza-Filho FJ. 2003. Microorganisms from canals of root-filled teeth with periapical lesions. *Int Endod J* 36(1): 1–11.

Pitt TL and Saunders NA. 2000. Molecular bacteriology: a diagnostic tool for the millennium. *J Clin Pathol* 53(1): 71–75.

Post JC, Aul JJ, White GJ, Wadowsky RM, Zavoral T, Tabari R, Kerber B, Doyle WJ, and Ehrlich GD. 1996. PCR-based detection of bacterial DNA after antimicrobial treatment is indicative of persistent, viable bacteria in the chinchilla model of otitis media. *Am J Otolaryngol* 17(2): 106–11.

Power EG. 1996. RAPD typing in microbiology—a technical review. *J Hosp Infect* 34(4): 247–65.

Rantakokko-Jalava K, Nikkari S, Jalava J, Eerola E, Skurnik M, Meurman O, Ruuskanen O, Alanen A, Kotilainen E, Toivanen P, and Kotilainen P. 2000. Direct amplification of rRNA genes in diagnosis of bacterial infections. *J Clin Microbiol* 38(1): 32–39.

Raoult D, Fournier PE, and Drancourt M. 2004. What does the future hold for clinical microbiology? *Nat Rev Microbiol* 2(2): 151–59.

Rappe MS and Giovannoni SJ. 2003. The uncultured microbial majority. *Annu Rev Microbiol* 57: 369–94.

Relman DA. 1997. Emerging infections and newly-recognised pathogens. *Neth J Med* 50(5): 216–20.

Relman DA. 1999. The search for unrecognized pathogens. *Science* 284(5418): 1308–10.

Renesto P, Crapoulet N, Ogata H, La Scola B, Vestris G, Claverie JM, and Raoult D. 2003. Genome-based design of a cell-free culture medium for *Tropheryma whipplei*. *Lancet* 362(9382): 447–49.

Riesenfeld CS, Schloss PD, and Handelsman J. 2004. Metagenomics: Genomic analysis of microbial communities. *Annu Rev Genet* 38: 525–52.

Rôças IN, Baumgartner JC, Xia T, and Siqueira JF, Jr. 2006. Prevalence of selected bacterial named species and uncultivated phylotypes in endodontic abscesses from two geographic locations. *J Endod* 32(12): 1135–38.

Rôças IN, Hulsmann M, and Siqueria JF, Jr. 2008. Microorganisms in root canal-treated teeth from a German population. *J Endod* 34(8): 926–31.

Rôças IN, Jung IY, Lee CY, and Siqueira JF, Jr. 2004a. Polymerase chain reaction identification of microorganisms in previously root-filled teeth in a South Korean population. *J Endod* 30(7): 504–8.

Rôças IN and Siqueira JF, Jr. 2002. Identification of *Dialister pneumosintes* in acute periradicular abscesses of humans by nested PCR. *Anaerobe* 8: 75–78.

Rôças IN and Siqueira JF, Jr. 2005a. Detection of novel oral species and phylotypes in symptomatic endodontic infections including abscesses. *FEMS Microbiol Lett* 250(2): 279–85.

Rôças IN and Siqueira JF, Jr. 2005b. Occurrence of two newly named oral treponemes—*Treponema parvum* and *Treponema putidum*—in primary endodontic infections. *Oral Microbiol Immunol* 20(6): 372–75.

Rôças IN and Siqueria JF, Jr. 2008. Root canal microbiota of teeth with chronic apical periodontitis. *J Clin Microbiol* 46(11): 3599–606.

Rôças IN and Siqueira JF, Jr. 2005c. Species-directed 16S rRNA gene nested PCR detection of *Olsenella* species in association with endodontic diseases. *Lett Appl Microbiol* 41(1): 12–16.

Rôças IN, Siqueira JF, Jr, Aboim MC, and Rosado AS. 2004b. Denaturing gradient gel electrophoresis analysis of bacterial communities associated with failed endodontic treatment. *Oral Surg Oral Med Oral Pathol Oral Radiol Endod* 98(6): 741–49.

Rôças IN, Siqueira JF, Jr, Andrade AF, and Uzeda M. 2003. Oral treponemes in primary root canal infections as detected by nested PCR. *Int Endod J* 36(1): 20–26.

Rôças IN, Siqueira JF, Jr, and Santos KR. 2004c. Association of *Enterococcus faecalis* with different forms of periradicular diseases. *J Endod* 30(5): 315–20.

Rôças IN, Siqueira JF, Jr, Santos KR, and Coelho AM. 2001. "Red complex" (*Bacteroides forsythus*, *Porphyromonas gingivalis*, and *Treponema denticola*) in endodontic infections: a molecular approach. *Oral Surg Oral Med Oral Pathol Oral Radiol Endod* 91(4): 468–71.

Rodriguez-Valera F. 2002. Approaches to prokaryotic biodiversity: a population genetics perspective. *Environ Microbiol* 4: 628–33.

Rolph HJ, Lennon A, Riggio MP, Saunders WP, MacKenzie D, Coldero L, and Bagg J. 2001. Molecular identification of microorganisms from endodontic infections. *J Clin Microbiol* 39(9): 3282–89.

Sabeti M, Simon JH, and Slots J. 2003a. Cytomegalovirus and Epstein–Barr virus are associated with symptomatic periapical pathosis. *Oral Microbiol Immunol* 18(5): 327–28.

Sabeti M and Slots J. 2004. Herpesviral–bacterial coinfection in periapical pathosis. *J Endod* 30(2): 69–72.

Sabeti M, Valles Y, Nowzari H, Simon JH, Kermani-Arab V, and Slots J. 2003b. Cytomegalovirus and Epstein–Barr virus DNA transcription in endodontic symptomatic lesions. *Oral Microbiol Immunol* 18(2): 104–8.

Saito D, de Toledo Leonardo R, Rodrigues JLM, Tsai SM, Hofling JF, and Gonçalves RB. 2006. Identification of bacteria in endodontic infections by sequence analysis of 16S rDNA clone libraries. *J Med Microbiol* 55(1): 101–7.

Sakamoto M, Huang Y, Umeda M, Ishikawa I, and Benno Y. 2002. Detection of novel oral phylotypes associated with periodontitis. *FEMS Microbiol Lett* 217(1): 65–69.

Sakamoto M, Rôças IN, Siqueira JF, Jr, and Benno Y. 2006. Molecular analysis of bacteria in asymptomatic and symptomatic endodontic infections. *Oral Microbiol Immunol* 21(2): 112–22.

Sakamoto M, Siqueira JF, Jr, Rôças IN, and Benno Y. 2007. Bacterial reduction and persistence after endodontic treatment procedures. *Oral Microbiol Immunol* 22(1): 19–23.

Sakamoto M, Siqueira JF, Jr, Rôças IN, and Benno Y. 2008. Molecular analysis of the root canal microbiota associated with endodontic treatment failures. *Oral Microbiol Immunol* 23(4): 275–81.

Sakamoto M, Umeda M, and Benno Y. 2005. Molecular analysis of human oral microbiota. *J Periodontal Res* 40(3): 277–85.

Sambrook J and Russell DW. 2001. *Molecular Cloning: A Laboratory Manual*. Cold Spring Harbor, NY: Cold Spring Harbor Laboratory Press.

Schena M, Shalon D, Davis RW, and Brown PO. 1995. Quantitative monitoring of gene expression patterns with a complementary DNA microarray. *Science* 270(5235): 467–70.

Sedgley C, Nagel A, Dahlen G, Reit C, and Molander A. 2006. Real-time quantitative polymerase chain reaction and culture analyses of *Enterococcus faecalis* in root canals. *J Endod* 32: 173–77.

Seol JH, Cho BH, Chung CP, and Bae KS. 2006. Multiplex polymerase chain reaction detection of black-pigmented bacteria in infections of endodontic origin. *J Endod* 32(2): 110–14.

Sharma S, Radl V, Hai B, Kloos K, Mrkonjic Fuka M, Engel M, Schauss K, and Schloter M. 2007. Quantification of functional genes from procaryotes in soil by PCR. *J Microbiol Methods* 68(3): 445–52.

Siqueira JF, Jr, Jung IY, Rôças IN, and Lee CY. 2005a. Differences in prevalence of selected bacterial species in primary endodontic infections from two distinct geographic locations. *Oral Surg Oral Med Oral Pathol Oral Radiol Endod* 99(5): 641–47.

Siqueira JF, Jr, and Lopes HP. 1999. Mechanisms of antimicrobial activity of calcium hydroxide: a critical review. *Int Endod J* 32(5): 361–69.

Siqueira JF, Jr, and Rôças IN. 2002. *Dialister pneumosintes* can be a suspected endodontic pathogen. *Oral Surg Oral Med Oral Pathol Oral Radiol Endod* 94(4): 494–98.

Siqueira JF, Jr, and Rôças IN. 2003a. *Bacteroides forsythus* in primary endodontic infections as detected by nested PCR. *J Endod* 29(6): 390–93.

Siqueira JF, Jr, and Rôças IN. 2003b. Detection of *Filifactor alocis* in endodontic infections associated with different forms of periradicular diseases. *Oral Microbiol Immunol* 18(4): 263–65.

Siqueira JF, Jr, and Rôças IN. 2003c. PCR-based identification of *Treponema maltophilum*, *T amylovorum*, *T medium*, and *T lecithinolyticum* in primary root canal infections. *Arch Oral Biol* 48(7): 495–502.

Siqueira JF, Jr, and Rôças IN. 2003d. PCR methodology as a valuable tool for identification of endodontic pathogens. *J Dent* 31(5): 333–39.

Siqueira JF, Jr, and Rôças IN. 2003e. Positive and negative bacterial associations involving *Dialister pneumosintes* in primary endodontic infections. *J Endod* 29(7): 438–41.

Siqueira JF, Jr, and Rôças IN. 2003f. *Pseudoramibacter alactolyticus* in primary endodontic infections. *J Endod* 29(11): 735–38.

Siqueira JF, Jr, and Rôças IN. 2003g. Polymerase chain reaction detection of *Propionibacterium propionicus* and *Actinomyces radicidentis* in primary and persistent endodontic infections. *Oral Surg Oral Med Oral Pathol Oral Radiol Endod* 96(2): 215–22.

Siqueira JF, Jr, and Rôças IN. 2004a. Nested PCR detection of *Centipeda periodontii* in primary endodontic infections. *J Endod* 30(3): 135–37.

Siqueira JF, Jr, and Rôças IN. 2004b. Polymerase chain reaction-based analysis of microorganisms associated with failed endodontic treatment. *Oral Surg Oral Med Oral Pathol Oral Radiol Endod* 97(1): 85–94.

Siqueira JF, Jr, and Rôças IN. 2004c. *Treponema* species associated with abscesses of endodontic origin. *Oral Microbiol Immunol* 19(5): 336–39.

Siqueira JF, Jr, and Rôças IN. 2005a. Exploiting molecular methods to explore endodontic infections: Part 1—current molecular technologies for microbiological diagnosis. *J Endod* 31(6): 411–23.

Siqueira JF, Jr, and Rôças IN. 2005b. Exploiting molecular methods to explore endodontic infections: Part 2—redefining the endodontic microbiota. *J Endod* 31(7): 488–98.

Siqueira JF, Jr, and Rôças IN. 2005c. Uncultivated phylotypes and newly named species associated with primary and persistent endodontic infections. *J Clin Microbiol* 43(7): 3314–19.

Siqueira JF, Jr, and Rôças IN. 2006. *Catonella morbi* and *Granulicatella adiacens*: new species in endodontic infections. *Oral Surg Oral Med Oral Pathol Oral Radiol Endod* 102: 259–64.

Siqueira JF, Jr, Rôças IN, Alves FR, and Santos KR. 2004a. Selected endodontic pathogens in the apical third of infected root canals: a molecular investigation. *J Endod* 30(9): 638–43.

Siqueira JF, Jr, Rôças IN, Andrade AF, and de Uzeda M. 2003. *Peptostreptococcus micros* in primary endodontic infections as detected by 16S rDNA-based polymerase chain reaction. *J Endod* 29(2): 111–13.

Siqueira JF, Jr, Rôças IN, Baumgartner JC, and Xia T. 2005b. Searching for Archaea in infections of endodontic origin. *J Endod* 31(10): 719–22.

Siqueira JF, Jr, Rôças IN, Cunha CD, and Rosado AS. 2005c. Novel bacterial phylotypes in endodontic infections. *J Dent Res* 84(6): 565–69.

Siqueira JF, Jr, Rôças IN, de Uzeda M, Colombo AP, and Santos KR. 2002a. Comparison of 16S rDNA-based PCR and checkerboard DNA–DNA hybridisation for detection of selected endodontic pathogens. *J Med Microbiol* 51(12): 1090–96.

Siqueira JF, Jr, Rôças IN, Favieri A, Oliveira JC, and Santos KR. 2001a. Polymerase chain reaction detection of *Treponema denticola* in endodontic infections within root canals. *Int Endod J* 34(4): 280–84.

Siqueira JF, Jr, Rôças IN, Favieri A, and Santos KR. 2000a. Detection of *Treponema denticola* in endodontic infections by 16S rRNA gene-directed polymerase chain reaction. *Oral Microbiol Immunol* 15(5): 335–37.

Siqueira JF, Jr, Rôças IN, Moraes SR, and Santos KR. 2002b. Direct amplification of rRNA gene sequences for identification of selected oral pathogens in root canal infections. *Int Endod J* 35(4): 345–51.

Siqueira JF, Jr, Rôças IN, Oliveira JC, and Santos KR. 2001b. Molecular detection of black-pigmented bacteria in infections of endodontic origin. *J Endod* 27(9): 563–66.

Siqueira JF, Jr, Rôças IN, and Rosado AS. 2004b. Investigation of bacterial communities associated with asymptomatic and symptomatic endodontic infections by denaturing gradient gel electrophoresis fingerprinting approach. *Oral Microbiol Immunol* 19(6): 363–70.

Siqueira JF, Jr, Rôças IN, and Rosado AS. 2005d. Application of denaturing gradient gel electrophoresis (DGGE) to the analysis of endodontic infections. *J Endod* 31(11): 775–82.

Siqueira JF, Jr, Rôças IN, Souto R, de Uzeda M, and Colombo AP. 2000b. Checkerboard DNA–DNA hybridization analysis of endodontic infections. *Oral Surg Oral Med Oral Pathol Oral Radiol Endod* 89(6): 744–48.

Siqueira JF, Jr, Rôças IN, Souto R, de Uzeda M, and Colombo AP. 2002c. *Actinomyces* species, streptococci, and *Enterococcus faecalis* in primary root canal infections. *J Endod* 28(3): 168–72.

Siqueira JF, Jr, Rôças IN, Souto R, Uzeda M, and Colombo AP. 2001c. Microbiological evaluation of acute periradicular abscesses by DNA–DNA hybridization. *Oral Surg Oral Med Oral Pathol Oral Radiol Endod* 92(4): 451–57.

Sjögren U, Figdor D, Persson S, and Sundqvist G. 1997. Influence of infection at the time of root filling on the outcome of endodontic treatment of teeth with apical periodontitis. *Int Endod J* 30(5): 297–306.

Slots J. 1986. Rapid identification of important periodontal microorganisms by cultivation. *Oral Microbiol Immunol* 1(1): 48–57.

Slots J. 2005. Herpesviruses in periodontal diseases. *Periodontol 2000* 38: 33–62.

Socransky SS, Gibbons RJ, Dale AC, Bortnick L, Rosenthal E, and MacDonald JB. 1963. The microbiota of the gingival crevice in man. 1. Total microscopic and viable counts and counts of specific organisms. *Arch Oral Biol* 8: 275–80.

Socransky SS, Haffajee AD, Cugini MA, Smith C, and Kent RL, Jr. 1998. Microbial complexes in subgingival plaque. *J Clin Periodontol* 25(2): 134–44.

Socransky SS, Smith C, Martin L, Paster BJ, Dewhirst FE, and Levin AE. 1994. "Checkerboard" DNA–DNA hybridization. *Biotechniques* 17(4): 788–92.

Song Y, Liu C, McTeague M, and Finegold SM. 2003. 16S ribosomal DNA sequence-based analysis of clinically significant Gram-positive anaerobic cocci. *J Clin Microbiol* 41: 1363–69.

Spangberg LS. 2006. Infatuated by Enterococci. *Oral Surg Oral Med Oral Pathol Oral Radiol Endod* 102(5): 577–78.

Spiegelman D, Whissell G, and Greer CW. 2005. A survey of the methods for the characterization of microbial consortia and communities. *Can J Microbiol* 51(5): 355–86.

Stanley HR. 1977. Importance of the leukocyte to dental and pulpal health. *J Endod* 3(9): 334–41.

Stevenson BS, Eichorst SA, Wertz JT, Schmidt TM, and Breznak JA. 2004. New strategies for cultivation and detection of previously uncultured microbes. *Appl Environ Microbiol* 70(8): 4748–55.

Strous M. 2007. Data storm. *Environ Microbiol* 9(1): 10–11.

Suau A, Bonnet R, Sutren M, Godon JJ, Gibson GR, Collins MD, and Dore J. 1999. Direct analysis of genes encoding 16S rRNA from complex communities reveals many novel molecular species within the human gut. *Appl Environ Microbiol* 65(11): 4799–807.

Sunde PT, Olsen I, Gobel UB, Theegarten D, Winter S, Debelian GJ, Tronstad L, and Moter A. 2003. Fluorescence in situ hybridization (FISH) for direct visualization of bacteria in periapical lesions of asymptomatic root-filled teeth. *Microbiology* 149(Pt 5): 1095–102.

Sunde PT, Tronstad L, Eribe ER, Lind PO, and Olsen I. 2000. Assessment of periradicular microbiota by DNA–DNA hybridization. *Endod Dent Traumatol* 16(5): 191–96.

Sundqvist G. 1992. Associations between microbial species in dental root canal infections. *Oral Microbiol Immunol* 7(5): 257–62.

Sundqvist G and Figdor D. 2003. Life as an endodontic pathogen. Ecological differences between the untreated and root-filled root canals. *Endod Top* 6(1): 3–28.

Sundqvist G, Figdor D, Persson S, and Sjogren U. 1998. Microbiologic analysis of teeth with failed endodontic treatment and the outcome of conservative re-treatment. *Oral Surg Oral Med Oral Pathol Oral Radiol Endod* 85(1): 86–93.

Sundqvist G, Johansson E, and Sjogren U. 1989. Prevalence of black-pigmented bacteroides species in root canal infections. *J Endod* 15(1): 13–19.

Tang Y-W, Procop GW, and Persing DH. 1997. Molecular diagnostics of infectious diseases. *Clin Chem* 43: 2021–38.

Tang YW, Ellis NM, Hopkins MK, Smith DH, Dodge DE, and Persing DH. 1998. Comparison of phenotypic and genotypic techniques for identification of unusual aerobic pathogenic Gram-negative bacilli. *J Clin Microbiol* 36: 3674–79.

Tanner A, Lai C-H, and Maiden M. 1992. Characteristics of oral Gram-negative species. In: Slots J and Taubman MA (eds), *Contemporary Oral Microbiology and Immunology*. St Louis, MO: Mosby, 299–341.

Tardif G, Sulavik MC, Jones GW, and Clewell DB. 1989. Spontaneous switching of the sucrose-promoted colony phenotype in *Streptococcus sanguis*. *Infect Immun* 57: 3945–48.

Tatakis DN and Kumar PS. 2005. Etiology and pathogenesis of periodontal diseases. *Dent Clin North Am* 49(3): 491–516.

Theron J and Cloete TE. 2000. Molecular techniques for determining microbial diversity and community structure in natural environments. *Crit Rev Microbiol* 26(1): 37–57.

Tyagi S and Kramer FR. 1996. Molecular beacons: probes that fluoresce upon hybridization. *Nat Biotechnol* 14(3): 303–8.

van Winkelhoff AJ, Carlee AW, and de Graaff J. 1985. *Bacteroides endodontalis* and others black-pigmented *Bacteroides* species in odontogenic abscesses. *Infect Immun* 49: 494–98.

Vianna ME, Conrads G, Gomes BPFA, and Horz HP. 2006a. Identification and quantification of archaea involved in primary endodontic infections. *J Clin Microbiol* 44: 1274–82.

Vianna ME, Horz HP, Gomes BP, and Conrads G. 2005. Microarrays complement culture methods for identification of bacteria in endodontic infections. *Oral Microbiol Immunol* 20(4): 253–58.

Vianna ME, Horz HP, Gomes BP, and Conrads G. 2006b. *In vivo* evaluation of microbial reduction after chemomechanical preparation of human root canals containing necrotic pulp tissue. *Int Endod J* 39: 484–92.

Vickerman MM, Brossard KA, Funk DB, Jesionowski AM, and Gill SR. 2007. Phylogenetic analysis of bacterial and archaeal species in symptomatic and asymptomatic endodontic infections. *J Med Microbiol* 56(Pt 1): 110–18.

Wade W. 2002. Unculturable bacteria—the uncharacterized organisms that cause oral infections. *J R Soc Med* 95(2): 81–83.

Wade WG. 2004. Non-culturable bacteria in complex commensal populations. *Adv Appl Microbiol* 54: 93–106.

Waltimo T, Trope M, Haapasalo M, and Orstavik D. 2005. Clinical efficacy of treatment procedures in endodontic infection control and one year follow-up of periapical healing. *J Endod* 31(12): 863–66.

Wang RF, Cao WW, and Cerniglia CE. 1996. PCR detection and quantitation of predominant anaerobic bacteria in human and animal fecal samples. *Appl Environ Microbiol* 62(4): 1242–47.

Ward DM, Weller R, and Bateson MM. 1990. 16S rRNA sequences reveal numerous uncultured microorganisms in a natural community. *Nature* 345(6270): 63–65.

Watts A and Paterson C. 1990. Detection of bacteria in histological sections of the dental pulp. *Int Endod J* 23(1): 1–12.

Welsh J and McClelland M. 1990. Fingerprinting genomes using PCR with arbitrary primers. *Nucleic Acids Res* 18(24): 7213–18.

Whelen AC and Persing DH. 1996. The role of nucleic acid amplification and detection in the clinical microbiology laboratory. *Annu Rev Microbiol* 50: 349–73.

Wicher K, Abbruscato F, Wicher V, Collins DN, Auger I, and Horowitz HW. 1998. Identification of persistent infection in experimental syphilis by PCR. *Infect Immun* 66(6): 2509–13.

Williams JM, Trope M, Caplan DJ, and Shugars DC. 2006. Detection and quantitation of *Enterococcus faecalis* by real-time PCR (qPCR), reverse transcription-PCR (RT-PCR), and cultivation during endodontic treatment. *J Endod* 32(8): 715–21.

Woese CR. 1987. Bacterial evolution. *Microbiol Rev* 51(2): 221–71.

Woese CR. 2000. Interpreting the universal phylogenetic tree. *Proc Natl Acad Sci USA* 97(15): 8392–96.

Xia T, Baumgartner JC, and David LL. 2000. Isolation and identification of *Prevotella tannerae* from endodontic infections. *Oral Microbiol Immunol* 15(4): 273–75.

Xu J. 2006. Microbial ecology in the age of genomics and metagenomics: concepts, tools, and recent advances. *Mol Ecol* 15(7): 1713–31.

Zambon JJ and Haraszthy VI. 1995. The laboratory diagnosis of periodontal infections. *Periodontol 2000* 7: 69–82.

Zhang K, Martiny AC, Reppas NB, Barry KW, Malek J, Chisholm SW, and Church GM. 2006. Sequencing genomes from single cells by polymerase cloning. *Nat Biotechnol* 24: 680–86.

Zoletti GO, Siqueira JF, Jr, and Santos KR. 2006. Identification of *Enterococcus faecalis* in root-filled teeth with or without periradicular lesions by culture-dependent and independent approaches. *J Endod* 32(8): 722–26.

# Chapter 6

# Extraradicular Endodontic Infections

*Leif Tronstad*

## 6.1 Introduction

The importance of bacteria as an etiologic factor for pulpal and periapical inflammation as expressed in the literature has varied over the years. However, presently it is understood and accepted that when the root canal is opened to the mouth, usually by caries or trauma, the pulp will become inflamed and necrotize as a result of bacteria entering the root canal from the oral cavity (see Chapters 3 and 4). With time, the inflammation will spread from the pulp through apical foramina and lateral canals beyond the apex of the tooth. At first, only the periodontal ligament will be involved in the periapical reaction. However, subsequently there will be resorption of root cementum, dentin, and alveolar bone so that all tissues of the periodontium will be affected. This condition is termed periradicular, periapical or, most commonly, apical periodontitis. The inflammatory process may lead to considerable bone loss, sometimes encompassing large areas of the alveolar process. Quite commonly a fistulous tract develops from the inflamed area to a body surface, in most instances to the oral vestibule (Tronstad 2008).

The prevalence of apical periodontitis in 35- to 45-year-olds is found to be 30–40% and is increasing with increasing age (Eriksen et al. 2002; Friedman 2002). Based on a comparison of results of studies from several European countries, it seems that apical periodontitis is more prevalent than severe marginal periodontitis (Eriksen et al. 2002; Skudutyte-Rysstad and Eriksen 2006). The disease, therefore, is of concern both for the individual, the dental profession, and the society.

## 6.2 Classical view of endodontic infections

Traditionally, it has been held that in teeth, with asymptomatic apical periodontitis, the infecting microorganisms are harbored in the root canal system and tubules of the root dentin, whereas the periapical lesion is free of bacteria (Andreasen and Rud 1972; Langeland et al. 1977; Nair et al. 1990; Sundqvist et al. 1998). It has been assumed that the defense systems mobilized by the periapical inflammation will eliminate the bacteria from the root canal which invade the periapex, and the well-known quote by Kronfeld (1939) that "a granuloma is not an area in which bacteria live, but a place in which they are destroyed" is still widely considered to be valid. Thus, the objective of treating a tooth with apical periodontitis is to eliminate the infection of the root canal system and the root, clean and prepare the root canal, and fill the canal bacteria tight so that it will not be reinfected (Tronstad 2008). Periapical repair will then normally occur. The inflammation subsides, new cementum forms in the resorptive areas of the root, and new bone develops where the bone was resorbed. In a period of 3–24 months, the apical periodontium is usually restored (Strindberg 1956; Kerekes and Tronstad 1979; Sjögren et al. 1997) (Fig. 6.1).

When a periapical lesion resolves following successful elimination of the root canal infection, it is rather impossible to ascertain whether the lesion in fact were free of bacteria or not prior to the treatment. Still, it is not unlikely that bacteria were present in the lesion, but without having established themselves as the cause of the periapical inflammation. Rather, they were in the tissue transiently until they were killed by the host defense systems. However, new bacteria will be released from biofilms in the root canal system more or less continuously, and, under the right conditions and with the right bacteria present, foci of extraradicular infection may be established on a permanent basis in persistent periapical lesions (Tronstad and Sunde 2003).

## 6.3 Exacerbations

In endodontics, an exacerbation is a flare-up of an asymptomatic apical periodontitis, resulting in a periapical abscess. It may occur spontaneously or

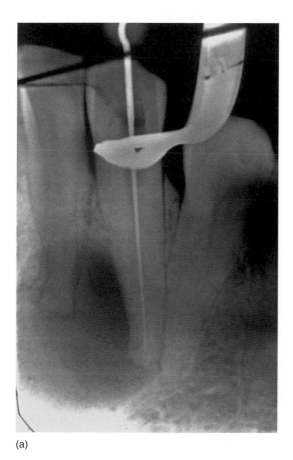

(a)

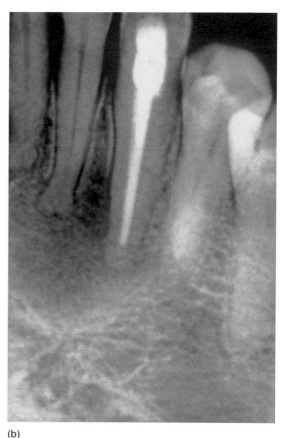

(b)

**Fig. 6.1** (a) Radiograph of mandibular canine with large periapical radiolucent lesion. The tooth is nonvital and the diagnosis is asymptomatic apical periodontitis. Therapy is nonsurgical endodontic treatment. (b) Postoperative radiograph after 6 months. Periapical healing is evident and the periapical lesion has filled in nicely with new bone.

following instrumentation of the root canal, and may be localized to the area of the infected tooth or be diffused with the form of a facial cellulitis (Tronstad 2008). A prerequisite for the development of an exacerbation is the introduction of bacteria from the root canal into the periapical tissues. The bacteria must have the means to enter and invade the periapex, and they have to be in sufficient numbers and have the capacity to overwhelm the local host defense. The infection is polymicrobial and is typically dominated by anaerobic organisms from the root canal (see Chapter 3). There is some evidence that the presence of black-pigmented anaerobes such as *Porphyromonas gingivalis*, *Porphyromonas endodontalis*, and *Prevotella intermedia* together with *Peptostreptococcus* spp. in the root canal flora favors the development of periapical abscesses (Sundqvist et al. 1979; Oguntebi et al. 1982; van Winkelhoff et al. 1985; Yoshida et al. 1987; Gomes et al. 1996). However, this information derives from culture-based studies that may not fully cover the complexity of the infection (see Chapter 4).

The most important cause of the symptoms in conjunction with an exacerbation is elevated pressure in the tissues. The objective of the treatment should therefore be to obtain drainage of pus through the root canal and/or by incision of a fluctuant abscess (Fig. 6.2). Normally, this results in relief from the pain. However, additional antibiotic therapy may be indicated if drainage of pus is not achieved, if the patient's general health is poor, and for any dramatic spreading of the infectious process (see Chapter 10). When the acute symptoms have subsided, the root canal treatment of the tooth continues in the normal fashion. The extraradicular bacteria apparently are eliminated, and it is not to be expected that an exacerbation has a negative influence on the final outcome of endodontic treatment (Fig. 6.2).

## 6.4 Success rate of endodontic treatment

The success rate of endodontic treatment is excellent (see Chapter 17). For nonvital teeth with root canal infection, but without radiographically visible apical periodontitis, a success rate of 90–95% is within reach. This is the same success rate as for pulpectomy treatment of vital teeth without root canal infection (Kerekes and Tronstad 1979; Sjögren et al. 1990). However, the success rate of teeth with apical periodontitis is reported to be 75–80%, or about 20%

lower than for vital teeth or nonvital teeth without periapical radiolucencies (Strindberg 1956; Kerekes and Tronstad 1979).

Thus, apical periodontitis strongly influences the outcome of endodontic treatment, and about 2 of every 10 teeth with this diagnosis will not respond to accepted therapy including effective antibacterial treatment of the root canal system and the root. The nonresponding teeth are usually referred to as treatment-resistant or refractory cases, and it has been claimed that when they do not respond to treatment it is because that not all roots can be rendered bacteria free (Byström et al. 1985; Sjögren et al. 1997). This may well be, but the fact that almost all teeth without apical periodontitis, but with a comparable root canal infection respond to treatment, suggests that this is not the main reason. Rather one must suspect that the nonresponding periapical lesions harbor bacteria that have established themselves within the lesions, conceivably as biofilm-like structures, maintaining the apical periodontitis (Tronstad et al. 1987). Clinically, this is supported by the observation that occasionally refractory cases may heal following systemic antibiotic therapy without further treatment of the teeth (Tronstad and Petersson 1995; Tronstad 2008).

## 6.5 Extraradicular endodontic infections

### 6.5.1 Results of culture-based studies

Over the years microbiological sampling has now and again been performed in the treatment of refractory apical periodontitis in our clinics. Our findings show that when proper endodontic treatment is ineffective, microorganisms are usually recovered from fistulous tracts or from surgically removed periapical lesions. The bacteria seemingly maintain the infectious disease process extraradicularly, independent of the root canal and the root (Fig. 6.3). In controlled studies, Happonen et al. (1985; Happonen 1986) by means of immunocytochemical methods have demonstrated the presence of *Actinomyces* spp. and *Propionibacterium propionicum* in asymptomatic periapical lesions refractory to endodontic treatment. Further studies have confirmed that these bacteria may invade and live in the granulation tissue outside the root (O'Grady and Reade 1988; Sjögren et al. 1988). Our group demonstrated with anaerobic cultivation that anaerobic and facultative anaerobic bacteria are able to survive in

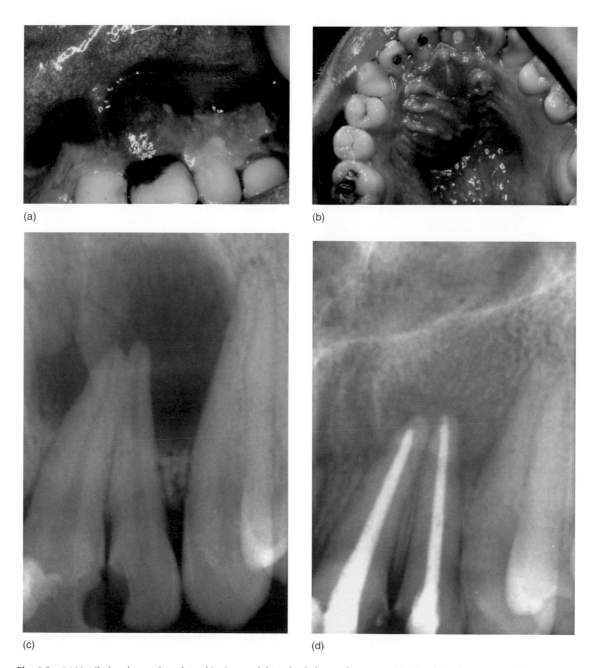

**Fig. 6.2** (a) Vestibular abscess from lateral incisor and (b) palatal abscess from central incisor following unattended traumatic injury to the teeth. Both teeth are nonvital. (c) Radiograph showing large radiolucent lesions from both teeth. The diagnosis is apical periodontitis with abscess (both teeth) and the therapy is nonsurgical endodontic treatment of both teeth. Drainage of the palatal abscess was obtained through the root canal of the central incisor, and the patient was quickly asymptomatic. The root canal treatment of both teeth then followed in the normal manner. (d) Postoperative radiograph after 6 months showing healing of the periapical lesions of both teeth.

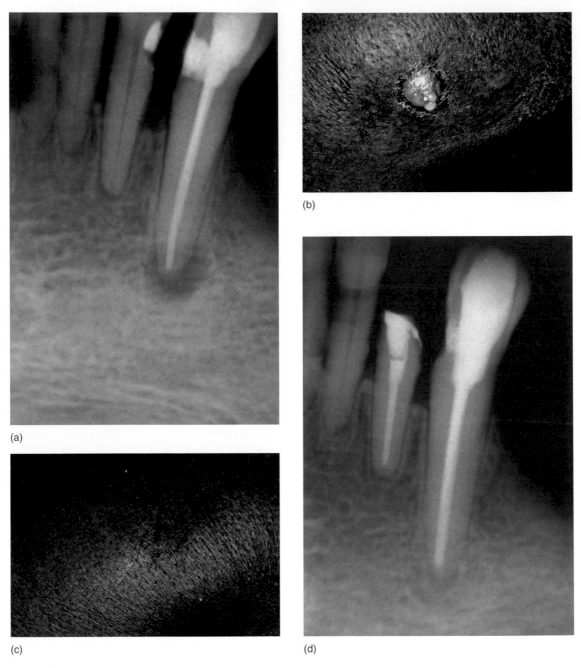

(a)

(b)

(c)

(d)

**Fig. 6.3**    (a) Postoperative radiograph of mandibular canine with small periapical radiolucent lesion. The diagnosis was asymptomatic apical periodontitis and the tooth was root filled after instrumentation and 3 weeks of calcium hydroxide in the canal. (b) After 5 weeks, the patient returned with an extraoral fistula on the chin coming from the canine. Nonsurgical retreatment was ineffective, and a microbial sample was taken from the fistulous tract. In accordance with the microbiological findings, the patient was given phenoxymethylpenicillin and metronidazole for 10 days. (c) The fistula closed during this period and the tooth was again root filled. (d) Final postoperative radiograph after 3 months. The periapical lesion has healed.

periapical inflammatory lesions (Tronstad et al. 1987, 1990a). The bacteria known from studies on root canal infection including *Porphyromonas* spp. and *Prevotella* spp. as well as Gram-positive rods and cocci were commonly found. In teeth where the endodontic therapy had been compromised and was grossly inadequate, for instance, after repeated but inadequate antibiotic treatment, multiple openings and closings of the root canal, or inadequate periapical surgical treatment, enteric and environmental bacteria and yeast were isolated. In such instances, organisms such as *Escherichia coli*, *Bacteroides fragilis*, *Pseudomonas aeruginosa*, *Enterobacter*, *Chlostridium*, *Proteus*, and *Klepsiella* species and yeast were recovered (Fig. 6.4). The presence of these and other mainly nonoral microorganisms may also suggest that blood-borne infection of the periapical lesion takes place.

These findings were generally supported in a number of subsequent cultivation studies (Iwy et al. 1990; Wasfy et al. 1992; Wayman et al. 1992; Vigil et al. 1997; Abou-Rass and Bogen 1998). Still, the results were met with skepticism and refusal to accept that multiple root canal bacteria might be able to invade, survive, and establish themselves in the periapical lesion, maintain the periapical inflammation, and be the cause of failure of endodontic treatment (Nair et al. 1990; Nair 1997; Sundqvist et al. 1998; Bergenholtz and Spångberg 2004). These authors consider the positive cultures obtained in the above studies to be due to poor case selection or contamination during microbial sampling, and maintain that the concept of the sterile apical granuloma still must be regarded as valid.

The question of extraradicular infection is of considerable clinical importance. Clinically, any spontaneous formation of a periapical abscess bears evidence that bacteria from the root canal are able to invade the periapical lesion when the conditions are right (van Winkelhoff et al. 1985). It is also well established that most microorganisms are able to adapt over time to live in many different environments, and that their numbers, rapid fluctuations, and amenability to genetic change give them effective tools for adaptation (Edwards 2000). They have a number of strategies for overcoming host innate and adaptive immune responses (Hornhef et al. 2002), and in fact can establish lifelong chronic infections in their hosts (Marshall and Warren 1984; Dunn et al. 1997; Kolltveit et al. 2002; Young at al 2002; da Silva et al. 2003, 2006) (Fig. 6.5). Also, recent evidence suggests the involvement of herpes viruses in the etiopathogenesis of apical periodon-

titis (see Chapter 8; Sabeti et al. 2003; Slots and Hames 2003). The viruses may cause the release of tissue destructive cytokines and the initiation of cytotoxic and immunopathologic events. The immune impairment and tissue changes resulting from the virus infection may then aid bacteria in invading and surviving in the periapical lesion.

However, the question of whether bacterial samples taken from surgically removed periapical lesions unavoidably will be contaminated by the indigenous oral flora is important, and was addressed in a methodological study (Sunde et al. 2000a). Thirty patients referred for surgical treatment of root-filled teeth with asymptomatic apical periodontitis were treated with apicoectomies using marginal and submarginal incisions. Before incision, the gingiva and mucosa were washed with 0.2% chlorhexidine gluconate. Bacterial samples were taken from the mucosa immediately before reflecting the flap, and from the exposed alveolar bone and the periapical lesion immediately after. All samples were cultured anaerobically on all-purpose and selected media.

In the marginal incision group, bacteria were recovered from the mucosa in 12 of 15 patients (80%). Bacterial growth was observed in all samples from the alveolar bone (100%) while the periapical lesions gave bacterial growth in 11 of 15 patients (73%). The microorganisms cultivated from the mucosa and the alveolar bone differed from the microorganisms recovered from the periapical lesions except in two patients where *Actinomyces odontolyticus* and *Veillonella* sp. were recovered from both the exposed alveolar bone and the periapical lesion. The biochemical/enzymatic profiles of these two species strains were identical, suggesting the possibility of either direct or circulatory translocation of the bacteria.

In the submarginal incision group, bacteria were cultured from the mucosa in 11 of 15 patients (73%). Three samples from the alveolar bone (20%) and 10 samples from the periapical lesions (67%) gave positive growth. The microorganisms cultivated from the mucosa and the exposed alveolar bone differed from the bacterial species recovered from the periapical lesions except in one patient where *Propionibacterium acnes* were cultured from both the exposed bone and the periapical lesion. However, the biochemical/enzymatic profiles of these strains were different, suggesting that a translocation had not occurred.

Thus, contamination of the periapical lesion during reflection of the flap and the microbial sampling

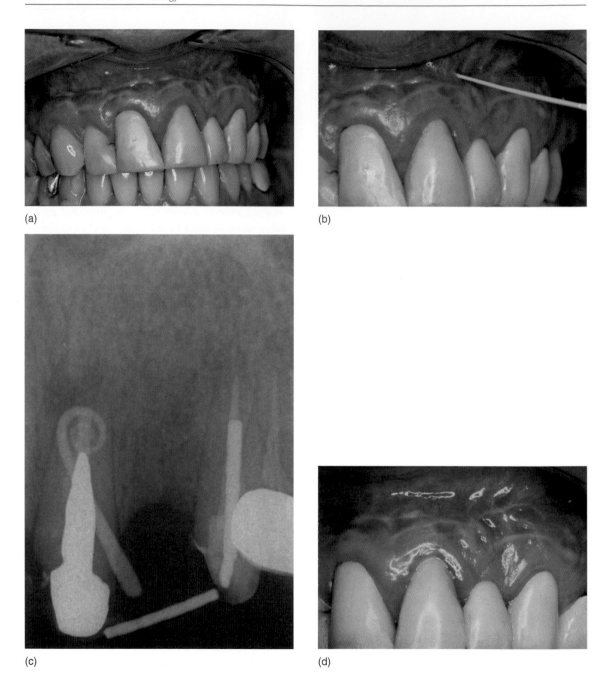

(a)

(b)

(c)

(d)

**Fig. 6.4**    (a) Patient with multiple vestibular scarring following six apicoectomies in the anterior region over an 8-year period. (b) A fistula in the region of the lateral incisor has persisted for the entire 8-year period. (c) Radiograph with gutta-percha cone in the fistulous tract that appears to originate from the central incisor. In addition to the apicoectomies, the lateral incisor has been extracted. (d) A microbial sample was taken from the fistulous tract. The cultivation results suggested that the infection was due to a pure culture of *Pseudomonas aeruginosa*. Ciprofloxacin for 4 weeks was prescribed and the fistula closed during this period and remained closed at the 2-year control.

**Fig. 6.5**  Scanning electron micrograph from stenotic lesion of aortic valve. Co-aggregating microorganisms with different morphologies are seen. Bacteria involved in endodontic infections have been recovered from lesions in the heart. (From Kolltveit et al. 2002.)

procedures was a very minor or rather a nonexisting problem in this study, suggesting that, when care is taken, meaningful bacterial samples may be obtained from the lesions during the surgical operation. Microorganisms known from studies of the root canal flora (da Silva et al. 2002) were recovered from the periapical lesions in 21 of 30 patients, and the results clearly supported the cultivation studies referred to above.

A finding that might be of some importance was the apparent translocation of bacteria into the surgical site when the incision was made through the sulcus where the bacteria are especially difficult to reach by oral antiseptics (Rindom-Schiøtt et al. 1970).

## 6.6 Molecular analyses of extraradicular infection

It is to be suspected that the extraradicular bacterial flora in refractory cases is more complex than that has been shown in the culture-based studies. Thus, 60% of the oral flora has been found to be uncultivable (Paster et al. 2006), and in studies of the root canal flora, many more bacterial species have been identified with molecular methods than in cultivation studies (Rolph et al. 2001; da Silva et al. 2002; Munson et al. 2002). Thus, a study on extraradicular endodontic infections was performed using the method developed by Socransky et al. (1994) for hybridizing large num-

bers of DNA samples against large numbers of DNA probes on a single support membrane, the so-called checkerboard DNA–DNA hybridization technique (Sunde et al. 2000b). This technique does not require bacterial viability although initially the probe bacteria must be cultured. It has higher sensitivity than culturing, and amplification of the bacterial DNA is not necessary for identification. Thirty-four patients referred for surgical treatment of root-filled teeth with asymptomatic apical periodontitis were treated with apicoectomies. In 17 patients, a marginal incision was made to expose the periapical lesion and in 17 patients a submarginal incision was used. The sampling procedures outlined in the previous methodological study were carefully followed (Sunde et al. 2000a). The 40 DNA probes used by Socransky et al. (1998) in their studies of the periodontal flora were used in this study as well.

Bacterial DNA was identified in all samples from the periapical lesions. Sterile transport medium without sample gave no hybridization signals. The number of species per lesion varied between 26 and 39 (mean 33.7 ± 3.3) in the marginal incision group and between 11 and 34 (mean 21.3 ± 6.3) in the submarginal group. Thus, significantly more bacterial species were found in the periapical lesions of treatment-resistant teeth with the DNA–DNA hybridization technique (11–39) than with cultivation (0–10). Interestingly, the number between 11 and 39, or on the average 20–30 species, coincides well with the number of microorganisms found in biofilms in the healthy oral cavity (Aas et al. 2005), in infected root canals (da Silva et al. 2002), and subgingivally in patients with active periodontal disease (Socransky et al. 1998). It was also confirmed that more bacteria are found when the sulcus is included in the flap, suggesting that a direct and/or circulatory translocation of bacteria to the surgical site may occur with this surgical approach.

In the submarginal incision group, DNA of *Fusobacterium nucleatum*, *Peptococcus micros*, *Actinomyces israelii*, *Actinomyces viscosus*, *Campylobacter rectus*, *P. gingivalis*, *Tannerella forsythia*, *Treponema socranskii* ssp. *socranskii*, *Aggregatibacter actinomycetemcomitans*, and *Streptococcus* spp. was present in more than 70% of the lesions (Fig. 6.6). Other designated endodontopathogens such as *Treponema denticola* (60%), *A. odontolyticus* (50%), *P. endodontalis* (50%), and *Eikenella corrodens* (40%) were commonly present. In 20–30% of the lesions, *Campylobacter gracilis*, *Eubacterium nodatum*,

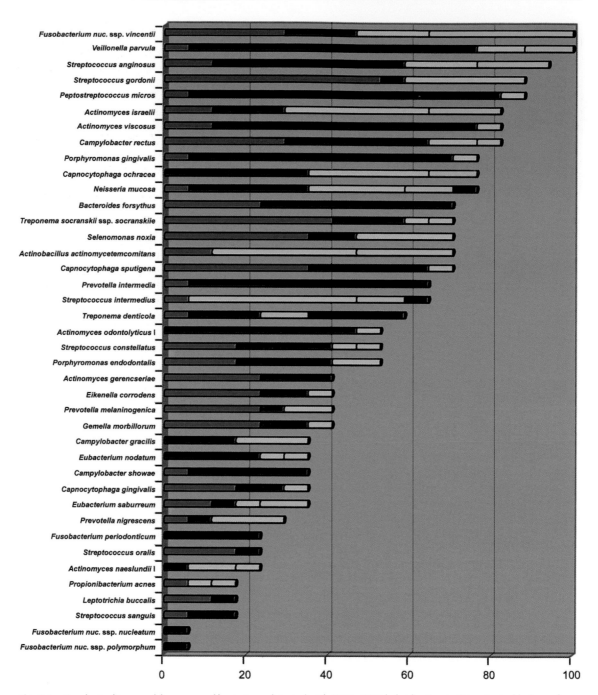

**Fig. 6.6**    Bar chart of type and frequency of bacteria as detected with DNA–DNA hybridization in 17 periapical lesions of asymptomatic teeth following submarginal incision and sampling from lesions. The length of the bars indicates the percentage of lesions colonized. (From Sunde et al. 2000b.)

*Campylobacter showae*, *Prevotella nigrescens*, and *Actinomyces naeslundii* were found. *Streptococcus*, *Enterococcus*, and *Staphylococcus* species were present as well. These findings were confirmed by the results of a parallel checkerboard DNA–DNA hybridization study carried out by a different group using the same study design and the same 40 DNA probes (Gatti et al. 2000).

The association of bacteria in mixed infections is not random. With regard to oral bacteria, this is best known from studies on dental plaque where six closely associated groups or complexes of bacterial species are recognized (Socransky et al. 1998). A "red complex" consisting of *T. forsythia*, *P. gingivalis*, and *T. denticola* is one group found in patients with active disease. An "orange complex" including *C. rectus*, *C. showae*, *E. nodatum*, *F. nucleatum*, *P. intermedia*, *P. nigrescens*, and *Parvimonas micra* is important for disease progression as well. A third, "blue group" includes the *Actinomyces*, and a fourth, "green group" consists of *Capnocytophaga* species, *A. actinomycetemcomitans*, *E. corrodens*, and *C. concisus*. The streptococci make up the fifth, "yellow group," and the sixth, "purple group" comprises *Veillonella parvula* and *A. odontolyticus*. Clearly, the periapical lesion is different from the periodontal pocket, and at present it is not known whether the findings in subgingival plaque are totally valid for endodontic infections. However, it should be noted that the red complex bacteria were together present in 60% of the periapical lesions. Also, *F. nucleatum* which is known to play an important role in established root canal infections (Sundqvist 1992) was present in 100% of the periapical lesions. This species co-aggregates with most oral bacteria, including strains of *P. gingivalis*, *T. denticola*, *A. actinomycetemcomitans*, *P. intermedia*, *Eubacterium* species, *Selenomonas* species, and, importantly, *Actinomyces* species. The important red and orange complexes contain most of these co-aggregating species. *P. endodontalis*, which in combination with other root canal bacteria is known to cause transmissible infection in guinea pigs (Sundqvist et al. 1979), was present in 50% of the lesions. Interestingly, combinations of the same bacteria, but without *P. endodontalis*, did not cause transmissible infection in the same experiments. Still, these and other noninfective bacteria may play an important role in maintaining the infection by providing nutrients for the principal pathogens and by synthesizing and degrading extracellular polysaccharides in the biofilms. *A. actinomycetemcomitans* is a designated periodontal pathogen (World Workshop on Periodontology 1996), and in the era of culture-based studies it was not regarded as being part of the endodontic flora (Sundqvist 1994). However, in a study using the checkerboard DNA–DNA hybridization technique, *A. actinomycetemcomitans* was recovered from more than 90% of the root canals (da Silva et al. 2002). This microorganism possesses a number of virulence factors, including a leukotoxin that targets and destroys host immune cells (Hendersen et al. 2003). The organism takes advantage of its high adhesiveness and is capable of rapid invasion and spread through eukaryotic cells (Meyer et al. 1996). It has been isolated from several nonoral infections including brain abscess (Renton et al. 1996), endocarditis (Brouqui and Raoult 2001), pericarditis (Horowitz et al. 1987), and from aortic aneurysms (da Silva et al. 2005). Thus, it is not surprising that *A. actinomycetemcomitans* is isolated from most of the periapical lesions in this study. From the molecular studies, it appears that the microbiotas of infected root canals, periapical lesions of treatment-resistant teeth, and periodontal pockets of patients with active periodontal disease are very similar and much more so than what was understood in the era of cultivation studies (Socransky et al. 1998; Ximenez-Fyvie et al. 2000, 2006; Gatti et al. 2000; Rupf et al. 2000; da Silva et al. 2002; Sunde et al. 2000b).

## 6.7 Visualization of extraradicular microorganisms

Certain clinical observations, for instance, extraradicular infections most often do not respond to systemic antibiotic therapy, suggest that the bacteria may have formed biofilms in the granulomatous tissue (Costerton et al. 1999; Stewart and Costerton 2001). It should then be possible to visualize the extraradicular bacteria, and a number of studies have been performed for this purpose.

## 6.8 Biofilm on root surfaces

Since biofilm preferably forms onto a solid surface, an obvious location to look for extraradicular biofilm would seem to be root tips included in periapical lesions (Tronstad et al. 1990a). Ten root tips that were removed during surgical treatment of root-filled teeth with periapical lesions were examined, five teeth with the diagnosis asymptomatic apical periodontitis

**Fig. 6.7**   Scanning electron micrograph of bacterial biofilm on surface of root tip within periapical lesion of root-filled tooth with asymptomatic apical periodontitis. The biofilm is dominated by cocci and short rods in an extracellular matrix. (From Tronstad et al. 1990b.)

**Fig. 6.8**   Scanning electron micrograph of bacterial biofilm adjacent to apical foramen of root-filled tooth with asymptomatic apical periodontitis. Bacterial colonies are recognized within smooth and structureless extracellular material. (From Tronstad et al. 1990b.)

and five teeth with the diagnosis apical periodontitis with fistula. To the naked eye, the surgically removed root tips appeared denuded. When examined in the scanning electron microscope, they were covered by soft tissue with fibers and cells in various stages of degradation. A bacterial plaque or biofilm was seen in areas of the root surfaces between fibers and cells and in crypts and holes (Fig. 6.7). The biofilms contained varying amounts of an amorphous extracellular material, sometimes making it difficult to distinguish individual bacterial cells. Still, it could be seen that the biofilm was dominated by cocci and short rods. Filamentous and fibrillar forms were recognized as well.

An additional conspicuous finding was a smooth, structureless coating or layer at the apex of the root tip, seemingly outside the apical foramen that was not visible in any of the specimens (Fig. 6.8). This smooth layer that gave the appearance of burned sugar was seen in nine of the ten specimens. It was interpreted as extracellular material of a biofilm, since at higher magnification, a variety of bacterial forms were recognized in the smooth material. The bacteria had formed colonies and were embedded in the structureless film.

Thus, bacteria were observed on the outer surfaces of all root tips studied. The bacteria were well established, and as anticipated, had formed biofilms in many areas of the apical root surfaces. The biofilm provides a number of advantages to the colonizing bacteria. It furthers the development of an appropriate physicochemical environment for the various bacteria and facilitates processing and uptake of nutrients, cross-

feeding, and removal of potentially harmful metabolic products (Costerton 1999). A further advantage of the biofilm for the bacterial populations is that it offers its residents protection from competing microorganisms and from environmental factors such as host defense mechanisms and potentially toxic substances such as antiseptics and antibiotics (Gilbert et al. 1997; Donlan and Costerton 2002). Also, the biofilm gives the colonizing species an opportunity for genetic exchange (Hausner and Wuertz 1999). By appearance, two types of biofilm were noted on the surfaces of the root tips, and it may be speculated that especially the smooth, structureless material of the biofilm outside the apical foramen represents an extracellular material that might act as a highly effective diffusion barrier, for instance, against an antibiotic (Hoyle et al. 1990; Gilbert et al. 1997). Clearly, the observed biofilms enable extraradicular bacteria to survive and live outside the tooth, and more than likely, the biofilms are important for maintaining the periapical inflammatory process. Interestingly, no differences were found between the teeth with or without fistulas.

## 6.9 Biofilm granules in granulomatous tissue

Many oral bacterial species possess surface structures such as fimbriae and fibrils that aid in their attachment to a wide variety of surfaces including soft tissue components like cells and fibers and even to other bacteria (Socransky and Hafajee 2002). Therefore, biofilms may develop in periapical lesions of refractory teeth independent of the solid surfaces of the root tip. In this context, it seemed important that in cases of cervicofacial actinomycosis granules containing *A. israelii* have been observed (Thoma 1963; Najjar 2006). These granules often have a bright, yellow color, and because of this, in older literature are referred to as *sulfur granules*. It was felt that the sulfur granules, in fact, might represent bacteria in a biofilm setting unattached to a *solid* surface.

In the quest for unattached biofilms in refractory periapical lesions, long-standing lesions in teeth with apical periodontitis where the teeth were treated with the long-term calcium hydroxide method were examined. This is a time-consuming, but usually efficient method of root-canal disinfection (Tronstad 2008). It is used with success in so-called problem teeth, that is, teeth with large periapical lesions, progressive external root resorption, and where conventional endodontic treatment has failed. Thus, the method was used in this study in a serious effort to obtain bacteria-free roots. All patients, 26 patients with the diagnosis asymptomatic apical periodontitis and 10 patients with the diagnosis apical periodontitis with fistula, were treated for 6–7 months with calcium hydroxide (Sunde et al. 2002). At the end of the treatment period, there was no evidence of healing of the lesions as evaluated clinically and radiographically. Two of the patients complained of apical tenderness and one had ample exudation from the root canal. These patients received antibiotics systemically (phenoxymethylpenicillin). None of the fistulas had closed. Five of the patients with fistulas received systemic antibiotic treatment (one patient, phenoxymethylpenicillin; one patient, amoxicillin; and three patients, amoxicillin and metronidazole), but without apparent clinical effect. The patients were then scheduled for root filling and surgical removal of the periapical lesions (Fig. 6.9).

Granules that were considered to be the equivalent of the so-called sulfur granules were recovered from lesions in 9 of the 36 patients (25%) (Fig. 6.10). They were free in the granulomatous tissue and varied in size from 0.5 to 4 mm. The granules were present in numbers between 2 and more than 10 and varied in color between whitish gray, bright yellow, brownish, or brownish green. Some of the granules were soft whereas others were hard and appeared calcified to varying degrees.

By scanning electron microscopy, it was seen that the granules were tightly packed with microorganisms (Fig. 6.11). Rod-like organisms were prominent (Fig. 6.12), and spirochete-like bacteria were commonly seen (Figs 6.12 and 6.13). In many of the granules, an amorphous material was seen between the bacterial cells (Fig. 6.13). In the granules that clinically felt hard, this material, when examined with EDXA, showed high amounts of silicon and low amounts of calcium. Occasionally macrophages were observed on the surface of the granules, seemingly engulfing bacteria (Fig. 6.11).

By transmission electron microscopy of the granules, bacteria with a Gram-positive and Gram-negative cell wall were observed. Outside the cell wall a slime-like layer was often present. This layer was seen to envelop several bacterial cells (Fig. 6.14). Outer membrane vesicles were observed in close contact with the bacterial cell wall and were also spread out between cells. Macrophages were seen at the surface of the granules, usually with engulfed bacteria (Fig. 6.15).

The electron-microscopic findings that the granules consist of tightly packed bacteria with an interspersed extracellular material strongly suggest that the granules, in fact, represent biofilm that has developed in the periapical lesions, unattached to and independent of the root of the tooth. Granules from seven of the nine patients yielded bacteria by culture, and in five of the cases positive for growth *A. israelii*, *A. viscosus*, *A. meyeri*, and *A. naeslundii* were cultured. These organisms have surface structures that aid in their attachment to a wide variety of surfaces. Especially, the means of attachment of *A. naeslundii* have been extensively studied, and the presence of fimbriae associated with attachment to epithelial cells, inflammatory cells, and oral bacteria has been demonstrated (Whittaker et al. 1996; Hallberg et al. 1998). It appears, therefore, that not only *A. israelii*, but also other *Actinomyces* species are important colonizers in the biofilm granules in periapical lesions. This observation is in accord with modern understanding of biofilm formation in the periodontal pocket where *Actinomyces* are thought to be pioneer bacteria and part of the *scaffolding structure* of the biofilm (Ximenez-Fyvie et al. 2000, 2006).

In addition to *Actinomyces*, three to six other microbes were recovered from the granules (by cultivation): *P. acnes*, *P. propionicum*, *Peptostreptococcus prevotii*, *Gemella morbillorum*, *Clostridium*

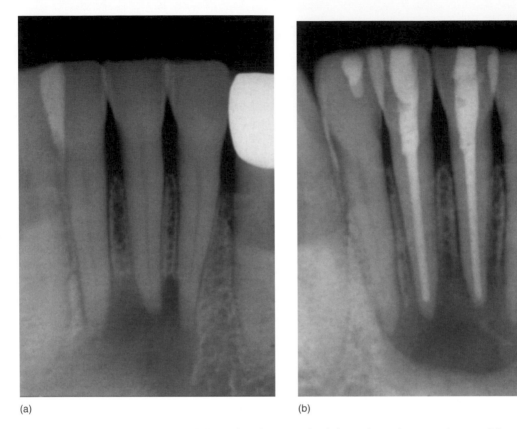

(a)                                              (b)

**Fig. 6.9**   (a) Radiograph of anterior mandibular teeth with periapical radiolucent lesion from central incisors following unattended traumatic injury to the region. The central incisors are nonvital and the diagnosis is asymptomatic apical periodontitis (both teeth) and the therapy is nonsurgical endodontic treatment of both teeth. (b) The treatment is ineffective (long-term calcium hydroxide treatment) and severe periapical exudation persists. Cultivation of bacterial samples from the root canal indicates that the infection is strongly dominated by *Staphylococcus aureus*. The bacteria present are sensitive to penicillin, and the patient receives phenoxymethylpenicillin and metronidazole for 10 days. (c) The periapical exudation has not stopped and the teeth are root filled and apicoectomized. (d) Postoperative radiograph after 3 months shows periapical healing.

*sordelli, Clostridium bifermentas, Leptotrichia buccalis, Staphylococcus chromogenes, Staphylococcus epidermidis, Vibrio metschnicovii*, and *Streptococcus* species. In the two granules not exhibiting *Actinomyces, Aerococcus viridans, Bacteroides ureolyticus, G. morbillorum, Capnocythophaga* species, *P. aeruginosa, Staphylococcus warneri*, and *S. oralis* were cultured. It cannot be known from this study whether or not biofilm granules may form without the participation of *Actinomyces*. We may have missed them by our cultivation procedures or they may possibly have died in the granules. However, many of the additional bacteria that were cultured are opportunistic pathogens and adept at forming biofilms as well (Socransky and Haffajee 2002).

Since this was a culture-based study, it is inevitable that not all bacteria present were recovered (see Chapter 4). An important example of this is that in the scanning electron microscope, spirochete-like organisms are commonly seen. The high occurrence of these and other strict anaerobic bacteria further strengthens the conclusion that the sulfur granules, in fact, are biofilm. The granules will have to contain microenvironments with a low redox potential since the anaerobes survive (Costerton 1999). This is a characteristic feature of a biofilm, and findings in studies with miniature electrodes have shown that oxygen may be completely consumed in some areas, leading to anaerobic niches in the film (Stewart and Costerton 2001).

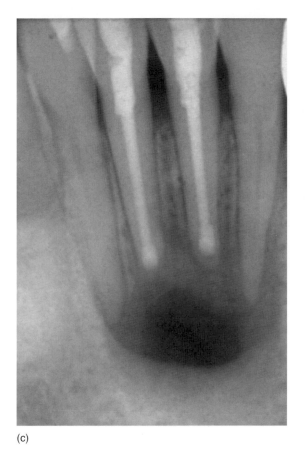

(c)

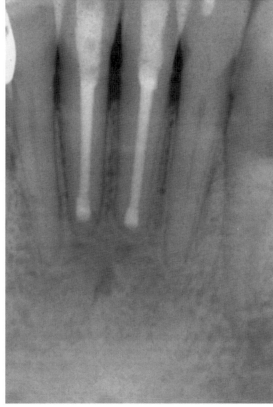

(d)

**Fig. 6.9**    *(Continued)*

In addition to the biofilm granules, samples from the periapical lesions as such were cultured as well. Thirty-five of the 36 lesions yielded microbial growth, and there was no statistical difference in the occurrence of microbial species between patients with or without fistulas. The flora of these lesions was somewhat different from the extraradicular flora of the studies referred to above in that the domination by Gram-positive organisms was clearer. Thus, 80% of the isolated bacteria were Gram-positive. Twenty-seven of the lesions (75%) contained organisms such as *Staphylococcus*, *Bacillus*, *Pseudomonas*, *Stenotrophomonas*, *Sphingomonas*, *Enterococcus*, *Enterobacter*, and *Candida* species. Of considerable clinical interest was the fact that the periapical samples from the eight patients having received systemic antibiotics before surgery showed microbial growth. In

five of these patients, *Enterococcus faecalis*, *Staphylococcus* species, *Pseudomonas* species, *Enterobacter cloacae*, and *Candida albicans* were found. In the two patients where the granules gave no growth, cultivation of the lesions showed *Sphingomonas paucimobilis* and *S. warneri* in one patient and *Stenotrophomonas maltophilia* in the other. One of these patients had taken amoxicillin and metronidazole before surgery and microbial sampling of the lesion.

## 6.10 Direct visualization of extraradicular bacteria

The visualization of mature bacterial biofilm on the external surfaces of root tips and in the form of biofilm granules in periapical granulomas has aided us in

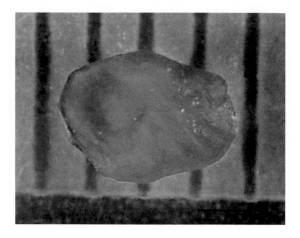

**Fig. 6.10** Biofilm granule (*sulfur granule*) from periapical lesion of tooth with asymptomatic, refractory apical periodontitis. The granule is soft, yellowish in color, and 3–4 mm in diameter. Three additional granules were recovered from the same lesion. (From Sunde et al. 2002.)

**Fig. 6.12** Scanning electron micrograph of cut surface of biofilm granule seen in Fig. 6.10. The granule consists of an abundance of bacteria. Rod-like organisms are prominent and spiral-formed bacteria are seen. (From Sunde et al. 2002.)

understanding the nature of extraradicular infection. However, bacteria have routinely been recovered also from the granulomatous tissue, and notably, most of the isolated bacteria are anaerobes. Recently, a fluorescence in situ hybridization method (FISH) has been developed whereby bacteria may be observed and even identified in the tissue of their natural environment (Moter and Göbel 2000). In formalin-fixed, plastic-embedded tissue sections, excellent conserva-

tion and visualization of bacteria have been achieved (Moter et al. 1998; Moter and Göbel 2000). In one study, the FISH technique was used to visualize and, with the probes available, identify bacteria directly within periapical lesions of asymptomatic, root-filled teeth (Sunde et al. 2003). The sections from the lesions were examined in a confocal laser scanning microscope that has become a valuable tool for obtaining high-resolution images and three-dimensional reconstruction of a variety of biological samples (Wecke et al. 2000). A probe, EUB 338, which is specific for

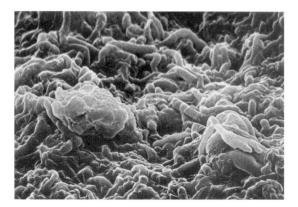

**Fig. 6.11** Scanning electron micrograph of surface area of biofilm granule seen in Fig. 6.10. Microorganisms that are tightly packed and glued together make up the outer boundary of the granule. Two macrophages are seen, seemingly engulfing bacteria. (From Sunde et al. 2002.)

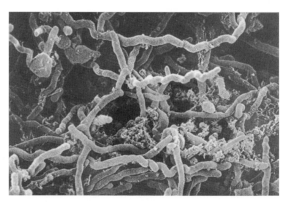

**Fig. 6.13** Scanning electron micrograph of cut surface of biofilm granule. In addition to mainly rod-like and spiral-formed bacteria, an amorphous material is seen between the cells. (From Sunde et al. 2002.)

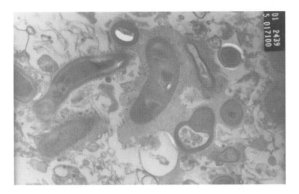

**Fig. 6.14** Transmission electron micrograph from biofilm granule. Gram-positive bacteria are seen. An extracellular material is enveloping several of the bacteria. (From Sunde et al. 2002.)

the domain *Bacteria*, was used to visualize the entire bacterial population in the specimens (Amann et al. 1990). In addition, a number of group-specific, genus-specific, and species-specific probes were applied (Loy et al. 2003).

With the universal probe EUB 338, bacteria were observed in 20 of 39 periapical lesions (three sections, 3 μm thick, from each lesion were examined) (Fig. 6.16). The bacteria were present in localized areas of the lesions, whereas other areas appeared to be free of bacteria. A variety of different bacterial morphotypes were seen: cocci, rods, spiral-, and spindle-shaped.

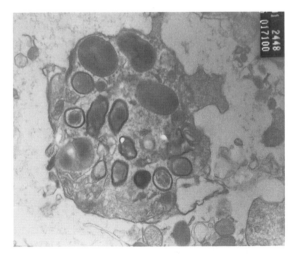

**Fig. 6.15** Transmission electron micrograph from biofilm granule. A macrophage with a variety of engulfed bacteria is seen. (From Sunde et al. 2002.)

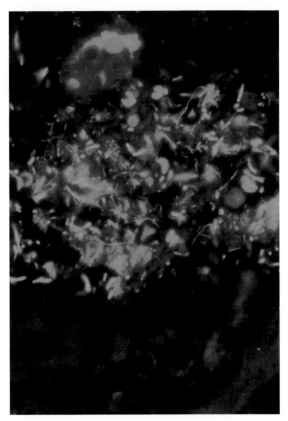

**Fig. 6.16** Fluorescence in situ hybridization using the *Bacteria*-specific probe EUB 338 on section from periapical granuloma. A large number of bacteria of different morphotypes are seen co-aggregating in limited area of tissue, possibly forming biofilm. (From Sunde et al. 2003.)

Organisms with different morphologies were seen to co-aggregate, forming small ecosystems, possibly biofilm, in the tissue (Fig. 6.16). Colonies of cocci were also seen, and the probe for the genus *Streptococcus* reacted specifically in three different lesions (Fig. 6.17). In some areas, looser groups of bacteria were present (Fig. 6.18), and single microorganisms, rods, and especially spirochete-like organisms were seen spread out in the tissue (Fig. 6.19). Hybridization with probes for *T. forsythia*, *P. gingivalis*, and *P. intermedia* gave positive signals in three different lesions (Fig. 6.20). This finding is consistent with the results of previous studies of endodontic infections using molecular techniques (Conrads et al. 1997; Gatti et al. 2000; Sunde et al. 2000b; Roças et al. 2001). One probe reacted specifically in one lesion, indicating the presence

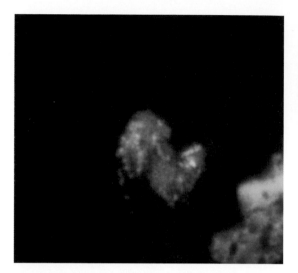

**Fig. 6.17** Fluorescence in situ hybridization on section from periapical granuloma. Simultaneous hybridization with the *Bacteria*-specific probe EUB 338 and the genus-specific probe for *Streptococcus* shows a mixed colony of streptococci (orange) and other cocci (green). (From Sunde et al. 2003.)

of *Treponema vincentii* and/or *T. vincentii*-related organisms. These organisms are associated with periodontal diseases (Choi et al. 1994; Dahle et al. 1996; Dewhirst et al. 2000), and may play a role in endodontic infections as well (Jung et al. 2000; da Silva et al. 2002; Siqueira and Roças 2003). An abundance of spirochete-like organisms of various sizes were detected with the EUB probe, including a distinct morphotype of a large curved bacterium, resembling the spirochete-like organism of 140 μm observed in the root canal (Dahle et al. 1993b) (Fig. 6.21). However, other than the signals from *T. vincentii* in one lesion, no specific signals could be obtained with the treponeme-specific probes, emphasizing the considerable genetic diversity of this group of organisms (Choi et al. 1994; Dewhirst et al. 2000).

The large number of spirochete-like organisms in the tissue was a striking observation. Motility and chemotaxis are major virulence factors of these organisms (Charon and Goldstein 2002), and spirochete species found in endodontic infections are known to be highly invasive to host tissues (Lux et al. 2001). They are known to penetrate endothelium (Dahle et al. 2003), and have been observed to invade surrounding tissues of sites with acute necrotizing and ulcerating gingivitis (Dewhirst et al. 2000). They may adapt to

**Fig. 6.18** Fluorescence in situ hybridization using the *Bacteria*-specific probe EUB 338 on section from periapical lesion. Bacteria of different morphotypes are visualized in the tissue. (From Sunde et al. 2003.)

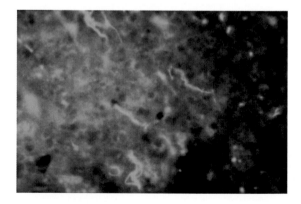

**Fig. 6.19** Fluorescence in situ hybridization using the *Bacteria*-specific probe EUB 338 on section from periapical lesion. Sole bacteria, mostly spirochete-like, are seen spread in the tissue.

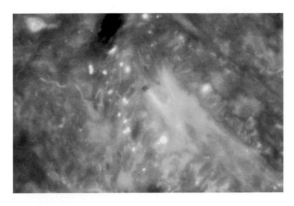

**Fig. 6.20** Fluorescence in situ hybridization on section from periapical lesion. Simultaneous hybridization with the *Bacteria*-specific probe EUB 338 and the species-specific probe for *P. gingivalis* localizes *P. gingivalis* cells (orange) in area of the tissue. (From Sunde et al. 2003.)

specific ecological niches that exclude other bacterial species (Canale-Parola 1978), and are generally found in the most advanced regions of infected tissues (Moter et al. 1998). Oral treponemes can adhere to cell surfaces and produce tissue-destructive enzymes (Dahle et al. 1993a), and have been shown to carry several putative virulence factors identical or similar to invasive organisms such as *Treponema pallidum*, the causative organism of syphilis (Centurion-Lara et al. 1999; Tzagaroulaki and Riviere 1999). Probably, the importance of treponemes in endodontic (and periodontal) infections is highly underestimated.

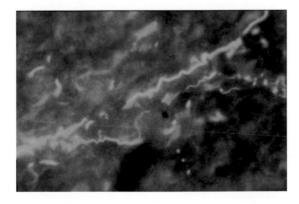

**Fig. 6.21** Fluorescence in situ hybridization using the *Bacteria*-specific probe EUB 338 on section from periapical lesion. A distinct morphotype of a large curved bacterium resembling the spirochete-like organism of 140 μm observed in the root canal is seen among other bacteria in the tissue.

The bright signals that were observed from the bacteria with the FISH technique result from high amounts of rRNA. This is evidence of physiological activity of the organisms at the time of sampling (Kemp et al. 1993; Wallner et al. 1993). Thus, the microorganisms seem to thrive in the tissue of the periapical lesions. The mechanisms by which the bacteria evade humoral and cellular immune responses are complex and not yet fully known (see Chapters 10 and 11). However, important is the fact that the outer surface molecules of the bacteria undergo antigenic variation so that the immune response mechanisms may not recognize or may not be able to clear the infectious agents (Hornhef et al. 2002; Underhill and Ozinsky 2002; LaFond et al. 2003). Persistent, even lifelong chronic infections may be established (Young et al. 2002).

## 6.11 Concluding remarks

More than 700 different species, of which over 50% have not yet been cultivated, have been detected in the oral cavity. The breadth and diversity of the oral flora in health and disease are continuously being investigated (see Chapter 1), and with the genetic methods that have become available in recent years, great progress has been made in understanding the nature of oral infections and assessing the organisms associated with disease. Thus, a new understanding of endodontic infections is slowly evolving due to the results of molecular and electron-microscopic studies.

Endodontic infections, including extraradicular infections, are polymicrobial in nature, and, interestingly, the flora is very similar to the flora of the periodontal pocket in patients with active periodontal disease. Even the numbers of infecting organisms are similar in that the same number of bacterial species is found in infected root canals and in periapical lesions as in plaque samples from patients with active periodontal disease. The red and orange complexes of bacteria regarded as important for the progression of periodontal disease apparently are important pathogens also in endodontic infections. No single pathogen of special importance has been identified. Still the role of *Actinomyces* needs to be emphasized. Traditionally, *Actinomyces* species, especially *A. israelii*, have been regarded as specific pathogens, causing the infectious disease actinomycosis in a similar manner to *T. pallidum* causing syphilis and *Mycobacterium tuberculosis* causing tuberculosis. As the evidence stands today, this may be incorrect. In periodontal research, it has been shown that *Actinomyces* are important colonizers and scaffold builders in dental plaque in both health and disease, but also that there is a significant decrease in the *Actinomyces* species and an increase in the proportion of members of the red and orange complexes in patients with active periodontal disease (Ximenez-Fyvie et al. 2000, 2006). In the biofilm granules observed in periapical lesions, *Actinomyces* species are never isolated as the sole organism, but with other bacteria normally found in endodontic and periodontal infections. These findings support the observations by Weese and Smith (1975) that actinomycotic infections are polymicrobial in nature. Thus, it may well be that the role of *Actinomyces* in oral infections, in fact, is limited to that of being pioneer bacteria and scaffold builders. With their apparent ability to survive under harsh conditions, and with their special surface structures enabling them to attach to a variety of surfaces, they may create conditions so that other bacteria (the real pathogens) are attracted to and may establish themselves at the site.

Microbiological and in situ observational evidence suggest that extraradicular infection is a common occurrence in asymptomatic teeth with apical periodontitis that do not respond to nonsurgical endodontic treatment. The presence of bacteria in the periapical lesions has been verified with bacterial cultivation, checkerboard DNA–DNA hybridization, fluorescence in situ hybridization, and electron microscopy. The bacteria are found in seemingly mature biofilms on the external surfaces of root tips included in the lesions, and in the form of biofilm granules free in the granulomatous tissue. With in situ hybridization, a variety of different bacterial morphotypes are also seen in looser groups in the tissue, and sole bacteria, especially spirochete-like organisms, are commonly observed between cells and fibers. More than 50% of the extraradicular bacteria are anaerobes, and with time, Gram-positive organisms will dominate the infection. In long-standing lesions, and when the root canal treatment has been grossly inadequate, nonoral bacteria may be recovered, especially *P. aeruginosa* and pseudomonas-like species. The pathogenic mechanisms used by the bacteria to cause disease may not yet be fully known. Still there can be little doubt that the extraradicular infection may maintain the apical periodontitis in treatment-resistant or refractory teeth independent of the infection of the root canal. As clinicians, we have to understand that an endodontic infection might not be limited to the root canal and the root, but may include the radiolucent periapical lesion as well.

## 6.12 References

Aas JA, Paster BJ, Stokes LN, Olsen I, and Dewhirst FE. 2005. Defining the normal flora of the oral cavity. *J Clin Microbiol* 43: 5721–32.

Abou-Rass M and Bogen G. 1998. Microorganisms in closed periapical lesions. *Int Endod J* 31: 39–47.

Amann R, Binder BJ, Olson RJ, Crisholm SW, Devereux R, and Stahl DA. 1990. Combination of 16S rRNA-targeted oligonucleotide probes with flow cytometry for analyzing mixed microbial populations. *Appl Environ Microbiol* 56: 1919–25.

Andreasen JO and Rud J. 1972. A histobacteriologic study of dental and periapical structures after endodontic surgery. *Int J Oral Surg* 1: 272–83.

Bergenholtz G and Spångberg L. 2004. Controversies in endodontics. *Crit Rev Oral Biol Med* 15: 99–114.

Brouqui P and Raoult D. 2001. Endocarditis due to rare and fastidious bacteria. *Clin Microbiol Rev* 14: 177–207.

Byström A, Claeson R, and Sundqvist G. 1985. The antibacterial effect of paramonochlorophenol, camphorated phenol and calcium hydroxide in the treatment of infected root canals. *Endod Dent Traumatol* 1: 170–75.

Canale-Parola E. 1978. Motility and chemotaxis of spirochetes. *Annu Rev Microbiol* 32: 69–99.

Centurion-Lara A, Castro C, Barrett L, Cameron C, Mostowfi M, Van Voorhis WC, and Lukehart SA. 1999. *Treponema pallidum* major sheath protein homologue TprK is a target of opsonic antibody and the protective immune response. *J Exp Med* 189: 647–56.

Charon NW and Goldstein SF. 2002. Genetics of motility and chemotaxis of a fascinating group of bacteria: the spirochetes. *Annu Rev Genet* 36: 47–73.

Choi B-K, Paster BJ, Dewhirst FE, and Göbel UB. 1994. Diversity of cultivable and uncultivable oral spirochetes from a patient with severe destructive periodontitis. *Infect Immun* 62: 1889–95.

Conrads G, Gharbia SE, Gulabivala K, Lambert F, and Shah HN. 1997. The use of a 16S rDNA directed PCR for the detection of endodontopathogenic bacteria. *J Endod* 23: 433–38.

Costerton JW. 1999. Introduction to biofilm. *Int J Antimicrob Agents* 11: 217–21.

Costerton JW, Stewart PS, and Greenberg EP. 1999. Bacterial biofilms: a common cause of persistent infections. *Science* 284: 1318–22.

Dahle UR, Sunde PT, and Tronstad L. 2003. Treponemes and endodontic infections. *Endod Top* 6: 160–70.

Dahle UR, Tronstad L, and Olsen I. 1993a. Spirochaetes in oral infections. *Endod Dent Traumatol* 9: 87–94.

Dahle UR, Tronstad L, and Olsen I. 1993b. Observation of an unusually large spirochete in endodontic infection. *Oral Microbiol Immunol* 8: 251–53.

Dahle UR, Tronstad L, and Olsen I. 1996. Characterization of new periodontal and endodontic isolates of spirochetes. *Eur J Oral Sci* 104: 41–47.

da Silva RM, Camargo SC, Debelian G, Eribe ER, Tronstad L, and Olsen I. 2002. DNA–DNA hybridization demonstrates a diverse endodontic microflora. *International Union of Microbiological Societies.* Paris: Proceedings of the World Congress, 33.

da Silva RM, Caugant DA, Eribe ERK, Aas JA, Lingaas PS, Geiran O, Tronstad L, and Olsen I. 2006. Bacterial diversity in aortic aneurysms determined by 16S ribosomal RNA gene analysis. *J Vas Surg* 44: 1055–60.

da Silva RM, Caugant DA, Lingaas PS, Geiran O, Tronstad L, and Olsen I. 2005. Detection of *Actinobacillus actinomycetemcomitans* but not bacteria of the red complex in aortic aneurysms by multiplex PCR. *J Periodontol* 76: 590–94.

da Silva RM, Lingaas PS, Geiran O, Tronstad L, and Olsen I. 2003. Multiple bacteria in aortic aneurysms. *J Vas Surg* 38: 1384–39.

Dewhirst DE, Tamer MA, Ericson RE, Lau CN, Levanos VA, Boches SK, Galvin JL, and Paster BJ. 2000. The diversity of periodontal spirochetes by 16S rRNA analysis. *Oral Microbiol Immunol* 15: 196–202.

Donlan RD and Costerton JW. 2002. Biofilms: survival mechanisms of clinically relevant microorganisms. *Clin Microbiol Rev* 15: 167–93.

Dunn BE, Cohen H, and Blaser MJ. 1997. *Helicobacter pylori.* *Clin Microbiol Rev* 10: 720–41.

Edwards DD. 2000. Enterococci attract attention of concerned microbiologists. *ASM News* 66: 540–45.

Eriksen HM, Kirkevang LL, and Petersson K. 2002. Endodontic epidemiology and treatment outcome: general considerations. *Endod Top* 2: 1–9.

Friedman S. 2002. Prognosis of initial endodontic therapy. *Endod Top* 2: 59–88.

Gatti JJ, Dobeck JM, Smith C, White RR, Socransky SS, and Skobe Z. 2000. Bacteria of asymptomatic periradicular endodontic lesions identified by DNA–DNA hybridization. *Endod Dent Traumatol* 16: 197–204.

Gilbert P, Das J, and Foley I. 1997. Biofilm susceptibility to antimicrobials. *Adv Dent Res* 11: 160–67.

Gomes BPFA, Lilley JD, and Drucker DB. 1996. Associations of endodontic symptoms and signs with particular combinations of specific bacteria. *Int Endod J* 29: 69–75.

Hallberg K, Holm C, Öhman U, and Strömberg N. 1998. *Actinomyces naeslundii* displays variant fimP and fimA fimbrial subunit genes corresponding to different types of acidic proline-rich protein and beta-linked galactosamine binding specificity. *Infect Immun* 66: 4403–10.

Happonen RP. 1986. Periapical actinomycosis: a follow-up study of 16 surgically treated cases. *Endod Dent Traumatol* 2: 205–9.

Happonen RP, Söderling E, Viander M, Linko-Kettunen L, and Pelliniemi LJ. 1985. Immunocytochemical demonstration of *Actinomyces* species and *Arachnia propionica* in periapical infections. *J Oral Pathol* 14: 405–13.

Hausner RA and Wuertz S. 1999. High rates of conjugation in bacterial biofilms as determined by quantitative in situ analysis. *Appl Environ Microbiol* 65: 3710–13.

Hendersen B, Nair SP, Ward JM, and Wilson M. 2003. Molecular pathogenicity of the oral opportunistic pathogen *Actinobacillus actinomycetemcomitans. Annu Rev Microbiol* 57: 29–55.

Hornhef MH, Wick MJ, Rehn M, and Nomark S. 2002. Bacterial strategies for overcoming host innate and adaptive immune responses. *Nat Immunol* 3: 1033–40.

Horowitz EA, Pugsley MP, Turbes PG, and Clark RB. 1987. Pericarditis caused by *Actinobacillus actinomycetemcomitans. J Infect Dis* 155: 152–53.

Hoyle BD, Jass J, and Costerton JW. 1990. The biofilm glycocalyx as a resistance factor. *J Antimicrob Chemother* 26: 1–6.

Iwy C, Macfarlane TW, Mackenzie D, and Stenhouse D. 1990. The microbiology of periapical granulomas. *Oral Surg Oral Med Oral Pathol* 69: 502–5.

Jung I-Y, Choi B-K, Kum K-Y, Yoo Y-J, Yoon T-C, Lee S-J, and Lee CY. 2000. Identification of oral spirochetes at the species level and their association with other bacteria in endodontic infections. *Oral Surg Oral Med Oral Pathol* 92: 329–34.

Kemp PF, Lee S, and LaRoche J. 1993. Estimating the growth rate of slowly growing bacteria from RNA content. *Appl Environ Microbiol* 59: 2594–601.

Kerekes K and Tronstad L. 1979. Long-term results of endodontic treatment performed with a standardized technique. *J Endod* 5: 83–90.

Kolltveit KM, Geiran O, Tronstad L, and Olsen I. 2002. Multiple bacteria in calcific aortic valve stenosis. *Microb Ecol Health Dis* 14: 110–17.

Kronfeld R. 1939. *Histopathology of the Teeth and Surrounding Structures.* Philadelphia: Lea and Febiger, 110.

LaFond RE, Centurion-Lara A, Godornes C, Rompalo AM, Van Voorhis WC, and Lukehart SA. 2003. Sequence diversity of *Treponema pallidum* subsp *pallidum* tprK in human syphilis lesions and rabbit-propagated isolates. *J Bacteriol* 185: 6262–68.

Langeland K, Block RM, and Grossman LI. 1977. A histopathologic and histobacteriologic study of 35 periapical endodontic surgical specimens. *J Endod* 3: 8–23.

Loy A, Horn M, and Wagner M. 2003. ProbeBase—an online resource for rRNA-targeted oligonucleotide probes. *Nucleic Acids Res* 31: 514–56.

Lux R, Miller JN, Park NH, and Shi W. 2001. Motility and chemotaxis in tissue penetration of oral epithelial cell layers by *Treponema denticola. Infect Immun* 69: 6276–83.

Marshall BJ and Warren JR. 1984. Unidentified curved bacilli in the stomach of patients with gastritis and peptic ulceration. *Lancet* 1: 1311–15.

Meyer DH, Lippmann JE, and Fives-Taylor PM. 1996. Invasion of epithelial cells by *Actinobacillus actinomycetemcomitans.* A dynamic, multistep process. *Infect Immun* 64: 2988–97.

Moter A and Göbel UB. 2000. Fluorescence in situ hybridization (FISH) for direct visualization of microorganisms. *J Microbiol Methods* 41: 85–112.

Moter A, Leist G, Rudolph R, Schrank K, Choi B-K, Wagner M, and Göbel UB. 1998. Fluorescence in situ hybridization shows spatial distribution of as yet uncultured treponemes in biopsies from digital dermatitis lesions. *Microbiology* 144: 2459–67.

Munson MA, Pitt-Ford T, Chong B, Weightman A, and Wade WG. 2002. Molecular and cultural analysis of the microflora associated with endodontic infections. *J Dent Res* 81: 761–66.

Nair PNR. 1997. Apical periodontitis: a dynamic encounter between root canal infection and host response. *Periodontology 2000* 13: 129–48.

Nair PNR, Sjögren U, Kahnberg KE, and Sundqvist G. 1990. Intraradicular bacteria and fungi in root-filled asymptomatic human teeth with therapy-resistant periapical lesions: a long term light and electron microscopic follow-up study. *J Endod* 16: 580–88.

Najjar T. 2006. *Actinomycosis*. Available at: http://www.emedicine.com/derm/topic767.htm (accessed February 12, 2007).

O'Grady JF and Reade PC. 1988. Periapical actinomycosis involving *Actinomyces israelii*. *J Endod* 14: 147–49.

Oguntebi B, Slee AM, Tanzer JM, and Langeland K. 1982. Predominant microflora associated with human dental periradicular abscesses. *J Clin Microbiol* 15: 964–66.

Paster BJ, Olsen I, Aas JA, and Dewhirst FE. 2006. The breadth of bacterial diversity in the human periodontal pocket and other oral sites. *Periodontology 2000* 42: 1–8.

Renton TF, Danks J, and Rosenfeld JV. 1996. Cerebral abscess complicating dental treatment. Case report and review of the literature. *Aust Dent J* 41: 12–15.

Rindom-Schiøtt C, Løe H, Børglum Jensen S, Kilian M, Davies RM, and Glavind K. 1970. The effect of chlorhexidine mouthrinses on the human oral flora. *J Periodontal Res* 5: 84–89.

Roças IN, Siqueira JF, Jr, Santos KR, and Coelho AM. 2001. "Red complex" (*Bacteroides forsythus, Porphyromonas gingivalis,* and *Treponema denticola*) in endodontic infections: a molecular approach. *Oral Surg Oral Med Oral Pathol* 91: 468–71.

Rolph HJ, Lennon A, Riggio MP, Saunders WP, MacKenzie D, and Coldero L. 2001. Molecular identification of microorganisms from endodontic infections. *J Clin Microbiol* 39: 3282–89.

Rupf S, Kannengiesser S, Merte K, Pfister W, Sigusch B, and Eschrich K. 2000. Comparison of profiles of key periodontal pathogens in periodontium and endodontium. *Endod Dent Traumatol* 16: 269–75.

Sabeti M, Simon JH, Nowsari H, and Slots J. 2003. Cytomegalovirus and Epstein-Barr virus active infection in periapical lesions of teeth with intact crowns. *J Endod* 29: 321–23.

Siqueira JF and Roças IN. 2003. PCR based identification of *Treponema maltophilum, T. amylovorum, T. medium,* and *T. lecithinolyticum* in primary root canal infections. *Arch Oral Biol* 48: 495–502.

Sjögren U, Figdor D, Persson S, and Sundqvist G. 1997. Influence of infection at the time of root filling on the outcome of endodontic treatment of teeth with apical periodontitis. *Int Endod J* 30: 297–306.

Sjögren U, Häglund B, Sundqvist G, and Wing K. 1990. Factors affecting the long-term results of endodontic treatment. *J Endod* 16: 498–504.

Sjögren U, Happonen RP, Kahnberg KE, and Sundqvist G. 1988. Survival of *Arachnia propionica* in periapical tissue. *Int Endod J* 21: 277–82.

Skudutyte-Rysstad R and Eriksen HM. 2006. Endodontic status amongst 35-year-old Oslo citizens and changes over a 30-year period. *Int Endod J* 39: 637–42.

Slots J and Hames HS. 2003. Herpesviruses in periapical pathosis: an etiopathogenic relationship? *Oral Surg Oral Med Oral Pathol* 96: 327–31.

Socransky SS and Haffajee AD. 2002. Dental biofilms: difficult therapeutic targets. *Periodontology 2000* 28: 12–55.

Socransky SS, Haffajee AD, Cugini MA, Smith C, and Kent RL, Jr. 1998. Microbial complexes in subgingival plaque. *J Clin Periodontol* 25: 134–44.

Socransky SS, Smith C, Martin L, Paster BJ, Dewhirst FE, and Levin AE. 1994. "Checkerboard" DNA–DNA hybridization. *Biotechniques* 17: 788–92.

Stewart PS and Costerton JW. 2001. Antibiotic resistance of bacteria in biofilms. *Lancet* 358: 135–38.

Strindberg LZ. 1956. The dependence of the results of pulp therapy on certain factors. An analytical study based on radiographic and clinical follow-up examinations. *Acta Odontol Scand* 14: 1–174.

Sunde PT, Olsen I, Debelian G, and Tronstad L. 2002. Microbiota of periapical lesions refractory to endodontic therapy. *J Endod* 28: 304–10.

Sunde PT, Olsen I, Göbel UB, Theegarten D, Winter S, Debelian G, Tronstad L, and Moter A. 2003. Fluorescence *in situ* hybridization (FISH) for direct visualization of bacteria in periapical lesions of asymptomatic root-filled teeth. *Microbiology* 149: 1095–102.

Sunde PT, Olsen I, Lind PO, and Tronstad L. 2000a. Extraradicular infection: a methodological study. *Endod Dent Traumatol* 16: 84–90.

Sunde PT, Tronstad L, Eribe ER, Lind PO, and Olsen I. 2000b. Assessment of periradicular microbiota by DNA-DNA hybridization. *Endod Dent Traumatol* 16: 191–96.

Sundqvist G. 1992. Associations between bacterial species in dental root canal infections. *Oral Microbiol Immunol* 7: 257–62.

Sundqvist G. 1994. Taxonomy, ecology, and pathogenicity of the root canal flora. *Oral Surg Oral Med Oral Pathol* 78: 522–30.

Sundqvist G, Eckerbom MI, Larsen ÅP, and Sjögren U. 1979. Capacity of anaerobic bacteria from necrotic pulp to induce purulent infections. *Infect Immun* 25: 685–93.

Sundqvist G, Figdor D, Persson S, and Sjögren U. 1998. Microbiologic analysis of teeth with failed endodontic treatment and the outcome of conservative re-treatment. *Oral Surg Oral Med Oral Pathol* 85: 86–93.

Thoma KH. 1963. *Oral Surgery*, 4th edn. Saint Louis: Mosby, 707–10, 764.

Tronstad L. 2008. *Clinical Endodontics*; 3rd edn. New York: Thieme.

Tronstad L, Barnett F, and Cervone F. 1990a. Periapical bacterial plaque in teeth refractory to endodontic treatment. *Endod Dent Traumatol* 6: 73–77.

Tronstad L, Barnett F, Riso K, and Slots J. 1987. Extraradicular endodontic infections. *Endod Dent Traumatol* 3: 86–90.

Tronstad L, Cervone F, Barnett F, and Slots J. 1990b. Further studies on extraradicular endodontic infections. *J Dent Res* 69: 299, IADR Abstract 1528.

Tronstad L and Petersson K. 1995. Endodontisk revisjonsbehandling. *Tandläkartidningen* 87: 161–70.

Tronstad L and Sunde PT. 2003. The evolving new understanding of endodontic infections. *Endod Top* 6: 57–77.

Tzagaroulaki E and Riviere G. 1999. Antibodies to *Treponema pallidum* in serum from subjects with periodontitis: relationship to pathogen-related oral spirochetes. *Oral Microbiol Immunol* 14: 375–78.

Underhill DM and Ozinsky A. 2002. Phagocytosis of microbes: complexity in action. *Annu Rev Immunol* 20: 825–52.

van Winkelhoff AJ, Carlee AW, and de Graaff J. 1985. *Bacteroides endodontalis* and other black pigmented *Bacteroides* species in odontogenic abscesses. *Infect Immun* 49: 494–97.

Vigil GV, Wayman BE, Dazey SE, Fowler CB, and Bradley DV. 1997. Identification and antibiotic sensitivity of bacteria isolated from periapical lesions. *J Endod* 23: 110–14.

Wallner G, Amann R, and Beisker W. 1993. Optimizing fluorescent in situ hybridization of suspended cells with rRNA-targeted oligonucleotide probes for the flow cytometric identification of microorganisms. *Cytometry* 14: 136–43.

Wasfy MO, McMahon KT, Minah GE, and Falkler WA. 1992. Microbial evaluation of periapical infections in Egypt. *Oral Microbiol Immunol* 7: 100–105.

Wayman BE, Murata SM, Almeida RJ, and Fowler CB. 1992. A bacteriological and histological evaluation of 58 periapical lesions. *J Endod* 18: 152–55.

Wecke J, Kersten T, Madela K, Moter A, Göbel UB, Friedman A, and Bernimoulin J-P. 2000. A novel technique for monitoring the development of bacterial biofilms in human periodontal pockets. *Science Direct* 191: 95–101.

Weese WC and Smith IM. 1975. A study of 57 cases of *Actinomycosis* over a 36-year period. *Arch Intern Med* 135: 1562–68.

Whittaker CJ, Klier CM, and Kolenbrander PE. 1996. Mechanisms of adhesion by oral bacteria. *Annu Rev Microbiol* 50: 513–52.

World Workshop on Periodontology. 1996. Consensus report for periodontal diseases. Pathogenesis and microbial factors. *Ann Periodontol* 1: 926–32.

Ximenez-Fyvie LA, Almaguer-Flores A, Jacobo-Soto V, Lara-Cordoba M, Moreno-Borjas J-Y, and Alcantara-Maruri E. 2006. Subgingival microbiota of periodontally untreated Mexican subjects with generalized aggressive periodontitis. *J Clin Periodontol* 33: 869–77.

Ximenez-Fyvie LA, Haffajee AD, and Socransky SS. 2000. Comparison of the microbiota of supra- and subgingival plaque in subjects in health and periodontitis. *J Clin Periodontol* 27: 648–57.

Yoshida M, Fukishima H, Yamamoto K, Ogawa K, Toda T, and Sagawa H. 1987. Correlation between clinical symptoms and microorganisms isolated from root canals of teeth with periapical pathosis. *J Endod* 13: 24–28.

Young D, Hussel T, and Dougan G. 2002. Chronic bacterial infection: living with unwanted guests. *Nat Immunol* 3: 1026–32.

# Chapter 7
# Virulence of Endodontic Bacterial Pathogens

*Christine M. Sedgley*

## 7.1 Pathogenicity and virulence

Humans have coevolved with a multitude of diverse microbial species that constitute a normal or commensal microflora. Indeed, it has been estimated that only 10% of the total number of cells present in the average human are actually "human," and the remainder are bacteria (Henderson et al. 1996). This flora populates the mucosal surfaces of the oral cavity, upper respiratory tract, gastrointestinal tract, urogenital tract, and the surface of the skin. In the healthy host, the absence of a constant state of inflammation indicates that a balance has developed between bacteria and the epithelia allowing both bacterial survival and prevention of the induction of inflammation that cause damage (Henderson and Wilson 1998).

### 7.1.1 Pathogenicity

In bacterial infections, a number of bacterial components are released, which induce the synthesis of cytokines that cause pathosis. Pathogenic microorganisms causing disease in a susceptible host must be able to adhere, colonize, survive, propagate, and invade, while at the same time evade host defense mechanisms. When the invading microorganisms are sufficiently pathogenic and the host is sufficiently com-

promised, damage can occur. Direct tissue damage caused by bacteria involves their products, such as enzymes, exotoxins, and metabolites. Indirect damage can be stimulated by the host immune response to various bacterial components such as lipopolysaccharide, peptidoglycan, lipoteichoic acid, fimbriae, proteins, capsule, and extracellular vesicles that can ultimately result in tissue destruction.

Traditionally, pathogenesis has often been described in terms of a predominantly microbial etiology, with little, if any, emphasis on host factors. This approach was in part influenced by Koch's postulates and landmark studies showing a direct cause-and-effect relationship between specific culturable species and diseases, for example, *Mycobacterium tuberculosis* and tuberculosis, and *Bacillus anthracis* and anthrax (Koch 1884). However, not all pathogens cause disease in all hosts and "non-pathogens" in healthy hosts can become pathogens in immunocompromised hosts. Also, the ability of bacteria to cause disease has been described in terms of the number of infecting bacteria, yet small numbers of highly virulent microorganisms can damage the immunocompromised host, whereas large numbers of low-virulence microorganisms may be countered by the healthy host. Further, it is now well established that viruses and fungi, as well as by-products of nonviable microorganisms, are also associated with pathogenesis (Walker et al. 2006).

### 7.1.2 Virulence

Microbial virulence generally refers to the degree of pathogenicity or disease-producing ability of a microorganism. In turn, pathogens are commonly distinguished from nonpathogens by their expression of intrinsic characteristics called virulence factors. In general terms, virulence factors enable a microorganism to establish itself on or within a host. The potential to cause disease is thereby enhanced, the response to which will be dependent on host–microbe interactions. Thus, multiple qualifiers can be applied to account for the status of the host as well as the pathogen. Virulence is therefore now understood to be multifactorial (Moine and Abraham 2004) with the susceptibility of the host playing a critical role (Casadevall and Pirofski 2001). Added to the complexity is that it may be difficult to distinguish virulence traits from common traits in commensals and opportunistic pathogens, traditionally considered low-virulence organisms (Casadevall and Pirofski 1999).

### 7.1.3 Pathogenicity and virulence of endodontic infections

Within the root canal system, variations in physicochemical factors could potentially influence the pathogenicity of bacteria. These include the availability of exogenous and endogenous nutrients, the degree of anaerobiosis (oxygen tension) and pH level, as well as the surfaces available for adherence and their characteristics (e.g., cells, dentin, medicament remnants, and root-filling materials). In periapical infections, there may be greater access to pathogens and their products by host immune factors.

## 7.2 Genetic aspects of bacterial virulence

Bacterial virulence factors are encoded by genes usually located on chromosomal DNA. The predominant means by which chromosomal genes are inherited is "vertical," in other words via replication, segregation, and cell division. Mechanisms of genetic variability between generations can include point mutations and genetic rearrangements.

Virulence factors may also be encoded by genes found on accessory genetic elements such as plasmids (see below). These can profoundly influence genome plasticity and evolution by allowing movement of genetic information both within and between species and conferring traits facilitating survival under atypical conditions. The dissemination of genes occurs via "horizontal" gene transfer, whereby genes may move between bacterial cells that are otherwise genetically unrelated.

Horizontal gene transfer of DNA in bacteria occurs by three basic methods: transformation, transduction, and conjugation. *Transformation* of bacteria involves the active uptake by a cell of extracellular DNA and its subsequent incorporation into the recipient genome. Natural transformation has been observed in the oral Gram-positive bacteria *Streptococcus mutans* in biofilms (Li et al. 2001), *Streptococcus gordonii* Challis (Wang et al. 2002), and *Streptococcus pneumoniae* (Morrison and Lee 2000). *Transduction* involves gene transfer whereby bacterial viruses (also termed "bacteriophages") carry genetic material to recipient cells. Bacteriophages have been isolated from *Actinobacillus* (now *Aggregatibacter*) *actinomycetemcomitans* (Haubek et al. 1997) and *Actinomyces* spp. (Yeung and Kozelsky 1997) in dental plaque and from *Enterococcus faecalis* in saliva samples (Bachrach et al. 2003). *Conjugation* is the most efficient horizontal gene transfer phenomenon in bacteria. The requirement for cell-to-cell contact distinguishes conjugation from transduction and transformation. DNA is transferred between cells that are in physical contact allowing unidirectional transfer of genetic information from donor to recipient. Conjugation can involve the crossing of species barriers and can also occur between bacteria and eukaryotic cells (Waters 2001). Chromosomal DNA segments, plasmids, and transposons (see below) can be transferred by conjugation (Fig. 7.1).

Horizontal gene transfer provides pathogens with the means to adapt rapidly, for example, by the acquisition of genes encoding virulence factors and antibiotic resistance (Clewell and Francia 2004). Some bacteria communicate and coordinate behavior via signaling molecules using a process called quorum sensing (Keller and Surette 2006), thereby modulating this process in a coordinated manner, once the bacterial population reaches a certain threshold.

Accessory genetic elements include plasmids, bacteriophages, transposons, insertion sequences, and pathogenicity islands (Fig. 7.2). *Plasmids* are

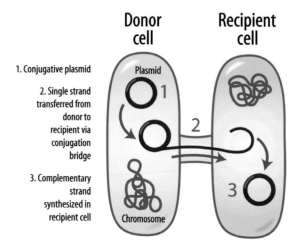

**Donor cell**

**Recipient cell**

1. Conjugative plasmid

2. Single strand transferred from donor to recipient via conjugation bridge

3. Complementary strand synthesized in recipient cell

Plasmid

Chromosome

**Fig. 7.1**    Conjugative plasmid transfer in Gram-negative bacteria.

extrachromosomal autonomously replicating elements important for bacterial adaptability and survival by the provision of functions that might not be encoded by the chromosome. Plasmids are found in bacteria (Clewell and Francia 2004), archaea (Brugger et al. 2002), and yeasts (Jayaram et al. 2004). They are typically covalently closed circular, double-stranded DNA molecules that replicate independently of chromosomal DNA and range in size from ~1 to >200 kilobase (kb) pairs (Sedgley and Clewell 2004). A simplified "map" of pAD1, a conjugative plasmid originally isolated from a clinical isolate of *E. faecalis*, is shown in Fig. 7.3. The copy number of plasmids in the bacterial cell is generally characteristic for the particular plasmid, and can range from >30 copies per chromosome for a small plasmid (e.g., <10 kb) to 1–2 copies for larger plasmids (e.g., >25 kb).

Some plasmids in *E. faecalis* conjugatively transfer copies of themselves from one bacterial cell to another using small peptides called sex pheromones as essential signals in the process (Clewell et al. 2002) (Fig. 7.4). The pheromones are secreted by a potential "recipient" cell, which "activates" the transfer system of a potential "donor" cell. The plasmid is then transferred from the donor to the recipient cell. Once the recipient cell has acquired the plasmid, it assumes a phenotype of the original donor and shuts down the production of endogenous pheromone. However, the transconjugants continue to produce pheromones specific for donors harboring different classes of plasmids.

From a clinical perspective, plasmids are particularly important because they are involved in the dissemination of antibiotic resistance, thereby enabling the survival of the strain, as well as a diverse range of products that may potentially contribute toward "virulence" that may adversely affect the host. These products include cytotoxins, adhesins, and certain metabolic enzymes. For example, cytolysins encoded by genes on plasmids in *E. faecalis*, often in association with clinical isolates (Ike et al. 1987; Huycke and Gilmore 1995), can lyse erythrocytes and other eukaryotic cells (Martinez and Baquero 2002).

Information relating to plasmids associated with endodontic microbiology appears to be limited to those associated with *Enterococcus* species. Interestingly, the *E. faecalis* MC4 strain used in monkey root canal infection studies (Moller et al. 2004) harbors a 130 kb conjugative, pheromone (cCF10)-responding plasmid, pAMS1, conferring chloramphenicol, streptomycin, and tetracycline resistances (Flannagan et al. 2008). Plasmid DNA was isolated from 25 to 33 endodontic enterococcal isolates (31 *E. faecalis* and 2 *Enterococcus faecium* strains), recovered from patients in Sweden, with up to four plasmids per strain (Sedgley et al. 2005b). Interestingly, several strains, which on initial screening appeared to be clones based on pulsed field gel electrophoresis analyses of total DNA, were actually shown to have distinct plasmid types (Fig. 7.5). Further, phenotypic studies showed that 16 of the 25 plasmid-positive strains exhibited a "clumping response" (characteristic of a response to pheromone) when exposed to a culture filtrate of a plasmid-free strain, suggesting the potential for conjugative transfer of genetic elements in these endodontic isolates. It is conceivable that if endodontic strains contain conjugative plasmids with genes that could enhance virulence during or after endodontic treatment, such properties might be transferrable to other strains remaining in the root canal system. Indeed, bidirectional transfer between *S. gordonii* and *E. faecalis* of an erythromycin-resistance gene on the conjugative plasmid pAM81 in root canals was shown in an ex vivo model (Sedgley et al. 2008).

*Transposons*, sometimes also called "jumping genes," are segments of DNA that can move ("jump") from one DNA molecule to another, for example, from the chromosome to a resident plasmid (Hayes 2003). Elements similar to the conjugative transposon Tn*916* were detected in 4 of 15 tetracycline-resistant bacteria isolated from root canals (Rossi-Fedele et al.

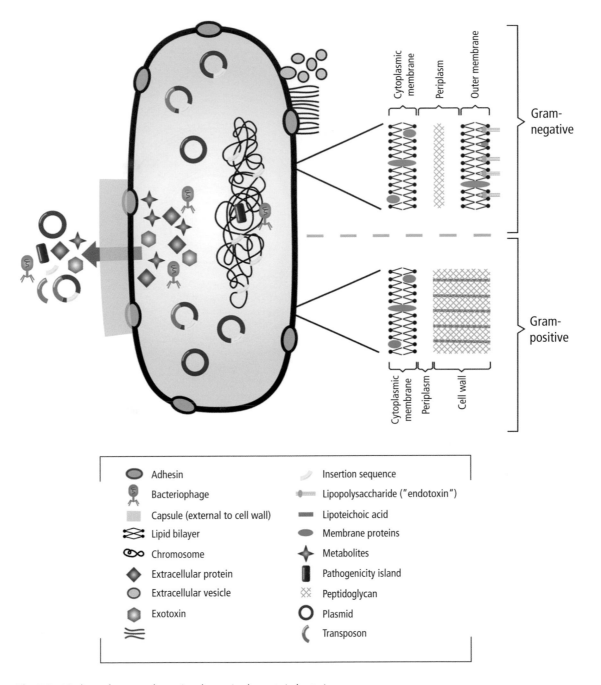

**Fig. 7.2** Virulence factors and associated genetic elements in bacteria.

Legend:

- Adhesin
- Bacteriophage
- Capsule (external to cell wall)
- Lipid bilayer
- Chromosome
- Extracellular protein
- Extracellular vesicle
- Exotoxin
- Insertion sequence
- Lipopolysaccharide ("endotoxin")
- Lipoteichoic acid
- Membrane proteins
- Metabolites
- Pathogenicity island
- Peptidoglycan
- Plasmid
- Transposon

Gram-negative:
- Cytoplasmic membrane
- Periplasm
- Outer membrane

Gram-positive:
- Cytoplasmic membrane
- Periplasm
- Cell wall

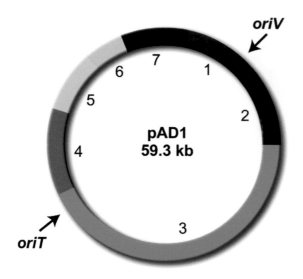

**Fig. 7.3**   Simplified map of the conjugative plasmid pAD1 originally isolated from *E. faecalis* DS16 based on Clewell et al. (2002). Segments are described according to the functions encoded by genes contained within (1) replication and maintenance; (2) regulation of pheromone response; (3) structural genes relating to conjugation; (4) unknown; (5) cytolysin biosynthesis; (6) unknown; (7) resistance to UV light; *oriV*, origin of replication; *oriT*, origin of transfer. (Reprinted from Sedgley and Clewell 2004.)

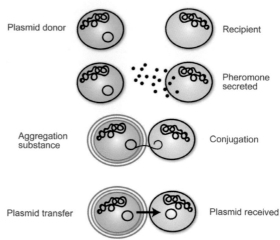

**Fig. 7.4**   Pheromone-initiated conjugative plasmid transfer. The pheromone induces the appearance of a surface adhesin (aggregation substance) that facilitates the attachment of the donor and recipient cells. Aggregates give rise to conjugal channels through which the plasmid is transferred from the donor to the recipient cell. (Adapted from Sedgley and Clewell 2004.)

2006). *Insertion sequences* are short DNA sequences that usually encode the ability to transpose but do not carry accessory genes like transposons do. Virulence genes can also be transferred via pathogenicity islands or horizontally transferable genomic islands that are located on the bacterial chromosome or may be a part of a plasmid (Coburn et al. 2007). More information is required about the role of accessory genetic elements encoding virulence genes in endodontic infections.

## 7.3 Virulence factors

It is well established that a diverse range of virulence factors can modulate bacterial adherence and invasion by the avoidance of host defense mechanisms or by indirectly or directly causing host damage. However, the specific application of this knowledge to endodontic bacterial pathogens is sparse at this time. Some with established or potential relevance to endodontic infections are listed in Table 7.1 and depicted in Fig. 7.2.

While both Gram-negative and Gram-positive bacteria have cytoplasmic membranes of similar lipid bi-layer structure, considerable structural differences are evident peripherally (Fig. 7.2). In addition, within the periplasm a number of important processes and regulatory functions occur that are essential to the viability and growth of the cell. Molecules within the periplasm provide a buffer against the changes occurring in the local surroundings (Oliver 1996). For many bacterial species, proinflammatory cytokine induction is a major virulence mechanism related to stimulation by certain components associated with the bacterial cell wall.

*Lipopolysaccharide* (LPS), also termed "endotoxin," is an integral component of the bacterial envelope of Gram-negative bacteria. It is composed of three parts: the lipid A portion of the molecule (the endotoxin component) serves as an anchor in the outer membrane, while the LPS core (an oligosaccharide) and side chain (the polysaccharide O-antigen) project from the surface.

LPS varies between different strains and has numerous biological effects. For example, LPS can activate the Hageman factor (factor XII) (Bjornson 1984), which in turn can activate several plasma protease cascades both directly and indirectly. Diversity in structure of the lipid A component may facilitate the evasion of recognition by toll-like receptor 4 (TLR4)

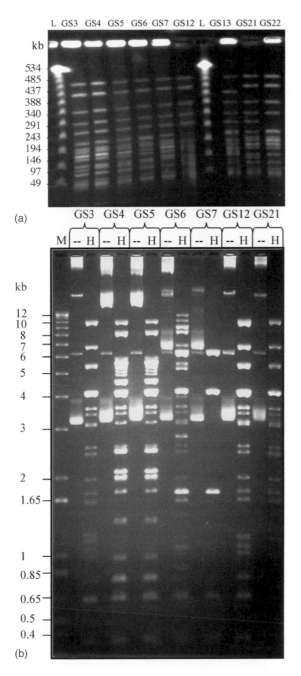

(a)

(b)

of immune cells. LPS is recognized by TLR4 and promotes the secretion of proinflammatory cytokines (Trent et al. 2006). For example, the expression of the proangiogenic vascular endothelial growth factor (VEGF) in odontoblast-like cells and macrophages was upregulated by LPS (Botero et al. 2003), in part mediated by TLR4 signaling (Botero et al. 2006). LPS can also trigger the mobilization of immunosurveillance mechanisms in the pulp. In vitro studies showed that LPS from clinical isolates activated complement (Horiba et al. 1992) and stimulated interleukin-1β release from human dental pulp cells in a time- and dose-dependent manner (Hosoya and Matsushima 1997). In postnatal human dental pulp stem cells, the expression of interleukin-8 was rapidly induced by LPS (Chang et al. 2005). LPS can also bind to CD14, a receptor that can be both soluble and membrane bound (Gioannini and Weiss 2007; Jerala 2007), with subsequent stimulation of inflammatory cytokines.

In the endodontic literature, endotoxin is the microbial virulence factor that has received the most attention. The presence of endotoxin was associated with periapical bone destruction in monkeys and cats (Dwyer and Torabinejad 1980; Dahlen et al. 1981). In humans, endotoxin was positively correlated with pulpal pain and periapical inflammation (Schein and Schilder 1975; Schonfeld et al. 1982). Similarly, endotoxin was more likely to be detected in root canals of symptomatic teeth compared with asymptomatic teeth (Horiba et al. 1991) and at a higher concentration (Jacinto et al. 2005).

*Peptidoglycan* (PG) is the major component of Gram-positive cell walls, providing cell wall shape and strength, while counteracting cytoplasmic osmotic pressure. PG consists of peptides and polysaccharides that form a homogeneous cross-linked layer outside the plasma membrane. In Gram-negative bacteria, the PG layer is considerably thinner. PG is released on cell lysis and can induce the upregulation of both proinflammatory and anti-inflammatory cytokines. For example, PG from *Staphylococcus aureus* induced IL-6 and IL-10 mRNA accumulation in monocytes and

**Fig. 7.5** (a) Total DNA and (b) plasmid DNA analysis of endodontic *E. faecalis*. (a) Pulsed field gel electrophoresis (PFGE) of *Sma*I-digested genomic DNA. Note similarities between GS3–GS7, GS12, and GS21. Reference standard, lambda phage DNA. (b) Plasmid analysis of the same *E. faecalis* isolates. Lane M, molecular size marker (1 kb Plus DNA Ladder, Invitrogen); --, undigested; H, digested with *Hin*DIII. Strain designations are shown above the lane designations. Isolates classified based on PFGE pattern as clonal, GS3, GS12, and GS21 are similar in plasmid content. GS4 and GS5 appear to be alike in plasmid content. GS6 and GS7 each contain two similar small plasmids; however, GS6 has two additional plasmids. (Adapted from Sedgley et al. 2005b.)

**Table 7.1**    Bacterial virulence factors that may contribute to endodontic infections.

|  | Gram-negative | Gram-positive | Effect |
|---|:---:|:---:|---|
| Structural components | | | |
| Lipopolysaccharide | + | − | Proinflammatory |
| Peptidoglycan | + | + | Proinflammatory |
| Lipoteichoic acid | − | + | Proinflammatory |
| Fimbriae | + | − | Adherence |
| Products | | | |
| Capsules | + | + | Protection |
| Extracellular vesicles | + | − | Secretory products |
| Exotoxins | + | + | Diverse |
| Extracellular proteins | + | + | Diverse |
| Short-chain fatty acids | + | + | Proinflammatory |
| Superoxide anions | + | + | Proinflammatory |

T cells, and stimulated IL-6 production in human dental pulp cells (Wang et al. 2000). PG-induced cytokine expression was mediated by TLR2 in fibroblasts (Yoshimura et al. 1999). The production of IL-6 was stimulated by PG from *Lactobacillus casei* in a time- and dose-dependent manner in human dental pulp cells (Matsushima et al. 1998). An adaptive immune response via macrophages may be facilitated by PG (Myhre et al. 2006). PG potency is strongly boosted in the presence of LPS (Wang et al. 2001; Myhre et al. 2006); this conceivably has broad implications for endodontic infections that are characteristically polymicrobial with significant numbers of Gram-negative anaerobes.

*Lipoteichoic acids* (LTAs) are an integral cell wall component of Gram-positive bacteria. They are amphiphilic polymers composed of glycerol phosphates and complex glycolipid. LTA can induce a diverse range of inflammatory diseases in animals (Costa et al. 2003) including nephritis, arthritis, septic shock, and multiorgan failure (Ginsburg 2002). The binding of Gram-positive bacteria to fibronectin in cell membranes and to neutrophils and lymphocytes may be facilitated by the lipid component of LTA (Courtney et al. 1981, 1988). In odontoblasts, LTA upregulated the expression of the cell-surface receptor TLR2 and the production of proinflammatory chemokines CCL2 and CXCL10 (Durand et al. 2006). When released on cell lysis, LTA can bind to target cells either specifically to TLRs and CD14 or nonspecifically to membrane phospholipids (Wang et al. 2000). Thereafter, LTA can interact with circulating antibodies and activate the complement cascade. While LTA and LPS share many pathogenic properties, LTA is the

less active, on a weight-for-weight basis (Myhre et al. 2006). LTA can trigger the release of many molecules from neutrophils and macrophages that include acid hydrolases, highly cationic proteinases, bactericidal cationic peptides, growth factors, reactive oxygen and nitrogen species, and cytotoxic cytokines (Ginsburg 2002). Alone or in combination, these molecules can subsequently amplify damage. LTA has both proinflammatory (Telles et al. 2003) and anti-inflammatory (Plitnick et al. 2001) effects. In macrophages and pulp cells, LTA can induce the expression of VEGF (Telles et al. 2003). However, LTA also inhibits the function of IL-2, an autocrine growth factor for T cells (Plitnick et al. 2001). It has been hypothesized that LTA may provide a selective advantage to Gram-positive bacteria by interfering with the immune response to infection (Hahn and Liewehr 2007).

*Fimbriae* are thin, filamentous macromolecules made of protein subunits of up to 10 nm in diameter and between approximately 100 nm and several micrometers in length. They are distinct from flagella that are longer and involved in cell motility. Fimbriae are involved in the attachment of bacteria to surfaces and interactions with other bacteria. Fimbriae are found on the surface of many Gram-positive and Gram-negative bacteria, for example, *Eikenella corrodens* (Hood and Hirschberg 1995), *Actinomyces israelii* (Figdor and Davies 1997), and *Actinomyces naeslundii* (Wu and Fives-Taylor 2001). There are several "types" of fimbriae. For example, type 1 fimbriae of uropathogenic *Escherichia coli* mediate adherence to urinary epithelium (Capitani et al. 2006). Type IV fimbriae, which can aggregate into bundles, have been detected on *E. corrodens* (Hood and Hirschberg 1995), a periodontal pathogen that has also been detected in root canal samples from teeth with acute periapical abscesses (Rocas and Siqueira 2006). The fimbriae of a periodontal clinical isolate of *Prevotella intermedia* were shown to induce hemagglutination (Leung et al. 1996). Fimbriae can also differentially facilitate bacterial adherence and invasion (Wu and Fives-Taylor 2001). For example, type II fimbriae of *Porphyromonas gingivalis* were more highly adherent to epithelial cells than type I fimbriae and thus had a greater potential to contribute toward virulence (Kato et al. 2007).

*Capsules* are generally composed of predominantly polysaccharides and form a well-organized layer, coating the outside of the cell wall. Microorganisms with capsules can evade or counteract host immune defenses by enabling avoidance of opsonin-mediated phagocytosis or the recognition of the underlying cell by complement and antibodies and subsequent phagocytosis (Abeyta et al. 2003). In Gram-negative black-pigmented bacteria, capsules were seen to facilitate the avoidance of, or survival after, phagocytosis (Sundqvist et al. 1982). Capsule formation by a pathogenic strain of *S. pneumoniae* was shown to facilitate efficient transfer from their initial site in a host, the lumenal mucus, to the epithelial surface, a capability not shared by capsule-deficient mutant strains (Nelson et al. 2007). Capsules can also provide protection of the microorganism against desiccation, bacterial viruses, and hydrophobic toxic materials (e.g., detergen).

*Extracellular vesicles* develop from evagination of the outer membrane of Gram-negative bacteria. They contain proteins and lipids derived from the periplasm and have an average diameter of 50–250 nm (Beveridge 1999). The contents of the vesicle are released into the extracellular environment where they can participate in a diverse array of activities involving both prokaryotic and eukaryotic cells (Beveridge 1999; Kuehn and Kesty 2005). These include bacterial adhesion, proteolytic activities, hemagglutination, and hemolysis (Kinder and Holt 1989). Extracellular vesicles can modulate interactions between adjacent bacteria. For example, vesicles from *P. gingivalis* induced aggregation among *Streptococcus* spp., *Fusobacterium nucleatum*, *A. naeslundii*, and *Actinomyces viscosus* (Kamaguchi et al. 2003). Extracellular vesicles can also provide their own "protection" by binding chlorhexidine (Grenier et al. 1995). Leukotoxin produced by *A. actinomycetemcomitans* (Kato et al. 2002) and the proteases Arg- and Lys-gingipain produced by *P. gingi-*

*valis* (Duncan et al. 2004) are specific virulence factors associated with vesicles produced by oral bacteria. The presence of outer membrane vesicles in lesions associated with teeth with refractory apical periodontitis was shown using transmission electron microscopy (Sunde et al. 2002).

*Exotoxins* are a diverse array of toxins secreted by a living microbial cell or released during cell lysis. They can target both eukaryotic cells and other microorganisms, as well as the extracellular matrix. Bacterial cytotoxins act on eukaryotic cells by targeting the cell cytostructure, either directly by modifying actin or indirectly by targeting regulators, in particular Rho GTPase regulators, which are essential for the functional integrity of the immune system (Aktories and Barbieri 2005). Exotoxins can trigger excessive and aberrant activation of T cells. Some exotoxins are extremely potent, for example, toxic shock syndrome toxin-1 (Dinges et al. 2000) and enterotoxins associated with food poisoning produced by certain strains of *S. aureus* (Balaban and Rasooly 2000) and pathogenic *E. coli* strains such as O157:H7 (Gyles 2007). The superantigens of *Streptococcus pyogenes* can result in multiorgan failure (Sriskandan et al. 2007). Synergism among pathogenic factors can also be involved as demonstrated in methicillin-resistant *S. aureus*, which causes hemorrhagic, necrotizing pneumonia by utilizing a bacterial toxin that destroys respiratory tissue and immune cells. The exotoxin works in synergy with other factors expressed by the strain: the leukotoxin Panton–Valentine leukocidin and the proinflammatory protein A in combination enhance strain virulence (Labandeira-Rey et al. 2007). Based on the elevated expression of proinflammatory cytokines by T cells obtained from periodontitis sites, it has been hypothesized that superantigens produced by periodontitis-associated bacteria may be contributory to the disease process (Sriskandan et al. 2007).

Bacterial toxins that target other microorganisms are called bacteriocins. These are protein or peptide "antibiotics" produced by some strains of Gram-positive and Gram-negative bacteria that are bacteriostatic or bactericidal to other, often closely related, bacterial strains as well as other species and genera. The production of bacteriocins may provide the producer strain with a selective advantage over other strains, especially those closely related to the bacteriocin-producing strain (Tomita et al. 1997). The capacity for bacteriocin production was shown in 14 of

33 *Enterococcus* species recovered from infected root canals (Sedgley et al. 2005b).

*Extracellular proteins* are produced by bacteria or released during bacterial cell lysis. Many of these diverse groups are enzymes with the potential to contribute to the spread of infection. The products may directly interact with TLRs to activate cells of the innate and adaptive immune systems, resulting in the production of cytokines. In monkeys, different strains recovered from infected root canals were shown to vary in their ability to produce different histolytic enzymes, several of which contributed to tissue disintegration, including hyaluronate lyase, chondroitin sulfatase, beta-glucuronidase, Dnase, and acid phosphatase (Dahlen et al. 1983). Posttreatment apical periodontitis may be associated with the presence of extracellular proteins produced by *S. gordonii*, *Streptococcus anginosus*, and *Streptococcus oralis* (Chavez de Paz et al. 2005). Endodontic isolates of *P. gingivalis* demonstrated evidence of the collagenase gene (Odell et al. 1999) that may explain why increased collagenase was found in association with larger periapical lesions (Hashioka et al. 1994).

In the enterococci, important extracellular virulence-related proteins are cytolysin, serine protease, gelatinase, aggregation substance, enterococcal surface protein (esp), Ace (an adhesin to collagen of *E. faecalis*), and EfaA (*E. faecalis* antigen) (Gilmore et al. 2002). The expression of the protein serine protease contributed to the binding of *E. faecalis* to dentin (Hubble et al. 2003). The production of gelatinase, observed in more than 70% of *E. faecalis* strains recovered from infected root canals (Sedgley et al. 2005b), was associated with extended survival in obturated teeth (Sedgley 2007).

### 7.3.1 Metabolic by-products

The process of metabolism results in a diverse array of metabolic by-products, many of which are released into the extracellular environment by living organisms or following lysis. The production of highly reactive free radicals and biologically toxic superoxide anion production is common among cells of the immune system and some bacterial species. For example, *E. faecalis* (Huycke et al. 1996) can produce extracellular superoxide that causes lysis of erythrocytes (Falcioni et al. 1981). Interspecies interactions may also be modulated by superoxides; extracellular superoxide production by *E. faecalis* enhanced its survival

in a mixed infection with *Bacteroides fragilis* in mice (Huycke and Gilmore 1997).

Short-chain fatty acids such as butyric and propionic acids are fermentation by-products of obligate anaerobes that can stimulate the release of cytokines (Niederman et al. 1997). In vitro investigations showed that penetration of butyric acid could occur in root canals of tooth sections obturated with gutta-percha and AH26 (Kersten and Moorer 1989). Short-chain fatty acids might play a role in the periapical infection process, perhaps via stimulation of monocyte IL-1β production, a cytokine associated with bone resorption (Eftimiadi et al. 1991), and increased T-lymphocyte cell apoptosis (Kurita-Ochiai et al. 2006).

## 7.4 Virulence associated with endodontic microorganisms and clinical implications

The relationship between the microflora in advancing caries and the histopathology of pulpitis involves irreversible tissue damage, healing, and repair that are modulated by both innate and adaptive immune responses (Hahn and Liewehr 2007). The microflora of carious dentin that has been strongly implicated in endodontic infections subsequent to pulpitis includes significant numbers of lactobacilli (Chhour et al. 2005) and Gram-negative bacteria (Martin et al. 2002). Many species recovered from infected root canals have also been identified as commensals in the oral cavity. The transition from oral "commensal" to root canal "pathogen" may reflect an innate ability to switch on "virulence genes" that enable survival and propagation in the root canal environment. On the other hand, the toll-like receptor response is tightly controlled to avoid an inflammatory response to commensals (Sirard et al. 2006).

*Prevotella* and *Porphyromonas* are genera of nonmotile Gram-negative obligately anaerobic rods belonging to the Bacteroidaceae family. They are sometimes generically identified as black-pigmented *Bacteroides* based on the fact that some species form brown or black pigments when cultured on blood-containing media. These species are frequently detected in root canal samples from symptomatic and asymptomatic root canal infections, and aspirates from acute periapical abscesses (Haapasalo et al. 1986; Siqueira and Rocas 2005). In vitro studies have shown that LPS from *Prevotella* and *Porphyromonas* stimulated IL-1β release (Hosoya and Matsushima 1997),

IL-8 expression (Chang et al. 2005), and VEGF expression (Botero et al. 2003). *P. intermedia* fimbriae induced hemagglutination activity in vitro (Leung et al. 1996). Supernatants of *Porphyromonas endodontalis*, *P. gingivalis*, and *P. intermedia* stimulated the expression of VEGF and IL-6 in human pulp fibroblasts (Yang et al. 2003, 2004).

*Fusobacterium* is a genus of Gram-negative anaerobic spindle-shaped rods or filaments belonging to the Bacteroidaceae family. They are nonmotile with the major end products of metabolism being butyric acid as well as lesser amounts of acetic, lactic, formic, and propionic acids. The release of inflammatory cytokines from neutrophils can be stimulated by butyric acid produced by *F. nucleatum* (Niederman et al. 1997). Virulence factors released by fusobacteria can stimulate numerous biological effects. LPS from *F. nucleatum* induced a rapid immune response when applied to pulp tissues in rats (Dahlen 1980) and produced a large array of biological effects in macrophage-like cells (U937 cells) by the upregulation of proinflammatory cytokines IL-1β, IL-6, TNF-α, and IL-8 (Grenier and Grignon 2006). Clinically, the recovery of *F. nucleatum* has been associated with the most severe flare-up pain and swelling (Chavez de Paz Villanueva 2002).

*Peptostreptococcus (Parvimonas)* is a genus of Gram-positive, asporogenous, anaerobic cocci from the Clostridiaceae family. The species *Peptostreptococcus micros* (also called *Parvimonas micra*) is commonly associated with periodontal disease and has been recovered from endodontic abscesses in children (Brook et al. 1981). *P. micros* can bind human plasminogen on their cell surface, which can be activated into plasmin. This activity, along with proteolytic capabilities, may facilitate dissemination of bacterial cells (Grenier and Bouclin 2006). Positive associations were described between *F. nucleatum* and *P. micros* in teeth with apical periodontitis (Sundqvist 1992), suggesting that this is a synergistic association that might enhance pathogenicity (Bolstad et al. 1996).

*Streptococcus* is a genus of Gram-positive, asporogenous, facultatively anaerobic, catalase-negative cocci, or coccoid bacteria. They are nonmotile and approximately 1 μm in diameter, occurring in pairs or chains. There are at least 50 species within the *Streptococcus* genus as determined by 16S rDNA gene sequencing (Facklam 2002). Several streptococcal species can form capsules. The cell walls of streptococci contain peptidoglycan (PG) and lipoteichoic acid (LTA).

Greater production of extracellular LTA by *S. mutans* was observed when grown in low pH (6.0 or 6.5) conditions at low dilutions (Jacques et al. 1979). On lysis of the bacterial cell (inducible by lysozyme, cationic peptides from leukocytes, or beta-lactam antibiotics), PG and LTA are released. These can bind to cell surface receptors and induce the release of proinflammatory cytokines. In mice, the production of proinflammatory cytokines was induced by extracellular products of *Streptococcus sanguis* and *Streptococcus mitis* (Takada et al. 1993). Contact with *Streptococcus salivarius* was associated with increased expression of IL-6, IL-8, and TNF-α by oral epithelial cells (Mostefaoui et al. 2004). Both teichoic acids and PG from *Streptococcus* spp. induced expression of TNF-α in human monocytes (Heumann et al. 1994). Extracellular proteins produced by root canal isolates of *S. gordonii*, *S. anginosus*, and *S. oralis* may contribute to posttreatment apical periodontitis (Chavez de Paz et al. 2005).

Streptococci have cell surface adhesins that facilitate binding to various substrates including dentin as well as other bacterial cells and epithelial cells (Jenkinson 1994). They also may recognize components present within dentinal tubules, such as collagen type I, which stimulates bacterial adhesion and intratubular growth. Cell surface adhesin proteins SspA and SspB, members of the antigen I/II family of streptococcal polypeptides, are involved in the binding and growth of streptococci into dentinal tubules (Love et al. 1997). *S. gordonii* has been shown to invade apical dentin up to 60 μm into dentinal tubules and up to 200 μm at the cervical and mid-root levels (Love 1996). *S. gordonii* amylase-binding protein A functions as an adhesin to amylase-coated hydroxyapatite (Rogers et al. 2001).

*Lactobacillus* is a genus of Gram-positive rods or coccobacilli, which grows under anaerobic, facultatively anaerobic, or microaerophilic conditions. Lactobacilli convert sugars, including lactose, to lactic acid. They are generally considered nonpathogens (Brouqui and Raoult 2001), apart from their association with dental caries (Brook 2003). Along with streptococci, lactobacilli are important microorganisms in the caries process comprising up to 50% of bacterial species in advanced carious lesions (Chhour et al. 2005). Oral *Lactobacillus* spp. bind to collagen type 1, the major collagen of dentin (McGrady et al. 1995). The expression of TNF-α through TLR2 in mouse immune cells was induced by *Lactobacillus* LTA (Matsuguchi et al. 2003). In teeth with apical periodontitis undergoing root canal treatment, *Lactobacillus* spp. (and *Olsenella uli*, a member of the *Olsenella* genus that was originally separated out from the *Lactobacillus* genus) predominated over other Gram-positive rods (Chavez de Paz et al. 2004).

*Enterococcus* is a genus of Gram-positive facultatively anaerobic coccoid bacteria. Enterococcal cells are ovoid and occur singly or in pairs or short chains, and can grow at temperatures ranging from 10°C–45°C. *E. faecalis* and *E. faecium* are the most common enterococcal species found in humans. *E. faecalis* is a common causative agent of infective endocarditis (Hill et al. 2007). Enterococci, predominantly *E. faecalis*, are frequently recovered from previously treated root canals (Rocas et al. 2004; Sedgley et al. 2006b). Interestingly, in monkey studies, the addition of *E. faecalis* to a four-strain collection (*S. anginosus*, *Peptostreptococcus anaerobius*, *Prevotella oralis*, and *F. nucleatum*) resulted in higher survival of the complete combination compared with the same bacterial combination without *E. faecalis* (Fabricius et al. 2006). When selected bacterial strains were inoculated into monkey root canals, *E. faecalis* was the only species to be reisolated from all 24 root canals as well as produce radiographic evidence of apical periodontitis after 8–12 months, including the canal where only *E. faecalis* was isolated (Moller et al. 2004). Similarly, *E. faecalis* can survive for extended periods in the root canal system in vitro (Sedgley et al. 2005a) and may be dependent on treatment conditions and phenotype (Sedgley 2007). *E. faecalis* was recovered from significantly more samples sealed with RoekoSeal than with AH-Plus or Roth's sealer, and viable counts were higher in association with the gelatinase-producing *E. faecalis* OG1RF compared to the gelatinase-deficient *E. faecalis* TX5128 (Sedgley 2007) (Fig. 7.6).

The persistence of *E. faecalis* in treated root canals has been attributed to resistance to the high pH of antimicrobial agents used during treatment, but the specific mechanisms are not clear. The ability for *E. faecalis* to survive over an extended period at pH 10 was associated with a 37-fold increase in gene transcripts of *ftsZ*, a gene involved in cell division (Appelbe and Sedgley 2007) (Fig. 7.7). Virulence factors identified in enterococci recovered from the oral cavity and infected root canals (Sedgley et al. 2004, 2005b, 2006a; Duggan and Sedgley 2007) are presented in Table 7.2. These include those with the potential to promote adaptation and survival in differ-

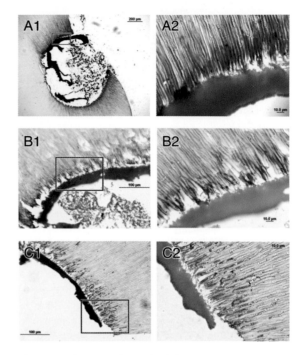

**Fig. 7.6** Infection of dentinal tubules 8 months after obturation with gutta-percha and RoekoSeal. Negative control, no bacteria (A1 and inset in A2), *E. faecalis* TX5128 (B1 and inset in B2) and *E. faecalis* OG1RF (C1 and inset in C2). Brown and Brenn stain. (From Sedgley 2007.)

ent environments: enterococcus surface protein (Esp), collagen-binding protein (Ace), and aggregation substance (AS), as well as factors that enable secretion of proteases (e.g., gelatinase) and toxins (e.g., cytolysin) (Sedgley et al. 2005b; Reynaud Af Geijersstam et al. 2007). Serine protease and Ace contribute to the ability of *E. faecalis* to bind to dentin (Hubble et al. 2003; Kowalski et al. 2006), and resistance to killing by human neutrophils is promoted by AS (Rakita et al. 1999).

However, despite their high prevalence and a capability for extended survival in the root canal system, a significant role for the species in the pathogenesis of human root canal infections has not been established. *E. faecalis* as a single species in monkey root canals caused only low-grade periapical reactions; only when present with other species within an "eight-strain collection" was lesion size larger (Fabricius et al. 1982). In addition, the recovery of *Enterococcus* spp. from previously treated root canals was more likely to be associated with a normal periapex (Kaufman et al.

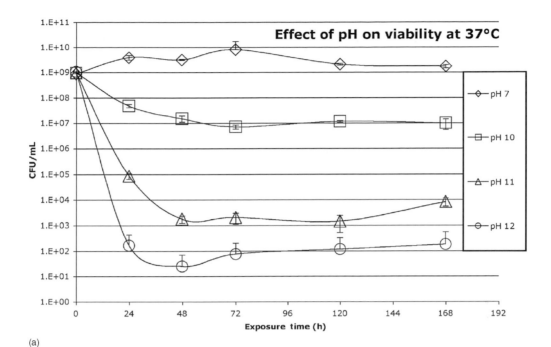

(a)

Fold-change* in level of gene transcripts compared to time zero at 37°C

|  | Incubation period (h) | Genes# dnaK | fba | ftsZ | GroEL | napA | pbp5 | tsf | tuf |
|---|---|---|---|---|---|---|---|---|---|
| pH 7 | 24 | 2 | −1 | 29 | −3 | −2 | −2 | −11 | −4 |
|  | 48 | −1 | −3 | 3 | −4 | −7 | −9 | −11 | −11 |
|  | 72 | −2 | −3 | −6 | −9 | −17 | −3 | −16 | −10 |
|  | 120 | −3 | −2 | −1 | −7 | −2 | −18 | −9 | −15 |
|  | 168 | −4 | −2 | −16 | −25 | −6 | −6 | −61 | −20 |
| pH 10 | 24 | 4 | −1 | 6 | 1 | −1 | 2 | −1 | −2 |
|  | 48 | −2 | −2 | 2 | −14 | −5 | −2 | −6 | −3 |
|  | 72 | 1 | 3 | 20 | −2 | 1 | 9 | 2 | 1 |
|  | 120 | 4 | 1 | 37 | 3 | 9 | 10 | 3 | −1 |
|  | 168 | −4 | −3 | 1 | −18 | −4 | 1 | −29 | −8 |
| pH 11 | 24 | 1 | −3 | −7 | −6 | −5 | −10 | −43 | −9 |
|  | 48 | −10 | −4 | −11 | −22 | −5 | −15 | −50 | −24 |
|  | 72 | −7 | −2 | −27 | −116 | −5 | −6 | −163 | −14 |
|  | 120 | −9 | −3 | −19 | −58 | −11 | −26 | −50 | −22 |
|  | 168 | −7 | −1 | 1 | −27 | −9 | −37 | −176 | −25 |

*A fold change of <−2 or >2 represents a real difference in levels of transcripts
#*dnaK,* protein refolding and degradation; *fba,* glycolysis and gluconeogenesis;
*ftsZ,* cell division; *GroEL,* protein refolding and degradation; *napA,* Na+/H+
antiporter; *pbp5,* peptidoglycan synthesis; *tsf* and *tuf,* protein synthesis

(b)

**Fig. 7.7** (a) Viable cell counts of *E. faecalis* after incubation at 37°C over various time periods at pH 7, 10, 11, and 12, and (b) corresponding level of gene transcripts at pH 7, 10, and 11 (with insufficient RNA being available from pH 12 samples for analyses). (Adapted from Appelbe and Sedgley 2007.)

**Table 7.2** Oral and endodontic *Enterococcus faecalis* with phenotypic and genotypic characteristics of virulence traits.[a]

| Source | n | Phenotypic tests | | | | Virulence genes[b] | | | | | | | |
|---|---|---|---|---|---|---|---|---|---|---|---|---|---|
| | | Gelatinase | Bacteriocin | Hemolysin | Pheromone response[c] | gelE | ef184/fsr | esp | asa | asa373 | ace | cylA | efaA |
| Endodontic | | | | | | | | | | | | | |
| *Primary* | 14 | 13 | 6 | 0 | 7 | 14 | 1 | 8 | 14 | 0 | 14 | 3 | 14 |
| *Retreatment* | 8 | 3 | 3 | 0 | 3 | 7 | 4 | 3 | 6 | 0 | 8 | 2 | 8 |
| *Unknown* | 11 | 8 | 4 | 0 | 6 | 11 | 5 | 8 | 11 | 0 | 11 | 1 | 11 |
| Combined | 33 | 24 | 13 | 0 | 16 | 32 | 10 | 19 | 31 | 0 | 33 | 6 | 33 |
| % | | 73 | 39 | 0 | 48 | 97 | 30 | 58 | 94 | 0 | 100 | 18 | 100 |
| Oral | | | | | | | | | | | | | |
| *Oral rinse* | 17 | 5 | 8 | 5 | 4 | 13 | 5 | 4 | 12 | 0 | 17 | 2 | 17 |
| *Tongue swab* | 3 | 0 | 2 | 2 | 1 | 2 | 1 | 2 | 3 | 0 | 3 | 2 | 3 |
| *Saliva* | 1 | 1 | 1 | 0 | 0 | 1 | 0 | 0 | 0 | 0 | 1 | 0 | 1 |
| Combined | 20 | 5 | 10 | 7 | 5 | 15 | 6 | 6 | 15 | 0 | 20 | 4 | 20 |
| % | | 25 | 50 | 35 | 25 | 75 | 30 | 30 | 75 | 0 | 100 | 20 | 100 |

[a] Combined from Sedgley et al. (2004, 2005b, 2006a) and Duggan and Sedgley (2007).
[b] gelE, gelatinase; ef184/fsr, gelatinase-negative phenotype; esp and ace, surface adherence factors; asa and asa373, aggregation substance; cylA, cytolysin activator; efaA, endocarditis antigen.
[c] Production of aggregation substance in response to E. faecalis pheromones.

2005). It has been hypothesized that the tissue damage found in periradicular infections that involve *E. faecalis* may be more to do with the host response instead of direct damage from bacterial products (Kayaoglu and Orstavik 2004).

*Actinomyces* is a genus of non-spore-forming Gram-positive bacteria occurring as rods, branched rods or filaments, or as rudimentary mycelia. All species can grow anaerobically. Many *Actinomyces* species are commensals in the oral cavity but can become opportunistic pathogens in humans and other mammals (Yeung 1999; Mardis and Many 2001). Occasionally, they cause actinomycosis. *A. israelii* has been implicated in osteoradionecrosis of the jaw (Happonen et al. 1983). *Actinomyces* species can be fimbriated or nonfimbriated. Higher cell surface interactive forces associated with fimbriated compared to nonfimbriated *Actinomyces* may contribute to modulation of their co-aggregation properties and adhesion (Tang et al. 2004). *A. israelii* and *A. naeslundii* injected into mice resulted in the production of suppurative lesions (Coleman and Georg 1969). *Actinomyces* species have been recovered from primary root canal infections (Borssen and Sundqvist 1981; Siqueira et al. 2002) and secondary root canal infections nonresponsive to conventional treatment (Borssen and Sundqvist 1981; Chavez de Paz et al. 2004). There are several case reports of *Actinomyces* species being isolated from persistent lesions following root canal filling (Figures and Douglas 1991; Sakellariou 1996), sometimes several years after completion of treatment (Sjogren et al. 1997; Hancock et al. 2001). This may be due in part to the ability of the branching filamentous organisms to evade elimination by host phagocytic cells (Figdor et al. 1992) or the formation of biofilms that overwhelm the host response.

*Propionibacterium* species are slow-growing, nonsporulating, Gram-positive anaerobic rods with propionic acid, an end product of fermentation. They are normal inhabitants of the skin and usually nonpathogenic (Roth and James 1988) but can be frequent contaminants of blood and body fluid cultures. The human cutaneous propionibacteria include *Propionibacterium acnes* and *Propionibacterium propionicum* (also called *Propionibacterium propionicus* and formerly *Arachnia propionica*). These species can be opportunistic pathogens, causing a diverse collection of infections that include acnes vulgaris (Vowels et al. 1995), central nervous system infections (Mory et al. 2005), and infective endo-

carditis (Delahaye et al. 2005). The species produce proinflammatory enzymes (lipases, neuraminidases, phosphatases, and proteases) with the capacity to contribute to direct damage to the host (Perry and Lambert 2006). The production of cytokines IL-1$\alpha$, IL-1$\beta$, IL-8, and TNF-$\alpha$ by monocytes was induced by *P. acnes* (Vowels et al. 1995). *P. propionicum* has been implicated as a causative agent of a disease process similar to actinomycosis (Happonen et al. 1985). The species has been cultured from deep layers of infected root canal dentin (Ando and Hoshino 1990) and has the ability to penetrate into dentinal tubules (Siqueira et al. 1996). In teeth with apical periodontitis undergoing root canal treatment, the most frequent *Propionibacterium* spp. species cultured was *P. propionicum* (Chavez de Paz et al. 2004).

*Yeasts* are unicellular fungi. The most clinically relevant yeasts belong to the large heterogeneous genus *Candida*, with members forming part of the commensal flora in many parts of the human body. *Candida* species can also be opportunistic pathogens as they have the ability to colonize and infect nearly all human tissues (Odds 1988). Yeasts can secrete proteases that allow degradation of human proteins. Secreted aspartyl proteinases (Sap proteins) at tissue lesion sites have been shown to directly contribute to *Candida albicans* pathogenicity (Naglik et al. 2004) by digesting molecules for nutritional purposes, disrupting host cell membranes for adhesion and invasion, and targeting immune cells for digestion or to avoid phagocytosis (Monod and Borg-von 2002). The ability to switch between phenotypes to allow adaptation to different ecological conditions and between the yeast and hyphae forms may also contribute to pathogenicity. Contact between oral epithelial cells and the hyphal form of *C. albicans* was associated with increased expression of IL-6, IL-8, and TNF-$\alpha$ by the epithelial cells (Mostefaoui et al. 2004). *C. albicans* can co-aggregate with some oral streptococci (Shen et al. 2005), a phenomenon that is enhanced when the yeast is in a starvation stage (Jenkinson et al. 1990). Yeast-like organisms were found in two of six specimens with therapy-resistant root canal infections (Nair et al. 1990). There was genotypic and phenotypic diversity among clinical strains of *C. albicans* recovered from infected root canals in Finland (Waltimo et al. 2001). These traits were similar in strains from other oral and nonoral sites, suggesting that *C. albicans* strains from infected root canals do not require unique characteristics to exist in the root canal environment (Waltimo et al. 2003). Fungi are discussed in greater detail in Chapter 9.

### 7.4.1 Combinations of microorganisms

It has been proposed that the combination of *F. nucleatum*, *Prevotella* spp., and *Porphyromonas* spp. may provide a risk factor for endodontic flare-ups by acting in synergy to increase the intensity of the periapical inflammatory reaction (Chavez de Paz Villanueva 2002). In mice, the combination of *F. nucleatum* and either *P. gingivalis* or *P. intermedia* induced more severe pathological subcutaneous lesions in mice in mixed, compared to pure culture (Baumgartner et al. 1992). Similarly, subcutaneous injections of the combination of *P. intermedia* and *P. micra* resulted in a more severe response in mice in combination rather than separately (Araki et al. 2004).

Specific interactions between streptococci and other bacteria may facilitate their invasion into dentin. For example, coinvasion of dentinal tubules by *P. gingivalis* and *S. gordonii*, but not with *S. mutans*, was facilitated by the streptococcal antigen I/II polypeptide (Love et al. 2000). Significant associations were reported to occur between specific combinations of species and clinical symptoms: swelling and the combination of *P. micra* and *Prevotella* spp., wet canals and the combinations of *Eubacterium* spp. with either *Prevotella* spp. or *Peptostreptococcus* spp., and pain and the combination of *Peptostreptococcus* spp. and *Prevotella* spp. (Gomes et al. 1996). In six of nine patients, the combination of *F. nucleatum* and *Streptococcus* spp. was associated with symptoms (Fouad et al. 2002). In root canals of teeth with apical periodontitis receiving treatment, there may be an association between *Lactobacillus* spp. and Gram-positive cocci (Chavez de Paz et al. 2004).

Endodontic infections are typically polymicrobial, and therefore interactions between different strains and species both antagonistic and synergistic would be expected. For example, the production of a bacteriocin by a producer strain that has an inhibitory activity against other strains may provide the producer with a selective advantage (Tomita et al. 1997; Riley and Wertz 2002), thereby modulating the infectious process. In contrast, virulence might be enhanced by synergistic interactions between species. For example, *P. micra* enhanced the pathogenicity of *Bacteroides melaninogenicus* (*Prevotella melaninogenica*) or *Bacteroides asaccharolyticus* (*Porphyromonas* spp.) in experimental infections in guinea pigs (Sundqvist

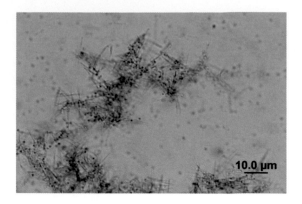

**Fig. 7.8**    Co-aggregation of *F. nucleatum* ATCC 10953 with *E. faecalis* E1. Co-aggregation assays were conducted according to the method of Cisar et al. (1979). Wet mounts of a combination of *F. nucleatum* and *E. faecalis* were stained with crystal violet. Images at 1,000× magnification were captured with an Olympus BX-51 microscope and DP-70 digital camera and software (Olympus, Melville, NY) (M. Chavarriaga and C.M. Sedgley, unpublished data).

et al. 1979). Other examples of beneficial interactions may utilize co-aggregation interactions (Cisar et al. 1979). Using confocal microscopy, co-aggregation interactions were observed in association with *Prevotella*, *Streptococcus*, and *Fusobacterium* species isolated from acute endodontic infections (Khemaleelakul et al. 2006). Co-aggregation interactions have also been observed between *F. nucleatum* and *E. faecalis* (Fig. 7.8) (M. Chavarriaga and C.M. Sedgley, unpublished data), and between *F. nucleatum* and *S. anginosus*, *P. anaerobius* and *P. oralis* (Johnson et al. 2006). *F. nucleatum* may provide a specific link or connection for other co-aggregating microorganisms (Kolenbrander et al. 1989). Interactions between genetically distinct bacteria are involved in the establishment and maintenance of biofilms (Kolenbrander 2000) and provide an accommodating environment for horizontal gene transfer (Sorensen et al. 2005).

*Biofilms* are sessile or pedunculated microbial communities composed of tower- or mushroom-shaped microcolonies, containing cells irreversibly attached to a substratum, an interface or each other. The matrix of extracellular polymeric substances is produced by cells within the biofilms (Donlan and Costerton, 2002). Interspersed between the microcolonies are channels separate from the external environment and through which fluids move by convection (Lawrence et al. 1991). The environmental conditions differ depending on location within the biofilm. For example, cells located more deeply in the biofilm are exposed to decreased oxygen tension, resulting in altered phenotypes in terms of growth rate and gene transcription that might facilitate certain virulence characteristics (ten Cate 2006). Although various irrigants and medicaments can disrupt biofilms in vitro (Distel et al. 2002; Clegg et al. 2006; Dunavant et al. 2006), the effectiveness of many antimicrobials is often impeded as a result of the slow metabolic rate of microorganisms in biofilms, as well as the extracellular matrix of the biofilm (Donlan and Costerton 2002; Sandoe et al. 2006).

Biofilms have been observed in the undebrided parts of the root canal system of surgically resected root apices (Nair et al. 2005). However, conditions under which biofilms might occur in infected root canals in vivo are not well understood (Svensater and Bergenholtz 2004) but may represent an important adaptive mechanism by which remaining bacteria survive treatment protocols (Chavez de Paz 2007). Formation of biofilms has been observed among bacteria from the mammalian gut, and the human urogenital tract (Fux et al. 2005), as well as being experimentally produced in root canals of extracted teeth with pure cultures of *E. faecalis* (Hubble et al. 2003). However, in contrast to strains from other clinical sources, in particular endocarditis strains (Sandoe et al. 2003; Mohamed et al. 2004), *E. faecalis* strains from oral and endodontic sources were not strong biofilm formers, and there were no significant associations between biofilm formation and the presence of the virulence determinants (genes or functional gene groups) *asa*, *cylA*, *esp*, and *gelE* (Duggan and Sedgley 2007).

Since biofilms are uniquely suited for horizontal gene transfer (Sorensen et al. 2005), they might provide an avenue for communication between species of relevance to endodontic infections, perhaps by facilitating plasmid transfer (Sedgley and Clewell 2004). More information is required on the role of biofilm communities in root canal infections, in particular in association with polymicrobial biofilms.

## 7.5 Concluding remarks and future directions

From the above, it is apparent that a diverse array of microbial virulence factors are potentially available in the infected root canal flora. It is reasonable to expect that the possession of certain virulence traits by one or a combination of species would provide a selective advantage over other species in the infected root canal in terms of survival and propagation of the infectious process. In particular, lipopolysaccharide (LPS) has been long recognized as a bacterial virulence factor of significance to endodontic infections. However, while these have the potential to modulate host–microbe interactions, apart from LPS, the specific role of these virulence factors and their relationship with clinical signs and symptoms at this time is still mostly speculative. More information is required about the identification and characterization of specific virulence factors as they pertain to human endodontic infections.

Nonetheless, there are some difficult barriers to overcome for progress in this field. The contribution of virulence factors associated with not-yet-cultivated organisms to the pathogenesis of endodontic infections is not at all clear. Considering that a large proportion of microorganisms from endodontic infections have not been cultivated, the development of methods to evaluate virulence properties in these endodontic bacterial pathogens is needed. In addition, the pathogenicity of specific bacterial species, or combinations of species identified in root canal infections in humans, has yet to be unequivocally established in controlled studies. From a practical standpoint, this may be difficult to establish considering the limitations of clinical sampling methods. Although intracanal paper point sampling has been the clinical standard for many decades, paper point samples hold only what is displaced from the canal into the paper point and do not provide information on inaccessible parts of the root canal. Further, the frequent recovery of the same species from both symptomatic and asymptomatic cases (Haapasalo et al. 1986; Baumgartner et al. 1999; Siqueira and Rocas 2005) makes interpretation difficult. This may indicate the capacity for species to modulate virulence under varying environmental conditions (Siqueira 2003) or alternatively that there may be different strains of the one species in root canal infections that express different virulence factors (Sedgley et al. 2005b; Reynaud Af Geijersstam et al. 2007). In addition, the available data thus far have focused on single species. There is a need for further investigation of virulence factors associated with polymicrobial endodontic infections existing as biofilm communities (Chavez de Paz 2007). Clinical studies are required that take the above factors into consideration as well as the potential modulatory role of individual host factors.

The design and implementation of such studies will likely present considerable challenges but have the potential to be rewarding beyond the scope of endodontics. In general terms, future gene expression studies may provide information on not only the viability but also the virulence of putative endodontic pathogens. A better understanding of the mechanisms by which the host regulates virulence gene expression by endodontic microorganisms may help explain different clinical presentations of endodontic infections and in turn ultimately contribute to the identification of therapeutic targets. In view of the fact that species recovered from root canal infections originate from the commensal oral flora, we may better understand how the host develops tolerance following exposure to commensal microorganisms that may influence the ability to counteract subsequent bacterial challenge (Medvedev et al. 2006). Other important and relevant areas of investigation could include how some bacteria can genetically vary their surface antigens, making them difficult targets for the immune system (Frank and Barbour 2006), and the ability of some intracellular bacteria to inactivate the killing mechanisms of phagocytic cells (Jansen and Yu 2006).

## 7.6 References

Abeyta M, Hardy GG, and Yother J. 2003. Genetic alteration of capsule type but not PspA type affects accessibility of surface-bound complement and surface antigens of *Streptococcus pneumoniae*. *Infect Immun* 71(1): 218–25.

Aktories K and Barbieri JT. 2005. Bacterial cytotoxins: targeting eukaryotic switches. *Nat Rev Microbiol* 3(5): 397–410.

Ando N and Hoshino E. 1990. Predominant obligate anaerobes invading the deep layers of root canal dentin. *Int Endod J* 23(1): 20–27.

Appelbe OK and Sedgley CM. 2007. Effects of prolonged exposure to alkaline pH on *Enterococcus faecalis* survival and specific gene transcripts. *Oral Microbiol Immunol* 22(3): 169–74.

Araki H, Kuriyama T, Nakagawa K, and Karasawa T. 2004. The microbial synergy of *Peptostreptococcus micros* and *Prevotella intermedia* in a murine abscess model. *Oral Microbiol Immunol* 19(3): 177–81.

Bachrach G, Leizerovici-Zigmond M, Zlotkin A, Naor R, and Steinberg D. 2003. Bacteriophage isolation from human saliva. *Lett Appl Microbiol* 36(1): 50–53.

Balaban N and Rasooly A. 2000. Staphylococcal enterotoxins. *Int J Food Microbiol* 61(1): 1–10.

Baumgartner JC, Falkler WA Jr, and Beckerman T. 1992. Experimentally induced infection by oral anaerobic microorganisms in a mouse model. *Oral Microbiol Immunol* 7(4): 253–56.

Baumgartner JC, Watkins BJ, Bae KS, and Xia T. 1999. Association of black-pigmented bacteria with endodontic infections. *J Endod* 25(6): 413–15.

Beveridge TJ. 1999. Structures of Gram-negative cell walls and their derived membrane vesicles. *J Bacteriol* 181(16): 4725–33.

Bjornson HS. 1984. Activation of Hageman factor by lipopolysaccharides of *Bacteroides fragilis, Bacteroides vulgatus*, and *Fusobacterium mortiferum*. *Rev Infect Dis* 6(Suppl 1): S30–33.

Bolstad AI, Jensen HB, and Bakken V. 1996. Taxonomy, biology, and periodontal aspects of *Fusobacterium nucleatum*. *Clin Microbiol Rev* 9(1): 55–71.

Borssen E and Sundqvist G. 1981. Actinomyces of infected dental root canals. *Oral Surg Oral Med Oral Pathol* 51(6): 643–48.

Botero TM, Mantellini MG, Song W, Hanks CT, and Nor JE. 2003. Effect of lipopolysaccharides on vascular endothelial growth factor expression in mouse pulp cells and macrophages. *Eur J Oral Sci* 111(3): 228–34.

Botero TM, Shelburne CE, Holland GR, Hanks CT, and Nor JE. 2006. TLR4 mediates LPS-induced VEGF expression in odontoblasts. *J Endod* 32(10): 951–55.

Brook I. 2003. Microbiology and management of endodontic infections in children. *J Clin Pediatr Dent* 28(1): 13–17.

Brook I, Grimm S, and Kielich RB. 1981. Bacteriology of acute periapical abscess in children. *J Endod* 7(8): 378–80.

Brouqui P and Raoult D. 2001. Endocarditis due to rare and fastidious bacteria. *Clin Microbiol Rev* 14(1): 177–207.

Brugger K, Redder P, She Q, Confalonieri F, Zivanovic Y, and Garrett RA. 2002. Mobile elements in archaeal genomes. *FEMS Microbiol Lett* 206(2): 131–41.

Capitani G, Eidam O, Glockshuber R, and Grutter MG. 2006. Structural and functional insights into the assembly of type 1 pili from *Escherichia coli*. *Microbes Infect* 8(8): 2284–90.

Casadevall A and Pirofski LA. 1999. Host–pathogen interactions: redefining the basic concepts of virulence and pathogenicity. *Infect Immun* 67(8): 3703–13.

Casadevall A and Pirofski L. 2001. Host–pathogen interactions: the attributes of virulence. *J Infect Dis* 184(3): 337–44.

Chang J, Zhang C, Tani-Ishii N, Shi S, and Wang CY. 2005. NF-kappaB activation in human dental pulp stem cells by TNF and LPS. *J Dent Res* 84(11): 994–98.

Chavez de Paz LE. 2007. Redefining the persistent infection in root canals: possible role of biofilm communities. *J Endod* 33(6): 652–62.

Chavez de Paz LE, Molander A, and Dahlen G. 2004. Grampositive rods prevailing in teeth with apical periodontitis undergoing root canal treatment. *Int Endod J* 37(9): 579–87.

Chavez de Paz LE, Svensater G, Dahlen G, and Bergenholtz G. 2005. Streptococci from root canals in teeth with apical periodontitis receiving endodontic treatment. *Oral Surg Oral Med Oral Pathol Oral Radiol Endod* 100(2): 232–41.

Chavez de Paz Villanueva LE. 2002. *Fusobacterium nucleatum* in endodontic flare-ups. *Oral Surg Oral Med Oral Pathol Oral Radiol Endod* 93(2): 179–83.

Chhour KL, Nadkarni MA, Byun R, Martin FE, Jacques NA, and Hunter N. 2005. Molecular analysis of microbial diversity in advanced caries. *J Clin Microbiol* 43(2): 843–49.

Cisar JO, Kolenbrander PE, and McIntire FC. 1979. Specificity of coaggregation reactions between human oral streptococci and strains of *Actinomyces viscosus* or *Actinomyces naeslundii*. *Infect Immun* 24(3): 742–52.

Clegg MS, Vertucci FJ, Walker C, Belanger M, and Britto LR. 2006. The effect of exposure to irrigant solutions on apical dentin biofilms in vitro. *J Endod* 32(5): 434–37.

Clewell DB and Francia MV. 2004. Conjugation in Gram-positive bacteria. *Plasmid Biology*. Washington, DC: American Society of Microbiology, 227–256.

Clewell DB, Francia MV, Flannagan SE, and An FY. 2002. Enterococcal plasmid transfer: sex pheromones, transfer origins, relaxases, and the *staphylococcus aureus* issue. *Plasmid* 48(3): 193–201.

Coburn PS, Baghdayan AS, Dolan GT, and Shankar N. 2007. Horizontal transfer of virulence genes encoded on the *Enterococcus faecalis* pathogenicity island. *Mol Microbiol* 63(2): 530–44.

Coleman RM and Georg LK. 1969. Comparative pathogenicity of *Actinomyces naeslundii* and *Actinomyces israelii*. *Appl Microbiol* 18(3): 427–32.

Costa ED, de Souza-Filho FJ, and Barbosa SV. 2003. Tissue reactions to a component of root canal system bacteria: lipoteichoic acid. *Braz Dent J* 14(2): 95–98.

Courtney H, Ofek I, Simpson WA, and Beachey EH. 1981. Characterization of lipoteichoic acid binding to polymorphonuclear leukocytes of human blood. *Infect Immun* 32(2): 625–31.

Courtney HS, Stanislawski L, Ofek I, Simpson WA, Hasty DL, and Beachey EH. 1988. Localization of a lipoteichoic acid binding site to a 24-kilodalton NH$_2$-terminal fragment of fibronectin. *Rev Infect Dis* 10(Suppl 2): S360–62.

Dahlen G. 1980. Immune response in rats against lipopolysaccharides of *Fusobacterium nucleatum* and *Bacteroides oralis* administered in the root canal. *Scand J Dent Res* 88(2): 122–29.

Dahlen G, Magnusson BC, and Moller A. 1981. Histological and histochemical study of the influence of lipopolysaccharide extracted from *Fusobacterium nucleatum* on the periapical tissues in the monkey *Macaca fascicularis*. *Arch Oral Biol* 26(7): 591–98.

Dahlen G, Wikstrom M, and Moller A. 1983. Production of histolytic enzymes by a combination of oral bacteria with known pathogenicity. *J Dent Res* 62(10): 1041–44.

Delahaye F, Fol S, Celard M, Vandenesch F, Beaune J, Bozio A, and de Gevigney G. 2005. *Propionibacterium acnes* infective endocarditis. Study of 11 cases and review of literature. *Arch Mal Coeur Vaiss* 98(12): 1212–28.

Dinges MM, Orwin PM, and Schlievert PM. 2000. Exotoxins of *Staphylococcus aureus*. *Clin Microbiol Rev* 13(1): 16–34.

Distel JW, Hatton JF, and Gillespie MJ. 2002. Biofilm formation in medicated root canals. *J Endod* 28(10): 689–93.

Donlan RM and Costerton JW. 2002. Biofilms: survival mechanisms of clinically relevant microorganisms. *Clin Microbiol Rev* 15(2): 167–93.

Duggan JM and Sedgley CM. 2007. Biofilm formation of oral and endodontic *Enterococcus faecalis*. *J Endod* 33(7): 815–18.

Dunavant TR, Regan JD, Glickman GN, Solomon ES, and Honeyman AL. 2006. Comparative evaluation of endodontic irrigants against *Enterococcus faecalis* biofilms. *J Endod* 32(6): 527–31.

Duncan L, Yoshioka M, Chandad F, and Grenier D. 2004. Loss of lipopolysaccharide receptor CD14 from the surface of human macrophage-like cells mediated by *Porphyromonas gingivalis* outer membrane vesicles. *Microb Pathog* 36(6): 319–25.

Durand SH, Flacher V, Romeas A, Carrouel F, Colomb E, Vincent C, Magloire H, Couble ML, Bleicher F, Staquet MJ, Lebecque S, and Farges JC. 2006. Lipoteichoic acid increases TLR and functional chemokine expression while reducing dentin formation in in vitro differentiated human odontoblasts. *J Immunol* 176(5): 2880–87.

Dwyer TG and Torabinejad M. 1980. Radiographic and histologic evaluation of the effect of endotoxin on the periapical tissues of the cat. *J Endod* 7(1): 31–35.

Eftimiadi C, Stashenko P, Tonetti M, Mangiante PE, Massara R, Zupo S, and Ferrarini M. 1991. Divergent effect of the anaerobic bacteria by-product butyric acid on the immune response: suppression of T-lymphocyte proliferation and stimulation of interleukin-1 beta production. *Oral Microbiol Immunol* 6(1): 17–23.

Fabricius L, Dahlen G, Holm SE, and Moller AJ. 1982. Influence of combinations of oral bacteria on periapical tissues of monkeys. *Scand J Dent Res* 90(3): 200–206.

Fabricius L, Dahlen G, Sundqvist G, Happonen RP, and Moller AJ. 2006. Influence of residual bacteria on periapical tissue healing after chemomechanical treatment and root filling of experimentally infected monkey teeth. *Eur J Oral Sci* 114(4): 278–85.

Facklam R. 2002. What happened to the streptococci: overview of taxonomic and nomenclature changes. *Clin Microbiol Rev* 15(4): 613–30.

Falcioni GC, Coderoni S, Tedeschi GG, Brunori M, and Rotilio G. 1981. Red cell lysis induced by microorganisms as a case of superoxide- and hydrogen peroxide-dependent hemolysis mediated by oxyhemoglobin. *Biochim Biophys Acta* 678(3): 437–41.

Figdor D and Davies J. 1997. Cell surface structures of *Actinomyces israelii*. *Aust Dent J* 42(2): 125–28.

Figdor D, Sjogren U, Sorlin S, Sundqvist G, and Nair PN. 1992. Pathogenicity of *Actinomyces israelii* and *Arachnia propionica*: experimental infection in guinea pigs and phagocytosis and intracellular killing by human polymorphonuclear leukocytes in vitro. *Oral Microbiol Immunol* 7(3): 129–36.

Figures KH and Douglas CW. 1991. Actinomycosis associated with a root-treated tooth: report of a case. *Int Endod J* 24(6): 326–29.

Flannagan SE, Clewell DB, and Sedgley CM. 2008. A "retrocidal" plasmid in *Enterococcus faecalis*: passage and protection. *Plasmid* 59(3): 217–30.

Fouad AF, Barry J, Caimano M, Clawson M, Zhu Q, Carver R, Hazlett K, and Radolf JD. 2002. PCR-based identification of bacteria associated with endodontic infections. *J Clin Microbiol* 40(9): 3223–31.

Frank SA and Barbour AG. 2006. Within-host dynamics of antigenic variation. *Infect Genet Evol* 6(2): 141–46.

Fux CA, Costerton JW, Stewart PS, and Stoodley P. 2005. Survival strategies of infectious biofilms. *Trends Microbiol* 13(1): 34–40.

Gilmore MS, Coburn PS, Nallapareddy SR, and Murray BE. 2002. Enterococcal virulence. *The Enterococci:*

*Pathogenesis, Molecular Biology, and Antibiotic Resistance.* Gilmore MS. Washington, DC: ASM Press, 301–54.

Ginsburg I. 2002. Role of lipoteichoic acid in infection and inflammation. *Lancet Infect Dis* 2(3): 171–79.

Gioannini TL and Weiss JP. 2007. Regulation of interactions of Gram-negative bacterial endotoxins with mammalian cells. *Immunol Res* 39(1–3): 249–60.

Gomes BP, Lilley JD, and Drucker DB. 1996. Associations of endodontic symptoms and signs with particular combinations of specific bacteria. *Int Endod J* 29(2): 69–75.

Grenier D, Bertrand J, and Mayrand D. 1995. *Porphyromonas gingivalis* outer membrane vesicles promote bacterial resistance to chlorhexidine. *Oral Microbiol Immunol* 10(5): 319–20.

Grenier D and Bouclin R. 2006. Contribution of proteases and plasmin-acquired activity in migration of *Peptostreptococcus micros* through a reconstituted basement membrane. *Oral Microbiol Immunol* 21(5): 319–25.

Grenier D and Grignon L. 2006. Response of human macrophage-like cells to stimulation by *Fusobacterium nucleatum* ssp. nucleatum lipopolysaccharide. *Oral Microbiol Immunol* 21(3): 190–96.

Gyles CL. 2007. Shiga toxin-producing *Escherichia coli*: an overview. *J Anim Sci* 85(13 Suppl): E45–62.

Haapasalo M, Ranta H, Ranta K, and Shah H. 1986. Black-pigmented *Bacteroides* spp. in human apical periodontitis. *Infect Immun* 53(1): 149–53.

Hahn CL and Liewehr FR. 2007. Relationships between caries bacteria, host responses, and clinical signs and symptoms of pulpitis. *J Endod* 33(3): 213–19.

Hancock HH III, Sigurdsson A, Trope M, and Moiseiwitsch J. 2001. Bacteria isolated after unsuccessful endodontic treatment in a North American population. *Oral Surg Oral Med Oral Pathol Oral Radiol Endod* 91(5): 579–86.

Happonen RP, Soderling E, Viander M, Linko-Kettunen L, and Pelliniemi LJ. 1985. Immunocytochemical demonstration of *Actinomyces* species and *Arachnia propionica* in periapical infections. *J Oral Pathol* 14(5): 405–13.

Happonen RP, Viander M, Pelliniemi L, and Aitasalo K. 1983. *Actinomyces israelii* in osteoradionecrosis of the jaws. Histopathologic and immunocytochemical study of five cases. *Oral Surg Oral Med Oral Pathol* 55(6): 580–88.

Hashioka K, Suzuki K, Yoshida T, Nakane A, Horiba N, and Nakamura H. 1994. Relationship between clinical symptoms and enzyme-producing bacteria isolated from infected root canals. *J Endod* 20(2): 75–77.

Haubek D, Willi K, Poulsen K, Meyer J, and Kilian M. 1997. Presence of bacteriophage Aa phi 23 correlates with the population genetic structure of *Actinobacillus actinomycetemcomitans*. *Eur J Oral Sci* 105(1): 2–8.

Hayes F. 2003. Transposon-based strategies for microbial functional genomics and proteomics. *Annu Rev Genet* 37: 3–29.

Henderson B, Poole S, and Wilson M. 1996. Microbial/host interactions in health and disease: who controls the cytokine network? *Immunopharmacology* 35(1): 1–21.

Henderson B and Wilson M. 1998. Commensal communism and the oral cavity. *J Dent Res* 77(9): 1674–83.

Heumann D, Barras C, Severin A, Glauser MP, and Tomasz A. 1994. Gram-positive cell walls stimulate synthesis of tumor necrosis factor alpha and interleukin-6 by human monocytes. *Infect Immun* 62(7): 2715–21.

Hill EE, Herijgers P, Claus P, Vanderschueren S, Herregods MC, and Peetermans WE. 2007. Infective endocarditis: changing epidemiology and predictors of 6-month mortality: a prospective cohort study. *Eur Heart J* 28(2): 196–203.

Hood BL and Hirschberg R. 1995. Purification and characterization of *Eikenella corrodens* type IV pilin. *Infect Immun* 63(9): 3693–96.

Horiba N, Maekawa Y, Abe Y, Ito M, Matsumoto T, and Nakamura H. 1991. Correlations between endotoxin and clinical symptoms or radiolucent areas in infected root canals. *Oral Surg Oral Med Oral Pathol* 71(4): 492–95.

Horiba N, Maekawa Y, Yamauchi Y, Ito M, Matsumoto T, and Nakamura H. 1992. Complement activation by lipopolysaccharides purified from Gram-negative bacteria isolated from infected root canals. *Oral Surg Oral Med Oral Pathol* 74(5): 648–51.

Hosoya S and Matsushima K. 1997. Stimulation of interleukin-1 beta production of human dental pulp cells by *Porphyromonas endodontalis* lipopolysaccharide. *J Endod* 23(1): 39–42.

Hubble TS, Hatton JF, Nallapareddy SR, Murray BE, and Gillespie MJ. 2003. Influence of *Enterococcus faecalis* proteases and the collagen-binding protein, Ace, on adhesion to dentin. *Oral Microbiol Immunol* 18(2): 121–26.

Huycke MM and Gilmore MS. 1995. Frequency of aggregation substance and cytolysin genes among enterococcal endocarditis isolates. *Plasmid* 34(2): 152–56.

Huycke MM and Gilmore MS. 1997. In vivo survival of *Enterococcus faecalis* is enhanced by extracellular superoxide production. *Adv Exp Med Biol* 418: 781–84.

Huycke MM, Joyce W, and Wack MF. 1996. Augmented production of extracellular superoxide by blood isolates of *Enterococcus faecalis*. *J Infect Dis* 173(3): 743–46.

Ike Y, Hashimoto H, and Clewell DB. 1987. High incidence of hemolysin production by *Enterococcus (Streptococcus) faecalis* strains associated with human parenteral infections. *J Clin Microbiol* 25(8): 1524–28.

Jacinto RC, Gomes BP, Shah HN, Ferraz CC, Zaia AA, and Souza-Filho FJ. 2005. Quantification of endotoxins in necrotic root canals from symptomatic and asymptomatic teeth. *J Med Microbiol* 54(Pt 8): 777–83.

Jacques NA, Hardy L, Knox KW, and Wicken AJ. 1979. Effect of growth conditions on the formation of extracellular lipoteichoic acid by *Streptococcus mutans* BHT. *Infect Immun* 25(1): 75–84.

Jansen A and Yu J. 2006. Differential gene expression of pathogens inside infected hosts. *Curr Opin Microbiol* 9(2): 138–42.

Jayaram M, Mehta S, Uzri D, and Velmurugan S. 2004. Segregation of the yeast plasmid: similarities and contrasts with bacterial plasmid partitioning. *Plasmid* 51(3): 162–78.

Jenkinson HF. 1994. Cell surface protein receptors in oral streptococci. *FEMS Microbiol Lett* 121(2): 133–40.

Jenkinson HF, Lala HC, and Shepherd MG. 1990. Coaggregation of *Streptococcus sanguis* and other streptococci with *Candida albicans*. *Infect Immun* 58(5): 1429–36.

Jerala R. 2007. Structural biology of the LPS recognition. *Int J Med Microbiol* 297(5): 353–63.

Johnson EM, Flannagan SE, and Sedgley CM. 2006. Coaggregation interactions between oral and endodontic *Enterococcus faecalis* and bacterial species isolated from persistent apical periodontitis. *J Endod* 32: 946–50.

Kamaguchi A, Nakayama K, Ichiyama S, Nakamura R, Watanabe T, Ohta M, Baba H, and Ohyama T. 2003. Effect of *Porphyromonas gingivalis* vesicles on coaggregation of *Staphylococcus aureus* to oral microorganisms. *Curr Microbiol* 47(6): 485–91.

Kato S, Kowashi Y, and Demuth DR. 2002. Outer membrane-like vesicles secreted by *Actinobacillus actinomycetemcomitans* are enriched in leukotoxin. *Microb Pathog* 32(1): 1–13.

Kato T, Kawai S, Nakano K, Inaba H, Kuboniwa M, Nakagawa I, Tsuda K, Omori H, Ooshima T, Yoshimori T, and Amano A. 2007. Virulence of *Porphyromonas gingivalis* is altered by substitution of fimbria gene with different genotype. *Cell Microbiol* 9(3): 753–65.

Kaufman B, Spangberg L, Barry J, and Fouad AF. 2005. *Enterococcus* spp. in endodontically treated teeth with and without periradicular lesions. *J Endod* 31(12): 851–56.

Kayaoglu G and Orstavik D. 2004. Virulence factors of *Enterococcus faecalis*: relationship to endodontic disease. *Crit Rev Oral Biol Med* 15(5): 308–20.

Keller L and Surette MG. 2006. Communication in bacteria: an ecological and evolutionary perspective. *Nat Rev Microbiol* 4(4): 249–58.

Kersten HW and Moorer WR. 1989. Particles and molecules in endodontic leakage. *Int Endod J* 22(3): 118–24.

Khemaleelakul S, Baumgartner JC, and Pruksakorn S. 2006. Autoaggregation and coaggregation of bacteria associated with acute endodontic infections. *J Endod* 32(4): 312–18.

Kinder SA and Holt SC. 1989. Characterization of coaggregation between *Bacteroides gingivalis* T22 and *Fusobacterium nucleatum* T18. *Infect Immun* 57(11): 3425–33.

Koch R. 1884. Die Aetiologie der Tuberkulose. *Mittheilingen aus dem Kaiserlichin Gesundheirsamte* 2: 1–88.

Kolenbrander PE. 2000. Oral microbial communities: Biofilms, interactions, and genetic systems. *Annu Rev Microbiol* 54: 413–37.

Kolenbrander PE, Andersen RN, and Moore LV. 1989. Coaggregation of *Fusobacterium nucleatum*, *Selenomonas flueggei*, *Selenomonas infelix*, *Selenomonas noxia*, and *Selenomonas sputigena* with strains from 11 genera of oral bacteria. *Infect Immun* 57(10): 3194–203.

Kowalski WJ, Kasper EL, Hatton JF, Murray BE, Nallapareddy SR, and Gillespie MJ. 2006. *Enterococcus faecalis* Adhesin, Ace, Mediates Attachment to Particulate Dentin. *J Endod* 32(7): 634–37.

Kuehn MJ and Kesty NC. 2005. Bacterial outer membrane vesicles and the host–pathogen interaction. *Genes Dev* 19(22): 2645–55.

Kurita-Ochiai T, Hashizume T, Yonezawa H, Ochiai K, and Yamamoto M. 2006. Characterization of the effects of butyric acid on cell proliferation, cell cycle distribution and apoptosis. *FEMS Immunol Med Microbiol* 47(1): 67–74.

Labandeira-Rey M, Couzon F, Boisset S, Brown EL, Bes M, Benito Y, Barbu EM, Vazquez V, Hook M, Etienne J, Vandenesch F, and Bowden MG. 2007. *Staphylococcus aureus* Panton-Valentine leukocidin causes necrotizing pneumonia. *Science* 315(5815): 1130–33.

Lawrence JR, Korber DR, Hoyle BD, Costerton JW, and Caldwell DE. 1991. Optical sectioning of microbial biofilms. *J Bacteriol* 173(20): 6558–67.

Leung KP, Fukushima H, Nesbitt WE, and Clark WB. 1996. *Prevotella intermedia* fimbriae mediate hemagglutination. *Oral Microbiol Immunol* 11(1): 42–50.

Li YH, Lau PC, Lee JH, Ellen RP, and Cvitkovitch DG. 2001. Natural genetic transformation of *Streptococcus mutans* growing in biofilms. *J Bacteriol* 183(3): 897–908.

Love RM. 1996. Regional variation in root dentinal tubule infection by *Streptococcus gordonii*. *J Endod* 22(6): 290–93.

Love RM, McMillan MD, and Jenkinson HF. 1997. Invasion of dentinal tubules by oral streptococci is associated with collagen recognition mediated by the antigen I/II family of polypeptides. *Infect Immun* 65(12): 5157–64.

Love RM, McMillan MD, Park Y, and Jenkinson HF. 2000. Coinvasion of dentinal tubules by *Porphyromonas gingivalis* and *Streptococcus gordonii* depends upon binding specificity of streptococcal antigen I/II adhesin. *Infect Immun* 68(3): 1359–65.

Mardis JS and Many WJ, Jr. 2001. Endocarditis due to *Actinomyces viscosus*. *South Med J* 94(2): 240–43.

Martin FE, Nadkarni MA, Jacques NA, and Hunter N. 2002. Quantitative microbiological study of human carious dentine by culture and real-time PCR: association of anaerobes with histopathological changes in chronic pulpitis. *J Clin Microbiol* 40(5): 1698–704.

Martinez JL and Baquero F. 2002. Interactions among strategies associated with bacterial infection: pathogenicity, epidemicity, and antibiotic resistance. *Clin Microbiol Rev* 15(4): 647–79.

Matsuguchi T, Takagi A, Matsuzaki T, Nagaoka M, Ishikawa K, Yokokura T, and Yoshikai Y. 2003. Lipoteichoic acids from *Lactobacillus strains* elicit strong tumor necrosis factor alpha-inducing activities in macrophages through Toll-like receptor 2. *Clin Diagn Lab Immunol* 10(2): 259–66.

Matsushima K, Ohbayashi E, Takeuchi H, Hosoya S, Abiko Y, and Yamazaki M. 1998. Stimulation of interleukin-6 production in human dental pulp cells by peptidoglycans from *Lactobacillus casei*. *J Endod* 24(4): 252–55.

McGrady JA, Butcher WG, Beighton D, and Switalski LM. 1995. Specific and charge interactions mediate collagen recognition by oral lactobacilli. *J Dent Res* 74(2): 649–57.

Medvedev AE, Sabroe I, Hasday JD, and Vogel SN. 2006. Tolerance to microbial TLR ligands: molecular mechanisms and relevance to disease. *J Endotoxin Res* 12(3): 133–50.

Mohamed JA, Huang W, Nallapareddy SR, Teng F, and Murray BE. 2004. Influence of origin of isolates, especially endocarditis isolates, and various genes on biofilm formation by *Enterococcus faecalis*. *Infect Immun* 72(6): 3658–63.

Moine P and Abraham E. 2004. Immunomodulation and sepsis: impact of the pathogen. *Shock* 22(4): 297–308.

Moller AJ, Fabricius L, Dahlen G, Sundqvist G, and Happonen RP. 2004. Apical periodontitis development and bacterial response to endodontic treatment. Experimental root canal infections in monkeys with selected bacterial strains. *Eur J Oral Sci* 112(3): 207–15.

Monod M and Borg-von ZM. 2002. Secreted aspartic proteases as virulence factors of *Candida* species. *Biol Chem* 383(7–8): 1087–93.

Morrison DA and Lee MS. 2000. Regulation of competence for genetic transformation in *Streptococcus pneumoniae*: a link between quorum sensing and DNA processing genes. *Res Microbiol* 151(6): 445–51.

Mory F, Fougnot S, Rabaud C, Schuhmacher H, and Lozniewski A. 2005. In vitro activities of cefotaxime, vancomycin, quinupristin/dalfopristin, linezolid and other antibiotics alone and in combination against *Propionibacterium acnes* isolates from central nervous system infections. *J Antimicrob Chemother* 55(2): 265–68.

Mostefaoui Y, Bart C, Frenette M, and Rouabhia M. 2004. *Candida albicans* and *Streptococcus salivarius* modulate IL-6, IL-8, and TNF-alpha expression and secretion by engineered human oral mucosa cells. *Cell Microbiol* 6(11): 1085–96.

Myhre AE, Aasen AO, Thiemermann C, and Wang JE. 2006. Peptidoglycan—an endotoxin in its own right? *Shock* 25(3): 227–35.

Naglik J, Albrecht A, Bader O, and Hube B. 2004. *Candida albicans* proteinases and host/pathogen interactions. *Cell Microbiol* 6(10): 915–26.

Nair PN, Henry S, Cano V, and Vera J. 2005. Microbial status of apical root canal system of human mandibular first molars with primary apical periodontitis after "one-visit" endodontic treatment. *Oral Surg Oral Med Oral Pathol Oral Radiol Endod* 99(2): 231–52.

Nair PN, Sjogren U, Krey G, Kahnberg KE, and Sundqvist G. 1990. Intraradicular bacteria and fungi in root-filled, asymptomatic human teeth with therapy-resistant periapical lesions: a long-term light and electron microscopic follow-up study. *J Endod* 16(12): 580–88.

Nelson AL, Roche AM, Gould JM, Chim K, Ratner AJ, and Weiser JN. 2007. Capsule enhances pneumococcal colonization by limiting mucus-mediated clearance. *Infect Immun* 75(1): 83–90.

Niederman R, Zhang J, and Kashket S. 1997. Short-chain carboxylic-acid-stimulated, PMN-mediated gingival inflammation. *Crit Rev Oral Biol Med* 8(3): 269–90.

Odds FC. 1988. *Candida and Candidosis: A Review and Bibliography*. London: Baillière Tindall-W.B. Saunders.

Odell LJ, Baumgartner JC, Xia T, and David LL. 1999. Survey for collagenase gene prtC in *Porphyromonas gingivalis* and *Porphyromonas endodontalis* isolated from endodontic infections. *J Endod* 25(8): 555–58.

Oliver D. 1996. Periplasm. In: Neidhardt FC (ed.), *Escherichia coli and Salmonella Cellular and Molecular Biology*, Vol. 1. Washington, DC: ASM Press, 88–103.

Perry AL and Lambert PA. 2006. *Propionibacterium acnes*. *Lett Appl Microbiol* 42(3): 185–88.

Plitnick LM, Jordan RA, Banas JA, Jelley-Gibbs DM, Walsh MC, Preissler MT, and Gosselin EJ. 2001. Lipoteichoic acid inhibits interleukin-2 (IL-2) function by direct binding to IL-2. *Clin Diagn Lab Immunol* 8(5): 972–79.

Rakita RM, Vanek NN, Jacques-Palaz K, Mee M, Mariscalco MM, Dunny GM, Snuggs M, Van Winkle WB, and Simon SI. 1999. *Enterococcus faecalis* bearing aggregation substance is resistant to killing by human neutrophils despite phagocytosis and neutrophil activation. *Infect Immun* 67(11): 6067–75.

Reynaud Af Geijersstam A, Culak R, Molenaar L, Chattaway M, Røslie E, Peciuliene V, Haapasalo M, and Shah HN. 2007. Comparative analysis of virulence determinants and mass spectral profiles of Finnish and Lithuanian endodontic *Enterococcus faecalis* isolates. *Oral Microbiol Immunol* 22(2): 87–94.

Riley MA and Wertz JE. 2002. Bacteriocins: evolution, ecology, and application. *Annu Rev Microbiol* 56: 117–37.

Rocas IN and Siqueira JF, Jr. 2006. Culture-independent detection of *Eikenella corrodens* and *Veillonella parvula* in primary endodontic infections. *J Endod* 32(6): 509–12.

Rocas IN, Siqueira JF, Jr, and Santos KR. 2004. Association of *Enterococcus faecalis* with different forms of periradicular diseases. *J Endod* 30(5): 315–20.

Rogers JD, Palmer RJ, Jr, Kolenbrander PE, and Scannapieco FA. 2001. Role of *Streptococcus gordonii* amylase-binding protein A in adhesion to hydroxyapatite, starch metabolism, and biofilm formation. *Infect Immun* 69(11): 7046–56.

Rossi-Fedele G, Scott W, Spratt D, Gulabivala K, and Roberts AP. 2006. Incidence and behaviour of Tn916-like elements within tetracycline-resistant bacteria isolated from root canals. *Oral Microbiol Immunol* 21(4): 218–22.

Roth RR and James WD. 1988. Microbial ecology of the skin. *Annu Rev Microbiol* 42: 441–64.

Sakellariou PL. 1996. Periapical actinomycosis: report of a case and review of the literature. *Endod Dent Traumatol* 12(3): 151–54.

Sandoe JA, Witherden IR, Cove JH, Heritage J, and Wilcox MH. 2003. Correlation between enterococcal biofilm formation in vitro and medical-device-related infection potential in vivo. *J Med Microbiol* 52(Pt 7): 547–50.

Sandoe JA, Wysome J, West AP, Heritage J, and Wilcox MH. 2006. Measurement of ampicillin, vancomycin, linezolid and gentamicin activity against enterococcal biofilms. *J Antimicrob Chemother* 57(4): 767–70.

Schein B and Schilder H. 1975. Endotoxin content in endodontically involved teeth. *J Endod* 1(1): 19–21.

Schonfeld SE, Greening AB, Glick DH, Frank AL, Simon JH, and Herles SM. 1982. Endotoxic activity in periapical lesions. *Oral Surg Oral Med Oral Pathol* 53(1): 82–87.

Sedgley C, Buck G, and Appelbe O. 2006a. Prevalence of *Enterococcus faecalis* at multiple oral sites in endodontic patients using culture and PCR. *J Endod* 32(2): 104–9.

Sedgley C, Nagel A, Dahlen G, Reit C, and Molander A. 2006b. Real-time quantitative polymerase chain reaction and culture analyses of *Enterococcus faecalis* in root canals. *J Endod* 32(3): 173–77.

Sedgley CM. 2007. The influence of root canal sealer on extended intracanal survival of *Enterococcus faecalis* with and without gelatinase production ability in obturated root canals. *J Endod* 33(5): 561–66.

Sedgley CM and Clewell DB. 2004. Bacterial plasmids in the oral and endodontic microflora. *Endod Topics* 9: 37–51.

Sedgley CM, Lee EH, Martin MJ, and Flannagan SE. 2008. Antibiotic resistance gene transfer between *Streptococcus gordonii* and *Enterococcus faecalis* in root canals of teeth ex vivo. *J Endod* 34(5): 570–74.

Sedgley CM, Lennan SL, and Appelbe OK. 2005a. Survival of *Enterococcus faecalis* in root canals ex vivo. *Int Endod J* 38(10): 735–42.

Sedgley CM, Lennan SL, and Clewell DB. 2004. Prevalence, phenotype and genotype of oral enterococci. *Oral Microbiol Immunol* 19(2): 95–101.

Sedgley CM, Molander A, Flannagan SE, Nagel AC, Appelbe OK, Clewell DB, and Dahlen G. 2005b. Virulence, phenotype and genotype characteristics of endodontic *Enterococcus* spp. *Oral Microbiol Immunol* 20(1): 10–19.

Shen S, Samaranayake LP, and Yip HK. 2005. Coaggregation profiles of the microflora from root surface caries lesions. *Arch Oral Biol* 50(1): 23–32.

Siqueira JF, Jr. 2003. Microbial causes of endodontic flare-ups. *Int Endod J* 36(7): 453–63.

Siqueira JF, Jr, De Uzeda M, and Fonseca ME. 1996. A scanning electron microscopic evaluation of in vitro dentinal tubules penetration by selected anaerobic bacteria. *J Endod* 22(6): 308–10.

Siqueira JF, Jr, and Rocas IN. 2005. Exploiting molecular methods to explore endodontic infections: Part 2—Redefining the endodontic microbiota. *J Endod* 31(7): 488–98.

Siqueira JF, Jr, Rocas IN, Souto R, de Uzeda M, and Colombo AP. 2002. *Actinomyces* species, streptococci, and *Enterococcus faecalis* in primary root canal infections. *J Endod* 28(3): 168–72.

Sirard JC, Bayardo M, and Didierlaurent A. 2006. Pathogen-specific TLR signaling in mucosa: mutual contribution of microbial TLR agonists and virulence factors. *Eur J Immunol* 36(2): 260–63.

Sjogren U, Figdor D, Persson S, and Sundqvist G. 1997. Influence of infection at the time of root filling on the outcome of endodontic treatment of teeth with apical periodontitis. *Int Endod J* 30(5): 297–306.

Sorensen SJ, Bailey M, Hansen LH, Kroer N, and Wuertz S. 2005. Studying plasmid horizontal transfer in situ: a critical review. *Nat Rev Microbiol* 3(9): 700–710.

Sriskandan S, Faulkner L, and Hopkins P. 2007. *Streptococcus pyogenes*: insight into the function of the streptococcal superantigens. *Int J Biochem Cell Biol* 39(1): 12–19.

Sunde PT, Olsen I, Debelian GJ, and Tronstad L. 2002. Microbiota of periapical lesions refractory to endodontic therapy. *J Endod* 28(4): 304–10.

Sundqvist G. 1992. Associations between microbial species in dental root canal infections. *Oral Microbiol Immunol* 7(5): 257–62.

Sundqvist G, Bloom GD, Enberg K, and Johansson E. 1982. Phagocytosis of *Bacteroides melaninogenicus* and *Bacteroides gingivalis* in vitro by human neutrophils. *J Periodontal Res* 17(2): 113–21.

Sundqvist GK, Eckerbom MI, Larsson AP, and Sjogren UT. 1979. Capacity of anaerobic bacteria from necrotic dental pulps to induce purulent infections. *Infect Immun* 25(2): 685–93.

Svensater G and Bergenholtz G. 2004. Biofilms in endodontic infections. *Endod Topics* 9: 27–36.

Takada H, Kawabata Y, Tamura M, Matsushita K, Igarashi H, Ohkuni H, Todome Y, Uchiyama T, and Kotani S. 1993. Cytokine induction by extracellular products of oral viridans group streptococci. *Infect Immun* 61(12): 5252–60.

Tang G, Yip HK, Samaranayake LP, Chan KY, Luo G, and Fang HH. 2004. Direct detection of cell surface interactive forces of sessile, fimbriated and non-fimbriated *Actinomyces* spp. using atomic force microscopy. *Arch Oral Biol* 49(9): 727–38.

Telles PD, Hanks CT, Machado MA, and Nor JE. 2003. Lipoteichoic acid up-regulates VEGF expression in macrophages and pulp cells. *J Dent Res* 82(6): 466–70.

ten Cate JM. 2006. Biofilms, a new approach to the microbiology of dental plaque. *Odontology* 94(1): 1–9.

Tomita H, Fujimoto S, Tanimoto K, and Ike Y. 1997. Cloning and genetic and sequence analyses of the bacteriocin 21 determinant encoded on the *Enterococcus faecalis* pheromone-responsive conjugative plasmid pPD1. *J Bacteriol* 179(24): 7843–55.

Trent MS, Stead CM, Tran AX, and Hankins JV. 2006. Diversity of endotoxin and its impact on pathogenesis. *J Endotoxin Res* 12(4): 205–23.

Vowels BR, Yang S, and Leyden JJ. 1995. Induction of proinflammatory cytokines by a soluble factor of *Propionibacterium acnes*: implications for chronic inflammatory acne. *Infect Immun* 63(8): 3158–65.

Walker L, Levine H, and Jucker M. 2006. Koch's postulates and infectious proteins. *Acta Neuropathol (Berl)* 112(1): 1–4.

Waltimo TM, Dassanayake RS, Orstavik D, Haapasalo MP, and Samaranayake LP. 2001. Phenotypes and randomly amplified polymorphic DNA profiles of *Candida albicans* isolates from root canal infections in a Finnish population. *Oral Microbiol Immunol* 16(2): 106–12.

Waltimo TM, Sen BH, Meurman JH, Orstavik D, and Haapasalo MP. 2003. Yeasts in apical periodontitis. *Crit Rev Oral Biol Med* 14(2): 128–37.

Wang BY, Chi B, and Kuramitsu HK. 2002. Genetic exchange between *Treponema denticola* and *Streptococcus gordonii* in biofilms. *Oral Microbiol Immunol* 17(2): 108–12.

Wang JE, Jorgensen PF, Almlof M, Thiemermann C, Foster SJ, Aasen AO, and Solberg R. 2000. Peptidoglycan and lipoteichoic acid from *Staphylococcus aureus* induce tumor necrosis factor alpha, interleukin 6 (IL-6), and IL-10 production in both T cells and monocytes in a human whole blood model. *Infect Immun* 68(7): 3965–70.

Wang JE, Jorgensen PF, Ellingsen EA, Almiof M, Thiemermann C, Foster SJ, Aasen AO, and Solberg R. 2001. Peptidoglycan primes for LPS-induced release of proinflammatory cytokines in whole human blood. *Shock* 16(3): 178–82.

Waters VL. 2001. Conjugation between bacterial and mammalian cells. *Nat Genet* 29(4): 375–76.

Wu H and Fives-Taylor PM. 2001. Molecular strategies for fimbrial expression and assembly. *Crit Rev Oral Biol Med* 12(2): 101–15.

Yang LC, Tsai CH, Huang FM, Liu CM, Lai CC, and Chang YC. 2003. Induction of interleukin-6 gene expression by pro-inflammatory cytokines and black-pigmented Bacteroides in human pulp cell cultures. *Int Endod J* 36(5): 352–57.

Yang LC, Tsai CH, Huang FM, Su YF, Lai CC, Liu CM, and Chang YC. 2004. Induction of vascular endothelial growth factor expression in human pulp fibroblasts stimulated with black-pigmented Bacteroides. *Int Endod J* 37(9): 588–92.

Yeung MK. 1999. Molecular and genetic analyses of Actinomyces spp. *Crit Rev Oral Biol Med* 10(2): 120–38.

Yeung MK and Kozelsky CS. 1997. Transfection of Actinomyces spp. by genomic DNA of bacteriophages from human dental plaque. *Plasmid* 37(2): 141–53.

Yoshimura A, Lien E, Ingalls RR, Tuomanen E, Dziarski R, and Golenbock D. 1999. Cutting edge: recognition of Gram-positive bacterial cell wall components by the innate immune system occurs via Toll-like receptor 2. *J Immunol* 163(1): 1–5.

# Chapter 8
# Herpesviruses in Endodontic Pathosis

*Mohammad Sabeti*

## 8.1 Introduction

Viruses are the simplest and smallest microorganisms that infect humans. They consist of either DNA or RNA surrounded by a protein coat termed as "capsid." The most commonly known viruses within oral cavity are the herpes viruses. They are the most important DNA viruses that cause oral disease in human. The hallmark of herpesvirus infections is immune impairment. Fig. 8.1 describes the infection process of herpesviruses.

To date, eight human herpesviruses have been identified, namely, herpes simplex virus type 1 and type 2 (HSV 1–2), varicella-zoster virus (VZV), Epstein–Barr virus (EBV), human cytomegalovirus (HCMV), human herpesvirus-6 (HHV-6), human herpesvirus-7 (HHV-7), and human herpesvirus-8 (HHV-8). Humans are the only source of infection for these eight herpesviruses.

Human herpesviruses are classified into three groups ($\alpha$, $\beta$, and $\delta$) based on details of tissue tropism, pathogenicity, and behavior in the laboratory (Table 8.1). In most individuals, primary infection by herpesviruses occurs early in life and exhibits few or no overt disease symptoms. Herpesviruses remain in infected hosts for a lifetime in a prolonged state of latency but retain their capacity for renewed or episodic reactivated replication. In the latent phase of infection, herpesviruses reside in the following cells:

1. Herpes simplex virus type 1 and type 2 in sensory nerve ganglia and monocytes
2. Epstein–Barr virus in B-lymphocytes and salivary gland tissue
3. Varicella-zoster in sensory nerve ganglia
4. Human cytomegalovirus in monocytes, macrophages, lymphocytes, and salivary gland tissue
5. Human herpesvirus-6 in lymphocytes and ductal epithelium of salivary gland
6. Human herpesvirus-7 in lymphocytes and salivary gland tissue
7. Human herpesvirus-8 in lymphocytes and macrophages

Reactivation of latent herpesviruses is involved in driving the pathological process of some types of symptomatic periapical disease. Physical trauma, stress, immunosuppression, immune dysfunction, and radiotherapy may trigger herpesviruses activation.

## 8.2 General description of herpesviruses

Membership in the family Herpesviridae is based on the structure of the virion (Roizman and Pellett 2001). The prototypical structure of herpesviruses consists

**Table 8.1** Classification of human herpesviruses.

| Herpesviruses | Herpes group |
| --- | --- |
| HSV-1 | α |
| HSV-2 | α |
| VZV | α |
| EBV | δ |
| HCMV | β |
| HHV-6 | β |
| HHV-7 | β |
| HHV-8 | δ |

of a double-stranded DNA genome ranging in size from 120 to 250 kilobase (kb) pairs encased within an icosahedral capsid and an amorphous proteinaceous tegument, which is surrounded by a lipid bilayer enve-

lope derived from the host cell membrane. Herpesviral replication takes place in the nucleus of the host cell and involves the expression of immediate-early, early, and late classes of genes. Late (structural) genes are expressed during the productive (lytic) phase of herpesviral infections. After primary exposure, herpesviruses establish latency in various host cell reservoirs, from which they may reactivate periodically (Sissons et al. 2002). Table 8.2 summarizes some of the common characteristics of herpesviruses.

Most herpesviruses are ubiquitous agents that are often acquired early in life and infect individuals from diverse geographical areas and economic background (Britt and Alford 1996; Rinckinson and Kueff 1996). Herpesvirus transmission occurs by intimate contact with infected secretions including saliva, blood, and

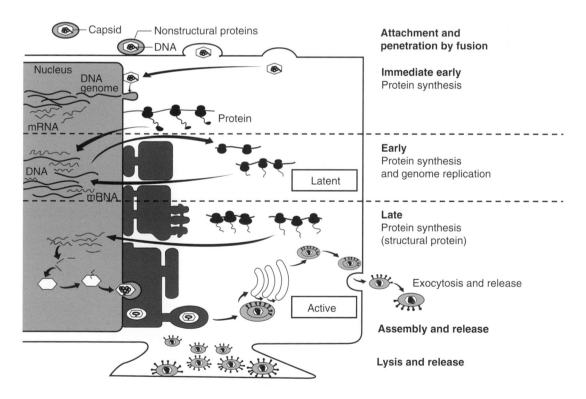

**Fig. 8.1** Schematic representation of the replication of herpesviruses. A virion initiates infection by fusion of the viral envelope with the plasma membrane after attachment to the cell surface. The capsid is transported to the nuclear pore, where viral DNA is released into the nucleus. Viral transcription and translation occur in three phases: immediate early, early, and late. Immediate early proteins shut off cell protein synthesis. Early proteins facilitate viral DNA replication. Late proteins are structural proteins of the virus, which form empty capsids. Viral DNA is packaged into preformed capsids in the nucleus. Viral glycoproteins and tegument protein patches in cellular membranes and capsids are enveloped. Virions are transported via endoplasmic reticulum and released by exocytosis or cell lysis. (Obtained from Slots et al. 2002, with the permission from the author.)

**Table 8.2**　Summary of common characteristics of herpesviruses.

A single double-stranded DNA molecule ranging from 120 to 250 kb pairs

An icosapentahedral capsid containing 162 capsomeres

An amorphous proteinaceous tegument, and surrounding the capsid and tegument

A lipid bilayer envelope derived from host cell membrane

Tissue tropism

The viral productive phase is followed by a latent phase in host cells, which ensures survival of the viral genome throughout the lifetime of the infected individual

Latent herpesvirus can undergo sporadic reactivation and reenter the productive phase

genital secretion (Gautheret-Dejean et al. 1977; Britt and Alford 1996; Ikuta et al. 2000). Acquisition of herpesviruses takes place from an early age and sometimes in uterus. A notable exception is herpesvirus-8 that is contracted in adulthood. Clinical manifestations of herpesvirus infections are highly diverse and range from mild or subclinical disease in most systemically healthy individuals to encephalitis, pneumonia, and other potentially lethal infections and various types of cancer including lymphoma, sarcoma, and carcinoma in immunocompromised hosts. Herpesviruses are the most important DNA viruses in oral pathology.

## 8.3 Human cytomegalovirus

Human cytomegalovirus is a ubiquitous agent that is often acquired early in life. Transmission of HCMV can happen prenatally from mother to infant. HCMV may be found in blood and in many body secretion including maternal milk, semen, and saliva (Gautheret-Dejean et al. 1977; Britt and Alford 1996). It is estimated that asymptomatic secretion shedding of HCMV into saliva, cervical secretions, semen, and breast milk occurs in 10–30% of infected individuals (Britt and Alford 1996). Most primary infections are asymptomatic. HCMV can infect and replicate in endothelial cells (Grefte et al. 1993a, b), ductal epithelial cells (Mocarski and Stinski 1979; Grundy 1990; Sinzger et al. 1995; Roizman and Pellett 2001), smooth muscle cells (Platcher et al. 1995; Sinzger et al. 1995; Mocarski 1996), and fibroblasts (Myerson et al. 1984; Platcher et al. 1995; Sinzger et al. 1995; Mocarski 1996; Roizman and Pellett 2001). HCMV

targets peripheral blood monocytes and lymphocytes during active infection (Myerson et al. 1984; Kapasi and Rice 1988; Dankner et al. 1990; Gerna et al. 1992; Meyer-Konig et al. 1995; Sinzger et al. 1995; Mocarski 1996). The site of latent HCMV is peripheral monocytes (Maciejowski et al. 1993; Taylor-Wiedeman et al. 1993) and may be found in bone marrow–derived progenitor cells (Maciejewski et al. 1992; Sinzger et al. 1995; Sinzger and Jahn 1996; Roizman and Pellett 2001). HCMV is emerging as an important opportunistic pathogen in immunocompromised person. HCMV is the most common life-threatening infection in HIV-infected patients (Griffiths and Emery 1997) and significant risk factor in transplant complications and rejections (Mutimer et al. 1997; Nowzari et al. 2003).

HCMV can cause serious infectious diseases. Cytomegalovirus infection is of great clinical significance in pregnant women, newborn infants with congenital or perinatal infection, immunosuppressed transplant patients, and individuals with AIDS. HCMV infection may be categorized under three clinical conditions: acquired HCMV infection, HCMV inclusion disease, and perinatal disease. Acquired HCMV infection can be observed in individuals with compromised immune system and patients who had tissue or organ transplants. HCMV infection can aggravate and enhance the immunosuppressiveness of HIV opportunistic infections. Necrotizing retinitis is a relatively common HCMV-induced complication in untreated HIV-infected individuals (Sweet 1999). HCMV may be a potential cause of salivary gland dysfunction in HIV patients. HCMV activation and resulting disease has become a major clinical problem in transplant recipients. HCMV infection is the most common reason for the transplant rejection, including bone marrow or stem cell grafts (Clark et al. 2003). HCMV inclusion disease is similar to infection mononucleosis. HCMV perinatal infection is the major cause of pregnancy complications and birth defects (Alford et al. 1979). HCMV-infected newborns may show low birth weight, jaundice, hepatosplenomegaly, skin rash, microcephaly, or chorioretinitis (Bale et al. 1990). HCMV infection is the cause of mental retardation and sensorineural deafness (Revello and Gerna 2004). Approximately one-third of newborns with symptomatic congenital HCMV infection born to mothers with recurrent HCMV infection or to mothers with primary HCMV infection during pregnancy may be premature (Boppana et al. 1999).

## 8.4 Epstein–Barr virus

EBV has two types: type 1 and type 2. The primary route of EBV acquisition is through salivary exchange in the oropharynx (Rickinson and Kieff 2001). EBV infection occurs in epithelial cells of oropharynx (Sixbey et al. 1984), and EBV DNA, RNA, and viral antigens have recently been detected in T-lymphocytes of Kawasaki-like disease (Kikuta et al. 1988), in nasal T-cell lymphomas (Tao et al. 1995), and in the epithelial cells of various human carcinomas (Raab-Traub et al. 1991; Luqmani et al. 2001). Latent EBV infection resides in B-lymphocytes (Klein 1989).

Epstein–Barr virus is a known cause of infectious mononucleosis and almost certainly plays a role in the etiology of nasopharyngeal carcinoma, Burkitt's lymphoma, and lymphoproliferative disorders in the presence of immunosuppression. Less certain is the role of Epstein–Barr virus in rheumatoid arthritis, Hodgkin's disease, and chronic fatigue syndrome. EBV may involve in development of aggressive types of non-Hodgkin lymphomas affecting gingiva (Yin et al. 1999). EBV is the main causative agent of infection mononucleosis, which is common clinical manifestation of a primary EBV infection in adolescents and young adults. Symptoms of this disease included fever, lymphadenopathy, malaise, and sore throat. Oral ulcers and multiple palatal petechiae have been reported. HIV-infected patients experience frequent EBV-2 infection and dual EBV1-EBV2 infections. EBV has also been implicated in multiple sclerosis and plays a role in development of oral hairy leukoplakia as evidenced by EBV replicating within epithelial cells (Walling et al. 2004a). EBV-encoded nuclear antigen (EBNA)-2 protein function (Walling et al. 2004b) and an EBV related decrease in oral epithelial Langerhans cells (Yamazaki and Nakajima 2004).

## 8.5 Herpes simplex virus type 1 and 2

Herpes simplex virus (HSV) infects keratinocytes monocytes, macrophages, and establishes latency in neurons (Laskaris 1996). Infected patients experience an initial primary infection followed with a latency period. HSV-1 mainly causes oral infections and HSV-2 anogenital infections. HSV-1 is responsible for most cases of herpetic gingivostomatitis with clinical presentation of multiple shallow ulcers throughout the keratinized and gland-bearing intraoral surface. Patient with latent herpes simplex infections develops episodes of recurrent oral herpes labialis characterized with occurrence of a cluster of vesicles and shallow ulcers localized to the lateral aspects of the lips. HSV-1 and HSV-2 are also implicated in recurrent erythema multiforme, Behcet's syndrome, some oral ulcers, and oral squamous carcinoma (Scully et al. 1991; Scott et al. 1997).

## 8.6 Varicella-zoster virus

Primary infection of varicella-zoster virus (VZV) (chickenpox) acquired during childhood that produces minor lesions throughout the oral cavity, whereas VZV reactivation in adults causes herpes zoster (shingles). Varicella is the initial infection of VZV and is a highly infectious disease transmitted by direct contact with lesions or by inhalation of infective droplets. Oral lesions include vesicles on the lips, and hard and soft palate (Millar and Troulis 1994; Miller 1996). Both primary and secondary VZV infection can produce gingival lesions (Laskaris 1996; Scully 1996). Following primary infection, VZV remains latent in the dorsal root ganglion cells for possibly later reactivation. Herpes zoster (shingles) develops from reactivation of a varicella infection. It involves the trigeminal nerve and forms ulcerated lesions with prominent red borders, resembling aphthous ulcers. Lesions are unilaterally distributed along the infected nerve (Millar and Troulis 1994; Miller 1996).

## 8.7 Human herpesvirus-6

Human herpesvirus-6 (HHV-6) exhibits tropism for $CD4^+$ T-lymphocytes (Ablashi et al. 1991) oral epithelium and gingival of periodontitis. It has also been reported to infect a wide variety of cell types (Ablashi et al. 1991). HHV-6 infects ductal epithelium of salivary glands and is isolated from saliva of most individuals (Yadav et al. 1997). HHV-6 mainly infects T-lymphocytes and occasionally B-lymphocytes (Lusso et al. 1988). Since EBV has been reported to be prevalent in preradicular pathosis, molecular interaction between HHV-6 and EBV may be important in the pathogenesis of periradicular infection. HHV-6 is frequently detected and reactivated in epithelial tumors of the oral cavity (Yadav et al. 1994; Parra and Slots 1996). HHV-6 may be involved in oral squamous carcinoma (Yadav et al. 1997). Yadav et al. studied 51 squamous cell carcinomas, 18 nonmalignant lesions, and 7 normal mucosa samples. HHV-6 was detected in 79% of malignancies, in 67% of lichen planus lesions and leukoplakia, but was absent in normal mucosa (Yadav et al. 1997). HHV-6 variant B was detected in 60% of the squamous carcinoma lesions (Yadav et al. 1997). HHV-6 is also frequently detected and reactivated in epithelial tumors of the oral cavity (Yadav et al. 1994; Parra and Slots 1996).

HHV-6 may be implicated in the pathogenesis of mononucleosis, pneumonia, meningitis, and encephalitis. It has been implicated as a cofactor of accelerated immunosuppression in HIV-infected individuals. It can proliferate $CD4^+$ and $CD8^+$ lymphocytes and natural killer cells, thereby increasing the severity of HIV infection (Shanavas et al. 1992). HHV-6 may be the cause of multiple sclerosis (Yoshikawa et al. 1992). HHV-6 is present in the nuclei of brain oligodendrocytes associated with multiple sclerosis plaques (Challoner et al. 1995).

## 8.8 Human herpesvirus-7 and -8

Salivary glands are a major site for a persistent and productive infection by human herpesvirus-7 (HHV-7) (Sada et al. 1996). HHV-7 infection occurs primarily in early childhood (Wyatt et al. 1991; Clark et al. 1993) and infectious virus is readily isolated from saliva (Hidaka et al. 1993). Studies have revealed that HHV-7 and HHV-6 are two beta-herpesviruses that are closely related and exhibit serologic cross-reactivity with each other (Levy 1997). HHV-7 infection usually occurs in childhood (Wyatt et al. 1991) and most adults are HHV-7 seropositive. HHV-7 is found in saliva (Hidaka et al. 1993), which presents the major mode of transmission, and is secreted for many years following initial infection (Takahashi et al. 1997). Minor labial salivary glands often harbor HHV-7 and may sometimes be the site of viral replication (Kimberlin 1998). In a study of more than 100 specimens from major salivary glands, Sada et al. detected HHV-7 in 100% of submandibular, in 85% of parotid and in 59% of minor lip salivary gland samples (Sada et al. 1996). HHV-7 has also been detected in periodontal pocket and gingival biopsy samples (Contreras et al. 2000).

HHV-8 is believed to be associated with development of Kaposi's sarcoma. The most commonly encountered angiosarcoma within oral cavity is a Kaposi's sarcoma that is associated with human immunodeficiency virus (HIV). Kaposi's sarcoma is rare in absence of HIV infection. HHV-8 DNA has also been identified in AIDS-related oral Kaposi's sarcoma (Flaitz et al. 1997) and body cavity–based non-Hodgkin's lymphomas, in Castleman's disease and in anti-immunoblastic lymphadenopathy (Luppi et al. 1994; Moore and Chang 1995; Kemeny et al. 1997). HHV-8 has also been detected in periodontal pocket and gingival biopsy samples (Contreras et al. 2000). Incidence of HHV-8 in 25% of the adult U.S. population and in about 8% of children has been reported (Lennette et al. 1996).

Kaposi's sarcoma is a unique form of angiosarcoma that occurs in elderly patients of mainly Mediterranean, Eastern European, or Middle Eastern descent, and are HIV-positive. It has a predilection for palate. HIV-associated Kaposi's sarcoma has become relatively prevalent and commonly seen on the skin and within oral cavity in 60% of patients and may later progress to extraoral sites (Flaitz et al. 1997), although the disease can also occur with dermal bullous pemphigoid in HIV-negative immunosuppressed patients (Gaspari et al. 1997). Immunosuppression serves to activate a latent HHV-8 infection Kaposi's sarcoma. It may become symptomatic in 25% of the patients (Di Alberti et al. 1996).

## 8.9 Association between herpesviruses and apical disease

Recent studies have investigated the occurrence of herpesviruses in periapical lesions (Sabeti et al. 2003a, b, c; Sabeti and Slots 2004). cDNA identification of genes transcribed late during the infectious cycle of herpesviruses was used to indicate herpesvirus-active infection (Sabeti et al. 2003a). The findings obtained from different studies with different sample of patients revealed a strong association of human cytomegalovirus and Epstein–Barr virus with symptomatic periapical lesions (Table 8.3). Herpes simplex virus infection demonstrated no relationship to periapical disease. Periapical lesions harboring cytomegalovirus/Epstein–Barr virus dual infection tended to show elevated occurrence of anaerobic bacteria, be symptomatic, and exhibit large size radiographic bone destruction (Sabeti and Slots 2004).

**Table 8.3**   Cytomegalovirus (CMV) and Epstein–Barr virus (EBV) active infection in periapical pathosis.

| Study | Total number of periapical lesions (sites) studied | Symptomatic lesions: number (%) infected[a] | Asymptomatic lesions (sites): number (%) infected | Large size lesions (5×7 mm or larger): number (%) infected | Small size lesions (sites): number (%) infected |
|---|---|---|---|---|---|
| Sabeti et al. (2003c) | 14 lesions, 2 healthy periapical sites | CMV+/EBV−: 5 (38%) EBV+/CMV−: 0 CMV+/EBV+: 8 (62%) CMV−/EBV−: 0 | CMV+/EBV−: 1 (33%) EBV+/CMV−: 0 CMV+/EBV+: 0 CMV−/EBV−: 2 (66%) | CMV+/EBV−: 1 (14%) EBV+/CMV−: 0 CMV+/EBV+: 6 (86%) CMV−/EBV−: 0 | CMV+/EBV−: 4 (44%) EBV+/CMV−: 0 CMV+/EBV+: 2 (22%) CMV−/EBV−: 3 (33%) |
| Sabeti et al. (2003a) | 5 lesions with calcified necrotic pulp | CMV+/EBV+: 5 (100%) | Not done | CMV+/EBV+: 5 (100%) | Not done |
| Sabeti et al. (2003b) | 14 lesions | CMV+/EBV−: 1 (14%) EBV+/CMV−: 0 CMV+/EBV+: 6 (86%) CMV−/EBV−: 0 | CMV+/EBV−: 0 EBV+/CMV−: 0 CMV+/EBV+: 1 (14%) CMV−/EBV−: 6 (86%) | CMV+/EBV−: 0 EBV+/CMV−: 0 CMV+/EBV+: 7 (58%) CMV−/EBV−: 5 (42%) | CMV+/EBV−: 1 (50%) EBV+/CMV−: 0 CMV+/EBV+: 0 CMV−/EBV−: 1 (50%) |
| Sabeti and Slots (2004) | 34 lesions | CMV+/EBV−: 6 (26%) EBV+/CMV−: 1 (4%) CMV+/EBV+: 16 (70%) CMV−/EBV−: 0 | CMV+/EBV−: 1 (9%) EBV+/CMV−: 0 CMV+/EBV+: 4 (36%) CMV−/EBV−: 6 (55%) | CMV+/EBV−: 3 (13%) EBV+/CMV−: 0 CMV+/EBV+: 19 (79%) CMV−/EBV−: 2 (8%) | CMV+/EBV−: 4 (40%) EBV+/CMV−: 1 (10%) CMV+/EBV+: 1 (10%) CMV−/EBV−: 4 (40%) |

[a]Symptomatic denotes swelling and pain.

Cytomegalovirus and Epstein–Barr virus in cooperation with specific bacterial species have also been associated with various types of advanced marginal periodontitis (Slots 2007) and several nonoral infectious diseases (Slots et al. 2002; Brogden et al. 2005). Most anaerobic bacteria were isolated from periapical lesions that showed HCMV/EBV dual infection were symptomatic or were large (Sabeti and Slots 2004). *Porphyromonas gingivalis/Porphyromonas endodontalis* were recovered only from symptomatic periapical lesions, supporting the notion that this group of organisms is capable of inducing acute endodontic infection (Sundqvist 1976). However, most of the symptomatic periapical lesions studied failed to yield black-pigmented anaerobic rods. Acute exacerbation of periapical disease may be caused by unique constellations of pathogenic bacteria or, alternatively, may result from a combination of herpesviral and bacterial causes. The latter possibility is consistent with the observed uniform presence of active herpesvirus infections in symptomatic periapical lesions and the proinflammatory potential of herpesviruses (Mogensen and Paludan 2001).

Herpesviruses possess several virulence factors of potential importance for periapical pathosis, including the ability to induce immune impairment (Michelson 1999; Boeckh and Nichols 2003) and subsequent overgrowth of pathogenic microorganisms (Sabeti and Slots 2004). In periodontitis, presence of subgingival HCMV or EBV is related to elevated bacterial load and occurrence of the periodontal pathogens *Porphyromonas gingivalis, Tannerella forsythensis, Dialister pneumosintes, Prevotella intermedia, Prevotella nigrescens, Treponema denticola,* and *Aggregatibacter actinomycetemcomitans* (Slots 2007). Herpesviruses seem also to cooperate with pathogenic bacteria in producing a variety of medical diseases, including inflammatory bowel disease, enterocolitis, esophagitis, pulmonary infections, sinusitis, acute otitis media, dermal abscesses, and pelvic inflammatory disease (Brogden et al., 2005). Additionally, herpesviruses may give rise to periapical pathosis by inducing cytokine and chemokine release from inflammatory and noninflammatory host cells (Mogensen and Paludan 2001; Sabeti et al. 2003a). Periapical sites having inadequate antiviral immune response may be particularly prone to tissue breakdown (Sabeti et al. 2003a). Viruses that infect mammals other than HCMV and EBV, alone or in cooperation with herpesviruses, may also play roles in the pathogenesis of pulpal and periapical pathosis (Elkins et al. 1994; Sigurdsson and Jacoway 1995).

## 8.10 Pathogenesis of herpesvirus-associated apical disease

Herpesviruses may cause disease as a direct result of viral infection and replication or as a result of virally induced impairment of the host defense. Herpesvirus-mediated pathogenicity may take place through several mechanisms, operating alone or in combination, and may involve both cellular and humoral host responses (Contreras and Slots 2000):

(1) Herpesviruses may cause direct cytopathic effects on periapical fibroblasts, endothelial cells, and bone cells, the results of which may be impaired tissue turnover and repair, and ultimately loss of tissue.

(2) HCMV and EBV may infect and alter functions of monocytes, macrophages, lymphocytes, and polymorphonuclear leukocytes. Impairment of host defense cells may predispose to overgrowth of endodontic pathogenic bacteria. Herpesvirus activation may induce significant immunosuppressive and immunomodulatory effects in periapical sites. Herpesviruses can trigger an array of host responses that include dysregulation of macrophages and lymphocytes and have a purpose to downregulate the antiviral host immune response (Boeckh and Nichols 2003). Host impairment includes silencing of natural killer cells, inhibition of apoptosis, and destruction of components of MHC class I and class II pathways within macrophages, markedly impairing their principal role in antigen presentation (Michelson 1999). In addition, HCMV encodes a unique homolog of interleukin (IL)-10, a Th2 cytokine that antagonizes Th1 responses, and its immunosuppressive properties may help HCMV circumvent detection and destruction by the host immune system (Kotenko et al. 2000). HCMV has also the ability to inhibit the expression of macrophage surface receptors for lipopolysaccharide, which impairs responsiveness to Gram-negative bacterial infections (Hopkins 1996).

(3) Herpesvirus infections elicit proinflammatory cytokine and chemokine release from inflammatory cells. Interleukin-1β and tumor necrosis factor-α are present in significant levels in periapical lesions (Lim et al. 1994; Wang et al. 1997; Kawashima and Stashenko 1999; Márton and Kiss 2000), and prostaglandin $E_2$ (PGE$_2$) concentration is higher in acute than in chronic periapical lesions (McNicholas et al. 1991). These inflammatory mediators, which are most likely produced locally by periapical macrophages (Miyauchi et al. 1996; Lin et al. 2000), are potent bone resorption-stimulating agents (Page et al. 1997; Márton and Kiss 2000). Previous studies have focused on lipopolysaccharide as an inductor of macrophage cytokine production (Page et al. 1997), but HCMV infection may possess higher potential to upregulate interleukin-1β and tumor necrosis factor-α gene expression in monocytes and macrophages. It might be that the relationship of macrophages and their products to periapical pathosis is in part due to HCMV-mediated cytokine release from periapical macrophages. EBV is a potent polyclonal B-lymphocyte activator, capable of inducing proliferation and differentiation of immunoglobulin-secreting cells. Periapical EBV infection may in part be responsible for the frequent occurrence of B cells in periapical lesions (Márton and Kiss 2000). Herpesvirus infections also affect cytokine networks (Mogensen and Paludan 2001). Cytokines and chemokines play important roles in the first line of defense against human herpesvirus infections and also contribute significantly to regulation of acquired immune responses. However, by a diverse array of strategies, herpesviruses are able to interfere with cytokine production or divert potent antiviral cytokine responses, which allow the viruses to survive throughout the lifetime of the host (Alcami and Koszinowski 2000; Tortorella et al. 2000). HCMV infection typically induces a proinflammatory cytokine profile, with production of IL-1β, IL-6, IL-12, tumor necrosis factor (TNF)-α, interferon (IFN)-α/β, and IFN-γ (Mogensen and Paludan 2001), and PGE$_2$ (Mocarski 2002). EBV infection stimulates the production of IL-1β, IL-1 receptor antagonist (IL-1Ra), IL-6, IL-8, IL-18, TNF-α, IFN-α/β, IFN-γ, monokine induced by IFN-γ (MIG), IFN-γ-inducible protein 10 (IP-10), and granulocyte-macrophage colony-stimulating factor (Mogensen and Paludan 2001). Proinflammatory activities normally serve a positive biological goal by aiming to overcome infection or invasion by infectious agents, but can also exert detrimental effects when a challenge becomes overwhelming or with a chronic pathophysiological stimulus. In an effort to counteract ongoing inflammation, the initial proinflammatory response triggers the release of anti-inflammatory mediators, such as transforming growth factor-β and IL-10 (Haveman et al. 1999). Also, viruses display great uniqueness when it comes to diverting the potent antiviral cytokine responses to their benefit (Tortorella et al. 2000). PGE$_2$, which is a key mediator of the periapical inflammatory response (Márton and Kiss 2000), increases rapidly in response to exposure of cells to HCMV, bacterial lipopolysaccharide, and the cytokines IL-1β and TNF-α (Sabeti and Slots 2004), and PGE$_2$ under certain circumstances may support HCMV replication (Takayama et al. 1996). Undoubtedly, a periapical HCMV infection can induce a

multiplicity of interconnected immunomodulatory reactions, and various stages of the infection may display different levels of specific inflammatory cells and mediators, underscoring the complexity of HCMV–host interactions in periapical disease.

(4) Herpesviruses may produce periapical tissue injury as result of immunopathologic responses. Th1 cells, which predominate periapical lesions (Kawashima and Stashenko 1999), are mediators of delayed-type hypersensitivity (Seymour et al. 1996). HCMV has the potential to induce cell-mediated immunosuppression by downregulating cell surface expression of major histocompatibility complex class I molecules, thereby interfering with cytotoxic T-lymphocyte recognition. EBV may induce proliferation of cytotoxic T-lymphocytes, the main purpose of which is to recognize and destroy virally infected cells, but may secondarily also inhibit various aspects of the immune response.

Control of herpesviral replication and prevention of pathosis depend on both innate and adaptive immune mechanisms. Antiviral antibodies may help control infectious virions, and cytotoxic T-lymphocytes play an important role in limiting the proliferation of herpesvirus-infected cells. The frequent presence of natural killer cells (CD8 T-lymphocytes) in chronic periapical lesions (Kettering and Torabinejad 1993; Márton and Kiss 2000) is consistent with an antiherpesviral host response. However, while antiherpesviral immune responses may be able to protect from disease, they are insufficient to eliminate reservoirs of persistent viral gene expression.

## 8.11 Model for herpesvirus-mediated apical disease

Fig. 8.2 describes an infectious disease model for the development of periapical pathosis based on herpesvirus–bacteria–host interactive responses. Herpesvirus infection of periapical sites may be important in a multistage pathogenesis by altering local host defenses. Initially, bacterial infection or mechanical trauma of the pulp causes inflammatory cells to enter pulpal and periapical tissues. In infected individuals, latent HCMV resides in periodontal macrophages and T-lymphocytes and latent EBV in periodontal B-lymphocytes (Contreras and Slots 2000). Reactivation of herpesviruses from latency may occur spontaneously or during periods of impaired host re-

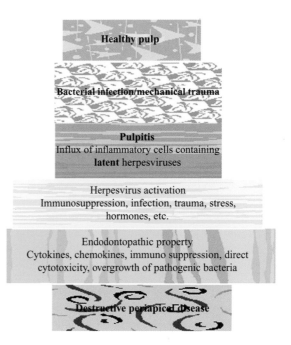

**Fig. 8.2** Herpesviruses in symptomatic endodontic pathosis.

sponse, resulting from immunosuppression, infection, physical trauma, hormonal changes, etc. Perhaps not coincidentally, various herpesvirus-activating factors are also known risk factors for periapical flare-ups (Walton and Fouad 1992; Imura and Zuolo 1995). Herpesviral activation leads to increased inflammatory mediator responses in macrophages and probably also in resident connective tissue cells within the periapical lesion. After reaching a critical viral load, activated macrophages and lymphocytes may trigger a cytokine/chemokine "storm" of IL-1β, TNF-α, IL-6, prostaglandins, interferons, and other multifunctional mediators, which in an enclosed area have the potential to propagate states of pain (Vane et al. 1998; Rittner et al. 2002; Rutkowski and DeLeo 2002; Zhu et al. 2002) and bone resorption (Wang et al. 1997; Kawashima and Stashenko 1999; De Jongh et al. 2003). Several of the herpesvirus-associated cytokines and chemokines are prominent in periapical lesions (McNicholas et al. 1991; Nair 1997; Wang et al. 1997; Lader and Flanagan 1998; Kawashima and Stashenko 1999; Márton and Kiss 2000; Radics et al. 2003). Herpesvirus-induced immune impairment may also cause an upgrowth of resident Gram-negative anaerobic bacteria (Sabeti and Slots 2004) whose lipopolysaccharide can induce cytokine and chemokine release from various mammalian cells and may act synergistically with HCMV in stimulating IL-1β gene transcription (Wara-aswapati et al. 2003). Moreover, in a vicious circle, triggering of cytokine responses may activate latent herpesviruses and in so doing further aggravate periapical disease.

In conclusion, endodontic inflammation can be initiated by a variety of infectious agents and is mediated by both cellular components, such as macrophages and leukocytes, and molecular components, including cytokines and chemokines, many of which possess pro- and/or anti-inflammatory properties, with harmful or beneficial effects. Reactivation of latent herpesviruses is involved in driving the pathological process of cases of symptomatic periapical disease. Herpesvirus reactivation and herpesviral–bacterial interactions may help explain various clinical characteristics of periapical infections. Alteration between prolonged periods of herpesvirus latency interrupted by periods of activation may partly be responsible for intermittent episodes of periapical disease flare-up. Frequent reactivation of periapical herpesviruses in some patients may result in rapid disease progression. Perhaps not coincidentally, herpesvirus-activating factors are also risk factors for acute endodontic disease (Torabinejad 1994). Detection of herpesvirus DNA in periapical lesions has brought a new dimension to our knowledge of periapical infections and calls for the inclusion of herpesviruses in studies on the pathogenesis of periapical pathosis, and the herpesviral findings may have future therapeutic relevance as well.

## 8.12 References

Ablashi DV, Salahuddin SZ, Josephs SF, Balachandran N, Krueger GR, and Gallo RC. 1991. Human Herpesvirus-6 (HHV-6). *In Vivo* 5(3): 193–99.

Alcami A and Koszinowski UH. 2000. Viral mechanisms of immune evasion. *Trends Microbiol* 8(9): 410–18.

Alford CA, Stagno S, and Pass RF. 1979. Natural history of perinatal cytomegalovirus infection. *Ciba Found Symp* 77: 125–47.

Bale JF, Miner L, and Petheram SJ. 1990. Congenital cytomegalovirus infection. *J Child Neurol* 5(2): 131–36.

Boeckh M and Nichols WG. 2003. Immunosuppressive effects of beta-herpesviruses. *Herpes* 10(1): 12–16.

Boppana SB, Fowler KB, Britt WJ, Stango S, and Pass RF. 1999. Symptomatic congenital cytomegalovirus infection in infants born to mothers with preexisting immunity to cytomegalovirus. *Pediatrics* 104(1): 55–60.

Britt WJ and Alford CA. 1996. Cytomegalovirus. In: Fields BN, Knipe DM, and Howley PM (eds), *Field Virology*, Vol. 2. Philadelphia: Lippincott-Raven Publisher, 2493–523.

Brogden KA, Guthmiller JM, and Taylor CE. 2005. Human polymicrobial infections. *Lancet* 365(9455): 253–55.

Challoner PB, Smith KT, Parker JD, MacLeod DL, Coulter SN, Rose TM, Schultz ER, Bennett JL, Garber RL, and Chang M. 1995. Plaque-associated expression of human herpesvirus 6 in multiple sclerosis. *Proc Natl Acad Sci USA* 92(16): 7440–44.

Clark DA, Emery VC, and Griffiths PD. 2003. Cytomegalovirus, human herpesvirus-6, and human herpesvirus-7 in hematological patients. *Semin Hematol* 40(2): 154–62.

Clark DA, Freeland ML, Mackie LK, Jarrett RF, and Onions DE. 1993. Prevalence of antibody to human herpesvirus 7 by age. *J Infect Dis* 168(1): 251–52.

Contreras A, Nowzari H, and Slots J. 2000. Herpesviruses in periodontal pocket and gingival biopsy samples. *Oral Microbiol Immunol* 15(1): 15–18.

Contreras A and Slots J. 2000. Herpesviruses in human periodontal disease. *J Periodontal Res* 35(1): 3–16.

Dankner WM, McCutchan JA, Richman DD, Hirata K, and Spector SA. 1990. Localization of human cytomegalovirus in peripheral blood leukocytes by in situ hybridization. *J Infect Dis* 161(1): 31–36.

De Jongh RF, Vissers KC, Meert TF, Booij LH, De Deyne CS, and Heylen RJ. 2003. The role of interleukin-6 in nociception and pain. *Anesth Analg* 96(4): 1096–103.

Di Alberti L, Teo CG, Porter S, Zakrzewska J, and Scully C. 1996. Kaposi's sarcoma herpesvirus in oral Kaposi's sarcoma. *Eur J Cancer* 32(1): 68–69.

Elkins DA, Torabinejad M, Schmidt RE, Rossi JJ, and Kettering JD. 1994. Polymerase chain reaction detection of human immunodeficiency virus DNA in human periradicular lesions. *J Endodon* 20(8): 386–88.

Flaitz CM, Jin YT, Hicks MJ, Nichols CM, Wang YW, and Su IS. 1997. Kaposi's sarcoma associated herpesvirus-like sequences (KSHV/HHV-8) in oral AIDS-Kaposi's sarcoma: A PCR and clinicopathologic study. *Oral Surg Oral Med Oral Pathol Oral Radiol Endod* 83(2): 259–64.

Gaspari AA, Marchese S, Powell D, Rady PL, and Tyring SK. 1997. Identification of HHV-8 in the skin lesions of Kaposi's sarcoma in an immunosuppressed patient with bullous pemphigoid. *J Am Acad Dermatol* 37(5): 843–47.

Gautheret-Dejean A, Aubin JT, Poirel L, Huraux JM, Nicola JC, and Rozenbaum W, Agut H. 1977. Detection of human betaherpesvirinae in saliva and urine from immunocompromised and immunocompetent subjects. *J Clin Microbiol* 35: 1600–603.

Gerna G, Zipeto D, Percivalle E, Parea M, Revello MG, Maccario R, Peri G, and Milanesi G. 1992. Human cytomegalovirus infection of the major leukocyte subpopulations and evidence for initial viral replication in polymorphonuclear leukocytes from viremic patients. *J Infect Dis* 166(6): 1236–44.

Grefte A, Blom N, Van Der Giessen M, van Son WJ, and The TH. 1993a. Ultrastructural analysis of circulating cytomegalic cells in patients with active cytomegalovirus infection: evidence for virus production and endothelial origin. *J Infect Dis* 168: 1110–18.

Grefte A, Van Der Giessen M, van Son WJ, and The TH. 1993b. Circulating cytomegalovirus (CMV)-infected endothelial cells in patients with an active CMV infection. *J Infect Dis* 167: 270–77.

Griffiths PD and Emery VC. 1997. Cytomegalovirus. In: Richman DD, Whitley RJ, and Hayden FG (eds), *Clinical Virology*. New York: Churchill Livingston, 445–70.

Grundy JE. 1990. Virologic and pathogenetic aspects of cytomegalovirus infection. *Rev Infect Dis* 12 (Suppl. 7): S711-19.

Haveman JW, Muller Kobold AC, Tervaert JW, Van Den Berg AP, Tulleken JE, Kallenberg CG, and The TH. 1999. The central role of monocytes in the pathogenesis of sepsis: consequences for immunomonitoring and treatment. *Neth J Med* 55(3): 132–41.

Hidaka Y, Liu Y, Yamamoto M, Mori R, Miyazaki C, Kusuhara K, Okada K, and Ueda K. 1993. Frequent isolation of human herpesvirus 7 from saliva samples. *J Med Virol* 40(4): 343–46.

Hopkins HA, Monick MM, and Hunninghake GW. 1996. Cytomegalovirus inhibits CD14 expression on human alveolar macrophages. *J Infect Dis* 174(1): 69–74.

Ikuta K, Satoh Y, Hoshikawa Y, and Sairenji T. 2000. Detection of Epstein–Barr virus in salivas and throat washings in healthy adults and children. *Microbes Infect* 2: 115–20.

Imura N and Zuolo ML. 1995. Factors associated with endodontic flare-ups: a prospective study. *Int Endod J* 28(5): 261–65.

Kapasi K and Rice GPA. 1988. Cytomegalovirus infection of peripherical blood mononuclear cells: effects of interleukin-1 and 2 production and responsiveness. *J Virol* 62(10): 3603–13.

Kawashima N and Stashenko P. 1999. Expression of bone-resorptive and regulatory cytokines in murine periapical inflammation. *Arch Oral Biol.* 44(1): 55–66.

Kemeny L, Gyulai R, Kiss M, Magy F, and Dobozy A. 1997. Kaposi's sarcoma-associated herpesvirus/human herpesvirus-8: a new virus in human pathology. *J Am Acad Dermatol* 37(1): 107–13.

Kettering JD and Torabinejad M. 1993. Presence of natural killer cells in human chronic periapical lesions. *Int Endod J.* 26(6): 344–47.

Kikuta H, Taguchi Y, Tomizawa K, Kojima K, Kawamura N, Ishizaka A, Sakiyama Y, Matsumoto S, Imai S, Kinoshita T, Koizumi S, Osato T, Kobayashi I, Hamada I, and Hirai K. 1988. Epstein–Barr virus genome positive T lymphocytes in a boy with EBV infection associated with Kawasaki-like disease. *Nature* 333(6172): 455–57.

Kimberlin DW. 1998. Human herpesvirus 6 and 7: identification of a newly recognized viral pathogens and their association with human disease. *Pediatr Infect Dis J* 17(1): 59–68.

Klein G. 1989. Viral latency and transformation: the strategy of Epstein–Barr virus. *Cell* 58(1): 5–8.

Kotenko SV, Saccani S, Izotova LS, Mirochnitchenko OV, and Pestka S. 2000. Human cytomegalovirus harbors its own unique IL-10 homolog (cmvIL-10). *Proc Natl Acad Sci USA* 97(4): 1695–700.

Lader CS and Flanagan AM. 1998. Prostaglandin E2, interleukin 1alpha, and tumor necrosis factor-alpha increase human osteoclast formation and bone resorption in vitro. *Endocrinology* 139(7): 3157–64.

Laskaris G. 1996. Oral manifestations of infectious diseases. *Dent Clin North Am* 40(2): 395–423.

Lennette ET, Blackbourn DJ, and Levy JA. 1996. Antibodies to human herpesvirus type 8 in the general population and in Kaposi's sarcoma patients. *Lancet* 348(9031): 858–61.

Levy JA. 1997. Three new human herpesviruses (HHV-6, 7, and 8). *Lancet* 349(9051): 558–63.

Lim GC, Torabinejad M, Kettering J, Linkhardt TA, and Finkelman RD. 1994. Interleukin 1-beta in symptomatic and asymptomatic human periradicular lesions. *J Endod.* 20(5): 225–27.

Lin SK, Hong CY, Chang HH, Chiang CP, Chen CS, Jeng JH, and Kuo MY. 2000. Immunolocalization of macrophages and transforming growth factor-beta 1 in induced rat periapical lesions. *J Endod* 26(6): 335–40.

Luppi M, Barozzi P, Maiorana A, Marasca R, and Torelli G. 1994. Human herpesvirus 6 infection in normal human brain tissue. *J Infect Dis* 169(4): 943–44.

Luqmani YA, Linjawi SO, and Shousha S. 2001. Detection of Epstein–Barr virus in gastrectomy specimens. *Oncol Rep* 8(5): 995–99.

Lusso P, Markham PD, Tschachler E, di Marzo Veronese F, Salahuddin SZ, Ablashi DV, Pahwa S, Krohn K, and Gallo RC. 1988. In vitro cellular tropism of human B-lymphotropic virus (human herpesvirus-6). *J Exp Med* 167(5): 1659–70.

Maciejowski JP, Bruening EE, Donahue RE, Mocarski ES, Young NS, and St Jeor SC. 1993. Infection of hematopoetic progenitor cells by human cytomegalovirus. *Virology* 195(2): 327–68.

Márton IJ and Kiss C. 2000. Protective and destructive immune reactions in apical periodontitis. *Oral Microbiol Immunol* 15(3): 139–50.

McNicholas S, Torabinejad M, Blankenship J, and Bakland L. 1991. The concentration of prostaglandin E2 in human periradicular lesions. *J Endod* 17(3): 97–100.

Meyer-Konig U, Serr A, and van Laer D, Kirste G, Wolff C, Haller O, Neumann-Haefelin D, and Hufert FT. 1995. Human cytomegalovirus immediate early and late transcripts in peripheral blood leukocytes: diagnostic value in renal transplant recipients. *J Infect Dis* 171(3): 705–9.

Michelson S. 1999. Human cytomegalovirus escape from immune detection. *Intervirology* 42(5–6): 301–7.

Millar EP and Troulis MJ. 1994. Herpes zoster of the trigeminal nerve: the dentist's role in diagnosis and management. *J Can Dent Assoc* 60(5): 450–53.

Miller CS. 1996. Viral infections in the immunocompromised patient. *Dermatol Clin* 14(2): 225–41.

Miyauchi M, Takata T, Ito H, Ogawa I, Kobayashi J, Nikai H, and Ijuhin N. 1996. Immunohistochemical detection of prostaglandins E2, F2 alpha, and 6-keto-prostaglandin F1 alpha in experimentally induced periapical inflammatory lesions in rats. *J Endod* 22(12): 635–37.

Mocarski ES, Jr. 1996. Cytomegalovirus and their replication. In: Fields BN, Knipe DM, and Howley PM (eds), *Field's Virology*, 3rd edn. Philadelphia: Lippiacott-Raven Publishers, 2447–92.

Mocarski ES, Jr. 2002. Virus self-improvement through inflammation: no pain, no gain. *Proc Natl Acad Sci USA* 99(2): 3362–64.

Mocarski ES and Stinski MF. 1979. Persistence of cytomegalovirus in human cells. *J Virol* 31(3): 761–75.

Mogensen TH and Paludan SR. 2001. Molecular pathways in virus-induced cytokine production. *Microbiol Mol Biol Rev* 65(1): 131–50.

Moore PS and Chang Y. 1995. Detection of herpesvirus-like DNA sequence in Kaposi's sarcoma in patients with and those without HIV infection. *N Engl J Med* 322(18): 1181–85.

Mutimer D, Mirza D, Shaw J, O'Donnell K, and Ellias E. 1997. Enhanced (cytomegalovirus) viral replication associated with septic bacterial complications in liver transplant recipients. *Transplantation* 63(10): 1411–15.

Myerson D, Hackman RC, Nelson JA, Ward DC, and McDougall JK. 1984. Widespread presence of histologically occult cytomegalovirus. *Hum Pathol* 15(5): 430–39.

Nair PN. 1997. Apical periodontitis: a dynamic encounter between root canal infection and host response. *Periodontology 2000* 13: 121–48.

Nowzari H, Jorgensen MG, Aswad S, Khan N, Osorio E, Safarian A, Shidban H, and Munroe S. 2003. Human cytomegalovirus-associated periodontitis in renal transplant patients. *Transplant Proc* 35(8): 2949–52.

Page RC, Offenbacher S, Schroeder HE, Seymour GJ, and Kornman KS. 1997. Advances in the pathogenesis of periodontitis: summary of developments, clinical implications and future directions. *Periodontol 2000*. 14: 216–48.

Parra B and Slots J. 1996. Detection of human viruses in periodontal pockets using polymerase chain reaction. *Oral Microbiol Immunol* 11: 289–93.

Platcher B, Sinzger C, and Jahn G. 1995. Cell types involved in replication and distribution of human cytomegalovirus. *Adv Virus Res* 46: 195–61.

Raab-Traub N, Rajadural P, Flynn K, and Lanier A. 1991. Epstein–Barr virus infection in carcinoma of the salivary gland. *J Virol* 65(12): 7032–36.

Radics T, Kiss C, Tar I, and Márton IJ. 2003. Interleukin-6 and granulocyte-macrophage colony-stimulating factor in apical periodontitis: correlation with clinical and histologic findings of the involved teeth. *Oral Microbiol Immunol* 18(1): 9–13.

Revello MG and Gerna G. 2004. Pathogenesis and prenatal diagnosis of human cytomegalovirus infection. *J Clin Virol* 29(2): 71–83.

Rickinson E and Kieff E. 2001. Epstein–Barr virus. In: Knipe DM and Howley PM (eds), *Field Virology*, 4th edn. Philadelphia: Lippincott, Williams & Wilkins, 2575–627.

Rinckinson AB and Kueff E. 1996. Epstein–Barr virus. In: Fields BN, Knipe DM, and Howely PM (eds) *Field's Virology*, 3rd edn. Philadelphia: Lippincott-Raven Publishers, 2397–450.

Rittner HL, Brack A, and Stein C. 2002. Pain and the immune system: friend or foe? *Der Anaesthesist* 51(5): 351–58 (German).

Roizman B and Pellett PE. 2001. The family Herpesviridae: a brief introduction. In: Knipe DM and Howley PM (eds), *Fields Virology*, 4th edn. Philadelphia, PA: Lippincott Williams & Wilkins, 2381–97.

Rutkowski MD and DeLeo JA. 2002. The role of cytokines in the initiation and maintenance of chronic pain. *Drug News Perspect* 15: 626–32.

Sabeti M and Slots J. 2004. Herpesviral–bacterial co-infection in periapical path osis. *J Endodont* 30(2): 69–72.

Sabeti M, Simon JH, Nowzari H, and Slots J. 2003a. Cytomegalovirus and Epstein–Barr virus active infection in periapical lesions of teeth with intact crowns. *J Endodont* 29(2): 321–23.

Sabeti M, Simon JH, and Slots J. 2003b. Cytomegalovirus and Epstein–Barr virus are associated with symptomatic periapical pathosis. *Oral Microbiol Immunol* 18(5): 327–28.

Sabeti M, Valles Y, Nowzari H, Simon JH, Kermani-Arab V, and Slots J. 2003c. Cytomegalovirus and Epstein–Barr virus DNA transcription in endodontic symptomatic lesions. *Oral Microbiol Immunol* 18(2): 104–8.

Sada E, Yasukawa M, and Ito C, Takeda A, Shiosaka T, Tanioka H, and Fujitaet S. 1996. Detection of human herpesvirus 6 and human herpesvirus 7 in the submandibular gland, parotid gland and lip salivary gland by PCR. *J Clin Microbiol* 34(9): 2320–21.

Scott DA, Coulter WA, Biagoni PA, O'Neill H, and Lamey PJ. 1997. Detection of herpes simplex type 1 shedding in the oral cavity by polymerase chain reaction and enzyme-linked immunosorbent assay at the prodromal stage of recrudecent herpes labialis. *J Oral Pathol Med* 26(7): 305–9.

Scully C. 1996. New aspects of oral viral diseases. In: Seifert G (ed.), *Oral Pathology: Actual Diagnostic and Prognostic Aspects. Current Topics in Pathology*, Vol. 90. Berlin: Springer Verlag, 29–96.

Scully C, Epstein J, Porter S, and Cox M. 1991. Viruses and chronic disorders involving the human oral mucosa. *Oral Surg Oral Med Oral Pathol* 72(5): 537–44.

Seymour GJ, Gemmell E, Kjeldsen M, Yamazaki K, Nakajima T, and Hara K. 1996. Cellular immunity and hypersensitivity as components of periodontal destruction. *Oral Dis*. 2(1): 96–101.

Shanavas KR, Vasudevan DM, Vijayakumar T, and Yadav M. 1992. Anti-HHV-6 antibodies in normal population and in cancer patients in India. *J Exp Pathol* 6(1–2): 95–105.

Sigurdsson A and Jacoway JR. 1995. Herpes zoster infection presenting as an acute pulpitis. *Oral Surg Oral Med Oral Pathol Oral Radiol Endod* 80(1): 92–95.

Sinzger C, Grefte A, Plachter B, Gouw ASH, Hauw TT, and Jahnet S. 1995. Fibroblasts, epithelial cells, endothelial cells and smooth muscle cells are major targets of human cytomegalovirus infection in lung and gastrointestinal tissues. *J Gen Virol* 76(4): 741–50.

Sinzger C and Jahn G. 1996. Human cytomegalovirus cell tropison and pathogenesis. *Intervirology* 39(5–6): 302–19.

Sissons JG, Bain M, and Wills MR. 2002. Latency and reactivation of human cytomegalovirus. *J Infect* 44: 73–77.

Sixbey JW, Nedrud JG, Rabb N, Hanes RA, and Pagano JS. 1984. Epstein–Barr virus replication in oropharyngeal epithelial cells. *N Engl J Med* 3109(19): 1225–30.

Slots J. 2007. Herpesviral–bacterial synergy in the pathogenesis of human periodontitis. *Curr Opin Infect Dis* 20(3): 278–83.

Slots J, Sugar C, and Kamma JJ. 2002. Cytomegalovirus periodontal presence is associated with subgingival *Dialister pneumosintes* and alveolar bone loss. *Oral Microbiol Immunol* 17(6): 369–74.

Sundqvist G. 1976. Bacteriological studies of necrotic dental pulps [thesis]. *Umeå University Odontological Dissertations*. No. 7. 5–94.

Sweet C. 1999. The pathogenicity of cytomegalovirus. *FEMS Microbiol Rev* 23(4): 457–82.

Takahashi Y, Yamada M, Nakamura J, Tsukazaki T, Padilla J, Kitamura T, Yoshida M, and Nii S. 1997. Transmission of herpesvirus 7 through multigenerational families in the same household. *Pediatr Infect Dis J* 16: 975–78.

Takayama S, Miki Y, Shimauchi H, and Okada H. 1996. Relationship between prostaglandin E2 concentrations in periapical exudates from root canals and clinical findings of periapical periodontitis. *J Endod.* 22(12): 677–80.

Tao Q, Ho FC, Loke SL, and Srivastava G. 1995. Epstein–Barr virus is located in tumor cells of nasal lymphomas of NK, T or B cell type. *Int J Cancer* 60(3): 315–20.

Taylor-Wiedeman J, Hayhurst GP, Sissons JG, and Sinclair JH. 1993. Polymorph nuclear cells are no sites of persistence of human cytomegalovirus in healthy individuals. *J Gen Virol* 74(2): 265–68.

Torabinejad M. 1994. Mediators of acute and chronic periradicular lesions. *Oral Surg Oral Med Oral Pathol* 78(4): 511–21.

Tortorella D, Gewurz BE, Furman MH, Schust DJ, and Ploegh HL. 2000. Viral subversion of the immune system. *Annu Rev Immunol* 18: 861–926.

Vane JR, Bakhle YS, and Botting RM. 1998. Cyclooxygenases 1 and 2. *Annu Rev Pharmacol Toxicol* 38: 97–120.

Walling DM, Etienne W, Ray AJ, Flaitz CM, and Nicholas CM. 2004a. Persistence and transition of Epsin–Barr virus genotypes in the pathogenesis of oral hairy leukoplakia. *J Infect Dis* 190(2): 387–95.

Walling DM, Ling PD, Gordadze AV, Montes-Walters M, Flaitz CM, and Nicholas CM. 2004b. Expression of Epstein–Barr virus latent gene in oral epithelium: determinants of the pathogenesis of oral hairy leukoplakia. *J Infect Dis* 190(2): 396–99.

Walton R and Fouad A. 1992. Endodontic interappointment flare-ups: a prospective study of incidence and related factors. *J Endod* 18(4): 172–77.

Wang CY, Tani-Ishii N, and Stashenko P. 1997. Bone-resorptive cytokine gene expression in periapical lesions in the rat. *Oral Microbiol Immunol.* 12(2): 65–71.

Wara-aswapati N, Boch JA, and Auron PE. 2003. Activation of interleukin 1β gene transcription by human cytomegalovirus: molecular mechanisms and relevance to periodontitis. *Oral Microbiol Immunol* 18(2): 67–71.

Wyatt LS, Rodriguez WJ, Balachandran N, and Frenkel N. 1991. Human herpesvirus 7: antigenic properties and prevalence in children and adults. *J Virol* 65(11): 6260–65.

Yadav M, Arivanathan M, Chandrashekran A, and Tan BS. 1997. Human herpesvirus-6 (HHV-6) DNA and virus-encoded antigen in oral lesions. Human herpesvirus-6 (HHV-6) DNA and virus-encoded antigen in oral lesions. *J Oral Pathol Med* 26(9): 393–401.

Yadav M, Chandrashekran A, Vasudevan DM, and Ablashi DV. 1994. Frequent detection of human herpesvirus-6 in oral carcinoma. *J Natl Cancer Inst* 86(23): 1793–94.

Yamazaki K and Nakajima T. 2004. Antigen specifity and T-cell clonality in periodontal disease. *Peirodontol 2000* 35: 75–100.

Yin HF, Jamlikhanova V, Okada N, and Takagi M. 1999. Primary natural killer/T-cells lymphomas of the oral cavity an aggressive neoplasms. *Virchows Arch* 435(4): 400–406.

Yoshikawa T, Nakashima T, Suga S, Asano Y, Yazaki T, Kimura H, Morishima T, Kondo K, and Yamanishi K. 1992. Human herpesvirus-6 DNA in cerebrospinal fluid of a child with exanthem subitum and meningoencephalitis. *Pediatrics* 89(5, Pt 1): 888–90.

Zhu H, Cong JP, Yu D, Bresnahan WA, and Shenk TE. 2002. Inhibition of cyclooxygenase 2 blocks human cytomegalovirus replication. *Proc Natl Acad Sci USA* 99(6): 3932–37.

# Chapter 9
# Fungi in Endodontic Infections

*Bilge Hakan Sen and B. Güniz Baksi*

## 9.1 General characteristics of fungi

Contrary to bacteria, which are prokaryotes, fungi are eukaryotic organisms. Most fungi are microscopic molds or yeasts. Molds are tangled masses of filaments of cells. Yeasts are typical unicellular fungi. Yeast cells have a cell wall, containing glucan, mannan, and chitin. Inside the cell wall, cell membrane, nucleus, a large vacuole, and membrane-bound organelles (mitochondria, endoplasmic reticulum) are other parts of a yeast cell.

The body of a fungus consists of tiny filaments called hyphae. Hyphae are tiny tubes filled with cytoplasm and nuclei. A mat of hyphae visible to the unaided eye is a mycelium. Some hyphae are divided by cross-section segments (walls) called septa. The septa have holes through which cytoplasm and organelles can move from segment to segment. In certain conditions, yeast cells can grow true hyphae or pseudo-

hyphae. (Please refer to Subsection 9.4.1 for more information.)

## 9.2 Oral yeasts and carriage

Most of the clinically important oral yeasts belong to the genus *Candida*. Taxonomy of *Candida* is as follows: kingdom, Fungi; phylum, Ascomycota; subphylum, Ascomycotina; class, Ascomycetes; order, Saccharomycetales; family, Saccharomycetaceae; genus, *Candida* (Waltimo et al. 2003b).

Among more than 300 cultivated microbial species or types in the oral cavity, there are many *Candida* species. These benign, commensal, or opportunistic species play important role in the development of oral as well as dental diseases. There are 150–200 species of *Candida*. *Candida albicans* is the most pathogenic type among the seven species most commonly found in the oral cavity (*C. albicans*, *C. glabrata*, *C. tropicalis*, *C. pseudotropicalis*, *C. guilliermondii*, *C. krusei*, and *C. parapsilosis*).

*C. albicans* is frequently isolated from the human mouth, yet few carriers develop clinical signs and symptoms of candidosis. The prevalence of the yeasts tends to be lower in healthy individuals than in hospital patients of any kind. Isolation rates of yeasts from healthy mouths in nine different studies range 2–37% as compared to 13–76% in nine other studies with hospital patients (Odds 1988).

## 9.3 Oral candidosis

Oral candidosis is an opportunistic infection associated with alteration in local and systemic defense mechanisms. Oral candidosis may clinically manifest itself in several different forms, which have been classified as pseudomembranous candidosis (thrush), erythematous (atrophic) candidosis, hyperplastic candidosis, and *Candida*-associated lesions (denture stomatitis, angular stomatitis, median rhomboid glossitis).

*C. albicans* is proved to be the most prevalent causative agent of oral candidosis when the hosts' physical and immunologic defenses have been undermined. In general terms, the severity and extent of *Candida* infections tend to increase with the number and severity of predisposing factors. Depressed host defenses (Odds 1988; Heimdahl and Nord 1990), endocrine disorders (Lamey et al. 1988; Darwazeh et al. 1991), mucosal lesions (Wilborn and Montes 1980; Krogh et al. 1987; Muzyka and Glick 1995), ill-fitting dentures (Olsen and Birkeland 1977; Samaranayake 1980; Budtz-Jorgensen et al. 1983; Beighton et al. 1990), poor oral and denture hygiene (Budtz-Jorgensen 1990) are components that markedly increase the hosts' susceptibility to oral candidosis. Nevertheless, it has been proposed that major increase in the incidence of candidosis over the past two to three decades can be attributed to iatrogenic causes. Use of broad-spectrum antibiotics, corticosteroids, drugs that induce neutropenia and xerostomia, psychoactive drugs and particularly immunosuppressive agents are established to be responsible for iatrogenic predisposition (Budzt-Jorgensen 1990; Heimdahl and Nord 1990; Oksala 1990; Peterson 1992; Narhi et al. 1993; Scully et al. 1994; Navazesh et al. 1995). Since the numbers of patients who receive immunosuppressive

therapy tend to increase annually, and since conditions such as AIDS add further to the number of "immunocompromised" hosts, increase in the incidence of oral candidosis is inevitable (Samaranayake 1992; Powderly et al. 1993).

## 9.4 Virulence factors and pathogenicity

*C. albicans* expresses a repertoire of activities that contribute to virulence. Total effect of many *Candida* factors leads to the establishment of infection in a suitably compromised host. Among these factors are the morphogenesis of *C. albicans* yeast cells to a filamentous growth, the production of phospholipases and secreted proteinases, and host cell recognition by cell surface adhesions (Calderone et al. 2000).

### 9.4.1 Morphogenesis and morphological transition

*C. albicans* and other types of *Candida* are aerobic yeasts that may reproduce in anaerobic conditions. This fungus has been demonstrated to grow in a number of morphological forms such as yeast (blastospore), true hyphae, pseudohyphae, and chlamydospores. The organism can grow in either yeast or hyphal form, or physically intermediate forms such as pseudohyphae. Yeast cells grow as round (sometimes oval), single cells and through the process of budding they give rise to colonies of physically separate cells. On the contrary, in hyphal growth form, an initial germ tube resembling a bud is extended into a long, unconstricted filament within which the individual cells are separated by septae. In between these two extremes, the fungus can exhibit a variety of growth forms that are referred to as pseudohyphae (Sudbery et al. 2004). Pseudohyphal growth demonstrates elongated cells connected in chains leading to filaments that resemble hyphae but made up of yeast-like individual cells. The elongation of buds in pseudohyphae can be so extreme that these filaments can superficially resemble hyphae. Due to this paradox, the term "filamentous" is adopted to refer both to pseudohyphae and hyphae in recent *Candida* literature (Kumamoto and Vinces 2005a).

Chlamydospores represent another functionally distinct cellular form, exhibiting complex combination of cell types. Chlamydospores are round, refractile spores with a thick cell wall. All growth patterns except chlamydospores show interconversion to each form of growth depending on the environmental conditions such as pH, temperature, and nutritional source.

Morphogenesis is believed to be important for virulence and has been the subject of many studies (Kobayashi and Cutler 1998; Brown 2002; Gow et al. 2002; Liu 2002). The potential role of hyphae formation in virulence has been reviewed in detail as well. Although there are contradictory reports regarding this subject, recent studies support the conclusion that the hyphal form is important for virulence. The hyphal tip is the site of apical secretion of enzymes that are able to degrade proteins, lipids, and other cellular components that further facilitate infiltration into tissues, presumably by liquefying the substrate in front of the advancing cell (Hube and Naglik 2001). The hyphae of pathogenic fungi also exhibit the phenomenon of contact sensing, or thigmotropism, which may enable them to navigate according to underlying surface topography and accordingly locate the points of weakened surface integrity, thereby gaining access to vulnerable sites for invasion (Gow et al. 2002). Although there is still no genetic basis to establish a role for yeast–hyphae morphogenesis as a virulence factor for *C. albicans*, the evidence has accumulated that formation of hyphae is one of the primary components of the overall virulence strategies of *C. albicans* (Kumamoto and Vinces 2005b). Recently, authors recommend evaluating the impact of morphogenesis at different disease stages such as colonization, penetration, dissemination, invasion, and necrosis (Gow et al. 2002). However the fundamental question of whether yeast or hyphal forms are more virulent is still remains to be answered.

### 9.4.2 Adherence

Adherence is the initial step in the process of tissue colonization and invasion. *C. albicans* has been shown to be significantly more adhesive than other *Candida* species, which rarely cause oral mucosal infections (Ray et al. 1984). Factors promoting the extent and strength of adherence to epithelial cells depend on the initial surface properties of both the organism and the substratum involved and can be influenced by several factors. These include factors related to yeast and host cells as well as the environmental factors affecting adhesion. Adherence of *C. albicans* to host tissue is considered as a crucial event in the

pathogenic process, and is a prerequisite for colonization and subsequent infection of the host. Adherence to host tissue is achieved by combination of specific and nonspecific mechanisms. Specific mechanisms include ligand–receptor interactions, while nonspecific mechanisms include electrostatic forces, aggregation, and cell surface hydrophobicity. It has been advocated that, although nonspecific interactions are involved in adherence, their overall contribution is less than that provided by specific mechanisms (Klotz 1994). An extensive body of literature exists on adherence of *C. albicans* to various oral mucosal (Willis et al. 2000; Blanco et al. 2006) and dental materials such as denture base acrylics and silicone-based resilient liner materials (Nevzatoglu et al. 2007), resilient denture-lining materials (Yilmaz et al. 2005), resin composite restorative dental materials (Maza et al. 2002), denture base materials (Radford et al. 1998, 1999), and orthodontic brackets (Gokdal et al. 2002); however, the mechanism of adherence to many cell types and surfaces remains unclear (Enache et al. 1996; Cotter and Kavanagh 2000). Knowledge of the mechanism by which *C. albicans* attaches itself to such surfaces may aid the development of treatment strategies that inhibit adherence of the fungus (San Millan et al. 1996). Many antifungal agents display an ability to retard adherence (Cotter and Kavanagh 2000). Many other methods that are nonantibiotic means of inhibiting adherence have been proposed. Among these methods were the disruption of surface-bound salivary protein, which promotes the adherence of *C. albicans* to surfaces (Nair and Samaranayake 1996), and monoclonal antibodies directed against extracellular matrix proteins collagen types I and IV (Cotter et al. 1998), the use of gelatin fragments for blocking the adherence of *C. albicans* to extracellular matrix proteins (Lee et al. 1996), and the use of antibodies and sugar amines to reduce the adherence to buccal epithelial cells (Collins-Lech et al. 1984; Lee et al. 1996). The change in hydrophobicity has also been reported to lead to a significant reduction in susceptibility to fungal adherence. A number of agents including the use of polyhexamethylene biguanides and quaternary ammonium compounds decrease the adhesion (Jones 1995; Schep et al. 1995). The removal of calcium by ethylene diamine tetraacetic acid (EDTA) and ethylene glycol tetraacetic acid (EGTA) is suggested to decrease the adherence of *C. albicans* to various extracellular matrix proteins (Klotz et al. 1993). Accordingly, the antifungal and fungicidal activity of calcium chelating or binding agents (particularly of EDTA) on *C. albicans* has been clearly demonstrated by Ates et al. (2005).

The importance of adherence of *C. albicans* to host tissue may be illustrated by its ability to adhere to various mucosal surfaces and to withstand forces that may lead to its removal, such as washing action of saliva/body fluids. Moreover, its ability to adhere to variety of oral surfaces including buccal epithelial cells, teeth, and saliva molecules as well as adherence to inert polymers and co-aggregation with several species of oral bacteria including *Streptococcus gordonii, S. mutans, S. oralis, S. sanguis, S. salivarus,* and *Actinomyces* species (Richards and Russell 1987; Branting et al. 1989; Jenkinson et al. 1990; Holmes et al. 1995; Millsap et al. 1998) makes its clearance a complex and multicomponent process.

### 9.4.3 Enzymes

The penetration of the surface epithelial cell by the candidal hyphae is probably due to an enzymatic process in combination with mechanical forces. The secreted aspartyl proteinases (SAPs) degrade many human proteins at lesion sites, such as albumin, hemoglobin, keratin, and secretory IgA (Hube et al. 1998). To date, nine different SAP genes have been identified in *C. albicans*. The proteolytic activity of SAPs has been associated with tissue invasion (Yang 2003). Other than SAPs the proteolytic enzymes include collagenase, glucosaminidases, acid and alkaline phosphatases, aminopeptidases, hyaluronidase, and chondroitin sulfatase, which act on the degradation of extracellular matrix proteins (Scully et al. 1994; Calderone and Fonzi 2001). Salivary proteins, including IgA, can be degraded by acidic proteinases of *Candida* particularly at low pH conditions (Samaranayake et al. 1994). It has been shown that a collagenolytic enzyme produced by *C. albicans* may digest the human dentine collagen (Kaminishi et al. 1986; Hagihara et al. 1988).

It has been shown that phospholipases are concentrated at the tips of fungal hyphae and localized in the vicinity of host cellular compartments where active invasion is occurring (Pugh and Cawson 1977; Ghannoum 2000). These enzyme activities were found in most *C. albicans* strains, but not in other less virulent *Candida* species (Samaranayake et al. 1984), and cause membrane damage to the host cells resulting in cell lysis (Ghannoum 2000).

#### 9.4.3.1 Evasion

In order to maintain *Candida* populations in the oral cavity, cells must grow and multiply at a rate at least equal to that of clearance. Cannon et al. (1995a) suggest that major factor influencing the balance among clearance, colonization, and candidosis is the interaction between *C. albicans* cells and the host defenses. Immune system defects are a major risk factor for candidosis. The presence of *Candida* species and the candidal overgrowth in the oral cavities of medically compromised patients have been demonstrated in many longitudinal studies (Arendorf and Walker 1979; Hauman et al. 1993; Swerdloff et al. 1993; Grimoud et al. 2003; Golecka et al. 2006; Li et al. 2006). *C. albicans* can evade host defenses as results of multiple mechanisms. Innate primary defense mechanisms play key roles in preventing yeast colonization of the oral cavity. Primary innate defenses include the epithelial barrier and anticandidal compounds of saliva such as lysozyme (Tobgi et al. 1988), histatins (Xu et al. 1991), lactoferrin (Nikawa et al. 1993), and calprotectin (Challacombe 1994). The major immunoglobulin in saliva is secretory IgA (SIgA), which aggregates yeasts and assists in clearance (Scully et al. 1994).

### 9.4.4 Biofilm formation

Biofilms are structured microbial communities that are attached to a surface. Microorganisms in biofilms are embedded within a matrix of extracellular polymers, and characteristically display a phenotype that is markedly different from planktonic cells (Douglas 2003). The first example of a biofilm to be recognized in medical systems was dental plaque on tooth surfaces. But, according to the estimates of National Institute of Health, more than 60% of microbial infections involve biofilms (Lewis 2001). These three-dimensional structures composed of yeast and hyphal cells embedded in an extracellular matrix constitute an important pitfall in the management of disseminated *Candida* infections because of their intrinsic resistance to almost all antifungals in clinical use. *Candida* biofilms are especially resistant to azoles and amphotericin B, but remain sensitive to the newly introduced echinocandins that target cell wall β-glucan biosynthesis (d'Enfert 2006).

Biofilm infections can be caused by a single microbial species or by a mixture of bacterial or fungal species (Costerton 1999). *C. albicans* has the ability to form biofilms on different surfaces, which is proposed to be one of the major reasons for its increased pathogenicity (Haynes 2001). Furthermore, the phenomenon of co-aggregation and co-adhesion between *Candida* and different bacteria and the effect of modulating factors such as saliva, sugars, and pH enhance the biofilm formation and colonization of oral mucosal and dental tissues (Jenkinson et al. 1990; Grimaudo et al. 1996).

Microbial biofilms are resistant to a variety of antimicrobial agents, including antibiotics, antiseptics, and industrial biocides. For example, when fungi exist in the biofilm form, they are five to eight times more resistant to clinically important antifungal agents such as amphotericin B, fluconazole, flucytosine, itraconazole, and ketaconazole than are planktonic cells (Hawser and Douglas 1995). The mechanisms of biofilm resistance to antimicrobial agents have not been fully understood, but several mechanisms have been suggested. Among these are (a) restricted penetration of drugs through the biofilm matrix, (b) phenotypic changes resulting from a decreased growth rate or nutritional limitation, and (c) expression of resistance genes induced by contact with a surface (Mah and O'Toole 2001; Donlan and Costerton 2002).

Bacteria are often found with *Candida* species in polymicrobial biofilms in vivo, and extensive interactions are demonstrated in these adherent populations (Douglas 2003). It has been reported that *Candida* resistance to fluconazole was enhanced in the presence of slime-producing staphylococci, but unaffected by the presence of slime-negative mutant (Douglas 2003).

### 9.4.5 Phenotypic switching

*C. albicans* is a very adaptable microorganism with the ability to survive in diverse and distinct anatomical sites. Micromorphological and physiological properties of *C. albicans* are rapidly modified in response to different growth conditions and environmental changes (Kennedy and Sandin 1988; Rams and Slots 1991; Soll 1992). Accordingly, pathogenicity may be increased after adaptation to the environment and phenotypic switching.

Many studies including healthy individuals have demonstrated strain specialization for particular anatomical sites (Soll et al. 1991; Hellstein et al. 1993; Kam and Xu 2002), and on the contrary, many studies have indicated the emergence of new, highly successful oral strains in the particular geographical locales (Tamura et al. 2001). In two consecutive studies, Hannula et al. (1997, 2001) revealed no difference in distribution of oral yeast species and of *C. albicans* phenotypes and genotypes between Finnish, American, and Turkish subjects obtained from the oral samples (periodontal pocket, oral mucosa, saliva). In an earlier study, Odds et al. (1983) also reported similar findings and found no significant differences among *C. albicans* phenotypes from different anatomical sources. However, in the same study they had found some differences among the phenotypes of strains from the different geographical areas. These and other studies emerged an interest in the phenotypic variability between strains and the developmental capacity for phenotypic variability within strains, which includes the capacity to differentiate between the yeast and hyphal forms, and the capacity to switch frequently and reversibly between general phenotypes that can be distinguished by colony morphology (Soll et al. 1994). Other than the differences in colony morphology, the general characteristics of switching in different strains were similar (Soll 1992; Soll et al. 1994) and included: (a) high- and low-frequency modes of spontaneous switching, (b) a basic original smooth phenotype, (c) reversibility and interconvertibility between phenotypes, (d) a limited number of predominant phenotypes, and (e) stimulation by low doses of ultraviolet irradiation.

Switching is associated with changes in micromorphology and physiological properties as well as a number of putative virulence traits. It therefore seems reasonable to suggest that switching may provide *C. albicans* and related infectious yeasts with the diversity that is expected such pervasive and successful pathogens. It has been proved that switching may provide an organism with the capacity to invade diverse body locations, evade the immune system, and/or change antibiotic resistance (Slutsky et al. 1985). Switching has been shown to occur not only in standard laboratory strains, but also in strains of *C. albicans* isolated from mouths of healthy patients. However, the fact that the genetic heterogeneity of strains increases with periodontal disease (Song et al. 2005) and HIV infection (Pizzo et al. 2002, 2005) was clearly demonstrated by recent studies. One study evaluated the *C. albicans* phenotypes and genotypes from infected root canals reported broad spectrum of heterogeneity of the strains, which was quite similar to previous reports from other oral and nonoral sources. Furthermore, the data implied that unusual strains of *C. albicans* are not involved in root canal infections. However, the systemic conditions of these patients were not described in this study (Waltimo et al. 2001).

## 9.5 Presence and pathogenicity of yeasts in different dental tissues

### 9.5.1 Yeasts and dental caries

Dental plaque has a diverse microbial population (Marsh 2004). Mutans streptococci and lactobacilli have been generally regarded as important microorganisms of dental plaque (Loesche 1986), while *Candida* is believed to be present temporarily as harmless saprophytic microorganisms. In addition to new microecological concepts of dental plaque and caries research, comprehensive studies on yeasts and their by-products have led to a renewed interest in *Candida* species and their possible role in the etiopathogenesis of dental caries.

*C. albicans* has a high acidogenic (Samaranayake et al. 1983, 1986; Odds 1988), aciduric and acidophilic (Odds 1988; Marchant et al. 2001), and cariogenic potential (Nikawa et al. 2003). It demonstrates a high affinity for uncoated (Nikawa et al. 2003) and coated hydroxylapatite (Cannon et al. 1995b; Nikawa et al. 1998) specifically through electrostatic interactions and dissolves hydroxylapatite to a greater extent (approximately 20-fold) when compared with *S. mutans*.

In addition, *C. albicans* also adheres to both denatured and intact collagen (Makihira et al. 2002a, b) and possesses collagenolytic activities (Kaminishi et al. 1986; Hagihara et al. 1988; Nishimura et al. 2002). Therefore, it is clear that the effect of *Candida* on dental hard tissues evolves in two ways. First, it may dissolve the inorganic material of dental hard tissues with its acidogenic properties and removes calcium. Second, it may attack the exposed collagen and causes dissolution of the organic material with its collagenolytic enzymes. However, the degree to which *Candida* may contribute to the pathogenesis of caries in vivo remains to be determined.

In a scanning electron-microscopic study, Sen et al. (1997a) have shown that *C. albicans* was able to adhere to normal or EDTA/NaOCl-treated enamel and cementum, demonstrating close adaptation (Fig. 9.1). The surfaces were covered with separate, but dense colonies. Different types of cell morphology including yeasts and hyphal structures were observed (Figs 9.2 and 9.3). Hyphae showed penetration into cracks and grew over the edges of the cavity in intimate contact. Yeast cells and hyphae were attached to the surfaces by strands of organic material (Fig. 9.4).

Taken together, with its acidogenic, collagenolytic, adhesive, and plaque-forming properties, *C. albicans* may contribute to the pathogenesis of caries.

*Candida* spp. have been frequently isolated from dental plaque in different clinical occasions (Hodson and Craig 1972; Brown et al. 1978, 1979; Beighton and

**Fig. 9.1** A colony of *C. albicans* consisting of yeast cells and hyphal extensions on untreated enamel surface. The extracellular material indicating dense cellular activity can be observed in the middle of the colony (original magnification 1,000×). (Courtesy of Bilge Hakan Sen, Kamran Safavi, and Larz Spangberg.)

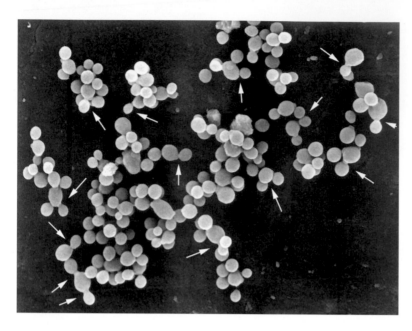

**Fig. 9.2** A colony of yeast cells on EDTA/NaOCl-treated cementum. The cells are in the stage of active budding (arrows) (original magnification 1,500×). (Courtesy of Bilge Hakan Sen, Kamran Safavi, and Larz Spangberg.)

Lynch 1995; de Carvalho et al. 2006; Eliasson et al. 2006). In addition to its presence in dental plaque, *Candida* has been also isolated from several forms of dental caries. Primary root carious lesions contain a high number of yeasts (Lynch and Beighton 1994; Beighton and Lynch 1995). Marchant et al. (2001) and de Carvalho et al. (2006) have demonstrated that there is significant association between the presence of *C. albicans* and early childhood caries. Sziegoleit et al. (1999) and Hossain et al. (2003) have also demonstrated that carious dentine has a high concentration of *Candida* spp., providing a significant ecological niche for the dissemination of these yeasts. Moreover, *Candida* spp. have been found to be closely related

**Fig. 9.3** A dense colony of *C. albicans* with numerous yeast cells on hyphal structures on untreated cementum surface (original magnification 550×). (Courtesy of Bilge Hakan Sen, Kamran Safavi, and Larz Spangberg.)

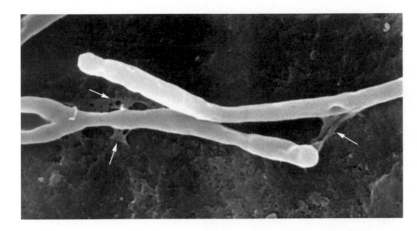

**Fig. 9.4** Two hyphal extensions showing close attachment to the enamel surface with secreted extracellular mucous material (arrows) (original magnification 5,000×). (Courtesy of Bilge Hakan Sen, Kamran Safavi, and Larz Spangberg.)

with postirradiation caries, particularly depending on the xerostomia after radiation therapy (Brown et al. 1978).

Oral presence of fungi in immunocompetent patients may not demonstrate a clinical pathology in most of the instances. However, there are indeed particular circumstances such as cancer or AIDS in which the immune system has been compromised. Damm et al. (1988) and Bunetel and Bonnaure-Mallet (1996) have reported that colonization of carious lesions by *Candida* may be responsible for development and recurrence of oral candidiasis in cancer patients. Jacob et al. (1998) have observed that *Candida* was frequently present (77%) in dental caries of HIV-infected people, and there was a significant association between clinical oral candidiasis and candidal colonization of carious dentin.

### 9.5.2 Dentin colonization and infection by yeasts

Colonization and penetration of dentin by microorganisms is considered as an important step for initiation and persistence of root canal infection. There are two main factors for invasion of dentin through dentinal tubules: colonizing species need to compete for space to adhere to the surface and sustain the infection, and for nutrient to survive. In addition, growth and survival in dentinal tubules can protect microorganisms from the effects of endodontic procedures such as instrumentation, irrigation, and disinfection.

There are a few in vitro studies showing infection of dentin by *Candida*. Sen et al. (1997b) investigated the interaction of *C. albicans* with smear-free root canal walls and the growth patterns of this fungus in relation to radicular dentin in a nutritionally stressed medium. The diameter of dentinal tubules after smear layer removal was 2.0–4.5 μm. While the size of the yeast cells was ranging from 2 to 5 μm, the width of growing hyphae was 1.5–2.0 μm and the length was extending as long as 100 μm. The canal walls were covered with yeast cells and hyphal structures forming dense, but separate, colonies (Fig. 9.5). Not only germ tubes, hyphae, and pseudohyphae, but also yeast cells showed evidence of penetrating into dentinal tubules (Fig. 9.6) and budding in the tubules, representing growth potential and active penetration. It was proposed that this contact-sensing (thigmotropism) ability of *Candida* supported dentinal invasion.

Contact sensing is a well-known property of neurons and other cells growing in close contact to surfaces (Bourett et al. 1987; Clark et al. 1990). This ability is actually a habituation of organisms to grow on the surfaces of solid materials or tissues (Sherwood et al. 1992; Gow et al. 1994). With this invasive affinity to dentin, *C. albicans* has been considered as a "dentinophilic" microorganism. The emergence of budding cells on previously formed pseudohyphae after 15 days in the nutritionally stressed medium with no addition of any type of sugars suggests that *C. albicans* can use dentin as a source of nutrition (Sen et al. 1997b).

In a parallel study from the same group (Sen et al. 1997a), colonization pattern of *C. albicans* was investigated in dentin cavities with or without smear layer. When there was no smear layer, a network of yeast cells and branching hyphae with bud clusters were present (Fig. 9.7). However, no dense colony was

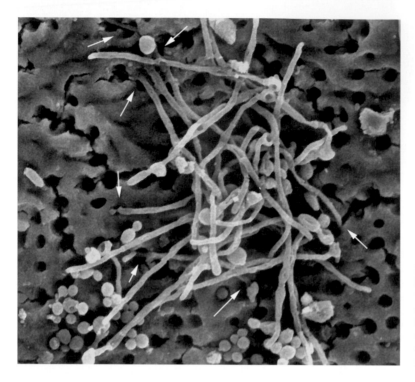

**Fig. 9.5**   A colony of yeast cells and hyphal structures grown on EDTA/NaOCl-treated radicular dentin. Note the penetration of hyphae (arrows) into dentinal tubules (original magnification 1,100×). (Courtesy of Bilge Hakan Sen, Kamran Safavi, and Larz Spangberg.)

observed at the base of the cavity or on the cavity walls, and dentin was still visible. Few hyphae demonstrated penetration into dentinal tubules. On the other hand, in the presence of smear layer, there was a dense mass of yeast cells and hyphae forming a thick biofilm layer

at the base and the walls of the cavity (Figs 9.8 and 9.9a). Fungal cells and mycelia were in close contact with the smear layer and its particles (Fig. 9.9b).

In a follow-up study, Sen et al. (2003) developed a reproducible, quantitative model for microbial

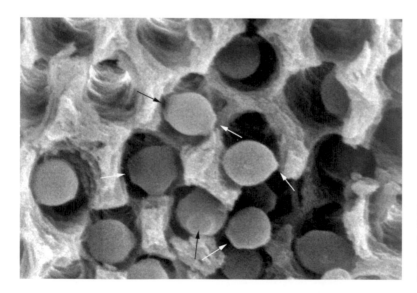

**Fig. 9.6**   Yeast cells migrating into the dentinal tubules. There are bud scars (arrows) on some of the cells (original magnification 5,500×). (Courtesy of Bilge Hakan Sen, Kamran Safavi, and Larz Spangberg.)

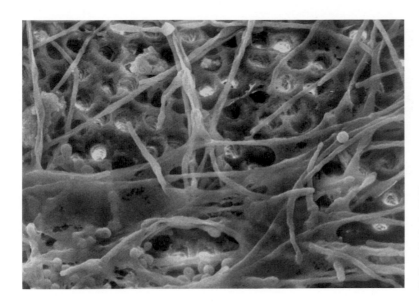

**Fig. 9.7** A colony of yeast cells and hyphae at the base of a smear-free dentin cavity. Note that part of the dentin is coated by an extracellular material (original magnification 1,500×). (Courtesy of Bilge Hakan Sen, Kamran Safavi, and Larz Spangberg.)

adhesion using a colorimetric method. After preparation of dentin disks, smear layer was either left intact or removed through EDTA/NaOCl irrigation. It was determined that presence of smear layer increased the adhesion of *C. albicans* to human dentin. Smear layer is composed of organic and inorganic material (Sen et al. 1995b). It has been previously observed that *Candida* has a specific affinity for dentinal collagen (Kaminishi et al. 1986; Hagihara et al. 1988) and type I collagen significantly increases adhesion of *Candida* (Makihira et al. 2002a, b). Furthermore, calcium ions control *Candida* morphogenesis (Holmes et al. 1991) and the adherence potential of *C albicans* to various extracellular matrix proteins (Klotz et al. 1993). Therefore, they have hypothesized that the increase in adhesion of *C. albicans* to smeared dentin is due to the presence of available resources of exposed collagen and calcium ions as a source of growth

**Fig. 9.8** A dense mass of yeast and hyphae at the base of a smeared dentinal cavity (original magnification 550×). (Courtesy of Bilge Hakan Sen, Kamran Safavi, and Larz Spangberg.)

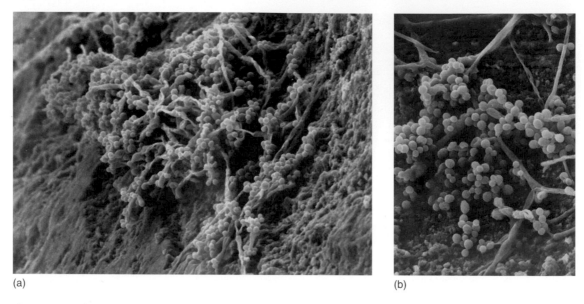

(a)                                                                      (b)

**Fig. 9.9**   (a) A dense colony of *C. albicans* particularly consisting of yeast cells on the smeared cavity wall (original magnification 750×). (b) Note that the yeast cells and hyphae are in close adaptation to the smeared surface (original magnification 1,500×). (Courtesy of Bilge Hakan Sen, Kamran Safavi, and Larz Spangberg.)

and adhesion. EDTA/NaOCl irrigation decreases the organic and the inorganic content of dentin. Hence, *C. albicans* loses its affinity to this poor substrate and shows less attachment to smear-free dentin. This finding is in accordance with the previous observational scanning electron microscopic (SEM) study (Sen et al. 1997a).

In an in vitro study by Waltimo et al. (2000a), penetration of *C. albicans* and *Enterococcus faecalis* into dentinal tubules was comparatively investigated through macroscopic and microscopic examination. *C. albicans* growth was weak and observed in 6 of 12 test systems during the 30-day incubation. Both microorganisms were capable of penetrating the 2 mm thick dentin discs. However, histological sections showed that continuous penetration of yeast cells and hyphae was limited to 60 μm in a few tubules and single yeast cells were found in low numbers throughout the dentin specimens. On the other hand, *E. faecalis* penetrated all dentin slices in a short period (1–5 days). While the infection was heavy in some of the tubules, majority of the tubules were free of bacterial cells. The authors concluded that the difference in penetration capacity of both microorganisms was mainly due to the different cell sizes of the species.

Siqueira et al. (2002a) investigated the colonization pattern of five fungal species—*C. albicans, C. glabrata, C. guillermondii, C. parapsilosis*, and *Saccharomyces cerevisiae*—by scanning electron microscopy. Regardless of the species, main growth form was single or budding yeast cells, but hyphal forms were not observed. While *C. albicans* colonized most of the specimens, the other fungal species showed separate or no colonization on the radicular dentin. *C. albicans* demonstrated different patterns of penetration into dentinal tubules. However, the majority of the tubules remained free of fungal cells.

Infection pattern of *C. albicans* on radicular dentin was similar in these in vitro studies (Sen et al. 1997a, b; Waltimo et al. 2000a; Siqueira et al. 2002a). *C. albicans* demonstrated slight to dense colonization and less or no penetration. However, there are case reports or clinical studies presenting heavy yeast infection of carious and normal dentin (Kinirons 1983; Damm et al. 1988; Jacob et al. 1998). A thick biofilm layer of *C. albicans* presenting different growth forms was always observed on exposed dentin. Hyphal extensions were detected almost in all dentinal tubules penetrating toward the pulpal space. Yeast cells were also shown in root canals in situ (Eidelman et al. 1978; Sen et al.

1995a). The root canal walls were covered by dense masses of yeast cells and hyphal structures. In addition, dentinal tubules were totally filled with hyphae (Eidelman et al. 1978). It is apparent that there is a distinct difference between in vitro and in vivo infection patterns of *C. albicans*. The reason for less penetration and colonization in vitro may be the use of irrigation solutions to remove the smear layer during experimental procedures. As stated previously, NaOCl and EDTA (or other acids) render dentin a deficient substrate, depleting available organic and inorganic content of dentin. To test this hypothesis, the colonization patterns of *C. albicans* on chemically treated or untreated radicular dentin were investigated by scanning electron microscopy (Turk et al. 2008). The colonization on untreated dentin surfaces was so dense that the yeast colonies could not be identified clearly. A dense mass of yeast cells forming a thick layer of biofilm was observed. On the other hand, the colonization was not extensive on treated dentin specimens and there were separate colonies. Few hyphal structures penetrated into dentinal tubules. The untreated dentin was considered to present available nutrition for *Candida* since it did not receive any chemical treatment. As observed in this study, *Candida* has an ability to change its colonization pattern depending on the condition of dentin surfaces. This fact should be considered in the future in vitro studies.

Most of our current knowledge about the colonization pattern of *Candida* originates from research using chemically modified dentin, grown in nutrient-rich media under optimal conditions. Therefore, extrapolation of the results from such studies (Sen et al. 1997a, b; Waltimo et al. 2000a; Siqueira et al. 2002a) to the in vivo situation may be highly misleading. Care should be taken when developing microbial models to evaluate adherence, colonization, and penetration patterns of *Candida* on root canal dentin.

### 9.5.3 Yeast infection of periodontal tissues

The interaction between the microorganisms present in the periodontal ligament and the host's immunologic system is the cause of periodontal diseases (Kornman et al. 1997; Azuma 2006). Although the cause-and-effect relationship of specific pathogens to periodontal disease is not fully established, some particular bacterial species are reported to be risk factors in periodontitis (Darveau et al. 1997; Ezzo and Cutler 2003). In addition to these bacteria, yeasts have also been isolated in many cases, suggesting a possible role of these microorganisms in the pathogenesis of periodontitis (Slots et al. 1988; Rams et al. 1990; Listgarten et al. 1993; Dahlen and Wikstrom 1995; Reynaud et al. 2001). The incidence of yeasts in relation with periodontal diseases is reported to be as high as approximately 15–24% in several studies (Slots et al. 1988; Najzar-Fleger et al. 1992; Reynaud et al. 2001; Järvensiu et al. 2004). However, they are not found to exceed 10% of the total viable count (Dahlen and Wikstrom 1995).

Certain immunocompromising conditions such as HIV infection, cytotoxic treatment, and broad use of antibiotics may cause superinfection of periodontal structures with yeasts (Peterson et al. 1987; Rams et al. 1990; Odden et al. 1994). Presence of yeasts in subgingival areas is a frequent finding in HIV-related periodontitis (62% of the subjects; 55% of the sites) (Zambon et al. 1990). Chattin et al. (1999) compared the presence of microorganisms at the periodontal sites of HIV-positive or HIV-negative subjects. The average cell numbers of *C. albicans* were significantly higher in samples of the HIV-positive group.

Overgrowth of yeasts in periodontal tissues has been demonstrated after systemic antibiotics therapy. González et al. (1987) found yeast cells invading gingival connective tissue in juvenile periodontitis. Large numbers of yeast cells were observed particularly after treatment with spiramycin. They concluded that antibiotics might favor overgrowth of yeasts. Another interesting finding in this study was that the researchers were unable to grow yeasts in Sabouraud agar (which is a specific medium for fungi) although they demonstrated their presence electron microscopically. Rams et al. (1990) also reported that systemic doxycycline therapy caused more than 10-fold increase in subgingival numbers of *Enterobacter aerogenes*, *Escherichia coli*, staphylococci, and *C. albicans*.

Yeasts can be actively responsible in the pathogenesis of the tissue breakdown in periodontal diseases. Odden et al. (1994) presented dense neutrophil infiltration of gingival epithelium and numerous mitoses as a reaction to candidal invasion. Järvensiu et al. (2004) studied the extent of candidal penetration into gingival tissues by immunochemistry and periodic acid-Schiff (PAS) staining. They found that hyphal germination started in the gingival pocket and *Candida* was typically present in the outer layers of the plaque, acting as a barrier between the host's immune system and the inner layers of microbial biofilm. Hyphae were also present deep in the connective tissue of periodontium, indicating candidal penetration and attachment.

According to the findings of these studies, it seems reasonable to consider yeasts not only as a frequently occurring microorganism in the periodontal microbial flora, but also as a part of the periodontal disease process.

### 9.5.4 Yeasts in root canals

#### 9.5.4.1 Primary root canal infections

The infection caused by microorganisms colonizing the necrotic pulpal tissues and root canal dentin is defined as primary root canal infection (Siqueira 2002). In general, primary infections are mixed and predominated by facultative or obligate anaerobic bacteria, depending on the microenvironmental changes and stresses as discussed in previous chapters.

Since fungi were not isolated in initial microbial flora of most root canal infections (Haapasalo 1989;

Sundqvist et al. 1989), they were not usually reported to be a common member of the microbial population isolated from primary endodontic infections (Sundqvist 1994). Even though the yeasts are present in the original sample, they may not be found on the plates selected for culturing because they have low colony-forming unit (CFU) number as compared with bacteria (Peciuliene et al. 2001; Waltimo et al. 2004a). In addition, they may be frequently considered as a contaminant particularly from the air (Waltimo et al. 2004a). However, there are multiple reports demonstrating presence of yeasts in infected root canals using either culturing, light and electron microscopy, or molecular techniques (Table 9.1).

The incidence of yeasts cultured from primary endodontic infections varies within a wide range. When a selective media had not been used, the incidence was reported to be as low as 0.5–10%. When the studies using selective medium such as Sabouraud's agar or broth are reevaluated, the occurrence of yeasts falls in the range of 1.9–61.5%. Najzar-Fleger et al. (1992)

**Table 9.1**    Literature relating to prevalence of yeasts isolated from root canals with primary endodontic infections.

| Authors | Number of samples (or teeth) | Method/culture | Prevalence of yeasts in root canals (%) |
|---|---|---|---|
| MacDonald et al. (1957) | 46 | Dextrose broth Thioglycolate broth | 2.2 |
| Leavitt et al. (1958) | 154 | Trypticase soy broth | 10 |
|  | 76 | Dextrose broth | 5 |
| Hobson (1959) | 98 | Nutrient broth Robertson's broth | 0.6 |
| **Jackson and Halder (1963)** | **214** | **Sabouraud's broth** | **26** |
| **Wilson and Hall (1968)** | **263** | **Robertson's meat broth Sabouraud's broth** | **1.9** |
| Goldman and Pearson (1969) | 563 | Trypticase soy broth Blood agar | 0.5 |
| Slack (1975) | 560 | Nutrient broth Robertson's broth | 5.2 |
| **Najzar-Fleger et al. (1992)** | **292** | **Sabouraud's agar** | **55** |
| Sen et al. (1995b) | 10 | SEM | 40 |
| Debelian et al. (1995) | 26 | Trypticase soy agar | 3.8 |
| Baumgartner et al. (2000) | 24 | PCR | 20.8 |
| **Lana et al. (2001)** | **27** | **Sabouraud's agar** | **7.4** |
| **Akdeniz et al. (2002)** | **13** | **Sabouraud's broth** | **61.5** |
| **Egan et al. (2002)** | **35** | **Sabouraud's agar** | **5.7** |
| Siqueira et al. (2002a) | 50 | PCR | 2 |
| Siqueira et al. (2002c) | 15 | SEM | 6.6 |
| **Ferrari et al. (2005)** | **25** | **Sabouraud's agar** | **4** |

The rows shown in bold indicate the studies in which a selective medium has been used for culturing of yeasts.

clearly demonstrated that the incidence was increased when Sabouraud's dextrose agar was used for cultivation, instead of nonselective blood agar.

Even though *Candida* spp., particularly *C. albicans*, have been isolated in most of the studies (Jackson and Halder 1963; Najzar-Fleger et al. 1992; Baumgartner et al. 2000; Lana et al. 2001; Egan et al. 2002), presence of other yeasts has also been reported. *S. cerevisiae* was recovered from both the root canal and the patient's blood samples taken during and after endodontic therapy (Debelian et al. 1995, 1997). Egan et al. (2002) investigated the relative prevalence and diversity of yeasts in saliva and root canals of teeth associated with apical periodontitis from the same patients. In addition to *C. albicans* and *C. sake*, *Rodotorula mucilaginosa* was isolated from the root canals. They also found that the presence of yeasts in root canals was significantly associated with their presence in saliva.

Fungi have also been observed electron microscopically in root canals associated with primary endodontic infections (Sen et al. 1995a; Siqueira et al. 2002b). Sen et al. (1995a) investigated the topography of the microbial flora of infected root canals. They observed that the root canals were heavily infected by cocci and rods in six teeth (Fig. 9.10). In the other four teeth, there were no bacteria, but yeast colonies were observed throughout the length of the root canals (Figs 9.11 and 9.12).

In addition to culturing and electron-microscopic methods, polymerase chain reaction (PCR) assay has

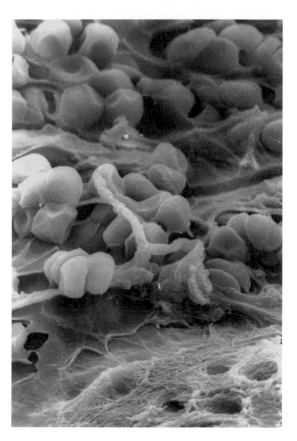

**Fig. 9.11** Yeast cells on the root canal wall in the middle third (original magnification 1,500×). (Courtesy of Bilge Hakan Sen, Beyser Piskin, and Tijen Pamir.)

**Fig. 9.10** Bacteria (arrows) in dentinal tubules approximately 100 µm from the root canal wall in the middle third of root (original magnification 2,000×). (Courtesy of Bilge Hakan Sen, Beyser Piskin, and Tijen Pamir.)

been used for detection of yeasts in root canals. Baumgartner et al. (2000) used PCR primers specific for the 18 S ribosomal RNA gene of *C. albicans* and stated that the sensitivity of PCR for detection of DNA of this yeast was $10^{-4}$ ng of DNA. The presence of *C. albicans* was detected in 5 of 24 (20.8%) samples. It was concluded that yeasts might be involved in root canal infections more often than previously believed. On the other hand, Siqueira et al. (2002c) used PCR along with species-specific primers and found fungi only in 1 of 50 specimens. The difference may be originated from using different primers as well as samples isolated from different geographical locations (Rocas et al. 2006; Machado de Oliveira et al. 2007).

There are case reports showing that pure cultures of *C. albicans* were isolated from primary endodontic infections associated with apical periodontitis

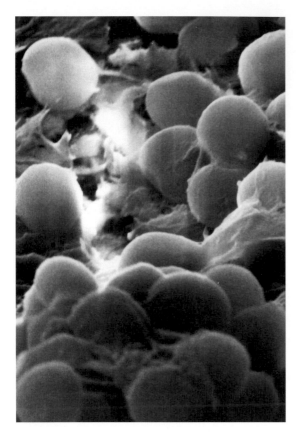

**Fig. 9.12** Yeast cells on the root canal wall in the apical third. Note that the cells are attached to each other and the dentinal wall with numerous organic strands (original magnification 5,000×). (Courtesy of Bilge Hakan Sen, Beyser Piskin, and Tijen Pamir.)

(Eidelman et al. 1978; Matusow 1981; Kinirons 1983; Damm et al. 1988).

### 9.5.4.2 Secondary root canal infections

This type of endodontic infection is caused by microorganisms that have not been in the root canal previously, but have penetrated into the endodontic space during treatment, between appointments, or after the endodontic treatment (Siqueira 2002).

As stated by Waltimo et al. (2003b), yeasts may be either present in low numbers or not present at all in primary endodontic infections. During endodontic procedures, they may reach higher percentages in the total cultivable flora or new yeast species may penetrate into the root canal system. Lana et al. (2001) investigated the microbial status of intact root canals with necrotic pulps. *C. tropicalis* and *S. cerevisiae* were recovered from two root canals (7.4%) before endodontic procedures were initiated. After instrumentation, irrigation with 2.5% NaOCl and disinfection with calcium hydroxide, these yeasts were not present in the root canal; however, *C. guilliermondii* and *C. parapsilosis* were recovered in the second and third collections, respectively. It is very likely that the latter species gained access to the root canal due to poor isolation or cavity seal. Pinheiro et al. (2003) demonstrated a significant association between coronally unsealed teeth and *Candida* spp. Wilson and Hall (1968) reported the prevalence of yeasts in primary endodontic infections as 1.9%. In the subsequent second and third visits, this incidence was increased to 6.8%. When they examined the cases with positive yeast culture, they observed that either the temporary restorations had defects or a very long period of time (3–4 months) had elapsed between the two visits (four cases). Jackson and Halder (1963) determined the presence of yeasts as high as 26% at the initial visit of endodontic therapy. At the subsequent visits after using chloramphenicol as an intermediate dressing, they isolated yeasts from the teeth that had been negative at the initial culture. They concluded that use of an antibacterial agent might favor the overgrowth of yeasts in the root canal.

### 9.5.4.3 Persistent root canal infections

After gaining access into the root canal system, the microorganisms may survive against harsh conditions such as intracanal procedures, disinfection, and obturation, and cause persistent infections. In addition to the establishment of presence of yeasts in primary and secondary infections in previous studies, there are considerable data that yeasts can also take part in the root canal microbiota of failed endodontic treatments.

The incidence of yeasts cultured from persistent endodontic infections is reported to be 2.9–22.2%. As stated previously, the use of a selective medium significantly affects the prevalence in root canals (Table 9.2). When a selective medium is used, their incidence is increased from 2.9–8.3% to 6.8–18%.

Yeasts are isolated either as pure cultures or together with other bacteria in endodontic therapy-resistant cases (Siren et al. 1997; Waltimo et al. 1997; Sundqvist et al. 1998; Peciuliene et al. 2001; Siqueira

**Table 9.2**  Literature relating to prevalence of yeasts isolated from root canals with persistent endodontic infections.

| Authors | Number of culture-positive samples (or teeth) | Method/culture | Prevalence of yeasts in root canals (%) |
|---|---|---|---|
| Nair et al. (1990) | 9 | LM and TEM | 22.2 |
| **Siren et al. (1997)** | **70** | **TSBV agar** | **12.9** |
| **Waltimo et al. (1997)** | **692** | **TSBV agar** | **6.8** |
| Sundqvist et al. (1998) | 24 | Thioglycolate | 8.3 |
| Hancock et al. (2001) | 34 | Agar | 2.9 |
| **Peciuliene et al. (2001)** | **33** | **Sabouraud's agar** | **18.2** |
| **Cheung and Ho (2001)** | **18** | **Sabouraud's agar** | **11.1** |
| **Egan et al. (2002)** | **25** | **Sabouraud's agar** | **16** |
| Pinheiro et al. (2003) | 51 | Nonselective | 3.9 |
| Siqueira and Roças (2004) | 22 | PCR | 9 |

The rows shown in bold indicate the studies in which a selective medium has been used for culturing of yeasts.
LM, light microscopy; TEM, transmission electron microscopy; TSBV, tryptic soy-serum-bacitracin-vancomycin.

and Roças 2004). Waltimo et al. (1997) isolated 48 fungal strains from 47 microbial samples, representing 7% of the 692 culture-positive samples. While *C. albicans* was the most common isolate, *C. glabrata*, *C. guilliermondii*, *C. inconspicua*, and *Geotrichium candidum* were also isolated. Fungi were demonstrated as pure cultures in six cases. Therefore, it was advocated that they had a definite pathogenic role in developing apical periodontitis. In most of the fungi-positive cases, facultative Gram-positive bacteria such as α- and nonhemolytic *Streptococcus* species were present. On the other hand, Gram-negative isolates were found randomly.

Peciuliene et al. (2001) reported the prevalence of *C. albicans* in root-filled teeth with chronic apical periodontitis as high as 18%, and this fungus was recovered from the root canals 50% with *E. faecalis* and 50% with other bacteria. However, yeasts constituted only <1% of the total cultivable microbial flora. They have concluded that the ecology in the root canals that were inadequately filled may favor the particular growth of *E. faecalis* and *C. albicans*. Adib et al. (2004) identified the cultivable microbial flora in root-filled teeth with persistent apical periodontitis and evident coronal microleakage. In addition to the presence of Gram-positive facultative anaerobes (75%), six strains of *Candida* were isolated from coronal dentin, root dentin, and root gutta-percha samples in three of eight subjects.

Apart from culture studies, the presence of fungi in failed root canal therapy has been shown using either microscopic or molecular biological techniques.

Nair et al. (1990) analyzed the apical part of nine therapy-resistant cases with light and electron microscopes. Two specimens revealed yeast-like microorganisms in the root canal and apical foramen. There were numerous budding yeasts, indicating that they were in active proliferation. They demonstrated fungi as a potential nonbacterial, microbial cause of nonhealed apical periodontitis. Siqueira and Roças (2004) studied root canal samples taken from 22 root-filled teeth with persistent periradicular lesions by means of PCR. While *E. faecalis* was the most prevalent bacterial species (77%), *C. albicans* was isolated in 9% of the samples.

### 9.5.5 Extraradicular yeast infections

Generally speaking, bacteria are located within the root canals of teeth with apical periodontitis and can occasionally penetrate into the periapical tissues in clinical conditions such as acute apical periodontitis, acute apical abscess, periapical actinomycosis, and osteomyelitis (Nair 1997; Dahlen 2002). On the other hand, many studies have shown that microorganisms can survive even in chronic apical pathologies of treated or untreated endodontic cases (Haapasalo et al. 1987; Tronstad et al. 1987, 1989; Iwu et al. 1990; Kiryu et al. 1994; Gatti et al. 2000). However, the attention has been focused on bacteria in these studies and any information related to presence of yeasts in periapical tissues has not been reported.

Nair et al. (1990) demonstrated electron microscopically that yeasts were present in the apical part of root canals, but not extended to the periapical tissues. On the other hand, in a study investigating root surfaces of teeth with chronic apical periodontitis using SEM, Lomcali et al. (1996) reported that there was a multilayered microbial plaque in most of the specimens and yeast cells were harbored in resorption lacunae.

Baumgartner et al. (2000) studied 19 samples of aseptic aspirates of cellulitis/abscesses of endodontic origin by means of PCR. They found that all samples were negative for the presence of *C. albicans* DNA. Waltimo et al. (2003a) investigated refractory periapical granulomas for the presence of *Candida* spp. using three different methods. They were able to extract DNA from 68 to 103 paraffin-embedded samples. PCR products demonstrated possible occurrence of *Candida* spp. in 18 of those 68 DNA-positive samples. However, PAS staining and immunohistological examination did not confirm this finding, and all PCR-positive samples were found to be negative for presence of *Candida*. Further sequencing of the PCR products revealed that the sequences were not typical for *Candida* spp. Therefore, they emphasized both the importance of primer selection in molecular microbiological studies and the use of sequencing to confirm the findings. Sunde et al. (2002) examined the periapical microbiota of 36 teeth with refractory apical periodontitis. Fifty-one percent of the bacterial strains were anaerobic and Gram-positive species had predominance (79.5%). In addition to bacterial species, they cultivated *C. albicans* in two patients. Even though the number of lesions with *Candida* was very small, this finding was in contradiction with the findings of Baumgartner et al. (2000) and Waltimo et al. (2003a).

Several case reports have also demonstrated the presence of yeasts in periapical tissues (Eidelman et al. 1978; Matusow 1981; McManners and Samaranayake 1990). Eidelman et al. (1978) showed that *C. albicans* was present in root canals and periapical granuloma of a patient suffering from chronic urticaria. Histological examinations revealed that the granuloma exhibited invasive *Candida* infection. The complete cure of the patient was achieved only after the extraction of *Candida*-infected teeth and curettage of the granuloma.

Damm et al. (1988) presented a patient with carcinoma ex-pleomorphic adenoma. Radicular and cervical soft tissues of extracted teeth of this patient were examined histologically using periodic acid-Schiff, Gomori methenamine silver, and Brown and Brenn stains. The microscopic analysis revealed that fibrous connective tissues contained mixed inflammatory cellular infiltrate with large areas of necrosis and abscess formation. The tissues were invaded by numerous colonies of yeast cells together with pseudohyphae, which were morphologically consistent with *C. albicans*.

Debelian et al. (1995, 1997) reported the presence of *S. cerevisiae* in the root canal and blood sample of a patient undergoing endodontic therapy. With the aid of phenotyping and genotyping tests, it was found that the root canal and blood isolates were identical. It was concluded that the root canal was the source of the blood isolate and fungemia occurred unintentionally during endodontic therapy. It was interesting to note that the patient had been an alcoholic for 20 years and alcoholism could be a predisposing factor for yeast infections (Bardwell et al. 1986; Trowbridge et al. 1999).

As seen in these cases, extraradicular presence of yeasts seems to be associated mostly with immunocompromised state of the patients.

## 9.6 Antifungal activity of endodontic irrigating solutions and disinfectants

During the 1950s and 1960s, local use of antibiotics had been extensively used in endodontic treatment. Most of the antibiotic pastes contained an antifungal component, particularly nystatin or sodium caprylate. It is clearly evident that there used to be a clinical approach to antifungal treatment of infected root canals. Since the 1970s, the popularity of local use of antibiotics in endodontics decreased mainly due to the high risks of producing resistant microorganisms and host sensitization. Since then, antifungal considerations in endodontic therapy have received little attention.

If there is a possible yeast infection in the root canal, use of common antibacterial solutions or disinfectants, which may have limited antifungal activities, may favor the overgrowth of yeasts between the visits (Sen et al. 1999). As stated previously, single *Candida* spp. may cause persistent root canal infections in vivo (Waltimo et al. 1997). Therefore, in addition to antibacterial effectiveness, it is desirable that root canal irrigating solutions or disinfectants should have antifungal capacity as well.

### 9.6.1 Sodium hypochlorite

Sodium hypochlorite (NaOCl) is perhaps the most widely used irrigant in endodontics today. It has a broad-spectrum of antimicrobial activity against bacteria, bacteriophages, spores, yeasts, and viruses (Mentz 1982). Its particular antifungal properties have been investigated in several studies with different methodologies (Ayhan et al. 1999; Waltimo et al. 1999a; Ferguson et al. 2002; Radcliffe et al. 2004).

Ferguson et al. (2002) reported that NaOCl was very effective against *C. albicans* having an MIC (minimumal inhibitory concentration) less than 10 µg/mL. Different concentrations of NaOCl ranging from 0.5 to 5% showed complete antifungal activity in a range from 10 s to 30 min (Smith and Wayman 1986; Harrison et al. 1990; Radcliffe et al. 2004; Vianna et al. 2004). According to Radcliffe et al. (2004), over 9 million CFU were decreased to below the limit of detection even with 0.5% NaOCl within a 10 s contact time. Waltimo et al. (1999a) used filter paper method to determine the susceptibilities of seven strains of *C. albicans* to endodontic disinfectants. Among a group that included iodine potassium-iodide (IKI), chlorhexidine acetate, and calcium hydroxide, NaOCl was the most effective disinfectant. Both 5 and 0.5% concentrations showed complete killing in 30 s. However, 0.05 and 0.005% concentrations were not effective even after 24 h.

Sen et al. (1999) investigated the antifungal effects of 0.12 chlorhexidine (CHX), 1 and 5% NaOCl on *C. albicans* biofilms grown on the radicular dentin walls and the effect of smear layer on antifungal capacity of these solutions. In the presence of smear layer, antifungal activity was not observed in 1, 5, and 30 min groups of any solution. Only 1 h treatment groups for all solutions were effective. When the smear layer was absent, 5% NaOCl alone started to exert its antifungal properties after 30 min. It is apparent that both smear layer and biofilms of *C. albicans* delayed or stopped the antifungal capacity of NaOCl and CHX and it took more time to reach the complete antifungal effect. This finding was different from that in other studies, which used direct exposure tests. Even though the direct exposure test is a practical laboratory method, the clinical relevancy of the methodology should be taken into consideration with the knowledge of complex root canal anatomy, polymicrobial root canal infections, and biofilm formation (Orstavik and Haapasalo 1990; Spratt et al. 2001; Clegg et al. 2006).

When embedded in a biofilm, microorganisms become resistant to antimicrobial agents that are active against the planktonic form of the same organisms in suspension (Evans and Holmes 1987). During irrigation of root canals with microbial biofilms, the first layer of microorganisms will be directly affected by a relatively high concentration of the irrigating or disinfecting solution. On the other hand, the extracellular matrix of the biofilm will stop the solution from affecting the deeper layers, which may contain still viable microorganisms. Endodontic instrumentation may disrupt the biofilm and expose its embedded microorganisms to irrigating/disinfecting solutions. However, it is also well-known that endodontic files cannot touch all root canal walls (Wu et al. 2002, 2003), and microorganisms may still remain in the root canal and dentinal tubules after instrumentation and irrigation (Moodnik et al. 1976; Nair et al. 2005).

Clegg et al. (2006) developed a good model to study the effect of biofilm in root canal disinfection. Polymicrobial biofilms were grown on apical dentin and the effect of different irrigating solutions was evaluated. It was concluded that 6% NaOCl was the only irrigant capable of both disrupting the biofilm physically and rendering bacteria nonviable. On the other hand, Ruff et al. (2006) evaluated antifungal efficacy of several irrigants as a final rinse after developing biofilms of *C. albicans* in root canals. On 1 min application, 6% NaOCl could not destroy all yeast cells, but lowered the growth potential considerably from $100 \times 10^3$ to $0.1 \times 10^3$.

### 9.6.2 Ethylene diamine tetraacetic acid

EDTA has been mainly used to increase the efficacy of root canal instrumentation (Fraser 1974; Hülsmann et al. 2002; Lim et al. 2003). It reacts with calcium ions in dentin and forms soluble calcium chelates (Fraser 1974). It is also highly effective in removing smear layer from root canal walls (Sen et al. 1995b; Torabinejad et al. 2002). In addition, EDTA does not only remove smear layer, but also removes the microorganisms entrapped in the smear layer (Bystrom and Sundqvist 1985; Yoshida et al. 1995). However, EDTA itself has not been considered as a particular antimicrobial agent and has no effect on gram-positive species (Heling and Chandler 1998). Orstavik and Haapasalo (1990) demonstrated that EDTA did not have any disinfecting action, but rather increased antibacterial effects of other agents by opening the dentinal tubules. On the other hand, some antibacterial activity of EDTA has been reported against streptococci and staphylococci (Masillamoni et al. 1981; Root et al. 1988). According to Haapasalo et al. (2005), EDTA extracts surface proteins of microorganisms by interacting with the metallic ions from the cell envelope and leads to microbial death only if the microorganisms are exposed to EDTA for a long time. EDTA alone or in combination with other antimicrobial agents is used for disruption and eradication of biofilm layers of several microorganisms such as *Pseudomonas aeruginosa*, *Staphylococcus aureus*, *Staphylococcus epidermidis*, and *C. albicans* in catheters or polycarbonate chips (Kite et al. 2004; Percival et al. 2005; Banin et al. 2006). EDTA has been found to be effective even on microorganisms cultivated on these artificial surfaces that contain no +2 ions. Its activity may be more evident on biofilms grown on organoinorganic surfaces such as dentin, particularly affecting the substrate as well.

From an endodontic point of view, antifungal properties of EDTA were first evaluated by Sen et al. (2000) in an agar diffusion study. They observed that EDTA had the highest antifungal activity among a variety of common antifungal drugs and endodontic irrigating solutions including nystatin, ketoconazole, NaOCl, and CHX. In addition, it was demonstrated that the root canal isolate of *C. albicans* was more susceptible to all disinfectants when compared with the oral cavity isolate. This potential antifungal activity of EDTA is considered interesting since it does not affect protein and DNA synthesis of *C. albicans* (Gil et al. 1994). However, there may be two particular ways for EDTA to show its antifungal activity: first, anti-colonization; second, antigrowth effects. According to Odds (1988), there is a direct relation between adherence and colonization capacity of *C. albicans*. The presence of calcium ions in the medium or environment has a critical role in the control of morphogenesis of *C. albicans* (Holmes et al. 1991) and its adherence capacity to various extracellular matrix proteins (Klotz et al. 1993). Hence, EDTA may prevent binding of *C. albicans* to proteins by chelating calcium ions. In addition to anti-colonization effect, EDTA may also reduce the growth potential of *C. albicans*. By removing calcium ions from the cell walls, it causes collapses in the cell wall and inhibits the enzyme reactions (Pugh and Cawson 1980).

Antifungal effect of EDTA has been confirmed by other studies (Grawehr et al. 2003; Ates et al. 2005). The effects of different calcium-chelating or calcium-binding agents on *C. albicans* were evaluated by determining minimum inhibitory (MIC) and fungicidal concentrations (MFC) of the solutions (Ates et al. 2005). Ketoconazole and EDTA showed the highest antifungal and fungicidal activities followed by titanium tetrafluoride. EGTA and sodium fluoride demonstrated weak activity against *C. albicans*. Even though EGTA is considered as a specific calcium ion chelator, it did not demonstrate an antifungal activity similar to that of EDTA. Since calcium-binding capacities of EDTA and EGTA are similar (Ueno et al. 1982), high antifungal activity of EDTA may be related to chelating not only calcium ions, but also other divalent ions such as magnesium, manganese, and zinc. These divalent ions are also considered important for the growth and morphogenesis of *C. albicans* (Bedell and Soll 1979; Holmes et al. 1991; Sohnle et al. 2001). In addition to the chelating activity, EDTA inhibits a pericellular metalloenzyme of *C. albicans*, which presents a cell-associated collagenolytic property (Nishimura et al. 2002).

In contrast to the above-mentioned studies presenting clear antifungal activity of EDTA, Ruff et al. (2006) demonstrated in a root infection model that a final rinse with 1 mL of 17% EDTA at a contact time of 1 min was not sufficient to remove *C. albicans* from the root canals. It is probable that EDTA interacts rapidly with root canal dentin through chelation mechanisms, reaches a saturation point with excessive calcium ions, and becomes ineffective against *C. albicans*. Accordingly, Grawehr et al. (2003) have reported that microbial growth inhibition of EDTA is strongly reduced when it is preincubated with dentin powder. This may be the explanation for the weak antibacterial effect of EDTA in infected dentin tubules (Orstavik and Haapasalo 1990; Heling and Chandler 1998). However, this problem may be solved in a clinical situation by increasing the volume and contact time of EDTA used for irrigating root canals.

Because of its limited antibacterial effectiveness, EDTA may not be recommended as a routine working solution during endodontic treatment by itself. However, due to its strong antifungal and biofilm-disrupting properties, copious irrigation with EDTA may be advised particularly in persistent root canal infections or in root canals of medically compromised patients who are prone to oral candidosis.

### 9.6.3 Chlorhexidine

CHX is one of the most commonly used biocide in antiseptic products in general medicine and dentistry. It has a wide antimicrobial spectrum and is effective against both Gram-positive and Gram-negative bacteria as well as yeasts. It permeates the cell wall or outer membrane and attacks the cytoplasm of bacteria or inner membrane or the yeast plasma membrane (McDonnell and Russell 1999). When *Candida* is exposed to CHX, both macroscopic and microscopic profound effects on the viability and structural integrity of *Candida* are observed (MacNeill et al. 1997). Nucleoproteins coagulate with inhibition budding and cell wall changes with possible escape of cytoplasmic components through the cell membrane (Bobichon and Bouchet 1987). *Candida* grown in the presence of CHX is more susceptible to spheroplasting, indicating further that the antiseptic affects the cell surface composition of *Candida* (McCourtie et al. 1986).

CHX has long-term antimicrobial properties due to its unique ability to bind to organic and inorganic dental tissues (Rolla et al. 1970; Parsons et al. 1980). It also inhibits the initial adherence of yeasts and other microorganisms efficiently, and perhaps further accumulation and biofilm formation (Waltimo et al. 2004b). Although it has strong antimicrobial properties with relatively low toxicity (Chang et al. 2001; Tanomaru Filho et al. 2002), its activity is mainly pH dependent and is considerably reduced by organic materials (Russell and Day 1993).

Use of CHX as an irrigating solution in endodontics has been suggested by Delany et al. (1982). Later, its gel form has been proposed as an intracanal dressing (Siqueira et al. 1997; Gomes et al. 2003). Even though the literature about the antibacterial effect of CHX is abundant, the scientific exploration of its antifungal effects in endodontics has been growing at a slower pace in recent years.

Ferguson et al. (2002) reported that MIC of CHX gluconate needed to inhibit the growth of *C. albicans* was less than 0.63 µg/mL. In a broth dilution test, both liquid and gel formulations of 2% CHX gluconate eliminated *C. albicans* in 15 s (Vianna et al. 2004). Similarly, Gomes et al. (2006) demonstrated that 2% CHX gluconate gel eliminated *C. albicans* in 15 s in a direct contact test. Waltimo et al. (1999a) evaluated the in vitro susceptibility of seven *C. albicans* strains to four disinfectants and their combinations, using filter paper method. They found that 0.5% CHX acetate killed all yeast cells within 5 min. Tenfold dilution of this solution became effective within 1 h and there was a slight variation in susceptibility of the strains to this dilution. It was also stated in this study that combinations of the disinfectants were equally or less effective than the more effective component. Sena et al. (2006) evaluated the effect of 2% CHX in liquid and gel forms against *C. albicans* biofilms grown on cellulose nitrate membranes. The contact time required to achieve negative cultures for CHX ranged between 30 s and 30 min, depending on the formulation type and mechanical agitation. The liquid form and mechanical agitation lowered the effective contact time.

It is apparent that CHX is very effective against *C. albicans* in direct contact tests or after growing biofilms on artificial membranes. However, these methods may not be clinically relevant. Considering the fact that organic matter, which is abundant in the root canal environment, reduces the activity of CHX, it may be expected that antimicrobial performance of CHX will be affected considerably. Haapasalo et al. (2000) demonstrated that the effect of CHX was reduced by the presence of dentin. Similarly, Portenier et al. (2002) showed that CHX was strongly inhibited by organic content of dentin as well as heat-killed cells of *E. faecalis* and *C. albicans*. However, dentin pretreated with EDTA showed only slight inhibition in this study. Loss of antimicrobial activity of CHX by bovine serum albumin had also been reported previously (Portenier et al. 2001).

These inhibition studies may help to explain relatively poor performance of CHX in the root canal. Sen et al. (1999) reported that antifungal activity of 0.12% started in 1 h regardless the presence or absence of smear layer. Menezes et al. (2004) infected the root canals with *C. albicans* for 7 days and observed that 2% CHX irrigation following instrumentation could not prevent further growth of *C. albicans*. In another tooth infection model, antifungal activity of 2% CHX gel was evaluated on biofilms of *C. albicans* in root canals (Ercan et al. 2006). It took 7 days for complete inhibition of *C. albicans* growth. In contrast to these studies, it was demonstrated that a final rinse of 2% CHX with 1 min application reached a negative culture in the root canals having biofilms of *C. albicans*. However, it should be strongly emphasized that no attempt was made in this study to neutralize CHX left in the root canals.

There is substantial evidence that CHX has a good potential of antifungal effects in general. However, its activity may be limited in the root canals by the presence of high amount of organic structures. In addition, CHX does not have any tissue-dissolving activity (Okino et al. 2004) and cannot remove smear layer (Yamashita et al. 2003) and cannot disrupt biofilms (Clegg et al. 2006). In this regard, use of CHX can be modified with the aid of other irrigating solutions. After completion biomechanical instrumentation with NaOCl as a working solution, smear layer is removed with sequential use of EDTA and NaOCl. Hence, the amount of organic structures present in the root canal and dentin walls is greatly reduced and CHX solution is used as a final rinse or gel form as a dressing. This may create an optimal environment for CHX to present its potent antimicrobial properties and bind to dentin to release back in a longer period.

### 9.6.4 MTAD

MTAD (a mixture of tetracycline isomer (doxycycline), an acid (citric acid), and a detergent (Tween 80); Biopure, Tulsa Dentsply, Tulsa, OK) is relatively a new product. Torabinejad et al. (2003a, b) have demonstrated that Biopure MTAD is an effective solution to remove smear layer without causing any erosion in dentinal tubules. MTAD was found to be very effective in eradicating *E. faecalis* (Shabahang and Torabinejad 2003; Shabahang et al. 2003; Torabinejad et al. 2003c). However, its activity was lower than EDTA, NaOCl, and their combinations (Dunavant et al. 2006; Baumgartner et al. 2007).

The antifungal effect of MTAD is not clear and has been studied in only one study so far. In a tooth model, it was demonstrated that a final rinse of 5 mL of Biopure MTAD at a contact time of 5 min was not effective against biofilm of *C. albicans* (Ruff et al. 2006). Doxycycline, which is the active antimicrobial ingredient in Biopure MTAD, demonstrates considerable antimicrobial activity against a wide spectrum of Gram-positive and Gram-negative microorganisms, but not against fungi (Liu et al. 2002). It has been previously reported that a well-recognized potential disadvantage of tetracycline administration is the induction of opportunistic infections such as candidiasis (Rams et al. 1990). In addition, MacNeill et al. (1997) observed that a solution of tetracycline hydrochloride at a concentration of 3.0 mg/mL caused a heavy and constant uniform growth of *C. albicans* during the study period of 10 days. The other component of MTAD that is citric acid also does not have any an-

tifungal activity (Smith and Wayman 1986). With the current knowledge of poor antifungal properties of both tetracycline and citric acid, it is not surprising to find out that Biopure MTAD is not effective against *C. albicans*. From the clinical point of view, it may be concluded that MTAD should be carefully used in persistent root canal infections that may have yeasts already in the microflora or in the root canals of the patients who already have oral candidosis.

### 9.6.5 Calcium hydroxide

Calcium hydroxide (CH) has been routinely used as an interappointment medicament in endodontics for many years. In cases of endodontic infections, a CH dressing for at least 1 week is advocated for use in root canals. However, it has limited effectiveness in eliminating bacteria from human root canal when assessed by culture techniques (Peters et al. 2002; Zerella et al. 2005; Sathorn et al. 2007). After isolation of yeasts particularly from root canals with secondary or persistent infections, there has been some interest in the antifungal capacity of calcium hydroxide preparations, in the last decade.

Waltimo et al. (1999b) demonstrated that saturated aqueous solutions of CH were ineffective against clinically important *Candida* spp. Even though they used a direct contact test, it required 16 h to kill 99.9% of the CFUs of *C. albicans* strains. When compared to *E. faecalis*, all *Candida* spp. showed either equally high or higher resistance to aqueous CH solution. In a parallel study from the same group (Waltimo et al. 1999a), they saturated irrigating solutions with CH powder and observed that these solutions were always more effective than CH alone. Similarly, Ferguson et al. (2002) reported ineffectiveness of saturated CH solution against *C. albicans*. Because *C. albicans* is able to survive in a wide range of pH, alkalinity of CH solutions may not have any effect on this organism. In addition, saturated CH solution readily presents calcium ions necessary for growth and morphogenesis of *Candida* (Holmes et al. 1991; Klotz et al. 1993). On the other hand, it has been reported that antifungal effectiveness of CH is increased when used as a paste (Ferguson et al. 2002). Siqueira et al. (2003) demonstrated that CH/glycerin paste started to disinfect the dentin specimens infected with *C. albicans* in 2 days and reached complete disinfection in 7 days.

Efforts have been made to increase antifungal properties of CH paste. In this regard, mixing it with other disinfectant solutions, particularly with CHX, has been proposed. However, the outcomes of this approach are somewhat contradictory. While Siqueira et al. (2003) demonstrated that CH/CHX combination was ineffective in disinfecting *C. albicans*–infected dentin specimens even after 7 days, Haenni et al. (2003) did not observe any additive antimicrobial effect of CHX, IKI, and NaOCl when mixed with CH powder. In contrast to these findings, Ercan et al. (2006) and Gomes et al. (2006) reported that CH mixed with 2% CHX solution or gel considerably improved antimicrobial effectiveness of CH in comparison to CH/sterile water combination. Zerella et al. (2005) studied potential interactions of CH with CHX in a preliminary study before the main clinical research. They found that there was a significant loss (>99%) of CHX when mixed with CH. However, despite this potential loss of CHX, the antimicrobial effectiveness of this mixture was as effective as CHX alone on *E. faecalis*. Therefore, it was considered that the combined effect of CH and CHX might have a clinical value. In the clinical study, CH/CHX mixture more effectively eliminated microorganisms from the root canal in comparison to CH/sterile water combination. However, this difference was not statistically significant due to the low number of subjects in the study. Clinical performance of CH/CHX combination seems to be promising; however, further in vitro studies and larger clinical studies are needed to elucidate whether CHX has an additive effect when mixed with CH.

## 9.6.6 Other antifungal measures

### 9.6.6.1 Effect of endodontic procedures

Biomechanical preparation of root canals cannot completely eliminate the bacteria from the root canals. The same concept may also be true for the presence of yeasts in various types of root canal infections.

Lana et al. (2001) analyzed the root canals with pulp necrosis microbiologically before and after endodontic procedures. At the first collection, yeasts (*C. tropicalis*, *S. cerevisiae*) were recovered from two root canals having intact pulp chambers. In the second collection taken after endodontic instrumentation, irrigation, and disinfection of 1 week, *C. guilliermondii* was isolated in one canal. After sealing the root canals without any dressing for another week, *C. parapsilosis* was present in another root canal.

In another clinical study (Ferrari et al. 2005), twenty-five single-rooted teeth with pulp necrosis, intact pulp chamber, and periradicular lesions were used to detect enterococci, enteric bacteria, and yeasts before (first sampling) and after (second sampling) endodontic instrumentation. While 0.5% NaOCl was the working solution, 10 mL EDTA was used for final irrigation. After sealing with temporary restoration for 1 week, third sample was obtained. At the first microbial sample, there was only one root canal infected by yeasts (*C. albicans*). After biomechanical instrumentation, yeasts were no longer detected. However, two root canals (*C. glabrata* and *C. magnoliae*) contained yeast cells after 7 days without intracanal dressing.

In an ex vivo study (Menezes et al. 2004), the teeth were contaminated with *C. albicans*. After biomechanical preparation up to size 50 K file was accomplished with sterile saline, the root canals were irrigated with 2.5% NaOCl or 2% CHX. Following the first sampling, the root canals were filled with Sabouraud's broth and were sealed with temporary cement for 1 week. Then, a second microbial sampling was performed. While the root canals were negative for *C. albicans* at the first sampling, the yeast cells were present at the second sampling. The results showed that *C. albicans* was able to recolonize the root canals after instrumentation and irrigation with 2.5% NaOCl or 2% CHX.

In contrast to the result of the above-mentioned studies, Peciuliene et al. (2001) did not recover yeasts after endodontic procedures, including instrumentation and irrigation with 2.5% NaOCl and 17% EDTA. Due to low number of *Candida*-positive cases, they stated that it was too early to draw a conclusion that yeasts were very sensitive to endodontic biomechanical instrumentation.

It is a fact that yeasts may be present in a wide range of percentage in various types of endodontic infections. There are two possible ways for their persistence in the root canal even after endodontic procedures. First, they may be very low in numbers at the beginning, but survive against endodontic treatment modalities and show considerable growth afterward. Second, new species may gain access to the root canal due to poor sealing and ineffective disinfectants. It is a requirement henceforth that both irrigating solutions and disinfectants should have a good antifungal capacity in addition to their antibacterial properties, since the microbial ecology of the root canals contains both bacterial and fungal species.

### 9.6.6.2 Antifungal agents

It has been emphasized in recent reports that the oral cavity may act as a reservoir of resistant yeast isolates and cause persistent yeast infections (Berthold et al. 1994; Kuriyama et al. 2005). Considering the fact that *Candida* species are highly resistant to calcium hydroxide (Waltimo et al. 1999a, b) and that the other endodontic disinfectants may be inactivated by the presence of organic structures such as dentin, serum albumin, and dead microbial cells (Haapasalo et al. 2000; Portenier et al. 2001, 2002), possible presence of yeasts in the root canals becomes a more pronounced issue. Therefore, use of specific antifungal agents may be considered after routine endodontic procedures in case of particular yeast infections of root canals. However, there is limited information about the effects of classical antifungal agents against yeasts isolated from root canals.

Sen et al. (2000) compared antifungal effects of endodontic irrigating solutions or disinfectants with classical antifungal agents, nystatin, and ketoconazole in an agar diffusion study. They found that both antifungal agents were superior to all solutions or disinfectants, except EDTA. In a subsequent study (Ates et al. 2005) determining MIC or MFC, ketoconazole was the most effective agent against three strains of *C. albicans*. While the root canal isolate was the most sensitive strain in the first study, there was no difference among the same strains in a follow-up study. It was concluded that the type of antimicrobial susceptibility test might affect the results of the studies when evaluating the antifungal capacity of the agents or resistance of the strains.

Waltimo et al. (2000b) determined MIC values of amphotericin B, 5-fluorocytosine, and three azole-based agents (fluconazole, miconazole, and clotrimazole) against 70 *C. albicans* strains isolated from either persistent cases of apical periodontitis or marginal periodontitis. All isolates were susceptible to amphotericin B and 5-fluorocytosine with an MIC below 1 μg/mL, and there was no difference between endodontic and periodontal isolates. However, their susceptibility to the three azoles varied. Fluconazole was mostly effective on endodontic isolates. Two periodontal and one endodontic strains presented azole cross-resistance. According to Waltimo et al. (2004a), fungal antibiotics should be used only for the treatment of acute endodontic cases after a substantial microbiological diagnosis is made.

## 9.7 Conclusions

During standard endodontic treatment procedures, most of the bacteria may be removed from the root canals, but fungi may survive. Then, they may demonstrate an overgrowth due to their opportunistic character. They may also gain access to the root canals during or after root canal therapy.

The incidence of yeasts in primary, secondary, or persistent endodontics infections has been reported to be in the range of 0.5–61.5%. However, it should be kept in mind that all of these studies have been accomplished in systemically healthy dental subjects. Considering that the presence of yeasts in oral and dental tissues of immunocompromised patients is relatively higher than that of the normal population, it may be proposed that their incidence will also be higher in the root canals of these patients. At this point, we should continue searching for new remedies not only against bacteria, but also against fungi in endodontics.

## 9.8 References

Adib V, Spratt D, Ng Y-L, and Gulabivala K. 2004. Cultivable microbiological flora associated with persistent periapical diseases and coronal leakage after root canal treatment: a preliminary study. *Int Endod J* 37(8): 542–51.

Akdeniz BG, Koparal E, Sen BH, Ates M, and Denizci AA. 2002. Prevalence of *Candida albicans* in oral cavities and root canals of children. *J Dent Child* 69(3): 289–92.

Arendorf TM and Walker DM. 1979. Oral candidal populations in health and disease. *Br Dent J* 147(10): 267–72.

Ates M, Akdeniz BG, and Sen BH. 2005. The effect of calcium chelating or binding agents on *Candida albicans*. *Oral Surg Oral Med Oral Pathol Oral Radiol Endod* 100(5): 626–30.

Ayhan H, Sultan N, Cirak M, Ruhi MZ, and Bodur H. 1999. Antimicrobial effects of various endodontic irrigants on selected microorganisms. *Int Endod J* 32(2): 99–102.

Azuma M. 2006. Fundamental mechanisms of host immune responses to infection. *J Periodontal Res* 41(5): 361–73.

Banin E, Brady KM, and Greenberg EP. 2006. Chelator-induced dispersal and killing of *Pseudomonas aeruginosa* cells in a biofilm. *Appl Environ Microbiol* 72(3): 2064–69.

Bardwell A, Hill DW, Runyon BA, and Koster FT. 1986. Disseminated macronodular cutaneous candidiasis in chronic alcoholism. *Arch Intern Med* 146(2): 385–86.

Baumgartner JC, Johal S, and Marshall JG. 2007. Comparison of the antimicrobial efficacy of 1.3% NaOCl/BioPure MTAD to 5.25% NaOCl/15% EDTA for root canal irrigation. *J Endod* 33(1): 48–51.

Baumgartner JC, Watts C, and Xia T. 2000. Occurrence of *Candida albicans* in infection of endodontic origin. *J Endod* 26(12): 695–98.

Bedell GW and Soll DR. 1979. Effects of low concentrations of zinc on the growth and dimorphism of *Candida albicans*: evidence for zinc-resistant and -sensitive pathways for mycelium formation. *Infect Immun* 26(1): 348–54.

Beighton D, Hellyer PH, and Heath MR. 1990. Associations between salivary levels of mutans streptococci, lactobacilli, yeasts and black-pigmented *Bacteroides* spp. and dental variables in elderly dental patients. *Arch Oral Biol* 35(Suppl): 173S–5S.

Beighton D and Lynch E. 1995. Comparison of selected microflora of plaque and underlying carious dentine associated with primary root caries lesions. *Caries Res* 29(2): 154–58.

Berthold P, Stewart J, Cumming C, Decker S, MacGregor R, and Malamud D. 1994. *Candida* organisms in dental plaque from AIDS patients. *J Infect Dis* 170(4): 1053–54.

Blanco MT, Morales JJ, Lucio L, Perez-Giraldo C, Hurtado C, and Gomez-Garcia AC. 2006. Modification of adherence to plastic and to human buccal cells of *Candida albicans* and *Candida dubliniensis* by a subinhibitory concentration of itraconazole. *Oral Microbiol Immunol* 21(1): 69–72.

Bobichon H and Bouchet P. 1987. Action of chlorhexidine on budding *Candida albicans*: scanning and transmission electron microscopic study. *Mycopathologia* 100(1): 27–35.

Bourett T, Hoch HC, and Staples RC. 1987. Association of the microtubule cytoskeleton with the thigmotropic signal for appressorium formation in *Uromyces*. *Mycologia* 79(4): 540–45.

Branting C, Sund ML, and Linder LE. 1989. The influence of *Streptococcus mutans* on adhesion of *Candida albicans* to acrylic surfaces *in vitro*. *Arch Oral Biol* 34(5): 347–53.

Brown AJP. 2002. Morphogenetic signalling pathways in *Candida albicans*. In: Calderone RA (ed), *Candida and Candidiasis*. Washington, DC: American Society for Microbiology Press, 95–106.

Brown LR, Dreizen S, Daly TE, Drane TE, Handler S, Riggan LJ, and Johnston DA. 1978. Interrelations of oral microorganisms, immunoglobulins, and dental caries following radiotherapy. *J Dent Res* 57(9–10): 882–93.

Brown LR, Mackler BF, Levy BM, Wright TE, Handler SF, Moylan JS, Perkins DH, and Keene HJ. 1979. Comparison of the plaque microflora in immunodeficient and immunocompetent dental patients. *J Dent Res* 58(12): 2344–52.

Budtz-Jorgensen E. 1990. Etiology, pathogenesis, therapy, and prophylaxis of oral yeast infections. *Acta Odontol Scand* 48(1): 61–69.

Budtz-Jorgensen E, Theilade E, and Theilade J. 1983. Quantitative relationship between yeast and bacteria in denture-induced stomatitis. *Scand J Dent Res* 91(2): 134–42.

Bunetel L and Bonnaure-Mallet M. 1996. Oral pathoses caused by *Candida albicans* during chemotherapy: update on development mechanisms. *Oral Surg Oral Med Oral Pathol Oral Radiol Endod* 82(2): 161–65.

Bystrom A and Sundqvist G. 1985. The antibacterial action of sodium hypochlorite and EDTA in 60 cases of endodontic therapy. *Int Endod J* 18(1): 35–40.

Calderone R, Suzuki S, Cannon R, Cho T, Boyd D, Calera J, Chibana H, Herman D, Holmes A, Jeng HW, Kaminishi H, Matsumoto T, Mikami T, O'Sullivan JM, Sudoh M, Suzuki M, Nakashima Y, Tanaka T, Tompkins GR, and Watanabe T. 2000. *Candida albicans*: Adherence, signaling and virulence. *Med Mycol* 38(Suppl 1): 125–37.

Calderone RA and Fonzi WA. 2001. Virulence factors of *Candida albicans*. *Trends Microbiol* 9(7): 327–35.

Cannon RD, Holmes Ar, Mason AB, and Monk BC. 1995a. Oral *Candida*: clearance, colonization, or candidiasis? *J Dent Res* 74(5): 1152–61.

Cannon RD, Nand AK, and Jenkinson HF. 1995b. Adherence of *Candida albicans* to human salivary components adsorbed to hydroxylapatite. *Microbiology* 141(Pt 1): 213–19.

Challacombe SJ. 1994. Immunologic aspects of oral candidiasis. *Oral Surg Oral Med Oral Pathol* 78(2): 202–10.

Chang YC, Huang FM, Tai KW, and Chou MY. 2001. The effect of sodium hypochlorite and chlorhexidine on cultured human periodontal ligament cells. *Oral Surg Oral Med Oral Pathol Oral Radiol Endod* 92(4): 446–50.

Chattin BR, Ishihara K, Okuda K, Hirai Y, and Ishikawa T. 1999. Specific microbial colonizations in the periodontal sites of HIV-infected subjects. *Microbiol Immunol* 43(9): 847–52.

Cheung GSP and Ho MWM. 2001. Microbial flora of root canal-treated teeth associated with asymptomatic periapical radiolucent lesions. *Oral Microbiol Immunol* 16(6): 332–37.

Clark P, Connolly P, Curtis AS, Dow JA, and Wilkinson CD. 1990. Topographical control of cell behaviour: II. Multiple grooved substrata. *Development* 108(4): 635–44.

Clegg MS, Vertucci FJ, Walker C, Belanger M, and Britto LR. 2006. The effect of exposure to irrigant solutions on apical dentin biofilms *in vitro*. *J Endod* 32(5): 434–37.

Collins-Lech C, Kalbfleisch JH, Franson TR, and Sohnle PG. 1984. Inhibition by sugars of *Candida albicans* adherence to human buccal mucosal cells and corneocytes *in vitro*. *Infect Immun* 46(3): 831–34.

Costerton JW. 1999. Introduction to biofilm. *Int J Antimicrob Agents* 11(3–4): 217–21.

Cotter G and Kavanagh K. 2000. Adherence mechanisms of *Candida albicans*. *Br J Biomed Sci* 57(3): 241–49.

Cotter G, Weedle R, and Kavanagh K. 1998. Monoclonal antibodies directed against extracellular matrix proteins reduce the adherence of *Candida albicans* to HEp-2 cells. *Mycopathologia* 141(3): 137–42.

d'Enfert C. 2006. Biofilms and their role in the resistance of pathogenic *Candida* to antifungal agents. *Curr Drug Targets* 7(4): 465–70.

Dahlen G. 2002. Microbiology and treatment of dental abscesses and periodontal-endodontic lesions. *Periodontology 2000* 28: 206–39.

Dahlen G and Wikstrom M. 1995. Occurrence of enteric rods, staphylococci and *Candida* in subgingival samples. *Oral Microbiol Immunol* 10(1): 42–46.

Damm DD, Neville BW, Geissler RH, White DK, Drummond JF, and Ferretti GA. 1988. Dentinal candidiasis in cancer patients. *Oral Surg Oral Med Oral Pathol* 65(1): 56–60.

Darveau RP, Tanner A, and Page RC. 1997. The microbial challenge in periodontitis. *Periodontology 2000* 14: 12–32.

Darwazeh AM, MacFarlane TW, McCuish A, and Lamey PJ. 1991. Mixed salivary glucose levels and candidal carriage in patients with diabetes mellitus. *J Oral Pathol and Med* 20(6): 280–83.

De Carvalho FB, Silva DS, Hebling J, Spolidorio LC, and Solidorio DM. 2006. Presence of mutans streptococci and *Candida* spp. in dental plaque/dentine of carious teeth and early childhood caries. *Arch Oral Biol* 51(11): 1024–28.

Debelian GJ, Olsen I, and Tronstad L. 1995. Bacteremia in conjunction with endodontic therapy. *Endod Dent Traumatol* 11(3): 142–49.

Debelian GJ, Olsen I, and Tronstad L. 1997. Observation of *Saccharomyces cerevisia* in blood of patient undergoing root canal treatment. *Int Endod J* 30(5): 313–17.

Delany GM, Patterson SS, Miller CH, and Newton CW. 1982. The effect of chlorhexidine gluconate irrigation on the root canal flora of freshly extracted necrotic teeth. *Oral Surg Oral Med Oral Pathol* 53(5): 518–23.

Donlan RM and Costerton JW. 2002. Biofilms: survival mechanisms of clinically relevant microorganisms. *Clin Microbiol Rev* 15(2): 167–93.

Douglas LJ. 2003. *Candida* biofilms and their role in infection. *Trends Microbiol* 11(1): 30–36.

Dunavant TR, Regan JD, Glickman GN, Solomon ES, and Honeyman AL. 2006. Comparative evaluation of endodontic irrigants against *Enterococcus faecalis* biofilms. *J Endod* 32(6): 527–31.

Egan MW, Spratt DA, Ng Y-L, Lam JM, Moles DR, and Gulabivala K. 2002. Prevalence of yeasts in saliva and root canals of teeth associated with apical periodontitis. *Int Endod J* 35(4): 321–29.

Eidelman D, Neuman I, Kuttin ES, Pinto MM, and Beemer AM. 1978. Dental sepsis due to *Candida albicans* causing urticaria: case report. *Ann Allergy* 41(3): 179–81.

Eliasson L, Carlen A, Almstahl A, Wikstrom M, and Lingstrom P. 2006. Dental plaque pH and micro-organisms during hyposalivation. *J Dent Res* 85(4): 334–38.

Enache E, Eskandari T, Borja L, Wadsworth E, Hoxter B, and Calderone R. 1996. *Candida albicans* adherence to a human oesophageal cell line. *Microbiology* 142(Pt 10): 2741–46.

Ercan E, Dalli M, and Dülgergil CT. 2006. *In vitro* assessment of the effectiveness of chlorhexidine gel and calcium hydroxide paste with chlorhexidine against *Enterococcus faecalis* and *Candida albicans*. *Oral Surg Oral Med Oral Pathol Oral Radiol Endod* 102(2): e27–31.

Evans RC and Holmes CJ. 1987. Effect of vancomycin hydrochloride on *Staphylococcus epidermidis* biofilm associated with silicone elastomer. *Antimicrob Agents Chemother* 31(6): 889–94.

Ezzo PJ and Cutler CW. 2003. Microorganisms as risk indicators for periodontal disease. *Periodontology 2000* 32: 24–35.

Ferguson JW, Hatton JF, and Gillespie MJ. 2002. Effectiveness of intracanal irrigants and medications against the yeast *Candida albicans*. *J Endod* 28(2): 68–71.

Ferrari PH, Cai S, and Bombana AC. 2005. Effect of endodontic procedures on enterococci, enteric bacteria and yeasts in primary endodontic infections. *Int Endod J* 38(6): 372–80.

Fraser JG. 1974. Chelating agents: their softening effect on root canal dentin. *Oral Surg Oral Med Oral Pathol* 37(5): 803–11.

Gatti JJ, Dobeck JM, Smith C, White RR, Socransky SS, and Skobe Z. 2000. Bacteria of asymptomatic periradicular endodontic lesions identified by DNA–DNA hybridization. *Endod Dent Traumatol* 16(5): 197–204.

Ghannoum MA. 2000. Potential role of phospholipases in virulence and fungal pathogenesis. *Clin Microbiol Rev* 13(1): 122–43.

Gil ML, Casanova M, and Martinez JP. 1994. Changes in the cell wall glycoprotein composition of *Candida albicans* associated to the inhibition of germ tube formation by EDTA. *Arch Microbiol* 161(4): 489–94.

Gokdal I, Kalkanci A, Pacal G, and Altug Z. 2002. *Candida* colonization on the surface of orthodontic brackets and the adhesion of these strains to buccal epithelial cells. *Mikrobiyol Bul* 36(1): 65–69.

Goldman M and Pearson AH. 1969. Postdebridement bacterial flora and antibiotic sensitivity. *Oral Surg Oral Med Oral Pathol* 28(6): 897–905.

Golecka M, Oldakowska-Jedynak U, Mierzwinska-Nastalska E, and Adamczyk-Sosinska E. 2006. *Candida*-associated denture stomatitis in patients after immunosuppression therapy. *Transplant Proc* 38(1): 155–56.

Gomes BP, Souza SF, Ferraz CC, Teixeira FB, Zaiza AA, Valdrighi L, and Souza-Filho FJ. 2003. Effectiveness of 2% chlorhexidine gel and calcium hydroxide against *Enterococcus faecalis* in bovine root dentine *in vitro*. *Int Endod J* 36(4): 267–75.

Gomes BPF, Vianna ME, Sena NT, Zaia AA, Ferraz CC, and Souza Filho FJ. 2006. *In vitro* evaluation of the antimicrobial activity of calcium hydroxide combined with chlorhexidine gel used as intracanal medicament. *Oral Surg Oral Med Oral Pathol Oral Radiol Endod* 102(4): 544–50.

González S, Lobos I, Guajardo A, Celis A, Zemelman R, Smith CT, and Saglie FR. 1987. Yeasts in juvenile periodontitis. Preliminary observations by scanning electron microscopy. *J Periodontol* 58(2): 119–24.

Gow NA, Brown AJ, and Odds FC. 2002. Fungal morphogenesis and host invasion. *Curr Opin Microbiol* 5(4): 366–71.

Gow NA, Perera TH, Sherwood-Higham J, Gooday GW, Gregory DW, and Marshall D. 1994. Investigation of touch-sensitive responses by hyphae of the human pathogenic fungus *Candida albicans*. *Scanning Microsc* 8(3): 705–10.

Grawehr M, Sener B, Waltimo T, and Zehnder M. 2003. Interactions of ethylenediamine tetraacetic acid with sodium hypochlorite in aqueous solutions. *Int Endod J* 36(6): 411–17.

Grimaudo NJ, Nesbitt WE, and Clark WB. 1996. Coaggregation of *Candida albicans* with oral *Actinomyces* species. *Oral Microbiol Immunol* 11(1): 59–61.

Grimoud AM, Marty N, Bocquet H, Andrieu S, Lodter JP, and Chabanon G. 2003. Colonization of the oral cavity by *Candida* species: risk factors in long-term geriatric care. *J Oral Sci* 45(1): 51–55.

Haapasalo HK, Siren EK, Waltimo TM, Orstavik D, and Haapasalo MP. 2000. Inactivation of local root canal medicaments by dentine: an *in vitro* study. *Int Endod J* 33(2): 126–31.

Haapasalo M. 1989. Bacteroides spp. in dental root canal infections. *Endod Dent Traumatol* 5(1): 1–10.

Haapasalo M, Endal U, Zandi H, and Coil MJ. 2005. Eradication of endodontic infection by instrumentation and irrigation solutions. *Endod Topics* 10(1): 77–102.

Haapasalo M, Ranta K, and Ranta H. 1987. Mixed anaerobic periapical infection with sinus tract. *Endod Dent Traumatol* 3(2): 83–85.

Haenni S, Schmidlin PR, Mueller B, Sener B, and Zehnder M. 2003. Chemical and antimicrobial properties of calcium hydroxide mixed with irrigating solutions. *Int Endod J* 36(2): 100–105.

Hagihara Y, Kaminishi H, Cho T, Tanaka M, and Kaita H. 1988. Degradation of human dentine collagen by an enzyme produced by the yeast *Candida albicans*. *Arch Oral Biol* 33(8): 617–19.

Hancock HH, Sigurdsson A, Trope M, and Moiseiwitsch J. 2001. Bacteria isolated after unsuccessful endodontic treatment in a North American population. *Oral Surg Oral Med Oral Pathol Oral Radiol Endod* 91(5): 579–86.

Hannula J, Dogan B, Slots J, Okte E, and Asikainen S. 2001. Subgingival strains of *Candida albicans* in relation to geographical origin and occurrence of periodontal pathogenic bacteria. *Oral Microbiol Immunol* 16(2): 113–18.

Hannula J, Saarela M, Alaluusua S, Slots J, and Asikainen S. 1997. Phenotypic and genotypic characterization of oral yeasts from Finland and the United States. *Oral Microbiol Immunol* 12(6): 358–65.

Harrison JW, Wagner GW, and Henry CA. 1990. Comparison of the antimicrobial effectiveness of regular and fresh scent Clorox. *J Endod* 16(7): 328–30.

Hauman CH, Thompson IO, Theunissen F, and Wolfaardt P. 1993. Oral carriage of *Candida* in healthy and HIV-seropositive persons. *Oral Surg Oral Med Oral Pathol* 76(5): 570–72.

Hawser SP and Douglas LJ. 1995. Resistance of *Candida albicans* biofilms to antifungal agents *in vitro*. *Antimicrob Agents Chemother* 39(9): 2128–31.

Haynes K. 2001. Virulence in *Candida* species. *Trends Microbiol* 9(12): 591–96.

Heimdahl A and Nord CE. 1990. Oral infections in immunocompromised patients. *J Clin Periodontol*, 17(7(Pt 2)): 501–3.

Heling I and Chandler NP. 1998. Antimicrobial effect of irrigant combinations within dentinal tubules. *Int Endod J* 31(1): 8–14.

Hellstein J, Vawter-Hugart H, Fotos P, Schmid J, and Soll DR. 1993. Genetic similarity and phenotypic diversity of commensal and pathogenic strains of *Candida albicans* isolated from the oral cavity. *J Clin Microbiol* 31(12): 3190–99.

Hobson P. 1959. An investigation into the bacteriological control of infected root canal. *Br Dent J* 20: 63–70.

Hodson JJ and Craig GT. 1972. The incidence of *Candida albicans* in the plaques of teeth of children. *Dent Pract Dent Rec* 22(8): 296–301.

Holmes AR, Cannon RD, and Jenkinson HF. 1995. Interactions of *Candida albicans* with bacteria and salivary molecules in oral biofilms. *J Ind Microbiol* 15(3): 208–13.

Holmes AR, Cannon RD, and Shepherd MG. 1991. Effect of calcium ion uptake on *Candida albicans* morphology. *FEMS Microbiol Lett* 61(2–3): 187–93.

Hossain H, Ansari F, Schulz-Weidner N, Wetzel WE, Chakraborty T, and Domann E. 2003. Clonal identity of *Candida albicans* in the oral cavity and the gastrointestinal tract of pre-school children. *Oral Microbiol Immunol* 18(5): 302–8.

Hube B and Naglik J. 2001. *Candida albicans* proteinases: Resolving the mystery of a gene family. *Microbiology* 147 (Pt 8): 1997–2005.

Hube B, Ruchel R, Monod M, Sanglard D, and Odds FC. 1998. Functional aspects of secreted *Candida* proteinases. *Adv Exp Med Biol* 436: 339–44.

Hülsmann M, Heckendorff M, and Schafers F. 2002. Comparative in-vitro evaluation of three chelator pastes. *Int Endod J* 35(8): 668–79.

Iwu C, MacFarlane TW, MacKenzie D, and Stenhouse D. 1990. The microbiology of periapical granulomas. *Oral Surg Oral Med Oral Pathol* 69(4): 502–5.

Jackson FL and Halder AR. 1963. Incidence of yeasts in root canals during therapy. *Br Dent J* 115: 459–60.

Jacob LS, Flaitz CM, Nichols CM, and Hicks MJ. 1998. Role of dentinal carious lesions in the pathogenesis of oral Candidiasis in HIV infections. *J Am Dent Assoc* 129(2): 187–94.

Järvensiu A, Hietanen J, Rautemaa R, Sorsa T, and Richardson M. 2004. *Candida* yeasts in chronic periodontitis tissues and subgingival microbial biofilms *in vivo*. *Oral Dis* 10(2): 106–12.

Jenkinson HF, Lala HC, and Shepherd MG. 1990. Coaggregation of *Streptococcus sanguis* and other streptococci with *Candida albicans*. *Infect Immun* 58(5): 1429–36.

Jones DS. 1995. The effects of sub-inhibitory concentrations of cationic, non-antibiotic, antimicrobial agents on the morphogenesis of *Candida albicans in vitro*. *Pharmacol Res* 12(12): 2057–59.

Kam AP and Xu J. 2002. Diversity of commensal yeasts within and among healthy hosts. *Diagn Microbiol Infect Dis* 43(1): 19–28.

Kaminishi H, Hagihara Y, Hayashi S, and Cho T. 1986. Isolation and characteristics of collagenolytic enzyme produced by *Candida albicans*. *Infect Immun* 53(2): 312–16.

Kennedy MJ and Sandin RL. 1988. Influence of growth conditions on *Candida albicans* adhesion, hydrophobicity and cell wall ultrastructure. *J Med Vet Mycol* 26(2): 79–92.

Kinirons MJ. 1983. Candidal invasion of dentine complicating hypodontia. *Br Dent J* 154(12): 400–401.

Kiryu T, Hoshino E, and Iwaku M. 1994. Bacteria invading periapical cementum. *J Endod* 20(4): 169–72.

Kite P, Eastwood K, Sugden S, and Percival SL. 2004. Use of *in vivo*-generated biofilms from hemodialysis catheters to test the efficacy of a novel antimicrobial catheter lock for biofilm eradication *in vitro*. *J Clin Microbiol* 42(7): 3073–76.

Klotz SA. 1994. The contribution of electrostatic forces to the process of adherence of *Candida albicans* yeast cells to substrates. *FEMS Microbiol Lett* 120(3): 257–62.

Klotz SA, Rutten MJ, Smith RL, Babcock SR, and Cunningham MD. 1993. Adherence of *Candida albicans* to immobilized extracellular matrix proteins is mediated by calcium-dependent surface glycoproteins. *Microb Pathog* 14(2): 133–47.

Kobayashi SD and Cutler JE. 1998. *Candida albicans* hyphal formation and virulence: is there a clearly defined role? *Trends Microbiol* 6(3): 92–94.

Kornman KS, Page RC, and Tonetti MS. 1997. The host response to the microbial challenge in periodontitis: assembling the players. *Periodontology 2000* 14: 33–53.

Krogh P, Holmstrup P, Thorn JJ, Vedtofte P, and Pindborg JJ. 1987. Yeast species and biotypes associated with oral leukoplakia and lichen planus. *Oral Surg Oral Med Oral Pathol* 63(1): 48–54.

Kumamoto CA and Vinces MD. 2005a. Contributions of hyphae and hypha-co-regulated genes to *Candida albicans* virulence. *Cell Microbiol* 7(11): 1546–54.

Kumamoto CA and Vinces MD. 2005b. Alternative *Candida albicans* lifestyles: growth on surfaces. *Annu Rev Microbiol* 59: 113–33.

Kuriyama T, Williams DW, Bagg J, Coulter WA, Ready D, and Lewis MA. 2005. *In vitro* susceptibility of oral *Candida* to seven antifungal agents. *Oral Microbiol Immunol* 20(6): 349–53.

Lamey PJ, Darwaza A, Fisher BM, Samaranayake LP, Macfarlane TW, and Frier BM. 1988. Secretor status, candidal carriage and candidal infection in patients with diabetes mellitus. *J Oral Pathol* 17(7): 354–57.

Lana MA, Ribeiro-Sobrinho AP, Stehling R, Garcia GD, Silva BK, Hamdan JS, Nicoli JR, Carvalho MA, and Farias Lde M. 2001. Microorganisms isolated from root canals presenting necrotic pulp and their drug susceptibility *in vitro*. *Oral Microbiol Immunol* 16(2): 100–105.

Leavitt JM, Irving JN, and Srugaevsky P. 1958. The bacterial flora of root canals as disclosed by a culture medium for endodontics. *Oral Surg Oral Med Oral Pathol* 11: 302–8.

Lee KK, Yu L, Macdonald DL, Paranchych W, Hodges RS, and Irvin RT. 1996. Anti-adhesin antibodies that recognize a receptor-binding motif (adhesintope) inhibit pilus/fimbrial-mediated adherence of *Pseudomonas aeruginosa* and *Candida albicans* to asialo-GM1 receptors and human buccal epithelial cell surface receptors. *Can J Microbiol* 42(5): 479–86.

Lewis K. 2001. Riddle of biofilm resistance. *Antimicrob Agents Chemother* 45(4): 999–1007.

Li SY, Yang YL, Chen KW, Cheng HH, Chiou CS, Wang TH, Lauderdale TL, Hung CC, and Lo HJ. 2006. Molecular epidemiology of long-term colonization of *Candida albicans* strains from HIV-infected patients. *Epidemiol Infect* 134(2): 265–69.

Lim TS, Wee TY, Choi MY, Koh WC, and Sae-Lim V. 2003. Light and scanning electron microscopic evaluation of Glyde File Prep in smear layer removal. *Int Endod J* 36(5): 336–43.

Listgarten MA, Lai CH, and Young V. 1993. Microbial composition and pattern of antibiotic resistance in subgingival microbial samples from patients with refractory periodontitis. *J Periodontol* 64(3): 155–61.

Liu H. 2002. Co-regulation of pathogenesis with dimorphism and phenotypic switching in *Candida albicans*, a commensal and a pathogen. *Int J Med Microbiol* 292(5–6): 299–311.

Liu Y, Ryan ME, Lee HM, Simon S, Tortora G, Lauzon C, Leung MK, and Golub LM. 2002. A chemically modified tetracycline (CMT-3) is a new antifungal agent. *Antimicrob Agents Chemother* 46(5): 1447–54.

Loesche WJ. 1986. Role of *Streptococcus mutans* in human dental decay. *Microbiol Rev* 50(4): 353–80.

Lomcali G, Sen BH, and Cankaya H. 1996. Scanning electron microscopic observations of apical root surfaces of teeth with apical periodontitis. *Endod Dent Traumatol* 12(2): 70–76.

Lynch E and Beighton D. 1994. A comparison of primary root caries lesions classified according to colour. *Caries Res* 28(4): 233–39.

MacDonald JB, Hare GC, and Wood AW. 1957. The bacteriologic status of the pulp chambers in intact teeth found to be nonvital following trauma. *Oral Surg Oral Med Oral Pathol* 10(3): 318–22.

MacNeill S, Rindler E, Walker A, Brown AR, and Cobb CM. 1997. Effects of tetracycline hydrochloride and chlorhexidine gluconate on *Candida albicans*. An *in vitro* study. *J Clin Periodontol* 24(10): 753–60.

Machado de Oliveira JC, Siqueira JF, Jr, Rocas IN, Baumgartner JC, Xia T, Peixoto RS, and Rosado AS. 2007. Bacterial community profiles of endodontic abscesses from Brazilian and USA subjects as compared by denaturing gradient gel electrophoresis analysis. *Oral Microbiol Immunol* 22(1): 14–18.

Mah TF and O'Toole GA. 2001. Mechanisms of biofilm resistance to antimicrobial agents. *Trends Microbiol* 9(1): 34–39.

Makihira S, Nikawa H, Tamagami M, Hamada T, Nishimura H, Ishida K, and Yamashiro H. 2002a. Bacterial and *Candida* adhesion to intact and denatured collagen *in vitro*. *Mycoses* 45(9–10): 389–92.

Makihira S, Nikawa H, Tamagami M, Hamada T, and Samaranayake LP. 2002b. Differences in *Candida albicans* adhesion to intact and denatured type I collagen *in vitro*. *Oral Microbiol Immunol* 17(2): 129–31.

Marchant S, Brailsford SR, Twomey AC, Roberts GJ, and Beighton D. 2001. The predominant microflora of nursing caries lesions. *Caries Res* 35(6): 397–406.

Marsh PD. 2004. Microbial ecology of dental plaque and its significance in health and disease. *Adv Dent Res* 8(2): 263–71.

Massillamoni CR, Kettering JD, and Torabinejad M. 1981. The biocompatibility of some root canal medicaments and irrigants. *Int Endod J* 14(2): 115–20.

Matusow RJ. 1981. Acute pulpa-alveolar cellulitis syndrome. III. Endodontic therapeutic factors and the resolution of a *Candida albicans* infections. *Oral Surg Oral Med Oral Pathol* 52(6): 630–34.

Maza JL, Elguezabal N, Prado C, Ellacuria J, Soler I, and Ponton J. 2002. *Candida albicans* adherence to resin-composite restorative dental material: Influence of whole human saliva. *Oral Surg Oral Med Oral Pathol Oral Radiol Endod* 94(5): 589–92.

McCourtie J, MacFarlane TW, and Samaranayake LP. 1986. A comparison of the effects of chlorhexidine gluconate, amphotericin B and nystatin on the adherence of *Candida* species to denture acrylic. *J Antimicrob Chemother* 17(5): 575–83.

McDonnell G and Russell AD. 1999. Antiseptics and disinfectants: activity, action, and resistance. *Clin Microbiol Rev* 12(1): 147–79.

McManners J and Samaranayake LP. 1990. Suppurative oral candidosis. Review of the literature and report of a case. *Int J Oral Maxillofac Surg* 19(5): 257–59.

Menezes MM, Valera MC, Jorge AO, Koga-Ito CY, Camargo CH, and Mancini MN. 2004. *In vitro* evaluation of the effectiveness of irrigants and intracanal medicaments on microorganisms within root canals. *Int Endod J* 37(5): 311–19.

Mentz TC. 1982. The use of sodium hypochlorite as a general endodontic medicament. *Int Endod J* 15(3): 132–36.

Millsap KW, Van Der Mei HC, Bos R, and Busscher HJ. 1998. Adhesive interactions between medically important yeasts and bacteria. *FEMS Microbiol Rev* 21(4): 321–36.

Moodnik RM, Dorn SO, Feldman MJ, Levey M, and Borden BG. 1976. Efficacy of biomechanical instrumentation: a scanning electron microscopic study. *J Endod* 2(9): 261–66.

Muzyka BC and Glick M. 1995. A review of oral fungal infections and appropriate therapy. *J Am Dent Assoc* 126(1): 63–72.

Nair PN. 1997. Apical periodontitis: a dynamic encounter between root canal infection and host response. *Periodontology 2000* 13: 121–48.

Nair PN, Henry S, Cano V, and Vera J. 2005. Microbial status of apical root canal system of human mandibular first molars with primary apical periodontitis after "one-visit" endodontic treatment. *Oral Surg Oral Med Oral Pathol Oral Radiol Endod* 99(2): 231–52.

Nair PNR, Sjögren U, Krey G, Kahnberg KE, and Sundqvist G. 1990. Intraradicular bacteria and fungi in root-filled, asymptomatic human teeth with therapy-resistant periapical lesions: a long-term light and electron microscopic follow-up study. *J Endod* 16(12): 580–88.

Nair RG and Samaranayake LP. 1996. The effect of oral commensal bacteria on candidal adhesion to denture acrylic surfaces. An *in vitro* study. *Acta Pathol Microbiol Immunol Scand* 104(5): 339–49.

Najzar-Fleger D, Filipovi D, Prpif G, and Kobler D. 1992. *Candida* in root canal in accordance with oral ecology. *Int Endod J* 25(1): 40–45.

Narhi TO, Ainamo A, and Meurman JH. 1993. Salivary yeasts, saliva, and oral mucosa in the elderly. *J Dent Res* 72(6): 1009–14.

Navazesh M, Wood GJ, and Brightman VJ. 1995. Relationship between salivary flow rates and *Candida albicans* counts. *Oral Surg Oral Med Oral Pathol Oral Radiol Endod* 80(3): 284–88.

Nevzatoglu EU, Ozcan M, Kulak-Ozkan Y, and Kadir T. 2007. Adherence of *Candida albicans* to denture base acrylics and silicone-based resilient liner materials with different surface finishes. *Clin Oral Investig* 11: 231–36.

Nikawa H, Hamada T, Yamashiro H, Murata H, and Subiwahjudi A. 1998. The effect of saliva or serum on *Streptococcus mutans* and *Candida albicans* colonization of hydroxylapatite beads. *J Dent* 26(1): 31–37.

Nikawa H, Samaranayake LP, Tenovuo J, Pang KM, and Hamada T. 1993. The fungicidal effect of human lactoferrin on *Candida albicans* and *Candida krusei*. *Arch Oral Biol* 38(12): 1057–63.

Nikawa H, Yamashiro H, Makihira S, Nishimura M, Egusa H, Furukawa M, Setijanto D, and Hamada T. 2003. *In vitro* cariogenic potential of *Candida albicans*. *Mycoses* 46(11–12): 471–78.

Nishimura M, Nikawa H, Yamashiro H, Nishimura H, Hamada T, and Embery G. 2002. Cell-associated collagenolytic activity by *Candida albicans*. *Mycopathologia* 153(3): 125–28.

Odden K, Schenck K, Koppang H, and Hurlen B. 1994. Candidal infection of the gingiva in HIV-infected persons. *J Oral Pathol Med* 23(4): 178–83.

Odds FC. 1988. *Candida and Candidosis*. London: Bailliere Tindall.

Odds FC, Abbott AB, Stiller RL, Scholer HJ, Polak A, and Stevens DA. 1983. Analysis of *Candida albicans* phenotypes from different geographical and anatomical sources. *J Clin Microbiol* 18(4): 849–57.

Okino LA, Siqueira EL, Santos M, Bombana AC, and Figueiredo JA. 2004. Dissolution of pulp tissue by aqueous solution of chlorhexidine digluconate and chlorhexidine digluconate gel. *Int Endod J* 37(1): 38–41.

Oksala E. 1990. Factors predisposing to oral yeast infections. *Acta Odontol Scand* 48: 71–74.

Olsen I and Birkeland JM. 1977. Denture stomatitis-yeast occurrence and the pH of saliva and denture plaque. *Scand J Dent Res* 85(2): 130–34.

Orstavik D and Haapasalo M. 1990. Disinfection by endodontic irrigants and dressings of experimentally infected dentinal tubules. *Endod Dent Traumatol* 6(4): 142–49.

Parsons GJ, Patterson SS, Miller CH, Katz S, Kafrawy AH, and Newton CW. 1980. Uptake and release of chlorhexidine by bovine pulp and dentin specimens and their subsequent acquisition of antibacterial properties. *Oral Surg Oral Med Oral Pathol* 49(5): 455–59.

Peciuliene V, Reynaud AH, Balciuniene I, and Haapasalo M. 2001. Isolation of yeasts and enteric bacteria in root-filled teeth with chronic apical periodontitis. *Int Endod J* 34(6): 429–34.

Percival SL, Kite P, Eastwood K, Murga R, Carr J, Arduino MJ, and Donlan RM. 2005. Tetrasodium EDTA as a novel central venous catheter lock solution against biofilm. *Infect Control Hosp Epidemiol* 26(6): 515–19.

Peters LB, van Winkelhoff AJ, Buijs JF, and Wesselink PR. 2002. Effects of instrumentation, irrigation and dressing with calcium hydroxide on infection in pulpless teeth with periapical bone lesions. *Int Endod J* 35(1): 13–21.

Peterson DE. 1992. Oral candidiasis. *Clin Geriatr Med* 8(3): 513–27.

Peterson DE, Minah GE, Overholser CD, Suzuki JB, DePaola LG, Stansbury DM, Williams LT, and Schimpff SC. 1987. Microbiology of acute periodontal infection in myelosuppressed cancer patients. *J Clin Oncol* 5(9): 1461–68.

Pinheiro ET, Gomes BPFA, Ferraz CCR, Sousa ELR, Teixeira FB, and Souza-Filho FJ. 2003. Microorganisms from canal of root-filled teeth with periapical lesions. *Int Endod J* 36(1): 1–11.

Pizzo G, Barchiesi F, Falconi Di Francesco L, Giuliana G, Arzeni D, Milici ME, D'Angelo M, and Scalise G. 2002. Genotyping and antifungal susceptibility of human subgingival *Candida albicans* isolates. *Arch Oral Biol* 47(3): 189–96.

Pizzo G, Giammanco GM, Pecorella S, Campisi G, Mammina C, and D'Angelo M. 2005. Biotypes and randomly amplified polymorphic DNA (RAPD) profiles of subgingival *Candida albicans* isolates in HIV infection. *New Microbiol* 28(1): 75–82.

Portenier I, Haapasalo H, Orstavik D, Yamauchi M, and Haapasalo M. 2002. Inactivation of the antibacterial activity of iodine potassium iodide and chlorhexidine digluconate against *Enterococcus faecalis* by dentin, dentin matrix, type-I collagen, and heat-killed microbial whole cells. *J Endod* 28(9): 634–37.

Portenier I, Haapasalo H, Rye A, Waltimo T, Orstavik D, and Haapasalo M. 2001. Inactivation of root canal medicaments by dentine, hydroxylapatite and bovine serum albumin. *Int Endod J* 34(3): 184–88.

Powderly WG, Robinson K, and Keath EJ. 1993. Molecular epidemiology of recurrent oral candidiasis in human immunodeficiency virus-positive patients: evidence for two patterns of recurrence. *J Infect Dis* 168(2): 463–66.

Pugh D and Cawson RA. 1977. The cytochemical localization of phospholipase in *Candida albicans* infecting the chick chorio-allantoic membrane. *Sabouraudia* 15(1): 29–35.

Pugh D and Cawson RA. 1980. Calcium, sequestering agents and nystatin-interactions on cell wall morphology and fungistasis of *Candida albicans*. *Sabouraudia* 18(2): 157–59.

Radcliffe CE, Potouridou L, Qureshi R, Habahbeh N, Qualtrough A, Worthington H, and Drucker DB. 2004. Antimicrobial activity of varying concentrations of sodium hypochlorite on the endodontic microorganisms *Actinomyces israelli*, *A. naeslundii*, *Candida albicans* and *Enterococcus faecalis*. *Int Endod J* 37(7): 438–46.

Radford DR, Challacombe SJ, and Walter JD. 1999. Denture plaque and adherence of *Candida albicans* to denture-base materials *in vivo* and *in vitro*. *Crit Rev Oral Biol Med* 10(1): 99–116.

Radford DR, Sweet SP, Challacombe SJ, and Walter JD. 1998. Adherence of *Candida albicans* to denture-base materials with different surface finishes. *J Dent* 26(7): 577–83.

Rams TE, Babalola OO, and Slots J. 1990. Subgingival occurrence of enteric rods, yeasts and staphylococci after systemic doxycycline therapy. *Oral Microbiol Immunol* 5(3): 166–68.

Rams TE and Slots J. 1991. *Candida* biotypes in human adult periodontitis. *Oral Microbiol Immunol* 6(3): 191–92.

Ray TL, Digre KB, and Payne CD. 1984. Adherence of *Candida* species to human epidermal corneocytes and buccal mucosal cells: correlation with cutaneous pathogenicity. *J Invest Dermatol* 83(1): 37–41.

Reynaud AH, Nygard-Østby B, Bøygard G-K, Eribe ER, Olsen I, and Gjermo P. 2001. Yeasts in periodontal pockets. *J Clin Periodontol* 28(9): 860–64.

Richards S and Russell C. 1987. The effect of sucrose on the colonization of acrylic by *Candida albicans* in pure and mixed culture in an artificial mouth. *J Appl Bacteriol* 62(5): 421–27.

Rocas IN, Baumgartner JC, Xia T, and Siqueira JF, Jr. 2006. Prevalence of selected bacterial named species and uncultivated phylotypes in endodontic abscesses from two geographic locations. *J Endod* 32(12): 1135–38.

Rolla G, Loe H, and Schiott CR. 1970. The affinity of chlorhexidine for hydroxyapatite and salivary mucins. *J Periodontal Res* 5(2): 90–95.

Root JL, McIntyre OR, Jacobs NJ, and Daghlian CP. 1988. Inhibitory effect of disodium EDTA upon the growth of *Staphylococcus epidermidis in vitro*: relation to infection prophylaxis of Hickman catheters. *Antimicrob Agents Chemother* 32(11): 1627–31.

Ruff ML, McClanahan SB, and Babel BS. 2006. *In vitro* antifungal efficacy of four irrigants as a final rinse. *J Endod* 32(4): 331–33.

Russell AD and Day MJ. 1993. Antibacterial activity of chlorhexidine. *J Hosp Infect* 25(4): 229–38.

Samaranayake LP. 1992. Oral mycoses in HIV infection. *Oral Surg Oral Med Oral Pathol* 73(2): 171–80.

Samaranayake LP, Geddes DA, Weetman DA, and MacFarlane TW. 1983. Growth and acid production of *Candida albicans* in carbohydrate supplemented media. *Microbios* 37(148): 105–15.

Samaranayake LP, Hughes A, Weetman DA, and MacFarlane TW. 1986. Growth and acid production of *Candida* species in human saliva supplemented with glucose. *J Oral Pathol* 15(5): 251–54.

Samaranayake LP and MacFarlane TW. 1980. An *in vitro* study of the adherence of *Candida albicans* to acrylic surfaces. *Arch Oral Biol* 25(8–9): 603–9.

Samaranayake LP, Raeside JM, and MacFarlane TW. 1984. Factors affecting the phospholipase activity of *Candida* species *in vitro*. *Sabouraudia* 22(3): 201–7.

Samaranayake YH, MacFarlane TW, Samaranayake LP, and Aitchison T. 1994. The *in vitro* proteolytic and saccharolytic activity of *Candida* species cultured in human saliva. *Oral Microbiol Immunol* 9(4): 229–35.

San Millan R, Ezkurra PA, Quindos G, Robert R, Senet JM, and Ponton J. 1996. Effect of monoclonal antibodies directed against *Candida albicans* cell wall antigens on the adhesion of the fungus to polystyrene. *Microbiology* 142(Pt 8): 2271–77.

Sathorn C, Parashos P, and Messer H. 2007. Antibacterial efficacy of calcium hydroxide intracanal dressing: a systematic review and meta-analysis. *Int Endod J* 40(1): 2–10.

Schep LJ, Jones DS, and Shepherd MG. 1995. Primary interactions of three quaternary ammonium compounds with blastospores of *Candida albicans* (MEN strain). *Pharmacol Res* 12(5): 649–52.

Scully C, El-Kabir M, and Samaranayake LP. 1994. *Candida* and oral candidosis: a review. *Crit Rev Oral Biol Med* 5: 125–57.

Sen BH, Akdeniz BG, and Denizci A. 2000. The effect of ethylenediamine-tetraacetic acid on *Candida albicans*. *Oral Surg Oral Med Oral Pathol Oral Radiol Endod* 90(5): 651–55.

Sen BH, Chungal NM, Liu H, and Fleischmann J. 2003. A new method for study in adhesion of *Candida albicans* to dentin in the presence or absence of smear layer. *Oral Surg Oral Med Oral Pathol Oral Radiol Endod* 96(2): 201–6.

Sen BH, Piskin B, and Demirci T. 1995a. Observation of bacteria and fungi in infected root canals and dentinal tubules by SEM. *Endod Dent Traumatol* 11(1): 6–9.

Sen BH, Safavi KE, and Spangberg LS. 1997a. Colonization of *Candida albicans* on cleaned human dental hard tissues. *Arch Oral Biol* 42(7): 513–20.

Sen BH, Safavi KE, and Spangberg LS. 1997b.Growth patterns of *Candida albicans* in relation to radicular dentin. *Oral Surg Oral Med Oral Pathol Oral Radiol Endod* 84(1): 68–73.

Sen BH, Safavi KE, and Spangberg LS. 1999. Antifungal effects of sodium hypochlorite and chlorhexidine in root canals. *J Endod* 25(4): 235–38.

Sen BH, Wesselink PR, and Turkun M. 1995b. The smear layer: a phenomenon in root canal therapy. *Int Endod J* 28(3): 141–48.

Sena NT, Gomes BP, Vianna ME, Berber VB, Zaia AA, Ferraz CC, and Souza-Filho FJ. 2006. *In vitro* antimicrobial activity of sodium hypochlorite and chlorhexidine against selected single-species biofilms. *Int Endod J* 39(11): 878–85.

Shabahang S, Pouresmail M, and Torabinejad M. 2003. *In vitro* antimicrobial efficacy of MTAD and sodium hypochlorite. *J Endod* 29(7): 450–52.

Shabahang S and Torabinejad M. 2003. Effect of MTAD on *Enterococcus faecalis*-contaminated root canals of extracted human teeth. *J Endod* 29(9): 576–79.

Sherwood J, Gow NA, Gooday GW, Gregory DW, and Marshall D. 1992. Contact sensing in *Candida albicans*: a possible aid to epithelial penetration. *J Med Vet Mycol* 30(6): 461–69.

Siqueira JF. 2002. Endodontic infections: concepts, paradigms and perspectives. *Oral Surg Oral Med Oral Pathol Oral Radiol Endod* 94(3): 281–93.

Siqueira JF and de Uzeda M, Jr. 1997. Intracanal medicaments: Evaluation of the antibacterial effects of chlorhexidine, metronidazole, and calcium hydroxide associated with three vehicles. *J Endod* 23(3): 167–69.

Siqueira JF, Rocas IN, Jr, Lopes HP, Elias CN, and de Uzeda M. 2002a. Fungal infection of the radicular dentin. *J Endod* 28(11): 770–73.

Siqueira JF and Roças IN. 2004. Polymerase chain reaction-based analysis of microorganisms associated with failed endodontic treatment. *Oral Surg Oral Med Oral Pathol Oral Radiol Endod* 97(1): 85–94.

Siqueira JF, Roças IN, and Lopes HP. 2002b. Pattern of microbial colonization in primary root canal infections. *Oral Surg Oral Med Oral Pathol Oral Radiol Endod* 93(2): 174–78.

Siqueira JF, Rocas IN, Moraes SR, and Santos KR. 2002c. Direct amplification of rRNA gene sequences for identification of selected oral pathogens in root canal infections. *Int Endod J* 35(4): 345–51.

Siqueira JF, Rocas IN, Lopes HP, Magalhaes FA, and Uzeda M. 2003. Elimination of *Candida albicans* infection of the radicular dentin by intracanal medications. *J Endod* 29(8): 501–04.

Siren EK, Haapasalo MP, Ranta K, Salmi P, and Kerosuo EN. 1997. Microbiological findings and clinical treatment procedures in endodontic cases selected for microbiological investigation. *Int Endod J* 30(2): 91–95.

Slack G. 1975. The resistance to antibiotics of microorganisms isolated from root canals. *Br Dent J* 18: 493–94.

Slots J, Rams TE, and Listgarten MA. 1988. Yeasts, enteric rods and pseudomonads in the subgingival flora of severe adult periodontitis. *Oral Microbiol Immunol* 3(2): 47–52.

Slutsky B, Buffo J, and Soll DR. 1985. High-frequency switching of colony morphology in *Candida albicans*. *Science* 230(4726): 666–69.

Smith JJ and Wayman BE. 1986. An evaluation of the antimicrobial effectiveness of citric acid as a root canal irrigant. *J Endod* 12(2): 54–58.

Sohnle PG, Hahn BL, and Karmarkar R. 2001. Effect of metals on *Candida albicans* growth in the presence of chemical chelators and human abscess fluid. *J Lab Clin Med* 137(4): 284–89.

Soll DR. 1992. High-frequency switching in *Candida albicans*. *Clin Microbiol Rev* 5(2): 183–203.

Soll DR, Galask R, Schmid J, Hanna C, Mac K, and Morrow B. 1991. Genetic dissimilarity of commensal strains of *Candida* spp. carried in different anatomical locations of the same healthy women. *J Clin Microbiol* 29(8): 1702–10.

Soll DR, Morrow B, Srikantha T, Vargas K, and Wertz P. 1994. Developmental and molecular biology of switching in *Candida albicans*. *Oral Surg Oral Med Oral Pathol* 78(2): 194–201.

Song X, Eribe ER, Sun J, Hansen BF, and Olsen I. 2005. Genetic relatedness of oral yeasts within and between patients with marginal periodontitis and subjects with oral health. *J Periodontal Res* 40(6): 446–52.

Spratt DA, Pratten J, Wilson M, and Gulabivala K. 2001. An *in vitro* evaluation of the antimicrobial efficacy of irrigants on biofilms of root canal isolates. *Int Endod J* 34(4): 300–307.

Sudbery P, Gow N, and Berman J. 2004. The distinct morphogenic states of *Candida albicans*. *Trends Microbiol* 12(7): 317–24.

Sunde PT, Olsen I, Debelian G, and Tronstad L. 2002. Microbiota of periapical lesions refractory to endodontic therapy. *J Endod* 28(4): 304–10.

Sundqvist G. 1994. Taxonomy, ecology, and pathogenicity of the root canal flora. *Oral Surg Oral Med Oral Pathol* 78(4): 522–30.

Sundqvist G, Figdor D, Persson S, and Sjögren U. 1998. Microbiological analysis of teeth with failed endodontic treatment and outcome of conservative re-treatment. *Oral Surg Oral Med Oral Pathol Oral Radiol Endod* 85(1): 86–93.

Sundqvist G, Johansson E, and Sjogren U. 1989. Prevalence of black-pigmented bacteroides species in root canal infections. *J Endod* 15(1): 13–19.

Swerdloff JN, Filler SG, and Edwards JE, Jr. 1993. Severe candidal infections in neutropenic patients. *Clin Infect Dis* 17(Suppl 2): S457–67.

Sziegoleit F, Sziegoleit A, and Wetzel WE. 1999. Effect of dental treatment and/or local application of amphotericin B to carious teeth on oral colonisation by *Candida*. *Med Mycol* 37(5): 345–50.

Tamura M, Watanabe K, Mikami Y, Yazawa K, and Nishimura K. 2001. Molecular characterization of new clinical isolates of *Candida albicans* and *C. dubliniensis* in Japan: analysis reveals a new genotype of *C. albicans* with group I intron. *J Clin Microbiol* 39(12): 4309–15.

Tanomaru Filho M, Leonardo MR, Silva LA, Anibal FF, and Faccioli LH. 2002. Inflammatory response to different endodontic irrigating solutions. *Int Endod J* 35(9): 735–39.

Tobgi RS, Samaranayake LP, and MacFarlane TW. 1988. *In vitro* susceptibility of *Candida* species to lysozyme. *Oral Microbiol Immunol* 3(1): 35–39.

Torabinejad M, Cho Y, Khademi AA, Bakland LK, and Shabahang S. 2003a. The effect of various concentrations of sodium hypochlorite on the ability of MTAD to remove the smear layer. *J Endod* 29(4): 233–39.

Torabinejad M, Handysides R, Khademi AA, and Bakland LK. 2002. Clinical implications of the smear layer in endodontics: a review. *Oral Surg Oral Med Oral Pathol Oral Radiol Endod* 94(6): 658–66.

Torabinejad M, Khademi AA, Babagoli J, Cho Y, Johnson WB, Bozhilov K, Kim J, and Shabahang S. 2003b. A new solution for the removal of the smear layer. *J Endod* 29(3): 170–75.

Torabinejad M, Shabahang S, Aprecio RM, and Kettering JD. 2003c. The antimicrobial effect of MTAD: an *in vitro* investigation. *J Endod* 29(6): 400–403.

Tronstad L, Barnett F, Riso K, and Slots J. 1987. Extraradicular endodontic infections. *Endod Dent Traumatol* 3(2): 86–90.

Tronstad L, Kreshtool D, and Barnett F. 1989. Microbiological monitoring and results of treatment of extraradicular endodontic infection. *Endod Dent Traumatol* 6(3): 129–36.

Trowbridge J, Ludmer LM, Riddle VD, Levy CS, and Barth WF. 1999. *Candida lambica* polyarthritis in a patient with chronic alcoholism. *J Rheumatol* 26(8): 1846–48.

Turk BT, Ates M, and Sen BH. 2008. The effect of treatment of radicular dentin on colonization patterns of *C. albicans*. *Oral Surg Oral Med Oral Pathol Oral Radiol Endod* 84(1): 68–73.

Ueno K, Imamura T, and Cheng KL. 1982. *CRC Handbook of Organic Analytical Reagents*. Boca Raton, FL: CRC Press.

Vianna ME, Gomes BP, Berber VB, Zaia AA, Ferraz CC, and de Souza-Filho FJ. 2004. *In vitro* evaluation of the antimicrobial activity of chlorhexidine and sodium hypochlorite. *Oral Surg Oral Med Oral Pathol Oral Radiol Endod* 97(1): 79–84.

Waltimo T, Haapasalo M, Zehnder M, and Meyer J. 2004a. Clinical aspects related to endodontic yeast infections. *Endod Topics* 9(1): 66–78.

Waltimo T, Kuusinen M, Järvensivu A, Väänänen A, Richardson M, Salo T, and Tjäderhane L. 2003a. *Candida* spp. in refractory periapical granulomas. *Int Endod J* 36(9): 643–47.

Waltimo T, Luo G, Samaranayake LP, and Vallittu PK. 2004b. Glass fibre-reinforced composite laced with chlorhexidine digluconate and yeast adhesion. *J Mater Sci-Mater Med* 15(2): 117–21.

Waltimo TM, Dassanayake RS, Orstavik D, Haapasalo MP, and Samaranayake LP. 2001. Phenotypes and randomly amplified polymorphic DNA profiles of *Candida albicans* isolates from root canal infections in a Finnish population. *Oral Microbiol Immunol* 16(2): 106–12.

Waltimo TM, Orstavik D, Siren EK, and Haapasalo MP. 1999a. *In vitro* susceptibility of *Candida albicans* to four disinfectants and their combinations. *Int Endod J* 32(6): 421–29.

Waltimo TM, Orstavik D, Siren EK, and Haapasalo MP. 2000a. *In vitro* yeast infection of human dentin. *J Endod* 26(4): 207–9.

Waltimo TM, Sen BH, Meurman JH, Orstavik D, and Haapasalo MP. 2003b. Yeasts in apical periodontitis. *Crit Rev Oral Biol Med* 14(2): 128–37.

Waltimo TM, Siren EK, Orstavik D, and Haapasalo MP. 1999b. Susceptibility of oral *Candida* species to calcium hydroxide *in vitro*. *Int Endod J* 32(2): 94–98.

Waltimo TM, Siren EK, Torkko HLK, Olsen I, and Haapasola MP. 1997. Fungi in therapy-resistant apical periodontitis. *Int Endod J* 30(2): 96–101.

Waltimo TMT, Orstavik D, Meurman JH, Samaranayeke LP, and Haapasalo MPP. 2000b. *In vitro* susceptibility of *Candida albicans* isolated from apical and marginal periodontitis to common antifungal agents. *Oral Microbiol Immunol* 15(4): 245–48.

Wilborn WH and Montes LF. 1980. Scanning electron microscopy of oral lesions in chronic mucocutaneous candidiasis. *JAMA* 244(20): 2294–97.

Willis AM, Coulter WA, Hayes JR, Bell P, and Lamey PJ. 2000. Factors affecting the adhesion of *Candida albicans* to epithelial cells of insulin-using diabetes mellitus patients. *J Med Microbiol* 49(3): 291–93.

Wilson MI and Hall J. 1968. Incidence of yeasts in root canals. *J Br Endod Soc* 2(4): 56–59.

Wu MK, Barkis D, Roris A, and Wesselink PR. 2002. Does the first file to bind correspond to the diameter of the canal in the apical region? *Int Endod J* 35(3): 264–67.

Wu MK, Van Der Sluis LW, and Wesselink PR. 2003. The capability of two hand instrumentation techniques to remove the inner layer of dentine in oval canals. *Int Endod J* 36(3): 218–24.

Xu T, Levitz SM, Diamond RD, and Oppenheim FG. 1991. Anticandidal activity of major human salivary histatins. *Infect Immun* 59(8): 2549–54.

Yamashita JC, Tanomaru Filho M, Leonardo MR, Rossi MA, and Silva LA. 2003. Scanning electron microscopic study of the cleaning ability of chlorhexidine as a root-canal irrigant. *Int Endod J* 36(6): 391–94.

Yang YL. 2003. Virulence factors of *Candida* species. *J Microbiol Immunol Infect* 36(4): 223–28.

Yilmaz H, Aydin C, Bal BT, and Ozcelik B. 2005. Effects of disinfectants on resilient denture-lining materials contaminated with *Staphylococcus aureus*, *Streptococcus sobrinus*, and *Candida albicans*. *Quintessence Int* 36(5): 373–81.

Yoshida T, Shibata T, Shinohara T, Gomya S, and Sekine I. 1995. Clinical evaluation of the efficacy of EDTA solution as an endodontic irrigant. *J Endod* 21(12): 592–93.

Zambon JJ, Reynolds HS, and Genco RJ. 1990. Studies of the subgingival microflora in patients with acquired immunodeficiency syndrome. *J Periodontol* 61(11): 699–704.

Zerella JA, Fouad AF, and Spangberg LS. 2005. Effectiveness of a calcium hydroxide and chlorhexidine digluconate mixture as disinfectant during retreatment of failed endodontic cases. *Oral Surg Oral Med Oral Pathol Oral Radiol Endod* 100(6): 756–61.

# Chapter 10

# Management of Aggressive Head and Neck Infections

*Domenick P. Coletti and Robert A. Ord*

## 10.1 Introduction

This chapter discusses the diagnosis and management of severe, life-threatening aggressive odontogenic infections of the head and neck. Intraoral soft tissue infections and bony infections, for example, osteomyelitis are not addressed. Early recognition by the oral health care professional of potentially severe infections and prompt management can prevent progression of the infection, with the later need for hospitalization and life-saving surgery. If not recognized and treated at an early stage, these infections may progress to cause airway obstruction, necrotizing infections with septic shock, or fatal outcomes from complications such as brain abscess, cavernous sinus thrombosis, or mediastinitis. Although medical advances in imaging, antibiotics, and critical care have increased our ability to treat infectious diseases, the outcome remains a balance between the virulence of the infectious agent and the immune response of the host. The development of bacterial strains resistant to multiple antibiotics (MRSA, VRE) "super bugs," the proliferation of bacteria causing necrotizing infections "flesh eating bacteria" (Deans 1994), and the increasing elderly population with multiple comorbidities and compromised immune system (AIDS, diabetes, renal failure, chemotherapy) have kept more pace than with these medical advances. Today, severe aggressive odontogenic infections remain an important cause of morbidity, mortality, and utilization of hospital resources.

## 10.2 Etiology and epidemiology

Although odontogenic sources, for example, periapical infections, pericoronitis, and periodontal abscesses account for most of the deep neck infections, they may also be secondary to facial fractures, penetrating neck wounds, skin lesions, for example, furuncles, folliculitis, infected sebaceous cysts, and salivary gland infections or sinusitis. In Wang et al.'s series, 157 of

250 maxillofacial infections (63%) were of odontogenic origin (Wang et al. 2005). Potter et al. (2002) found 69% of deep space infections to be of odontogenic origin and 11% peritonsillar (Riggio et al. 2007). In subcategorizing odontogenic infections, 65% were secondary to caries, 22% pericoronitis, and 22% were periodontal (some patients had more than one source), with 68% of cases originating in the mandibular third molars, followed by the mandibular bicuspids and molars (Flynn et al. 2006a, b). Wang et al.'s study also confirmed previous studies that most patients were adults, older than 18 years, and although 15% of patients were under 12 years old, severe odontogenic infections were rare in teenagers (2%) (Wang et al. 2005). Peters et al. reviewed 128 patients with only odontogenic maxillofacial infections and their average age was 35.8 years with a slight female predominance 54% (Peters et al. 1996). However, other authors have described more male patients (62%) and also a high incidence of minorities, African American 54%, Hispanic 22% (Flynn et al. 2006a, b). In children Dodson et al. showed that most had maxillary buccal infections in contrast to adults, who have more infections related to the mandible (Dodson et al. 1989, 1991).

Many publications have looked at these patients for comorbid conditions that may have immunosuppressive potential, with conflicting results reporting HIV/diabetes in only 8% (Flynn et al. 2006a, b) to immunocompromise of 24% (Peters et al. 1996) up to 64.5% (Wang et al. 2005). In this last series Wang et al. found that all 26 of their patients under the age of 18 years to be perfectly healthy; however, 131 patients over 18 years of age (64.2%) were compromised, 8 with diabetes, 35 alcohol addiction, 34 intravenous drug abuse, and 6 with HIV infection. This, however, may reflect the patient population of the study that was set in a large urban public hospital, rather than a potential specific predilection for developing these serious infections.

In necrotizing infections, such as necrotizing fasciitis, underlying medical compromise with diabetes, steroid or chemotherapy treatment, leukemia, renal failure or HIV is very frequent which may, at least in part, account for the increased mortality rate 20–73% (Lin et al. 2001). In one large series of head and neck infection cases, 89.4% had an underlying medical condition that included 72.3% diabetes, 19.1% acute or chronic renal failure, and 12% with underlying malignancy (Lin et al. 2001). A comprehensive review

article of necrotizing fasciitis concluded that 70% of adult cases have an underlying immune compromising disease (McGurk 2003). Odontogenic infection is less frequent (10.6%) (Lin et al. 2001) as an etiology as most cases result from trauma or surgery.

## 10.3 Microbiology

In most odontogenic infections there is a polymicrobial mixed aerobic and anaerobic infection with intraoral bacteria. Cultures show mixed infections with Gram-positive cocci and Gram-negative rods predominating. Wang et al. reported that in their 250 cases obtaining cultures and sensitivity reports did not appear to be clinically helpful as it failed to lead to any antibiotic or treatment change (Wang et al. 2005). Other authors recommend cultures for extensive or rapidly spreading infections, in immune compromised patients, recurrent infections, where there is no response to antibiotics, nosocomial infections, or necrotizing and gas-producing infections as opposed to purely empiric therapy (Jones and Cadelaria 2000). Lastly, others have described the use of polymerase chain reaction for microbiological detection at a molecular level; this methodology has been shown to detect some uncultivable species of bacteria (Riggio et al. 2007). The clinical applicability of these types of diagnostics is questionable, in a similar fashion to cultures; it does not presently appear as though molecular microbiology will change the initial management of these types of cases. However, there may be some benefit in assisting with the selection of a narrow-spectrum antibiotic after surgical therapy has been implemented. In Flynn et al.'s prospective study, a total of 45 different species of bacteria were isolated, both aerobic and anaerobic in their cultures of severe odontogenic infections. The most frequent pathogens isolated were *Prevotella* spp., *Streptococcus viridans* (which includes the *Streptococcus milleri* group), and *Peptostreptococcus* spp. (some species are now classified under *Parvimonas* spp.) (Flynn et al. 2006b). These findings agree with other recent publications since 1998 (Sakamoto et al. 1998; Kuriyama et al. 2000; Stephanopoulus and Kolokotronis 2004).

In Flynn et al.'s study a positive culture for *Peptostreptococcus* spp. was a negative predictor for abscess formation, but positively associated with a cellulitis. Unless patients were allergic or had a necrotizing infection, all were started on penicillin G, IV and subsequently 19% of isolated species were found to be penicillin resistant, and one or more of these resistant bacteria were found in 54% of cases. In addition, clindamycin-resistant strains were found in 17% of cases. Although all six patients who failed penicillin treatment were found to have one or more penicillin-resistant strains, and six out of ten (60%) patients with resistant bacteria were treatment failures. The authors point out that the identification of penicillin-resistant organisms was often delayed for up to 2 weeks, with clinical decisions already having been made (Flynn et al. 2006a).

Huang et al. analyzed a cohort of 56 patients with diabetes and 129 patients without diabetes presenting with deep neck infections (Huang et al. 2005). They found the expected bacterial cultures in the nondiabetic group, with *S. viridans* (43.7%), followed by *Peptostreptococcus* spp. and *Klebsiella pneumoniae*; however, the predominant pathogen in the diabetic group was *K. pneumoniae* (56.1%) followed by *S. viridans*, and this difference was highly significant ($p = 0.0004$). Because of the very predictable polymicrobial mixed infections, most patients are treated empirically on hospital admission with regimes that usually include a penicillin, clindamycin, and/or metronidazole. Reported antibiotic regimes are penicillin + metronidazole, or clindamycin in 60 and 20% of 157 patients, respectively (Wang et al. 2005), and penicillin G with or without oxacillin in 54% or clindamycin in 33% of patients (Flynn et al. 2006a), or cefotaxime + metronidazole + corticosteroids in one trial of antibiotics without surgical drainage (Mayor et al. 2001).

In necrotizing fasciitis, most infections particularly those of the limbs, perineum, and abdomen have been attributed to group A streptococci, which appears to have an increasing incidence (Kaul et al. 1997). However, although the classic disease is seen with group A streptococci alone or in combination with *Staphylococcus aureus*, this disease may also be polymicrobial with prominent anaerobic species present (Brook and Frazier 1995). In Lin et al.'s paper, *K. pneumoniae* was the most common organism followed by *S. aureus* and *Streptococcus* spp. Due to the severe systemic compromise and high mortality rate, triple antibiotic therapy is usually mandated. Regimes recommended include a penicillin, aminoglycoside, and anti-anaerobe (Lin et al. 2001), or gentamycin, metronidazole, and clindamycin (Flynn et al. 2006a).

## 10.4 Anatomy and pathogenesis of spread

Essentially, there are three patterns of spread seen in severe odontogenic infections (although a mixture of these types is frequently seen). These are abscess formation, cellulites, and necrotizing fasciitis. An abscess is a localized collection of pus and it is classically caused by *Staphylococcus* species that produce coagulase enzyme, which "walls off" infections. Cellulitis is an infection that spreads widely through the tissues, not usually producing much pus but permeating tissue spaces, classically caused by streptococci that produce collagenase and hyaluronidase to break down the ground substance of connective tissue and promote spread. In necrotizing fasciitis there is a rapid progressive liquefaction of subcutaneous fat and fascia with thrombosis of the subdermal veins. The skin becomes reddened, and then necrotic as its blood supply is lost. Although the underlying muscle is usually preserved, myositis can occur as can gas formation ("gas gangrene"). Extensive underlying skin necrosis may be present with comparatively little in the way of clinical signs.

The patterns of spread of the cellulitic infections will determine symptoms and signs and also whether the airway is liable to be compromised. Pus and cellulitic fluid will usually follow the path of least resistance and track along fascial planes in the neck and face. The direction of spread for odontogenic infections is therefore determined by the anatomical relationships of the tooth root to the jawbone and the muscle insertions into the jaw. As discussed earlier, the majority of severe odontogenic infections are related to the mandibular third molars, followed by the other molars and premolars, and this is the area that is focused on in this section. It should also be appreciated by the reader that most of these spaces or potential spaces are interconnected so that many of these infections involve multiple spaces. Additionally, fascial layers extending along the retropharyngeal/prevertebral and carotid/jugular sheath may allow infection to spread into the mediastinum and thorax, far from the head and neck.

Once pulpal necrosis has occurred, periapical infection is located within the medullary bone of the jaw. This infection will eventually perforate the cortical plate to gain access to the soft tissue spaces. In the mandibular molar region, the roots tend to be situated lingually and lingual perforation is more common. If the apex of the tooth is inferior to the attachment of the mylohyoid muscle and to the mylohyoid ridge (which is usually the case), pus can directly enter the submandibular space. If the apex is superior to the mylohyoid attachment then it will enter the sublingual space. Both these spaces are connected posteriorly at the free edge of the mylohyoid muscle. If the pus perforates buccally in the mandibular molar region and it is inferior to the buccinator muscle attachment, it can involve the buccal space. If the apex of the tooth is superior to the buccinator insertion, the pus will present intraorally as a vestibular abscess. (This can easily be drained intraorally and will not be discussed further.) Severe pericoronitis and periapical infection from mandibular third molars may track posteriorly into the masticator and lateral pharyngeal spaces. In the maxilla, buccal perforation from posterior teeth may enter the buccal space and from the canine, via the canine space, which can lead to orbital or intracranial involvement. Infections perforating the palatal side of the alveolus, for example, from lateral incisors or the palatal roots of molars are usually well confined by the thick bound down palatal mucosa of the hard palate and just require simple intraoral drainage.

In order to understand the presentation and management of these infections, some knowledge of the anatomy of these spaces is essential. A brief outline of their anatomy and intercommunications is given below, but for more details referral to a surgical text is recommended.

### 10.4.1 Submandibular space (Fig. 10.1)

The submandibular space lies inferior to the mandible and contains the submandibular gland, lymph nodes, the facial artery, and the anterior facial vein. The mandibular branch of the facial nerve lies superiorly and the lingual and hypoglossal nerve posteriorly and deeply. The medial/deep boundary is delineated by the mylohyoid muscle with contributions from the hypoglossus and styloglossus, while laterally and superficially are skin and platysma. Anteriorly is the anterior belly of the digastric muscle although the submandibular space communicates freely anteriorly with the submental space. Posteriorly are the posterior belly of digastric and the stylohyoid muscles. Superiorly is the lower border of the mandible and the pterygomasseteric sling and inferiorly the hyoid bone. The space communicates with the submental anteriorly, the sublingual space posterosuperiorly, and the lateral pharyngeal posteriorly.

### 10.4.2 Sublingual space (Fig. 10.2)

The sublingual space contains the sublingual gland, Wharton's duct, and the lingual nerve. Medially it is bounded by the genioglossus, geniohyoid, and styloglossus muscles, and laterally by the lingual plate of the mandible. Posteriorly it connects with the lateral pharyngeal space. Superiorly is the mucosa of the floor of mouth and inferiorly the mylohyoid muscle.

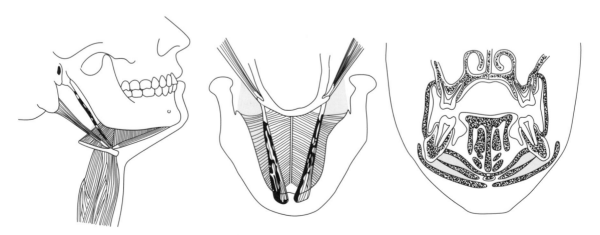

**Fig. 10.1**   Anatomical boundaries of the submandibular space. (Adapted from Hohl et al. (1983).)

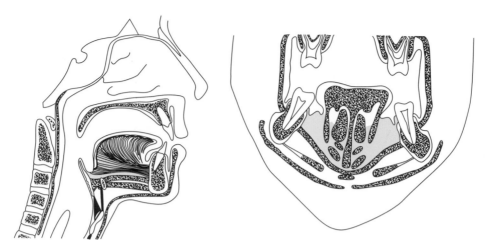

**Fig. 10.2** Anatomical boundaries of the sublingual space. (Adapted from Hohl et al. (1983).)

### 10.4.3 Submental space (Fig. 10.3)

The submental space is between the anterior bellies of the digastric muscles, superiorly bounded by the mylohyoid muscle, and inferiorly bounded by the platysma.

There is one named infection that is important to highlight which is Ludwig's angina (Figs 10.4a–b), which was first described by William Frederick von Ludwig in 1836. It is a bilateral infection of the submandibular, sublingual, and submental spaces; the term *angina* means "to constrict." Ironically, Dr Ludwig died at the age of 75 of an acute infection of

the neck. Some feel he succumbed to the very disease he is famous for describing.

### 10.4.4 Buccal space (Fig. 10.5)

The buccal space contains fat, branches of the facial nerve, the parotid duct, and the facial artery and vein. Medially lies the buccinator muscle and laterally skin. Anteriorly is the oral commissure with zygomaticus major and depressor anguli oris muscle, and the infraorbital space, while posteriorly is the pterygomandibular raphe. Superiorly lies the zygomatic arch

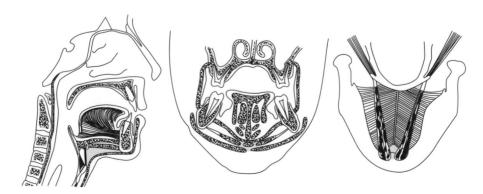

**Fig. 10.3** Anatomical boundaries of the submental space. (Adapted from Hohl et al. (1983).)

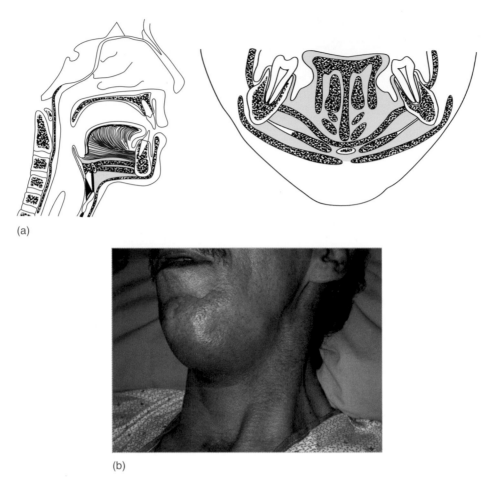

**Fig. 10.4** (a) Anatomical boundaries of the Ludwig's angina; (b) clinical example of Ludwig's angina. (Part (a), adapted from Hohl et al. (1983).)

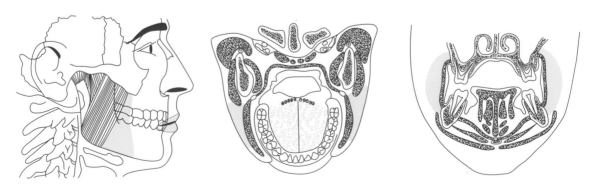

**Fig. 10.5** Anatomical boundaries of the buccal space. (Adapted from Hohl et al. (1983).)

and inferiorly the mandible. Infections can track to the submandibular and masticator spaces.

### 10.4.5 Lateral pharyngeal space (Figs 10.6a–b)

This is a pyramidal shaped space with its base superiorly and may be divided into an anterior and posterior compartment by the styloid muscles. It contains fat, lymph nodes, the carotid artery, and jugular vein as well as cranial nerves IX–XII. The medial boundary is the superior and middle constrictor muscles and laterally are the medial pterygoid and parotid glands. The base of the skull is superior and the hyoid bone inferior. Anteriorly lies the pterygomandibular raphe, but the space communicates freely with the submandibular and sublingual spaces, while posteriorly it communicates directly with the retropharyngeal space, which runs inferiorly to C6–T4 region where its fascia fuses with the alar fascia. If pus perforates the alar fascia, it enters the prevertebral (danger) space and can track along the whole length of the spinal column.

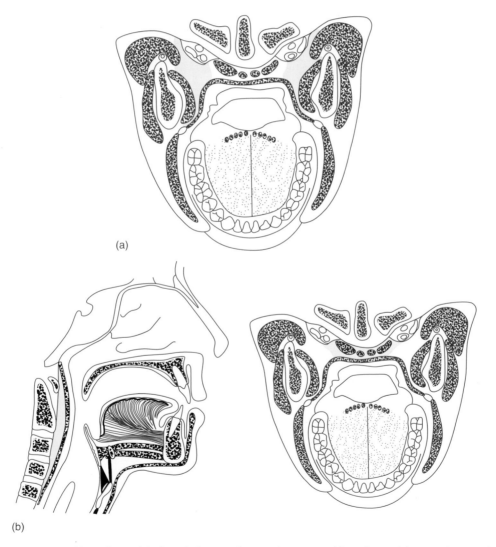

(a)

(b)

**Fig. 10.6** (a) Anatomical boundaries of the lateral pharyngeal space; (b) anatomical boundaries of the retropharyngeal space. (Adapted from Hohl et al. (1983).)

### 10.4.6 Masticator space

This space is composed of those spaces formed by the insertions of the masticatory muscles to the vertical ramus of the mandible, and it is subdivided into the masseteric, pterygomandibular, and temporal spaces:

#### 10.4.6.1 Masseteric space (Fig. 10.7a)

This small potential space is bounded medially by the buccal plate of the vertical ramus of the mandible and laterally by the masseter muscle. Superiorly it can pass into the infratemporal space, but inferiorly it is restricted by the attachment of the pterygomasseteric sling to the mandible.

#### 10.4.6.2 Pterygomandibular space (Fig. 10.7b)

This space is bounded medially by the medial pterygoid muscle and laterally by the lingual surface of the vertical ramus of the mandible. Superiorly lies the

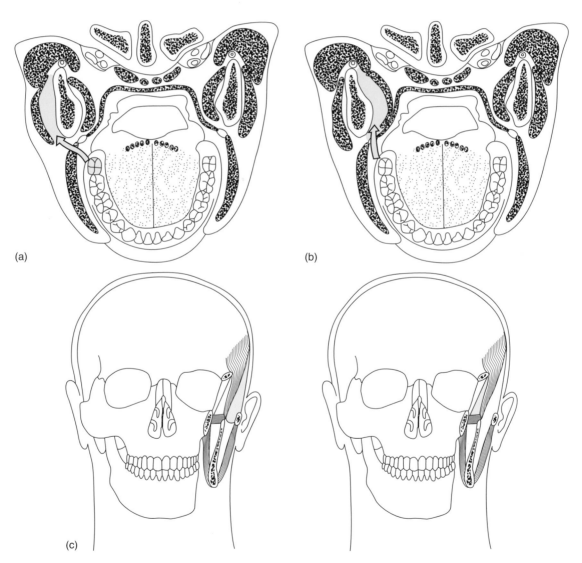

(a)

(b)

(c)

**Fig. 10.7**  (a) Anatomical boundaries of the submasseteric space; (b) anatomical boundaries of the pterygomandibular space; (c) anatomical boundaries of the superficial and deep temporal space. (Adapted from Hohl et al. (1983).)

lateral pterygoid muscle and inferior the pterygomasseteric sling. Anteriorly are the pterygomandibular raphe and the parotid gland, respectively.

### 10.4.6.3 Temporal space (Fig. 10.7c)

This is divided into the superficial and deep spaces. In the deep space the pus lies between the temporal surface of the skull medially and the temporalis muscle laterally, while in the superficial space the temporalis muscle is medial and the thick temporalis fascia lateral. These spaces communicate with the rest of the masticator spaces inferiorly via the infratemporal space and the attachment of the temporalis tendon to the coronoid process.

## 10.4.7 Canine space (Fig. 10.8)

This space contains terminal branches of the facial artery and vein (angular vein). Medially lies the nasal bone and laterally the buccal space. Anterior is the facial skin and posterior the maxilla, while inferiorly are the levator labii superiori, levator labii, and alaeque nasi muscles. The orbital septum is the superior boundary.

Lastly, one infectious process worth brief discussion for the sake of thoroughness is NOMA, otherwise known as *cancrum oris*, which in Latin means "to devour." NOMA is associated with a low virulent organism such as fusobacterium necrophorum (Paster et al. 2002); this infectious process, some feel, is an extensive form of ANUG (acute necrotizing ulcerative gingivitis). It typically begins with necrosis of the periodontium and then extends itself into the soft tissue of the face, hence its pretense "to devour" (Fig. 10.9). The common denominator in NOMA is malnutrition and immune suppression; in African NOMA there is also an association with AIDS, malaria, measles, and anemia. NOMA is rarely seen in western civilizations due to their low starvation rates; however, it may present in severely malnourished immune-compromised patients, such as those undergoing chemotherapy or transplantation, and as previously mentioned the AIDS population. NOMA is initially managed by nutritional supplementation, fluid resuscitation, blood transfusion, and soft-tissue debridement. Reconstructing these patients can be surgically challenging, especially if the functional units of the face are involved. Many reconstructive surgeons will follow the principles of the "reconstructive ladder;" which is as follows:

- Healing by secondary intention
- Skin grafting
- Local flaps
- Regional flaps
- Free tissue transfer

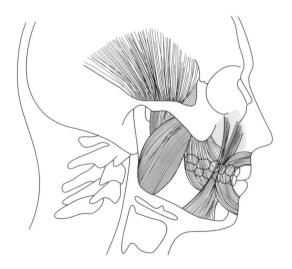
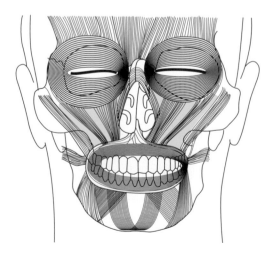

**Fig. 10.8**   Anatomical boundaries of the canine space. (Adapted from Hohl et al. (1983).)

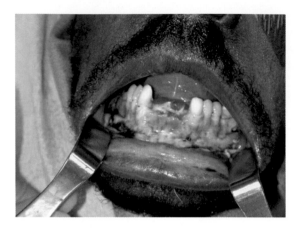

**Fig. 10.9**  Example of cancrum oris in an HIV patient.

# 10.5 Diagnosis

## 10.5.1 History

The diagnosis for patients with severe aggressive odontogenic infections will usually be evident but commences with a history and physical examination. Important areas to cover in the history will include precipitating factors (toothache or trauma), and specific symptoms that may alert the clinician to the fact that that the patient has impending airway obstruction. Symptoms that are very concerning are dysphagia and/or odynophagia, or difficulty in and/or pain with swallowing, or speaking with a muffled or "hot potato" voice quality. Patients with these symptoms represent true surgical emergencies and can quickly progress to a life-threatening situation. A complaint of inability to fully open the mouth is suggestive of pus causing spasm of the masticatory muscles. The examination is directed toward an overall systemic assessment of the patient and a good head and neck examination.

## 10.5.2 Systemic examination

Initial examination may reveal a patient sitting with his/her head held forward, drooling and using his/her accessory muscles of respiration. This is the worst case scenario with the patients leaning forward to protect their airway and unable to control their own saliva due to tongue edema. It can be predicted that even attempting to lay the patients flat to intubate them may precipitate complete obstruction. Fortunately the majority of patients do not present

at such an advanced stage. Fever and tachycardia may be marked especially in children, and this may exacerbate dehydration in patients with inability to swallow. These systemic manifestations of acute infection may destabilize preexisting diseases, for example, diabetes and cause hyperglycemia and ketosis. In necrotizing fasciitis, severe systemic effects may be seen; more than 50% have hypotension and 10–30% display one or more of the following: acute renal failure, coagulopathy, abnormal liver function, acute respiratory distress syndrome, abnormal liver function, or hemolytic anemia (McGurk 2003).

## 10.5.3 Head and neck examination

The classical signs of acute inflammation, swelling, redness, pain, and loss of function are usually present in cases of abscess or cellulitis but not always in necrotizing fasciitis, which can have gross necrosis with little in the way of obvious inflammatory signs. In an abscess the swelling is localized and painful and fluctuant to the touch. The overlying skin may be shiny or reddened and in advanced cases becomes thinned prior to spontaneous drainage of pus (Fig. 10.10). In cellulitis the swelling is diffuse, firm, and brawny to palpation. The overlying skin may be red and pitting edema can occur. In necrotizing fasciitis the skin is diffusely reddened over a wide area, and with progression becomes white and then black and necrotic. Blisters or bullae may occur and the skin is tender to palpation and crepitus due to gas-forming organisms may be elicited.

In abscess/cellulitis the swelling may be obvious on initial visual examination, for example, swelling of the cheek in buccal space infections or of the submandibular region in submandibular space infections. Considerable swelling may occur in the lateral pharyngeal/masticator space, which can only be detected on intraoral examination when the bulging of the pharyngeal wall, soft palate, and deviation of the uvula will be noted. However, in infections around the masticatory muscles severe trismus may occur, which will prevent a good intraoral exam and be a problem for airway access. Some space infections such as the masseteric have trismus out of all proportion to their other signs. Lastly, because odontogenic infections can be life-threatening due to airway loss, the practitioner must thoroughly evaluate swelling on the floor of mouth/tongue, and oropharynx (Fig. 10.11).

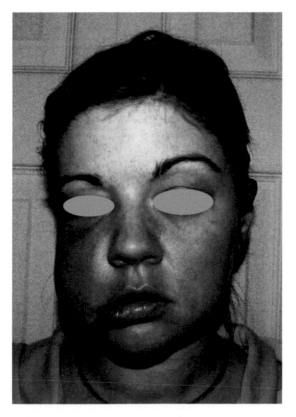

**Fig. 10.10** Clinical example of facial swelling and cellulitis of a canine and buccal space abscess.

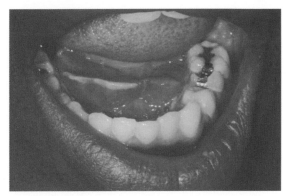

**Fig. 10.11** Noted floor of mouth edema and elevation of the tongue.

### 10.5.4 Laboratory investigations

Blood tests will usually show a leukocytosis with increased percentage of granulocytes and a left shift (i.e., an increase in immature forms as the body attempts to produce more granulocytes). In these infections the presence of a "normal" white cell count is clinically abnormal and usually indicates a deficient immune response. Blood glucose and electrolytes will be essential for diabetic patients and for suspected necrotizing fasciitis to diagnose underlying systemic problems.

### 10.5.5 Imaging

Plain radiographs such as a panoramic film will be useful to identify an odontogenic etiology. Soft tissue lateral views of the neck have been used to visualize increased prevertebral soft tissue thickening in retropharyngeal abscess (Haug et al. 1991). CT scanning is now the imaging modality mostly used to as-

sess severe neck infections (Fig. 10.12). Lazor et al. (1994) reviewed 38 patients with retro/parapharyngeal abscesses and found that intraoperative findings correlated with CT scan in 76.3% of cases, with a false-positive rate of 13.2% and a false-negative rate of 10.5%. Miller et al. (1999) compared clinical examination to contrast-enhanced CT in a blinded prospective trial with 35 patients to predict the presence of a drainable purulent infection in suspected deep neck abscess. Twenty-two patients had

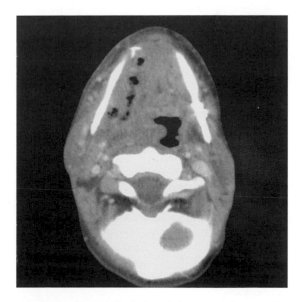

**Fig. 10.12** CT scan of right lateral pharyngeal abscess with tracheal deviation and gas-forming organisms.

purulent drainable collections. Clinical examination was 63% accurate with sensitivity of 55% and specificity of 73%, whereas CT was 77% accurate with sensitivity of 95% and specificity of 53%. If the two modalities were combined, accuracy was 89%, sensitivity 95%, and specificity 80%. The authors concluded that both clinical examination and contrast-enhanced CT were critical components in diagnosing these infections. Recently Smith et al. (2006) attempted to use Hounsfield units from their CT scan images to predict the presence of pus versus cellulitis and found this was unreliable. Due to their negative exploration rate of 25%, these authors emphasized the importance of clinical decision making for surgical drainage. Although the contrast-enhanced CT appears to be the accepted "gold standard" in imaging, ultrasound has also been advocated in children to assess abscess formation (Yeow et al. 2001; Duque et al. 2006).

Although CT scanning has a great deal of advantages for the clinician being able to show all the spaces of the face and neck to demonstrate involvement of areas that were not clinically suspected, visualizing spaces that cannot be directly examined in patients with severe trismus, and also showing areas such as the brain and mediastinum; when complications need to be ruled out, clinical assessment is paramount. Patients with impending airway obstruction should not be delayed from having an airway secured in the operating room (OR) to wait for a CT scan, nor should the surgeon forget that placing these patients in a supine position for scanning places them at risk for airway obstruction. The CT scanner is often in a remote area of the hospital and not ideally suited to deal with airway emergencies. Despite the valuable information that can be gained from imaging, the clinician has to balance the benefits against the risk in any case where airway is an issue.

## 10.6 Airway management

It is essential to secure the airway in patients with severe deep neck infections. This is frequently a challenging task due to generalized fascial swelling with edema of the tongue, pharynx, and larynx, often accompanied by severe trismus. When the cords can be visualized then standard oral endotracheal intubation is a safe way to proceed. However, in the presence of edema inappropriate manipulation of the airway may cause further swelling and bleeding, requiring emergent surgical intervention. In addition, a retropharyngeal or peritonsillar collection of pus may be ruptured by injudicious use of the laryngoscope with aspiration of purulent material with subsequent pulmonary or mediastinal complications. In most severe infections standard oral intubation is not possible so the choice of airway management will be either a fiberoptic intubation or a tracheotomy. In Potter et al.'s review of the literature, they found that in cases managed by otolaryngologists tracheotomy was the treatment of choice and in those managed primarily by Oral and Maxillofacial surgeons fiber optic intubation was preferred (Potter et al. 2002).

Ovassapian et al. report 26 patients with deep neck infections and 17 with Ludwig's angina who underwent awake fiber optic intubation. Three patients were intubated in the sitting position, 2 in Fowler's position, and 21 supine with head 10–15° raised. Twenty-five of the 26 intubations were successful. Postsurgery 7 patients were kept intubated and 5 underwent tracheotomy. Twelve patients remembered part of the procedure (Ovassapian et al. 2005). In Flynn et al.'s series (Flynn et al. 2006a, b) 49% underwent fiber optic intubation, 43% direct laryngoscopic intubation, and 8% no intubation; only one patient (3%) had a tracheotomy and this was the patient who required reoperation. Obviously awake fiberoptic intubation requires a skilled anesthesiologist and a surgeon on hand should intubation prove unsuccessful. The major advantage is avoiding the complications associated with tracheotomy; however, the airway is not secured and there is more patient discomfort with the patient requiring mechanical ventilation. In this setting it is often difficult to assess when the patient is suitable for extubation despite the presence of a cuff leak and ventilatory weaning parameters being met. Reintubation can be hazardous in these patients.

Tracheotomy also requires skill with these patients as swelling of the neck may obscure the usual landmarks. In the series by Potter et al., 34 patients underwent tracheotomy and 51 endotracheal intubation. One patient in the tracheotomy group had a cerebrovascular accident from rupture of a carotid aneurysm during the tracheotomy, and 2 patients in the intubation group died, one from an unplanned extubation and the other from laryngeal edema postextubation with inability to reintubate. Patients with tracheotomy had a shorter hospital stay, significantly less time in the intensive care unit and 60% less hospital cost (Potter et al. 2002). It was felt that tracheotomy provided better utilization of critical care resources. In true airway emergencies cricothyroidotomy is probably a quicker and safer choice for the nonspecialist surgeon to secure an airway, but again swelling and edema may make palpation of the cricoid and thyroid cartilages difficult.

## 10.7 Medical and surgical management

Obviously, initial therapy for these patients will be directed to their overall systemic condition. In the case of necrotizing fasciitis, fluid resuscitation for septic shock and control of hyperglycemic ketosis may be required prior to surgical intervention. Diabetics who are uncontrolled should also be stabilized as much as possible prior to surgery. The traditional management of these infections is reflected in the statement from the article, by Wang et al., that "early dental extractions, incision and drainage, coupled with intravenous antibiotic therapy is the most effective treatment" (Wang et al. 2005). Most surgeons would feel that pus is most effectively treated by surgical drainage (Fig. 10.13). Even in the case of cellulitis that is causing systemic signs and involving important fascial spaces, aggressive incision and drainage will help earlier resolution (Flynn 1991). This may be due to decompression of the space or changing the microenvironment

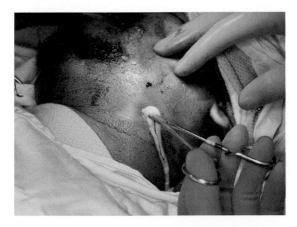

**Fig. 10.13**  Incision and drainage of left submandibular abscess.

to promote a more aerobic background. However, it is currently debated in the literature whether there is a place for nonsurgical management or radiologically guided aspiration, and for which types of patients it may be indicated.

It would appear from the literature that peritonsillar and parapharyngeal abscesses, especially those in children, may possibly be treated medically with intravenous antibiotics alone. Sichel et al. undertook a prospective study on patients with infection limited to the parapharyngeal space. Twelve patients presented with this diagnosis but 5 were excluded as other spaces were involved. Seven patients without systemic signs of shock were treated with intravenous amoxicillin/clavulanic acid for 9–14 days and all were cured without need for drainage. Six of the 7 cases were of pediatric patients (Sichel et al. 2002). In another prospective study of 34 patients with parapharyngeal abscesses (confirmed by contrast-enhanced CT scan), all of whom received antibiotics; 15 patients were surgically drained; and 19 cases had antibiotics alone or needle aspiration. Length of hospital stay was 8.2 days in the medical group and 11.6 days in the surgical group. One patient in the medical group developed mediastinitis. The conclusion was that parapharyngeal infections may localize and be effectively treated without drainage (Oh et al. 2007). However, in a review of 205 children with lateral neck infections, it was found that the clinical diagnosis corresponded with the radiological findings in only 73.6% of cases and ultrasound with the surgical findings in only 65.2%. It was also found that clinical assessment of lateral neck infections was poor underestimating suppuration, although the diagnosis of an abscess clinically correlated highly with surgical findings (Courtney et al. 2007). Therefore, diagnosis and selection of patients for conservative treatment or drainage is fraught with inaccuracy.

In peritonsillar abscesses catheter or needle drainage is the method of choice and ancillary steroids may reduce morbidity (Herzon and Martin 2006). Image-guided aspiration using either CT or ultrasound has been applied to deep neck infections. Poe et al. reported the use of CT-guided aspiration in a small series of 4 cases without complications (Poe et al. 1996). In 15 cases of unilocular neck abscess, 13 were successfully treated with needle or catheter aspiration, 2 of whom required re-aspiration. Two patients failed and required surgical drainage (Yeow et al. 2001). In a prospective controlled study of 14 patients with well-defined and/or unilocular abscesses, all were

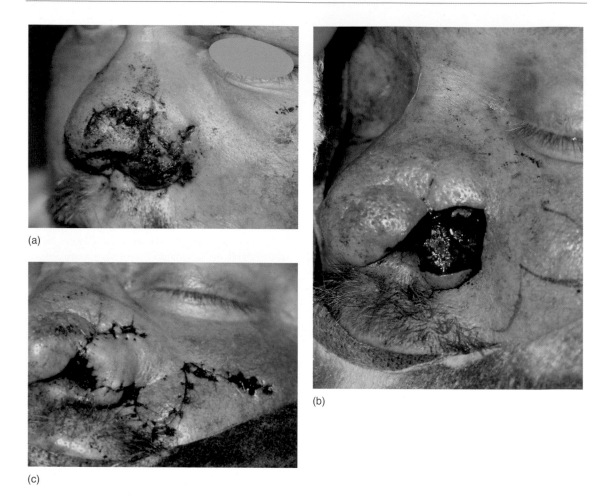

**Fig. 10.14**   (a) Alar defect of infected wound; (b) wound thoroughly debrided and a bilobed local flap marked; (c) rotation and closure of local flap.

successfully treated by ultrasound-guided drainage, and 8 patients had an indwelling catheter (Chang et al. 2005). There are a few trials that examine this method for space infections from purely odontogenic sources. One report described that the surgeon undertook aspiration drainage under ultrasound guidance of 11 patients with a submasseteric space infection. Eight of 11 (73%) required no surgical drainage. Clinically when mouth opening improved to 1.5 cm or more following aspiration drainage, cure was more likely (Al-Belasy 2005).

However, in most severe deep neck space infections from odontogenic causes, abscesses are not well defined and may occupy more than one space, and cellulitic processes are poorly confined with impending airway problems. In these circumstances aggressive drainage using large incision to allow adequate exploration of all spaces and breakdown of loculations with digital examination is preferred. Prior to drainage, needle aspiration may be performed to obtain a pus sample for culture and sensitivity in patients whom this information may be necessary, for example, immunocompromised patients. The principles of drainage include placing incisions in noninvolved skin in a site that allows dependant drainage and avoids important anatomic structures, for example, facial nerve branches. Blunt dissection by opening hemostats or finger dissection will break down loculi and allow wide

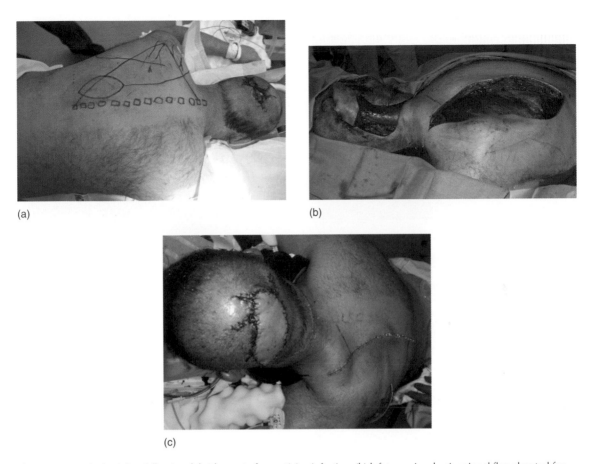

(a)

(b)

(c)

**Fig. 10.15**  (a) Scalp defect following debridement of necrotizing infection; (b) left trapezius dorsi regional flap elevated for scalp reconstruction; (c) inset flap reconstructing defect.

exploration of all spaces. Latex or rubber drains are placed to keep the wound open for drainage and to allow irrigation. In cases of cellulitis large incision to open the spaces widely and exploration of adjacent spaces with the placement of multiple drains is essential although cellulitic edema fluid is usually seen with little frank pus. Postdrainage there will initially be increased edema, and endotracheal tubes should not be removed when airway compromise is an issue. Wide drainage of these spaces will reduce the chance of missing loculations or involved spaces and the need for subsequent reoperation for patients whose fever and swelling are not resolving. An alternative approach with small incisions for Ludwig's angina in 113 patients has been published. In 62 there was extension to the parapharyngeal space and 32 had retropharyn-geal involvement. In this series, 33 patients had major complications such as mediastinitis, sepsis, or death, but the authors concluded that drainage using small incisions was safe and effective in Ludwig's angina (Bross-Soriano et al. 2004).

In necrotizing fasciitis aggressive surgical debridement is essential and delay is associated with an increased death rate. The area of reddened skin is usually delineated with a surgical marking pen so that postsurgical progression can be followed. The skin is incised and usually no bleeding is observed due to blood vessel thrombosis. The underlying fascia is necrotic and "dish water" pus is classic, although foul-smelling pus and gas may be obtained. The skin and fascia are excised radically until bleeding skin edges are observed. The underlying muscle is

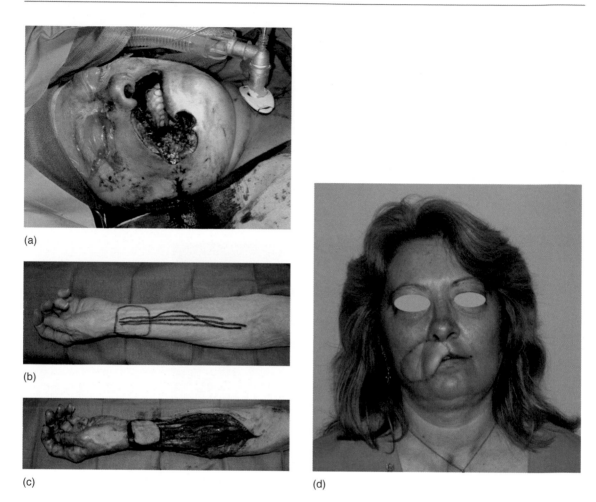

(a)

(b)

(c)

(d)

**Fig. 10.16**    (a) Upper lip debridement after a necrotizing infection; (b) example of a radial forearm free flap with the skin paddle based over the radial artery; (c) example of the raised flap prior to inset; (d) one month after upper lip reconstruction.

usually uninvolved; however, if it is involved, it is also vigorously debrided. The wound is packed open and irrigated with hydrogen peroxide and saline, with frequent dressing changes. Further visits to the OR are done daily as the disease declares itself and more necrotic skin is excised. The surgeon will usually have multiple OR sessions before the disease is stabilized. It is usually recommended waiting 7–10 days until healthy granulation tissue is present prior to reconstruction. If available, hyperbaric oxygen may be helpful in these cases, but aggressive surgical management is undoubtedly the primary treatment.

As previously stated, reconstruction is not considered until after the wounds are stabilized with signs of a healthy granulation tissue bed. In addition, these patients are typically debilitated as a result of sepsis, poor nutritional status, and diminished immunologic response. Sequencing reconstructive strategies for patients with aggressive head and neck infections (i.e., necrotizing fasciitis) can be a monumental challenge. These wounds can present with extensive tissue loss following aggressive debridement, and, therefore, may require composite (skin, muscle, bone, etc.) tissue replacement. The challenges of reconstruction for these aggressive head and neck infections include compromised airway support, impaired sensory and motor control, facial deformity, inability to control secretions, and impaired speech. The goals of reconstruction are similar to those in oncological head and neck surgery, which are restoration of function and form. In order to achieve these goals, the surgeon must understand the impact of the reconstruction ladder on the surgical and prosthetic phases of reconstruction. The stages ("rungs") of the ladder include primary closure, followed by skin grafts, local flaps, distant flaps, and finally free flaps. A true comprehension of the architectural hard and soft tissue defects helps the surgeon in developing a strategic approach to these difficult and challenging reconstructions. The differences between a graft and flap are based on the donor tissues blood supply; a graft depends on the recipient beds blood supply and lacks its own. A flap is independent of the recipient beds blood supply and carries its own. There are different types of flaps available for utilization to the surgeon. These depend on the size and complexity of the defect and the underlying comorbidities of the patient. A local flap transposes healthy tissue based on a *random* blood supply to the defect (Figs 10.14a–c); the regional flap rotates distant tissue based on an *axial* blood supply (blood vessel) to the defect site (Figs 10.15a–c); lastly the free flap transfers distant tissue based on an axial blood supply by the donor vessels are intentionally transected and then anastomosed to vessels in the recipient bed to restore blood flow (Figs 10.16a–d). The advantage of free tissue transfer it offers is a unique combination of vascularized tissues (bone, muscle, skin) in a single stage. The initial sequencing of reconstruction is focused on restoring oral lining, followed by soft tissue coverage of the face and then skeletal support. Prior to implementing any of the reconstruction options available, it is important the patient is medically stabilized; also, all the necrotic and infected tissues must be debrided with no signs of progression. Since some of these cases can result in a large surface area of skin loss, they are reconstructed analogous to a burn patient, with multiple skin-grafting procedures. However, other patients may have composite tissue loss and are suitable candidates for free tissue transfer. There is a population of patients that may not be suitable candidates for a free flap surgery due to medical comorbidities, and are more suitable for regional flaps. Because each option has its own inherent advantages and disadvantages, the surgeon can employ a combination of each

in order to achieve the best reconstruction possible. Some of these cases can be extraordinarily complex, and require well thought-out planning and sequencing of treatment.

## 10.8 References

Al-Belasy FA. 2005. Ultrasound-guided drainage of submasseteric space abscesses. *J Oral Maxillofac Surg* 63(1): 36–41.

Brook I and Frazier EH. 1995. Clinical and microbiological features of necrotizing fasciitis. *J Clin Microbial* 33: 2382–87.

Bross-Soriano D, Arrieta-Gomez JR, Prado-Valleros HS, chimelmitz-Idi J, and Jorba-Basave S. 2004. *Otolaryngol Head Neck Surg* 130(6): 712–17.

Chang KP, Chen YL, Hao SP, and Chen SM. 2005. Ultrasound-guided closed drainage for abscesses of the head and neck. *Otolaryngol Head Neck Surg* 132(1): 119–24.

Courtney MJ, Miteff A, and Mahadevan M. 2007. Management of pediatric lateral neck infections: Does the adage "... never let the sun go down on undrained pus..." hold true? *Int J Pediatr Otorhinolaryngol* 71(1): 95–100.

Deans M. 1994. Flesh eating bug scare. *Lancet* 343: 1418.

Dodson TB, Barton JA, and Kaban LB. 1991. Predictors of outcome in children hospitalized with maxillofacial infections a linear logistic model. *J Oral Maxillofac Surg* 49: 838–42.

Dodson TB, Perrot DH, and Kaban LB. 1989. Pediatric maxillofacial infections: A retrospective study of 113 patients. *J Oral Maxillofac Surg* 47: 327–30.

Duque CS, Guerra L, and Roy S. 2006. Use of intraoperative ultrasound for localizing difficult parapharyngeal space abscesses in children. *Int J Pediatr Otolaryngol* 71(3): 375–78.

Flynn TR. 1991. Odontogenic infections. *Oral Maxillofac Clin North Amer* 3: 311–29.

Flynn TR, Shanti RB, and Hayes C. 2006a. Severe odontogenic infections, Part 2: Prospective outcomes study. *J Oral Maxillofac Surg* 64: 1104–13.

Flynn TR, Shanti RB, Levi MH, Adams AK, Kraut RA, and Trieger N. 2006b. Severe odontogenic infections, Part 1: Prospective report. *J Oral Maxillofacial Surg* 64: 1093–1103.

Haug RH, Wible RT, and Lieberman J. 1991. Measurement standards for the prevertebral region in the lateral soft-tissue radiographs of the neck. *J Maxillofac Surg* 49: 1149–56.

Herzon FS and Martin AD. 2006. Medical and surgical treatment of peritonsillar, retropharyngeal, and parapharyngeal abscesses. *Curr Infect Dis Rep* 8(3): 196–202.

Huang T-T, Tseng F-Y, Liu T-C, Hsu C-J, and Chen Y-S. 2005. Deep neck infections in diabetic patients: Comparison of the picture and outcomes with nondiabetic patients. *Otolaryngol Head Neck Surg* 132: 943–47.

Jones JL and Cadelaria LM. 2000. Head and neck infections. In: Fonseca R (ed.), *Oral and Maxillofacial Infections in Oral and Maxillofacial Surgery*. WB Saunders Co., 81.

Kaul R, Mcgeer A, Low DE, Green K, Schwartz AE, and Simar AE. 1997. Population based surveillance for group A Streptococcal necrotizing fasciitis. *Am J Med* 103: 18–24.

Kuriyama T, Karasawa T, Nakagawa K, Saiki Y, Yamamoto E, and Nakamura S. 2000. Bacteriologic features and anti microbial susceptibility in isolates from odontogenic infections. *Oral Surg Oral Med Oral Path Oral Radiol Endod* 90: 600–608.

Lazor JB, Cunningham MJ, Eavey RD, and Weber AL. 1994. Comparison of computed tomography and surgical findings in deep neck infections. *Otolaryngol Head Neck Surg* 111: 746–50.

Lin C, Yeh F-L, Lin J-T, Ma H, Hwang C-H, Shen B-H, and Fang R-H. 2001. Necrotizing fascitis of the head and neck an analysis of 47 cases. *Plast Reconstr Surg* 107: 1684–93.

Mayor GP, Millán J, and Martínez-Vidal A. 2001. Is conservative treatment of deep space infections appropriate. *Head Neck* 23: 126–33.

McGurk M. 2003. Diagnosis and treatment of necrotizing fasciitis in the head and neck region. *Oral Maxillofac Surg Clin N Amer* 15: 59–67.

Miller WD, Furst IM, Sandor GK, and Keller MA. 1999. A prospective blinded comparison of clinical examination and computed tomography in deep neck infections. *Laryngoscope* 109: 1873–79.

Oh JH, Kim Y, and Kim CH. 2007. Parapharyngeal abscess: Comprehensive management protocol. *ORL J Otorhinolaryngol Relat Spec* 69(1): 37–42.

Ovassapian A, Tuncbilek M, Weitzel EK, and Joshi CW. 2005. Airway management in adult patients with deep neck infections: A case series and review of the literature. *Anesth Analg* 100(2): 585–89.

Paster BJ, Falkler, Jr, WA, Jr, Enwonwu CO, Idigbe EO, Savage KO, Levanos VA, Tamer MA, Ericson RL, Lau CN, and Dewhirst FE. 2002. Prevalent bacterial species and novel phylotypes in advanced noma lesions. *J Clin Microbiol* 40(6): 2187–91.

Peters ES, Fong B, Wormuth DW, and Sonis ST. 1996. Risk factors affecting hospital length of stay in patients with odontogenic maxillofacial infections. *J Oral Maxillofac Surg* 54: 1386–91.

Poe LB, Petro GR, and Matta I. 1996. Percutaneous CT-guided aspiration of deep neck abscesses. *Am J Neuroradiol* 17: 1359–63.

Potter JS, Herford AS, and Ellis E. 2002. Tracheotomy versus endotracheal intubation for airway management in deep neck space infections. *J Oral Maxillofac Surg* 60: 349–54.

Riggio MP, Aga H, Murray CA, Jackson MS, Lennon A, Hammersley N, and Bagg J. 2007. Identification of bacteria associated with spreading odontogenic infections by 16 S rRNA gene sequencing. *Oral Surg Oral Med Oral Pathol Oral Radiol Endod* 103(5): 610–17.

Sakamoto H, Kato H, Sato T, and Saski J. 1998. Semiquantitive bacteriology of closed odontogenic abscesses. *Bull Tokyo Dental Coll* 39: 103–7.

Sichel JY, Dano I, Hocwald E, Biron A, and Eliasher R. 2002. Nonsurgical management of parapharyngeal space infections: A prospective study. *Laryngoscope* 112(5): 906–10.

Smith JL II, Hsu JM, and Chang J. 2006. Predicting deep neck space abscess using computed tomography. *Am J Otolaryngol* 27(4): 244–47.

Stephanopoulus PK and Kolokotronis AE. 2004. The clinical significance of anaerobic bacteria in acute orofacial odontogenic infections. *Oral Surg Oral Med Oral Path Oral Radiol Endod* 98: 398–403.

Wang J, Ahani A, and Pogrel MA. 2005. A five year retrospective study of odontogenic maxillofacial infections in a large urban public hospital. *Int J Oral Maxillofac Surg* 34: 646–49.

Yeow K-M, Liao C-T, and Hao S-T. 2001. US-guided needle aspiration and catheter drainage as an alternative to open surgical drainage for uniloculated neck abscesses. *J Vasc Interv Radiol* 12: 585–94.

# Chapter 11

# Endodontic Infections and Pain

*Anibal Diogenes and Ken M. Hargreaves*

## 11.1 Introduction

Odontogenic pain is a prevalent condition. Epidemiological studies indicate that dental pain is reported by about 12–14% of the population, which translates to tens of millions of patients in the United States alone (Locker and Grushka 1987; Lipton et al. 1993). Since odontalgia is often directly related to microbial infection and resulting tissue inflammation, this indicates that one of the most common forms of acute pain are actually a consequence of endodontic infections. Research conducted in the last several decades has greatly increased our understanding of the pain system, and it is now widely appreciated to be a highly dynamic sensory system that rapidly alters its response properties due to conditions such as inflammation due to bacterial infection. Increased knowledge of the pain system and its properties contributes to improved clinical skills for diagnosis and treatment of odontogenic pain. In this chapter, we review the pain system and its interaction with microbial and inflammatory factors and use this knowledge base to make evidence-based recommendations for managing endodontic-related pain.

Although many patients may view pain from a binary perspective ("it hurts" vs "it doesn't hurt"), it is important to realize that this sensory system undergoes substantial dynamic alterations following tissue injury. This neuronal plasticity is manifested in three major pain symptoms: allodynia, hyperalgesia, and spontaneous pain. Allodynia is defined as a reduction in pain threshold where normally nonpainful stimuli can now elicit pain. Mechanical allodynia is a reduction in mechanical pain thresholds and classically is evaluated by percussing a tooth with a mirror handle. Studies of nearly 1,000 patients indicate that mechanical allodynia can occur in 57% of patients with a diagnosis of irreversible pulpitis (Owatz et al. 2007), indicating that this is an early feature of pulpal inflammation. New technologies for diagnosing mechanical allodynia have been introduced, which provide an actual measure of mechanical allodynia in painful teeth (i.e., in newtons of force) rather than the +/– outcome of the percussion test (Khan et al. 2007). Hyperalgesia is defined as an increase in the perceived magnitude of a noxious stimulus; in other words, a stimulus that is painful when applied to healthy tissue is perceived as being much more painful under conditions of hyperalgesia. This is classically evaluated by the response to cold stimulation of teeth, where the response becomes much more exaggerated for a diseased tooth than a normal tooth.

## 11.2 Biology of the pain system

Historically, many patients harbor a negative association between dentistry and pain. Moreover, clinicians often face the challenge of managing pain and suffering that occurs in their patients. It should be

213

appreciated that the oral cavity is exquisitely innervated by pain-sensing afferent neurons (nociceptors). Indeed, most studies indicate that all physiological stimuli applied to dental pulp results in a sensation of pain. Moreover, there is a large area of the cerebral cortex solely dedicated to interpreting sensory inputs from the orofacial structures (Penfield 1950).

Pain is a complex sensation with three major levels of possible interventions. First, it is initiated by activation, depolarization, and conduction of action potentials by nociceptor neurons. Second, the action potentials reach the trigeminal nucleus where considerable information processing and modulation occurs. Third, these action potentials are relayed to the thalamus and ultimately to the cerebral cortex where they are perceived as pain, with both affective (e.g., suffering) and sensory intensity components. Increased understanding of the pain system provides considerable clinical insight into strategies for adequate treatment. In this chapter, we address pain system and its relation to odontogenic infections.

Dental pulp and periradicular tissues are innervated primarily by sensory neurons, originating from the trigeminal ganglia where their cell bodies reside. Each neuronal cell body has one axon that divides into a peripheral projection, where it terminates into free nerve endings innervating a target tissue (e.g., pulp) and a central projection to the trigeminal brain stem nuclei complex, where it synapses with second order or projection neurons. These projection neurons ascend through the contralateral trigeminothalamic tract and synapse with higher-order neurons in the thalamic with final neuronal projections into the sensory cerebral cortex (Penfield 1950). Although the dental structure is one of the most heavily innervated tissues in the body with pain-sensing neurons (nociceptors), not all neurons in the dental pulp and periodontal ligament are nociceptors. There are postganglionic sympathetic fibers from the cervical sympathetic ganglia and other afferent low-threshold (A-β) trigeminal fibers responsible for detecting non-noxious tactile information (Hargreaves and Goodis 2002).

Nociceptive afferent nerve fibers in dental pulp and periodontal ligament (PDL) are normally composed of C-fibers (25–50%) and A-δ (25%) (Byers 1984; Mengel et al. 1992, 1993). Therefore, the dental pulp and PDL are mainly innervated by nociceptors and to a lesser extent by A-β and sympathetic fibers. The C-fibers and A-δ fibers innervating the dental pulp and PDL are typically sensitive to thermal, mechanical, and chemical stimuli (polymodal nociceptors). There are significant differences between C-fiber and A-δ nociceptors regarding anatomy (neuronal morphology and distribution) and conduction velocity.

Electrophysiological studies measuring the conduction velocity of sensory neurons form the basis of widely used nomenclature of these fibers. The C-fibers have a slower conduction velocity (<2 m/s) than A-δ fibers (≥2 m/s conduction velocity). Anatomically, C-fibers are unmyelinated and originate from small neurons and, in contrast, A-δ fibers are lightly myelinated and originate from small- to medium-diameter cell bodies in one of the three divisions of the trigeminal ganglion. Myelination provides axonal insulation, allowing action potentials to propagate faster in A-δ fibers. Although trigeminal neurons project their central terminals to different brain stem nuclei (the nuclei interpolaris, oralis, and caudalis), much of the nociceptive input is thought to be located in the outer layers of the nucleus caudalis (Dubner and Bennett 1983; Hargreaves and Milam 2001; Sessle 2005). This nucleus is also known as medullary dorsal horn due to its anatomical similarity to spinal dorsal horn. There is a subpopulation of afferent nociceptive neurons that are called peptidergic neurons because they express important neuropeptides such as calcitonin gene-related peptide (CGRP) and substance P. The role of these neuropeptides in the modulation of the nociceptive signaling is discussed below.

In the dental pulp, the greatest density of innervation is found in the coronal aspect at the level of the pulpal horns (Byers 1984). At these regions, nociceptors are presented as intertwined ramified free nerve endings forming a plexus structure (Luthman et al. 1992). In general, nociceptive innervation density decreases in the apical direction. A-δ fibers are known to traverse the odontoblastic layer, lose their myelination, and project approximately 0.1–0.2 mm far into the dentinal tubules, whereas C-fiber innervation is found deeper in the dental pulp. The innervation of the dentinal tubules by A-δ nociceptive fibers agrees qualitatively with the sharp and transient pain usually originated when the dentin is stimulated (e.g., dentinal sensitivity). According to the hydrodynamic theory, this pain is the result of fluid moving in the dentinal tubules due to thermal (cold drinks or dichlorodifluoromethane application during pulp vitality examination), drilling (during cavity preparation), or hypertonic (sweets) stimuli (Brannstrom and Johnson 1978). This theory has been substantiated by several studies showing that the movement of fluid in the tubules generates discharges in nociceptors and that closing of the dentinal tubules confers resistance to those noxious dentinal stimuli (Ahlquist et al. 1994; Andrew and Matthews 2000; Charoenlarp et al. 2007).

Besides the movement of fluid through the dentinal tubules and subsequent activation of mechanosensitive fibers, transdentinal cooling or heating could potentially activate C-fibers located deeper in the dental pulp. In order for this to happen in uninjured tissue, the temperature gradient needs to diffuse from the mineralized tissues and reach C-fibers at a noxious level sufficient for activating these high-threshold fibers. Indeed, in an experimental setting, heating of the tooth generated a sharp pain that subsided (consistent with A-δ fiber activation) followed by a pain characterized by a slow, burning pain (consistent with C-fiber activation) (Jyvasjarvi and Kniffki 1987). Importantly, a subset of C-fibers in the dental pulp expresses the transient receptor potential vanilloid type-1 channel (TRPV1), an ionotropic channel known to be activated by noxious temperatures ($>43°C$), low pH ($<5.5$), and selectively by capsaicin, the pungent ingredient in chilli peppers (Ikeda et al. 1997; Caterina et al. 1999; Morgan et al. 2005). In human subjects intradermal injection of capsaicin evoked a pain characterized as throbbing (slow) burning pain (Simone et al. 1989). Although C-fiber nociceptors likely encode this deep pulpal thermal nociception, perhaps their most important role is detection of inflammatory pain.

A myriad of inflammatory mediators are released on tissue damage. These inflammatory mediators are usually released in a cascade that comprises an "inflammatory soup" (Julius and Basbaum 2001; Cunha et al. 2005). Nociceptors express an array of receptors for inflammatory mediators such as the B2 receptor for bradykinin, TrkA receptor for nerve growth factor (NGF), prostaglandin receptors, and others (Julius and Basbaum 2001). Whenever the concentration of any given mediator is sufficient to bind and activate its receptors in the nociceptors, intracellular signaling pathways are triggered. Many intracellular signaling pathways are known to sensitize or activate nociceptors. For example, local elevation of prostaglandin E2 activates protein kinase A, leading to nociceptor sensitization, while increased tissue amounts of bradykinin activates phospholipase C and protein kinase C, leading to nociceptor activation (e.g., depolarization) (Nicol et al. 1992; Cesare et al. 1999). Some of these intracellular signaling pathways are involved in

phosphorylating receptors such as TRPV1, resulting in a decreased activation threshold (manifested as allodynia) or an augmented response to a suprathreshold stimulus (contributing to hyperalgesia). For example, bradykinin and prolactin trigger phosphorylation of TRPV1 reducing its temperature of activation from $43°C$ to as low as $36°C$ (Cesare et al. 1999; Diogenes et al. 2006). Thus, allodynia can be produced by inflammatory mediators that sensitize nociceptors. This reduced activation threshold could explain how minor, innocuous transdentinal heating of inflamed pulp might still activate TRPV1, leading to pain perception (Ahlberg 1978). More importantly, the lowering of the temperature of activation of TRPV1 to less than $37°C$ may lead to continuous activation of these channels contributing to spontaneous pain. Interestingly, the dull, throbbing pain sensation typically associated with activation of C nociceptors is often reported by patients with irreversible pulpitis. In many of these patients, cooling the tooth (i.e., drinking ice water) appears to promote a transient alleviation of the symptoms. It can be hypothesized that this alleviation could be due to the cooling of the pulp to temperatures below the threshold for activation of the sensitized TRPV1 receptor.

## 11.3 Mechanisms of pain due to endodontic infections

Odontogenic infection is one of the most prevalent infectious diseases that affect humans. When bacteria gain access to the pulp, a significant inflammatory response is initiated. As discussed elsewhere in this text, bacteria can induce tissue injury and inflammation through three general pathways. First, bacteria can release tissue-modifying agents such as enzymes capable of degrading host tissue elements (Sedgley et al. 2005; Reynaud af Geijersstam et al. 2007). Second, bacteria can stimulate the innate immune system by activation of pattern recognition receptors such as the toll-like receptors (TLR) (O'Neill 2004; Wadachi and Hargreaves 2006). And third, chronic bacterial infections can lead to activation of specific or adaptive immune response (Baumgartner and Falkler 1991).

These three pathways could lead to nociceptor activation by either indirect or direct mechanisms. First, it is known that bacteria trigger local host tissue responses including release of prostaglandins and cytokines, factors well recognized to sensitize or activate

**Table 11.1**   The toll-like receptor (TLR) family: An innate immune system capable of detecting microbial substances.

| TLR | Foreign substance |
| --- | --- |
| TLR-1/2 | Bacterial lipopeptides |
| TLR-2 | Bacterial lipopeptides, lipoteichoic acid from Gram-positive bacteria, zymosan from yeast |
| TLR-3 | Double-stranded RNA from viruses |
| TLR-4 | Endotoxin (lipopolysaccharide) from Gram-negative bacteria |
| TLR-5 | Flagellin |
| TLR-7/8 | Uridine-rich single-stranded RNA from viruses |
| TLR-9 | CpG DNA |
| TLR-10 | Unknown |
| TLR-11 | Uropathogenic bacteria |
| TLR-12/13 | Unknown |

Modified from Wickelgren I. 2006. *Science* 312: 186.

nociceptors via a paracrine action (Nakanishi et al. 1995; Hahn et al. 2000). Thus, certain strains of bacteria might indirectly stimulate nociceptors via release of host factors. However, recent research has suggested an alternative direct pathway by which bacterial-derived substances activate TLRs expressed on nociceptive neurons. The TLR family of receptors represents an ancient family of proteins capable of detecting a broad range of microbial substances (Table 11.1). Given the prevalence of Gram-negative organisms in odontogenic infections and clinical studies reporting an association with pain (Hashioka et al. 1992; Siqueira et al. 2004), we explored whether TLR-4 was expressed on trigeminal nociceptors (Wadachi and Hargreaves 2006).

The TLR-4 receptor detects lipopolysaccharides (LPS) derived from the surface of Gram-negative bacteria (Poltorak et al. 1998; Qi and Shelhamer 2005). As illustrated in Fig. 11.1, the activation of TLR4 is greatly enhanced by the participation of other membrane-anchored receptors that lack an intracellular domain, namely myeloid differentiation-2 receptor (MD2) and cluster differentiation-14 receptor (CD14). These two accessory proteins bind to LPS but are unable to trigger any intracellular signaling since they lack an intracellular domain. However, the MD2 receptor binds to LPS and to an extracellular domain in the TLR4 receptors itself. The formation of a TLR4/MD2 dimer and its LPS binding is essential for LPS signaling in immune cells. This interaction is believed to promote TLR4 activation by LPS and stabilize the molecular conformational change of the activated LPS/TLR4/MD2 complex (Fitzgerald

et al. 2004). On the other hand, the CD14 receptor has the capacity to bind to multiple LPS molecules, increasing the concentration of the ligand that is then presented to TLR4. Therefore, the formation of an LPS/CD14/TLR4/MD2 complex has been shown to evoke greater cellular responses than an LPS/TLR4/MD2 interaction alone. It is important to note that soluble LPS in the extracellular space is often bound to an LPS-binding protein (LPSB), which serves as a carrier protein, allowing better interaction of LPS with any of the members of the MD2/CD14/TLR4 complex. Thus, there are several extracellular proteins that serve to greatly amplify the ability of host cells to detect the presence of LPS.

LPS at very low concentrations promotes maximal activation of immune cells, particularly macrophages

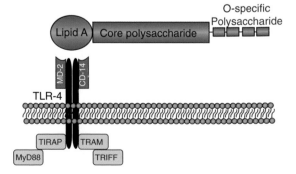

**Fig. 11.1**   Schematic illustration of the structure of lipopolysaccharides (LPS) and the pattern recognition receptor, TLR-4, which detects LPS and its associated intracellular signaling molecules.

and neutrophils. In these cells, LPS evokes an NFκB-dependent gene upregulation and release of cytokines such as tumor necrosis factor-α (TNF-α) and interleukin-1α, and chemotactic agents such as leukotriene B4 (Doyle and O'Neill 2006). On the release of these cytokines, the inflammatory reaction amplifies rapidly as infiltrating immune cells start releasing more inflammatory mediators that act in concert to promote the cardinal signs of inflammation: edema and pain.

As mentioned previously, nociceptors express the receptors for numerous inflammatory mediators, and activation of these receptors leads to the sensitization of these pain-sensing fibers (Byers et al. 1990b; Fried and Risling 1991; Goodis et al. 2000; Chang et al. 2006). The concentration of inflammatory mediators such as bradykinin, interleukin-1α, and NGF have been found to be dramatically increased in inflamed dental pulp as compared to asymptomatic control dental pulp (Nakanishi et al. 1995; Shimauchi et al. 1997; Wheeler et al. 1998; Lepinski et al. 2000). Some of these mediators such as bradykinin and prostaglandins evoke an acute sensitization of nociceptors by triggering local intracellular events (e.g., phosphorylation of channels).

Other mediators such as NGF bind to their receptors and are retrogradely transported from the periphery to the cell body located in the trigeminal ganglion. In the cell body NGF triggers the transcription of genes for ionotropic channels (e.g., transient receptor potential A1, TRPA1) and G–protein-coupled receptors that are anterogradely transported back to the periphery where they will be involved in an exacerbated nociceptive signaling (Obata et al. 2005; Diogenes et al. 2007a). Also, NGF promotes neuronal sprouting at the site of injury, increasing the local density of nociceptors (Byers et al. 1990a). This process requires hours to days to occur and has long-lasting effects. In fact, peripherally administered NGF is known to generate hyperalgesia and allodynia that last days to months in both humans and experimental animals (McMahon 1996; Mendell et al. 1999).

The activation of a peptidergic subset of nociceptors is often accompanied with a vasodilation reaction (Simone et al. 1989). This occurs because these neurons release vasoactive neuropeptides such as CGRP and substance P when depolarized. Certainly, the enhanced activation of these nociceptors in inflammatory conditions leads to neuropeptide release. For example, elevated levels of substance P have been found in the pulp of patients with irreversible pulpitis (Bowles et al. 2003). CGRP and substance P promote vasodilation and plasma extravasation, respectively, thus participating in the inflammatory process. The mechanism for neuronal-generated inflammation is called neurogenic inflammation and is often intertwined with classic inflammatory processes due to a positive feedback loop. Therefore, it is hard to determine which event physiologically started first. Although inflammation and neurogenic inflammation have different etiologies, they are often present together as part of an overall amplifying inflammatory response.

It has been suggested that odontoblasts play a monitoring and functional role in pulpal inflammation since they represent the first biological barrier, between mineralized tissues and the dental pulp, encountered by pathogens. Odontoblasts are highly specialized cells that express a "molecular repertoire" far greater than that needed for simple dentin production and deposition. Human odontoblasts express TLR4 and TLR2 known to be classically expressed in immune cells and therefore recognize cell wall components from Gram-negative and Gram-positive bacteria, respectively (Veerayutthwilai et al. 2007). The activation of TLR2 and TLR4 in human odontoblasts results in the upregulation of soluble factors such as TNF-α and IL-1β (Veerayutthwilai et al. 2007). Therefore, bacteria invading the pulp are most likely first detected by odontoblasts via activation of TLR4 and TLR2 receptors, with subsequent paracrine communication to nearby cells, including immune cells. Together, these cells then release inflammatory mediators leading to the indirect sensitization of nociceptors by bacteria (Fig. 11.2). Although a direct communication of odontoblasts and neurons has not yet been shown, it is curious that odontoblasts can induce membrane depolarization much like neurons. They express the TRPV1 ("capsaicin") receptor (Okumura et al. 2005) and tetrodotoxin-resistant sodium channels (Allard et al. 2006) that typically are expressed only in a nociceptive subset of sensory neurons. Thus, it is possible that stimuli such as noxious heat (via TRPV1) or LPS (via TLR4) directly excite odontoblasts, which in turn relay the excitation to adjacent sensory neurons. This is an important and exciting new question deserving additional research.

As mentioned above, the sensitization of nociceptors in infections is classically believed to be due to the release of inflammatory mediators from

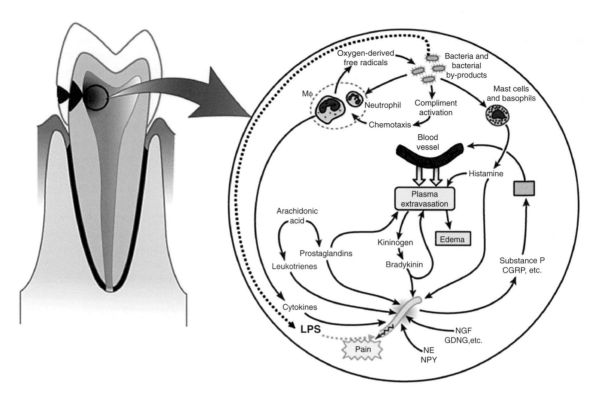

**Fig. 11.2** Schematic illustration of direct and indirect mechanisms by which bacteria can activate pulpal or periradicular nociceptors. (Modified and published with permission from Owatz et al. 2007.)

bacteria-activated immune cells. However, Wadachi and Hargreaves (2006) demonstrated, for the first time, that trigeminal nociceptors in the human dental pulp express both TLR4 and CD14 (Fig. 11.3). This novel finding raises the possibility that sensory neurons are able to "sense" the presence of Gram-negative bacteria.

More recent studies from our laboratory have demonstrated that LPS derived from *Porphyromonas gingivalis*, which is a bacterial species known to be involved in odontogenic infections, directly sensitizes TRPV1 responses in cultured rat trigeminal neurons (Diogenes et al. 2007b; Ferraz et al. 2007). We found that exposure of trigeminal neurons in culture to LPS acutely sensitized nociceptors significantly increasing the release of the neuropeptide CGRP when neurons were stimulated with capsaicin. These data have exciting clinical significance, since deeper carious lesions and primary periapical lesions are predominantly associated with Gram-negative bacteria. Moreover,

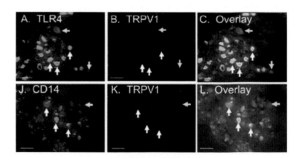

**Fig. 11.3** Evaluation of the expression patterns of TLR4 and CD14 in human trigeminal sensory neurons. White arrows depict examples of neurons expressing both markers for each row of three images, and yellow arrows depict examples of neurons that express one but not both markers. Human trigeminal neurons were evaluated for colocalization of TLR4 (Panel A), CD14 (Panel J), with a marker for the capsaicin-sensitive subclass of nociceptors (TRPV1, Panels B and C for TLR4, and Panels K and L for CD14). (Reproduced by permission from Wadachi and Hargreaves 2006.)

elevated LPS concentrations in infected root canal systems have been positively correlated with painful presentations of periradicular periodontitis (Jacinto et al. 2003).

The direct sensitization of nociceptors by LPS may explain why some immunocompromised patients also present devastatingly painful infections (Epstein 1988; Navarro et al. 1998). In these patients, there are decreased levels of inflammatory mediators, which normally would result in attenuated inflammatory pain. However, bacteria grow unchecked in these patients, elevating local LPS levels with the potential to sensitize nociceptors in the infected area. Another clinical consideration is that following endodontic treatment, LPS from dead bacteria needs to be adequately removed or neutralized since there is the potential for sensitization of nociceptors in the periradicular area of infection. More research is needed to investigate this novel bacteria–neuronal communication and the clinical implications of this direct sensitization of nociceptors by LPS.

The central circuitry at the level of nucleus caudalis is far from being a simple static "connector" that relays the original burst of action potential from the peripheral neuron to the cerebral cortex. Instead, there is significant processing of the nociceptive signal at this level. A process called "central sensitization" occurs when the primary afferent signal is amplified at this central terminal. Central sensitization represents a major component in inflammatory hyperalgesia and allodynia. Central sensitization is triggered by an increase in nociceptive peripheral afferent barrage (e.g., augmented responses due to inflammation or nerve injury). This constant primary afferent barrage evokes the release of neuropeptides (e.g., substance P) and neurotransmitters (e.g., glutamate) from central terminals of these neurons.

Glutamate activation of the glutamate NMDA (*N*-methyl-D-aspartate) receptor is an important mechanism in central sensitization. Interventions at this receptor in the central terminals induce either reduction or enhancement of central sensitization generated by repeated C-fiber discharge. For example, an NMDA antagonist reduces central sensitization, whereas brain-derived neurotrophic factor facilitates NMDA activation, thereby enhancing central sensitization (Woolf and Thompson 1991; Kerr et al. 1999).

Clinically, central sensitization can be observed in patients complaining of persistent pain even after the removal of inflamed peripheral tissue with endodontic

and restorative procedures. Studies have shown that the presence of preoperative pain increases the risk of postoperative pain (Torabinejad et al. 1994). Furthermore, preemptive analgesia during surgical procedures is associated with reduced postoperative pain in many situations (Woolf and Chong 1993; Reuben 2007).

## 11.4 Clinical strategies for treating pain due to endodontic infections

The management of the patient with pain due to an odontogenic infection involves both pharmacological and nonpharmacological therapies. Appropriate pharmacotherapy includes drugs that reduce pain perception or microbial infection. Pain perception can be reduced by the administration of local anesthetics or the use of analgesics. It is well recognized by clinicians that odontogenic pain patients often experience considerably reduced efficacy of local anesthetics, particularly in cases involving painful mandibular teeth. It has been speculated that this might be due, in part, to direct effects of prostaglandins on sensitizing neuronal voltage-gated sodium channels via activation of protein kinase A and subsequent protein phosphorylation (Hargreaves and Keiser 2002). This hypothesis suggests that preoperative administration of nonsteroidal anti-inflammatory drugs (NSAIDs) in patients who can tolerate this drug class might actually increase the efficacy of local anesthetics. Interestingly, two clinical trials have recently reported that oral analgesics do tend to increase the success of local anesthetics in odontogenic pain patients (Modaresi et al. 2006; Ianiro et al. 2007). In addition, several randomized clinical trials have indicated that the efficacy of anesthetizing painful mandibular teeth can be increased about twofold (from ~40 to ~85% of patients) by combining conventional inferior alveolar nerve block injection with intraosseous injection of local anesthetics (Nusstein et al. 1998; Parente et al. 1998; Nusstein et al. 2003; Bigby et al. 2006). However, it should be noted that none of these clinical trials evaluated the efficacy of intraosseous anesthetics in cases of pain due to pulpal necrosis with periradicular involvement. Indeed, at least some investigators have expressed concern about the efficacy and potential for side effects when intraosseous anesthetics are used in cases involving pulpal necrosis (Reader and Nusstein 2002). Thus, additional clinical trials in this area appear warranted.

Oral analgesics have been shown to be useful in treating odontogenic pain. In general, the NSAIDs have been shown to have efficacy for reducing odontogenic pain (Doroschak et al. 1999; Holstein et al. 2002; Keiser and Hargreaves 2002). Surveys of endodontists reveal that most clinicians would use NSAID drugs for pain scenarios, with ibuprofen 600 mg being the most common dosage (Mickel et al. 2006). Although NSAIDs are available in formulations combined with a narcotic drug, there are large and dose-related increases in the potential for adverse effects (Dionne 1999). Instead, there is increasing evidence that NSAIDs can be combined with acetaminophen to provide greater analgesia for treating odontogenic pain (Menhinick et al. 2004). This important clinical finding is bolstered by recent reports that acetaminophen is actually a prodrug that is converted in the brain to a drug with considerable activity for enhancing the endogenous cannabinoid analgesic system (Hogestatt et al. 2005; Bertolini et al. 2006; Ottani et al. 2006). Thus, important evidence from randomized controlled trials and from bench-top research indicates that moderate to severe pain can often be managed by combinations of an NSAID with acetaminophen.

Pharmacotherapy is also used to reduce microorganisms in infected root canal systems. Several drug classes have been used for disruption or elimination of microorganisms in the infected root canal systems or associated tissues. As described elsewhere in this text, common endodontic pathogens in primary endodontic infections include members of Gram-negative anaerobic species. Accordingly, oral antibiotics including the penicillins and clindamycin comprise appropriate and popular forms of antibiotics used by endodontists (Yingling et al. 2002; Baumgartner and Xia 2003). Although such surveys indicate appropriate usage in the majority of clinical scenarios, it should be noted that they also reveal that some clinicians prescribe these drugs up to 80 times per week. In general, antibiotics are not indicated for cases of irreversible pulpitis, where pain of inflammatory origin is dominant (Nagle et al. 2000; Keenan et al. 2006). In addition, most (Walton and Fouad 1992; Walton and Chiappinelli 1993; Fouad et al. 1996; Henry et al. 2001; Pickenpaugh et al. 2001), but not all (Torabinejad et al. 1994), placebo-controlled clinical trials indicate that antibiotics have no effect for reducing odontogenic pain. Moreover, a comprehensive evidence-based review of the literature came to the same conclusion

about the lack of efficacy of antibiotics for relieving pain of odontogenic origin (Fouad 2002). Instead of systemic antibiotics, the chemomechanical debridement procedures appear to provide effective postoperative pain control since even placebo-treated patients report an approximately 80% reduction in pain within 48 h of endodontic instrumentation procedures (Torabinejad et al. 1994).

The term "chemomechanical debridement" is of value since it reflects the notion that bulk removal of infected tissue combined with direct application of antimicrobial agents into infected root canal systems provides a useful strategy for achieving both short-term (pain) and long-term (healing) endodontic outcomes. Although mechanical debridement using rotary NiTi file systems can reduce bacterial load, most of the treated root canal systems still retain cultivable microorganisms when instrumentation is performed in the presence of sterile saline (Shuping et al. 2000). Instead, antimicrobial agents must be provided. Local antimicrobial therapy generally consists of delivery of compounds such as NaOCl, 2% chlorhexidine, calcium hydroxide, MTAD (Biopure$^\circledR$), or other agents into infected root canal systems (Gilad et al. 1999; Shuping et al. 2000; Torabinejad et al. 2003; Waltimo et al. 2005; Zerella et al. 2005; Hoelscher et al. 2006; Cook et al. 2007). However, despite the well-reported disinfecting action of these intracanal agents, they do not seem to improve postoperative pain (Torabinejad et al. 2005; Ehrmann et al. 2007). It can be hypothesized that this may be, at least in part, due to the inability of these intracanal disinfectants to neutralize bacterial biproducts (e.g., LPS) that have diffused into the periradicular tissues. In addition, it should be noted that ultrasonic delivery of NaOCl may improve local tissue debridement (Gutarts et al. 2005; Burleson et al. 2007).

Taken together, there is no one magic bullet for treating pain associated with odontogenic infections. Instead, local anesthetics, analgesics, and effective chemomechanical debridement provide an effective multidisciplinary strategy for the control of pain due to microbial infection.

## 11.5 References

Ahlberg KF. 1978. Influence of local noxious heat stimulation on sensory nerve activity in the feline dental pulp. *Acta Physiol Scand* 103(1): 71–80.

Ahlquist M, Franzen O, Coffey J, and Pashley D. 1994. Dental pain evoked by hydrostatic pressures applied to exposed dentin in man: A test of the hydrodynamic theory of dentin sensitivity. *J Endod* 20(3): 130–34.

Allard B, Magloire H, Couble ML, Maurin JC, and Bleicher F. 2006. Voltage-gated sodium channels confer excitability to human odontoblasts: Possible role in tooth pain transmission. *J Biol Chem* 281(39): 29002–10.

Andrew D and Matthews B. 2000. Displacement of the contents of dentinal tubules and sensory transduction in intradental nerves of the cat. *J Physiol* 529 (Pt 3): 791–802.

Baumgartner JC and Falkler WA, Jr. 1991. Biosynthesis of IgG in periapical lesion explant cultures. *J Endod* 17(4): 143–46.

Baumgartner JC and Xia T. 2003. Antibiotic susceptibility of bacteria associated with endodontic abscesses. *J Endod* 29(1): 44–47.

Bertolini A, Ferrari A, Ottani A, Guerzoni S, Tacchi R, and Leone S. 2006. Paracetamol: New vistas of an old drug. *CNS Drug Rev* 12(3–4): 250–75.

Bigby J, Reader A, Nusstein J, Beck M, and Weaver J. 2006. Articaine for supplemental intraosseous anesthesia in patients with irreversible pulpitis. *J Endod* 32(11): 1044–47.

Bowles WR, Withrow JC, Lepinski AM, and Hargreaves KM. 2003. Tissue levels of immunoreactive substance P are increased in patients with irreversible pulpitis. *J Endod* 29(4): 265–67.

Brannstrom M and Johnson G. 1978. The sensory mechanism in human dentin as revealed by evaporation and mechanical removal of dentin. *J Dent Res* 57(1): 49–53.

Burleson A, Nusstein JRA, and Beck M. 2007. The in vivo evaluation of hand/rotary/ultrasound instrumentation in necrotic, human mandibular molars. *J Endod* 33: 782–87.

Byers MR. 1984. Dental sensory receptors. *Int Rev Neurobiol* 25: 39–94.

Byers MR, Schatteman GC, and Bothwell M. 1990a. Multiple functions for NGF receptor in developing, aging and injured rat teeth are suggested by epithelial, mesenchymal and neural immunoreactivity. *Development* 109(2): 461–71.

Byers MR, Taylor PE, Khayat BG, and Kimberly CL. 1990b. Effects of injury and inflammation on pulpal and periapical nerves. *J Endod* 16(2): 78–84.

Caterina MJ, Rosen TA, Tominaga M, Brake AJ, and Julius D. 1999. A capsaicin-receptor homologue with a high threshold for noxious heat. *Nature* 398(6726): 436–41.

Cesare P, Dekker LV, Sardini A, Parker PJ, and McNaughton PA. 1999. Specific involvement of PKC-epsilon in sensitization of the neuronal response to painful heat. *Neuron* 23(3): 617–24.

Chang MC, Chen YJ, Tai TF, Tai MR, Li MY, Tsai YL, Lan WH, Wang YL, and Jeng JH. 2006. Cytokine-induced prostaglandin E2 production and cyclooxygenase-2 expression in dental pulp cells: Downstream calcium signalling via activation of prostaglandin EP receptor. *Int Endod J* 39(10): 819–26.

Charoenlarp P, Wanachantararak S, Vongsavan N, and Matthews B. 2007. Pain and the rate of dentinal fluid flow produced by hydrostatic pressure stimulation of exposed dentine in man. *Arch Oral Biol* 52(7): 625–31.

Cook J, Nandakumar R, and Fouad AF. 2007. Molecular- and culture-based comparison of the effects of antimicrobial agents on bacterial survival in infected dentinal tubules. *J Endod* 33(6): 690–92.

Cunha TM, Verri WA, Jr, Silva JS, Poole S, Cunha FQ, and Ferreira SH. 2005. A cascade of cytokines mediates mechanical inflammatory hypernociception in mice. *Proc Natl Acad Sci U S A* 102(5): 1755–60.

Diogenes A, Akopian AN, and Hargreaves KM. 2007a. NGF Up-regulates TRPA1: Implications for orofacial pain. *J Dent Res* 86(6): 550–55.

Diogenes A, Ferraz CCR, and Hargreaves KM. 2007b. Painful infections: Bacteria regulate nociceptors via TLR4 and CD14 receptors. *J Endod* 33(Abstract): 340.

Diogenes A, Patwardhan AM, Jeske NA, Ruparel NB, Goffin V, Akopian AN, and Hargreaves KM. 2006. Prolactin modulates TRPV1 in female rat trigeminal sensory neurons. *J Neurosci* 26(31): 8126–36.

Dionne RA. 1999. Additive analgesic effects of oxycodone and ibuprofen in the oral surgery model. *J Oral Maxillofac Surg* 57(6): 673–78.

Doroschak AM, Bowles WR, and Hargreaves KM. 1999. Evaluation of the combination of flurbiprofen and tramadol for management of endodontic pain. *J Endod* 25(10): 660–63.

Doyle SL and O'Neill LA. 2006. Toll-like receptors: From the discovery of NFkappaB to new insights into transcriptional regulations in innate immunity. *Biochem Pharmacol* 72(9): 1102–13.

Dubner R and Bennett GJ. 1983. Spinal and trigeminal mechanisms of nociception. *Annu Rev Neurosci* 6: 381–418.

Ehrmann EH, Messer HH, and Clark RM. 2007. Flare-ups in endodontics and their relationship to various medicaments. *Aust Endod J* 33: 119–30.

Epstein JB. 1988. The painful mouth. Mucositis, gingivitis, and stomatitis. *Infect Dis Clin North Am* 2(1): 183–202.

Ferraz CCR, Diogenes A, and Hargreaves KM. 2007. Potential mechanisms for exacerbated pain due to infection: NGF upregulates TLR4 and CD14 in trigeminal neurons. *J Endod* 33: 359.

Fitzgerald KA, Rowe DC, and Golenbock DT. 2004. Endotoxin recognition and signal transduction by the TLR4/MD2-complex. *Microbes Infect* 6(15): 1361–67.

Fouad AF. 2002. Are antibiotics effective for endodontic pain? *Endod Topics* 2: 52–66.

Fouad AF, Rivera EM, and Walton RE. 1996. Penicillin as a supplement in resolving the localized acute apical abscess. *Oral Surg Oral Med Oral Pathol Oral Radiol Endod* 81(5): 590–95.

Fried K and Risling M. 1991. Nerve growth factor receptor-like immunoreactivity in primary and permanent canine tooth pulps of the cat. *Cell Tissue Res* 264(2): 321–28.

Gilad JZ, Teles R, Goodson M, White RR, and Stashenko P. 1999. Development of a clindamycin-impregnated fiber as an intracanal medication in endodontic therapy. *J Endod* 25(11): 722–27.

Goodis HE, Bowles WR, and Hargreaves KM. 2000. Prostaglandin E2 enhances bradykinin-evoked iCGRP release in bovine dental pulp. *J Dent Res* 79(8): 1604–7.

Gutarts R, Nusstein J, Reader A, and Beck M. 2005. In vivo debridement efficacy of ultrasonic irrigation following hand-rotary instrumentation in human mandibular molars. *J Endod* 31(3): 166–70.

Hahn CL, Best AM, and Tew JG. 2000. Cytokine induction by Streptococcus mutans and pulpal pathogenesis. *Infect Immun* 68(12): 6785–89.

Hargreaves KM and Goodis HE. 2002. *Seltzer and Bender's Dental Pulp*. Carol Stream, IL: Quintessence Publishing Co.

Hargreaves KM and Keiser K. 2002. Local anesthetic failure in endodontics: Mechanisms and management. *Endod Topics* 1: 26–39.

Hargreaves KM and Milam S. 2001. Mechanisms of pain and analgesia. In: Dionne R and Phero J (eds), *Management of Pain and Anxiety in Dental Practice*. New York: Elsevier, 18–40.

Hashioka K, Yamasaki M, Nakane A, Horiba N, and Nakamura H. 1992. The relationship between clinical symptoms and anaerobic bacteria from infected root canals. *J Endod* 18(11): 558–61.

Henry M, Reader A, and Beck M. 2001. Effect of penicillin on postoperative endodontic pain and swelling in symptomatic necrotic teeth. *J Endod* 27(2): 117–23.

Hoelscher AA, Bahcall JK, and Maki JS. 2006. In vitro evaluation of the antimicrobial effects of a root canal sealerantibiotic combination against *Enterococcus faecalis*. *J Endod* 32(2): 145–47.

Hogestatt ED, Jonsson BA, Ermund A, Andersson DA, Bjork H, Alexander JP, Cravatt BF, Basbaum AI, and Zygmunt PM. 2005. Conversion of acetaminophen to the bioactive N-acylphenolamine AM404 via fatty acid amide hydrolase-dependent arachidonic acid conjugation in the nervous system. *J Biol Chem* 280(36): 31405–12.

Holstein A, Hargreaves KM, and Niederman R. 2002. Evaluation of NSAIDs for treating post-endodontic pain. A systematic review. *Endod Topics* 3: 3–13.

Ianiro SR, Jeansonne BG, McNeal SF, and Eleazer PD. 2007. The effect of preoperative acetaminophen or a combination of acetaminophen and Ibuprofen on the success of inferior alveolar nerve block for teeth with irreversible pulpitis. *J Endod* 33(1): 11–14.

Ikeda H, Tokita Y, and Suda H. 1997. Capsaicin-sensitive A delta fibers in cat tooth pulp. *J Dent Res* 76(7): 1341–49.

Jacinto RC, Gomes BP, Ferraz CC, Zaia AA, and Filho FJ. 2003. Microbiological analysis of infected root canals from symptomatic and asymptomatic teeth with periapical periodontitis and the antimicrobial susceptibility of some isolated anaerobic bacteria. *Oral Microbiol Immunol* 18(5): 285–92.

Julius D and Basbaum AI. 2001. Molecular mechanisms of nociception. *Nature* 413(6852): 203–10.

Jyvasjarvi E and Kniffki KD. 1987. Cold stimulation of teeth: A comparison between the responses of cat intradental A delta and C fibres and human sensation. *J Physiol* 391: 193–207.

Keenan JV, Farman AG, Fedorowicz Z, and Newton JT. 2006. A Cochrane systematic review finds no evidence to support the use of antibiotics for pain relief in irreversible pulpitis. *J Endod* 32(2): 87–92.

Keiser K and Hargreaves KM. 2002. Building effective strategies for the management of endodontic pain. *Endod Topics* 3: 92–104.

Kerr BJ, Bradbury EJ, Bennett DL, Trivedi PM, Dassan P, French J, Shelton DB, McMahon SB, and Thompson SW. 1999. Brain-derived neurotrophic factor modulates nociceptive sensory inputs and NMDA-evoked responses in the rat spinal cord. *J Neurosci* 19(12): 5138–48.

Khan AA, McCreary B, Owatz CB, Schindler WG, Schwartz SA, Keiser K, and Hargreaves KM. 2007. The development of a diagnostic instrument for the measurement of mechanical allodynia. *J Endod* 33(6): 663–66.

Lepinski AM, Hargreaves KM, Goodis HE, and Bowles WR. 2000. Bradykinin levels in dental pulp by microdialysis. *J Endod* 26(12): 744–47.

Lipton JA, Ship JA, and Larach-Robinson D. 1993. Estimated prevalence and distribution of reported orofacial pain in the United States. *J Am Dent Assoc* 124(10): 115–21.

Locker D and Grushka M. 1987. Prevalence of oral and facial pain and discomfort: Preliminary results of a mail survey. *Community Dent Oral Epidemiol* 15(3): 169–72.

Luthman J, Luthman D, and Hokfelt T. 1992. Occurrence and distribution of different neurochemical markers in the human dental pulp. *Arch Oral Biol* 37(3): 193–208.

McMahon SB. 1996. NGF as a mediator of inflammatory pain. *Philos Trans R Soc Lond B Biol Sci* 351(1338): 431–40.

Mendell LM, Albers KM, and Davis BM. 1999. Neurotrophins, nociceptors, and pain. *Microsc Res Tech* 45(4–5): 252–61.

Mengel MK, Jyvasjarvi E, and Kniffki KD. 1992. Identification and characterization of afferent periodontal C fibres in the cat. *Pain* 48(3): 413–20.

Mengel MK, Jyvasjarvi E, and Kniffki KD. 1993. Identification and characterization of afferent periodontal A delta fibres in the cat. *J Physiol* 464: 393–405.

Menhinick KA, Gutmann JL, Regan JD, Taylor SE, and Buschang PH. 2004. The efficacy of pain control following nonsurgical root canal treatment using ibuprofen or a combination of ibuprofen and acetaminophen in a randomized, double-blind, placebo-controlled study. *Int Endod J* 37(8): 531–41.

Mickel AK, Wright AP, Chogle S, Jones JJ, Kantorovich I, and Curd F. 2006. An analysis of current analgesic preferences for endodontic pain management. *J Endod* 32(12): 1146–54.

Modaresi J, Dianat O, and Mozayeni MA. 2006. The efficacy comparison of ibuprofen, acetaminophen-codeine, and placebo premedication therapy on the depth of anesthesia during treatment of inflamed teeth. *Oral Surg Oral Med Oral Pathol Oral Radiol Endod* 102(3): 399–403.

Morgan CR, Rodd HD, Clayton N, Davis JB, and Boissonade FM. 2005. Vanilloid receptor 1 expression in human tooth pulp in relation to caries and pain. *J Orofac Pain* 19(3): 248–60.

Nagle D, Reader A, Beck M, and Weaver J. 2000. Effect of systemic penicillin on pain in untreated irreversible pulpitis. *Oral Surg Oral Med Oral Pathol Oral Radiol Endod* 90(5): 636–40.

Nakanishi T, Matsuo T, and Ebisu S. 1995. Quantitative analysis of immunoglobulins and inflammatory factors in human pulpal blood from exposed pulps. *J Endod* 21(3): 131–36.

Navarro V, Meseguer V, Fernandez A, Medrano F, Saez JA, and Puras A. 1998. Psoas muscle abscess. Description of a series of 19 cases. *Enferm Infecc Microbiol Clin* 16(3): 118–22.

Nicol GD, Klingberg DK, and Vasko MR. 1992. Prostaglandin E2 increases calcium conductance and stimulates release of substance P in avian sensory neurons. *J Neurosci* 12(5): 1917–27.

Nusstein J, Kennedy S, Reader A, Beck M, and Weaver J. 2003. Anesthetic efficacy of the supplemental X-tip intraosseous injection in patients with irreversible pulpitis. *J Endod* 29(11): 724–28.

Nusstein J, Reader A, Nist R, Beck M, and Meyers WJ. 1998. Anesthetic efficacy of the supplemental intraosseous injection of 2% lidocaine with 1:100,000 epinephrine in irreversible pulpitis. *J Endod* 24(7): 487–91.

O'Neill LA. 2004. Immunology. After the toll rush. *Science* 303(5663): 1481–82.

Obata K, Katsura H, Mizushima T, Yamanaka H, Kobayashi K, Dai Y, Fukuoka T, Tokunaga A, Tominaga M, and Noguchi K. 2005. TRPA1 induced in sensory neurons contributes to cold hyperalgesia after inflammation and nerve injury. *J Clin Invest* 115(9): 2393–2401.

Okumura R, Shima K, Muramatsu T, Nakagawa K, Shimono M, Suzuki T, Magloire H, and Shibukawa Y. 2005. The odontoblast as a sensory receptor cell? The expression of TRPV1 (VR-1) channels. *Arch Histol Cytol* 68(4): 251–57.

Ottani A, Leone S, Sandrini M, Ferrari A, and Bertolini A. 2006. The analgesic activity of paracetamol is prevented by the blockade of cannabinoid CB1 receptors. *Eur J Pharmacol* 531(1–3): 280–81.

Owatz CB, Khan AA, Schindler WG, Schwartz SA, Keiser K, and Hargreaves KM. 2007. The incidence of mechanical allodynia in patients with irreversible pulpitis. *J Endod* 33(5): 552–56.

Parente SA, Anderson RW, Herman WW, Kimbrough WF, and Weller RN. 1998. Anesthetic efficacy of the supplemental intraosseous injection for teeth with irreversible pulpitis. *J Endod* 24(12): 826–28.

Penfield W and Ramussen G. 1950. *The Cerebral Cortex of Man*. New York: MacMillan.

Pickenpaugh L, Reader A, Beck M, Meyers WJ, and Peterson LJ. 2001. Effect of prophylactic amoxicillin on endodontic flare-up in asymptomatic, necrotic teeth. *J Endod* 27(1): 53–56.

Poltorak A, He X, Smirnova I, Liu MY, Van Huffel C, Du X, Birdwell D, Alejos E, Silva M, Galanos C, Freudenberg M, Ricciardi-Castagnoli P, Layton B, and Beutler B. 1998. Defective LPS signaling in C3 H/HeJ and C57 BL/10ScCr mice: Mutations in Tlr4 gene. *Science* 282(5396): 2085–88.

Qi HY and Shelhamer JH. 2005. Toll-like receptor 4 signaling regulates cytosolic phospholipase A2 activation and lipid generation in lipopolysaccharide-stimulated macrophages. *J Biol Chem* 280(47): 38969–75.

Reader A and Nusstein J. 2002. Local anesthesia for endontic pain. *Endod Topics* 3: 14–30.

Reuben SS. 2007. Chronic pain after surgery: What can we do to prevent it. *Curr Pain Headache Rep* 11(1): 5–13.

Reynaud af Geijersstam A, Culak R, Molenaar L, Chattaway M, Roslie E, Peciuliene V, Haapasalo M, and Shah HN. 2007. Comparative analysis of virulence determinants and mass spectral profiles of Finnish and Lithuanian endodontic *Enterococcus faecalis* isolates. *Oral Microbiol Immunol* 22(2): 87–94.

Sedgley CM, Molander A, Flannagan SE, Nagel AC, Appelbe OK, Clewell DB, and Dahlen G. 2005. Virulence, phenotype and genotype characteristics of endodontic *Enterococcus* spp. *Oral Microbiol Immunol* 20(1): 10–19.

Sessle BJ. 2005. Peripheral and central mechanisms of orofacial pain and their clinical correlates. *Minerva Anestesiol* 71(4): 117–36.

Shimauchi H, Takayama S, Miki Y, and Okada H. 1997. The change of periapical exudate prostaglandin E2 levels during root canal treatment. *J Endod* 23(12): 755–58.

Shuping GB, Orstavik D, Sigurdsson A, and Trope M. 2000. Reduction of intracanal bacteria using nickel-titanium rotary instrumentation and various medications. *J Endod* 26(12): 751–55.

Simone DA, Baumann TK, and LaMotte RH. 1989. Dose-dependent pain and mechanical hyperalgesia in humans after intradermal injection of capsaicin. *Pain* 38(1): 99–107.

Siqueira JF, Jr, Rocas IN, and Rosado AS. 2004. Investigation of bacterial communities associated with asymptomatic and symptomatic endodontic infections by denaturing gradient gel electrophoresis fingerprinting approach. *Oral Microbiol Immunol* 19(6): 363–70.

Torabinejad M, Cymerman JJ, Frankson M, Lemon RR, Maggio JD, and Schilder H. 1994. Effectiveness of various medications on postoperative pain following complete instrumentation. *J Endod* 20(7): 345–54.

Torabinejad M, Shabahang S, Aprecio RM, and Kettering JD. 2003. The antimicrobial effect of MTAD: An in vitro investigation. *J Endod* 29(6): 400–403.

Torabinejad M, Shabahang S, and Bahjri K. 2005. Effect of MTAD on postoperative discomfort: A randomized clinical trial. *J Endod* 31: 171–76.

Veerayutthwilai O, Byers MR, Pham TT, Darveau RP, and Dale BA. 2007. Differential regulation of immune responses by odontoblasts. *Oral Microbiol Immunol* 22(1): 5–13.

Wadachi R and Hargreaves KM. 2006. Trigeminal nociceptors express TLR-4 and CD14: A mechanism for pain due to infection. *J Dent Res* 85(1): 49–53.

Waltimo T, Trope M, Haapasalo M, and Orstavik D. 2005. Clinical efficacy of treatment procedures in endodontic infection control and one year follow-up of periapical healing. *J Endod* 31(12): 863–66.

Walton R and Fouad A. 1992. Endodontic interappointment flare-ups: A prospective study of incidence and related factors. *J Endod* 18(4): 172–77.

Walton RE and Chiappinelli J. 1993. Prophylactic penicillin: Effect on posttreatment symptoms following root canal treatment of asymptomatic periapical pathosis. *J Endod* 19(9): 466–70.

Wheeler EF, Naftel JP, Pan M, von Bartheld CS, and Byers MR. 1998. Neurotrophin receptor expression is induced in a subpopulation of trigeminal neurons that label by retrograde transport of NGF or fluoro-gold following tooth injury. *Brain Res Mol Brain Res* 61(1–2): 23–38.

Woolf CJ and Chong MS. 1993. Preemptive analgesia—treating postoperative pain by preventing the establishment of central sensitization. *Anesth Analg* 77(2): 362–79.

Woolf CJ and Thompson SW. 1991. The induction and maintenance of central sensitization is dependent on N-methyl-D-aspartic acid receptor activation: Implications for the treatment of post-injury pain hypersensitivity states. *Pain* 44(3): 293–99.

Yingling NM, Byrne BE, and Hartwell GR. 2002. Antibiotic use by members of the American Association of Endodon-tists in the year 2000: Report of a national survey. *J Endod* 28(5): 396–404.

Zerella JA, Fouad AF, and Spangberg LS. 2005. Effectiveness of a calcium hydroxide and chlorhexidine digluconate mixture as disinfectant during retreatment of failed endodontic cases. *Oral Surg Oral Med Oral Pathol Oral Radiol Endod* 100(6): 756–61.

# Chapter 12
# Systemic Antibiotics in Endodontic Infections

*J. Craig Baumgartner and John R. Smith*

## 12.1 Introduction

"Antibiotic" in Greek means against life. Antibiotics (antimicrobials) are synthetic or naturally occurring chemicals that either inhibit the growth or kill microbes. The modern concept of specific antimicrobials (magic bullets) to treat infections was formulated by Paul Ehrlich in 1906. However, the use of various agents to treat infections dates back to about 2,500 years ago. Physicians in China used soybean curd, soil/mudpacks, and plant products to treat their patients. Numerous antibiotics have been discovered in soil. It is likely that some of the Chinese soil/mudpacks contained molds or bacteria that produced antibiotics. The use of sulfonamides in the 1930s ushered in the modern era of antibiotic treatment of infections. However, the so-called golden age of antimicrobial therapy did not begin until the use of penicillin in 1941. Although Fleming discovered the antimicrobial properties of the mold penicillium in 1929, it was not until Florey and Chain mass produced penicillin that it became available in quantities needed for widespread clinical use. Since then, over 100 antibiotics have been used therapeutically, with dramatic clinical efficacy in treating infectious disease. How-ever, because of widespread and often promiscuous use of antibiotics, numerous microbes have developed resistance to previously effective antibiotics. In addition, many of the newer antibiotics have toxic side effects.

The efficacy of antibiotics is based on the principle of selective toxicity. Ideally, chemotherapeutic agents have selective toxicity for the infectious organisms, but not for the host cells. Many antibiotics in use do have toxic side effects for patients. The amount of toxicity is primarily determined by an antibiotic's mechanism of activity. There are two basic mechanisms of activity for antibiotics. The first mechanism of activity is blocking a function necessary to the microbe but not the host, and the second mechanism of activity is blocking a function necessary for both the microbe and the host. Penicillin is a prime example of an antibiotic that has selective toxicity for bacteria but not human cells. Penicillin inhibits cell wall production in bacteria but has no effect on mammalian cells, fungi, viruses, or protozoa. The efficacy of many antibiotics is based on varying degrees of selective toxicity between mammalian cells and microbes. For example, an antibiotic that inhibits protein synthesis may have a profound effect on bacteria but minimal affect on mammalian protein synthesis. In this chapter, the discussion is restricted to that of systemically administered antibiotics. A discussion of locally applied antibiotics is presented in Chapter 14.

## 12.2 Classification of antibiotics

Antibiotics may be classified according to mechanism of action. The mechanism of action is a factor that determines whether an antibiotic is bacteriostatic or bactericidal. Categories for the classification of antibiotics include antimetabolites, inhibitors of cell wall synthesis, inhibitors of protein synthesis, inhibitors of nucleic acid synthesis, and cell membrane agents (Yagiela et al. 1998).

An *antimetabolite* may be considered a competitive substrate because of its similarity to the normal substrate. The antibiotic binds to a specific enzyme to produce a nonfunctional complex. For example, sulfonamides are antibiotics that act as antimetabolites by inhibiting the production of folic acid in the bacterial cell.

Several antibiotics act by inhibition of cell wall synthesis with penicillin being the prototype. This is truly selective toxicity for bacteria because mammalian cells do not have cell walls. Interference with cell wall synthesis allows movement of water into the bacterial cell and extrusion of cellular contents.

Clindamycin and erythromycin are examples of antibiotics that bind to bacterial ribosomes and inhibit protein synthesis. They act by preventing growth of amino acid chains in the process of peptide elongation. Defective proteins are formed that are nonfunctional for cell growth and viability.

The mechanism of activity for metronidazole and fluoroquinolone derivatives is by inhibition of nucleic acid synthesis. For example, the fluoroquinolone derivatives block DNA gyrase in bacteria. Similar enzymes in mammalian cells are less sensitive to the effects of these antibiotics.

Antibiotics such as nystatin and amphotericin B used to treat fungal infections affect cell membranes. Damage to the cell membrane leads to changes in membrane permeability and inhibition of growth or death of the cell. Because of similarities between mammalian and fungal cell membranes, these antibiotics may cause cell damage to the mammalian cells.

## 12.3 Efficacy of antibiotics

Clinical success is based on making a correct diagnosis and, in the case of an infection, knowing what microbe(s) is producing the infection. Effective treatment requires the removal of the reservoir of infection. Although antibiotics may reduce the number of viable bacteria, ultimate healing is dependent on the host's innate and specific immune responses. Endodontic infections are polymicrobial and should be treated as such, understanding that polymicrobial endodontic infections are primarily strict anaerobic and facultative anaerobic bacteria are important in choosing an effective antibiotic. Bactericidal agents are preferred over bacteriostatic agents for several reasons. Bactericidal antibiotics provide a decrease in number of bacteria by causing cell death. They produce a more rapid clinical effect than bacteriostatic agents. Bactericidal antibiotics bind to the bacteria and produce an effect even after falling below a minimum inhibitory concentration (MIC). For example, penicillin binds to bacterial enzymes that result in delayed lysis of the cells after its tissue level has fallen below the MIC.

In general, it is best to use the narrowest spectrum of antibiotic to which the bacteria are susceptible. This produces less alteration of normal flora and selects for fewer resistant organisms. For example, penicillin V with a relatively narrow spectrum would have less effect on bacteria in the intestine than amoxicillin that has an extended spectrum.

An MIC must be reached at the site of infection for an antibiotic to be effective. The absorption of antibiotics is usually decreased in the presence of food, so antibiotics, in general, should be taken 1 h before or 2 h after the consumption of food. The calcium in milk may prevent the absorption of tetracycline antibiotics. In addition, antacids containing metal cations may decrease absorption of some antibiotics. Improper dosing is of concern. Inadequate dosage and inadequate duration may cause a failure of antibiotic therapy. Concentrations below the MIC of an antibiotic select for resistant bacteria. Likewise, inadequate duration may lead to recurrence of the infection. The antibiotic regimen should be continued for 2–3 days following remission of clinical signs and symptoms. Excessive dosage may produce toxic side effects depending on the antibiotic (e.g., aminoglycosides), while overdosage with nontoxic penicillin is not a problem.

Patient compliance is important for clinical success. Patients may not have expensive prescriptions filled or not follow the dosage schedule for various reasons including the occurrence of unpleasant side effects. Following the dosage schedule to maintain an MIC is important especially with bacteriostatic antibiotics. The dosage schedule is determined based on the normal rate of biotransformation and excretion. The dose is repeated at specific intervals to maintain minimal serum and tissue levels.

The effectiveness and toxicity of an antibiotic may be altered if its metabolism or excretion is altered or if the active metabolite is not distributed to the infected tissue. The penicillins, cephalosporins, aminoglycosides, and polymyxins are mostly excreted unchanged through the kidney, while other antibiotics are metabolized in the liver. The antibiotic and dosage may need to be altered for patients with liver or kidney disease. The blood supply to areas with abscesses or necrotic tissue may be compromised and prevent delivery of the antibiotic at an effective tissue level. Debridement of necrotic tissue and drainage of abscesses or cellulitis is important to improve circulation and deliver the antibiotic at an MIC. It is important to treat an infection as soon as possible when only a smaller number of bacteria are present and before it spreads to other tissues. Because periradicular abscesses are relatively avascular with an accumulation of purulence and necrotic cells, surgical intervention is needed to achieve an effective level of antibiotic. Antibiotics cannot be relied on to kill all the bacteria without the removal of the cause of the infection.

## 12.4 Host factors

Several host factors influence the efficacy and safety of antibiotic therapy. They include the patient's defense mechanisms (innate and specific immunity), age, allergies, pregnancy, and genetic determinants. Impairment of the patient's immune system may occur because of disease such as diabetes or acquired immune deficiency syndrome (AIDS). Even though antibiotic therapy may decrease the number of infecting bacteria, some patients may not be able to mount an immune response to recover. Age-related elimination of antibiotics must also be considered. Children and the elderly may require lower doses because of the inability to properly metabolize an antibiotic. The use of antibiotics during pregnancy or while breast-feeding is associated with some risk. For example, the use of tetracycline during the last half of pregnancy can cause hypoplasia of the teeth and bones of the fetus and even congenital cataracts. Patients with multiple allergies have a greater chance of having an allergy to antibiotics.

## 12.5 Bacterial resistance

Bacterial resistance to antibiotics may be natural or acquired. Natural resistance is present in the bacterial genome prior to contact with the antibiotic, whereas acquired resistance is the result of the selection of resistant organisms by the antibiotic. An example of natural resistance is that of aerobes and facultative anaerobes to metronidazole. Because of its mechanism of activity, it is only effective against strict anaerobic bacteria. Acquired resistance occurs as a result of selective pressure during antibiotic therapy. Some strains of bacteria with acquired resistance may be present before antibiotics are administered but many emerge during treatment. Antibiotic therapy does not cause bacterial mutation, but it kills or inhibits the susceptible bacteria and allows the resistant organisms to take over the ecosystem. The selective pressure of antibiotics in hospitals and other environments often favor the emergence of the resistant bacterial strains. The selected resistant strains then become the predominant bacterial species. Eventually some species acquire resistance to the extent that they are considered resistant to a specific antibiotic (e.g., *Staphylococcus aureus* resistant to penicillin G or methicillin).

The circular bacterial genome consists of about 2,000 genes. Bacterial resistance arises following a change in genetic makeup. These changes occur either because of spontaneous mutations or by horizontal transfer of genetic material. Once a bacterium has a gene, it is passed vertically to its daughter cells. Chromosomal mutations usually occur as stepwise changes. Usually, several changes are needed before there is a change in phenotypic expression. The presence of an antibiotic produces selective pressure with the most resistant strain eventually dominating the microbial mixture. For most bacteria, spontaneous mutations occur about once every $10^5$–$10^9$ cell divisions. To place this in perspective, realize that 1 mL of media turbid with bacteria contains $10^6$–$10^8$ bacteria. So there may be hundreds of mutants in 1 mL of turbid media. Low concentrations of antibiotic favor the selection of stepwise mutations for resistance to the antibiotic. High concentrations of antibiotic may still inhibit or kill first- or second-step resistant organisms. Thus, early treatment with antibiotics and drainage of purulent exudate containing large numbers of bacteria reduce the chance of selecting for resistant bacteria.

Once a strain of bacteria has resistance to antibiotics, it is possible for the gene to be horizontally transferred to other strains of the same species and to other species. These resistant strains can be transferred to other patients. So the selection of resistant bacteria in one patient may be detrimental to other patients in the future. There are four methods of horizontal transfer of genetic material. They are conjugation, transformation, transduction, and transposition. Conjugation is the transfer of DNA from one bacterium to another through sex pili. Transformation is the uptake of extracellular DNA and incorporation of the DNA into the recipient's chromosome where there is a similar base sequence. Transduction is the injection of genetic material by a virus (bacteriophage) into a bacterium. Transposition is the insertion of DNA transposons into a chromosome without the need for homology between the donor and the recipient cell. Conjugation has the greatest significance because multiple antibiotic-resistance genes may be transmitted. The extrachromosomal DNA segments transferred during conjugation are like other bacterial plasmids. They are known as resistance (R) factors. The R factors can replicate independently of the chromosomal DNA and may contain genes for resistance to several antibiotics. The resistance transfer factor is a segment of DNA that contains the genes needed for transfer of the R factor. In addition to R factors possibly coding for multiple drug resistance, they can be transferred from nonpathogenic bacteria to pathogenic bacteria. In one step, the R factor can produce resistant organisms from those previously susceptible to the antibiotics. There are now some strains of *Enterococcus*, *Pseudomonas*, and *Enterobacter* that are resistant to all clinically available antibiotics (Hardman et al. 2005).

Mechanisms of bacterial resistance include a change in surface receptors, extrusion of the antibiotic from the cell, the production of enzymes to alter the permeability of the bacterial cell to an antibiotic, or an enzyme to directly inactivate the antibiotic. The production of penicillinase is a classic example of the latter. To minimize the development of bacterial resistance, antibiotics should only be used when they are truly indicated. Antibiotics should be used systemically rather than locally, in adequate doses and duration, and when indicated based on susceptibility tests.

## 12.6 Antibiotic toxicities, allergies, and superinfections

Clinicians must always be aware that untoward reactions to the administration of antibiotics range from mild to fatal. The toxicity of an antibiotic is related to its effects on mammalian cells. The toxic reactions vary with each drug, the concentration in organ systems, and the person receiving the drug. Toxic reactions to antibiotics often involve the gastrointestinal (GI) tract, kidney, liver, central nervous system, and blood. Orally administered antibiotics may affect the GI tract by producing symptoms ranging from nausea and vomiting to diarrhea. These effects may occur associated with irritation of the GI mucosa (e.g., penicillin), damage to the GI cells (e.g., neomycin), or stimulation of endogenous receptors (e.g., erythromycin). Antibiotics that use the kidney for excretion may be toxic to the kidney. Antibiotics that may produce nephrotoxicity are tetracyclines, cephalosporins, aminoglycosides, and sulfonamides. Although some antibiotics are excreted unchanged by the kidney, many are biotransformed in the liver. After being metabolized in the liver, some antibiotics are excreted by biliary secretion. Antibiotics that may cause hepatotoxicity are tetracyclines, amphotericin B, and isoniazid. Large doses of penicillin G may produce a nonspecific irritation with neurotoxic manifestations. Aminoglycosides, griseofulvin, and tetracycline may be associated with neurotoxicity. Antibiotics may produce hematological changes. The dose-related effect of chloramphenicol on the bone marrow is an example of an adverse hematological change. Broad-spectrum penicillins and cephalosporins may be associated with bleeding disorders.

Although penicillins have little toxicity, they have the highest incidence of allergic reactions. Most antibiotics are relatively small molecules. To be allergenic the antibiotic molecule acts as a hapten to combine with endogenous proteins to form the antigens, producing the allergic reaction. Some antibiotics with large molecular structures (e.g., erythromycin estolate) are capable of acting as antigenic determinants without combining with an endogenous protein.

Cross-allergenicity within a class is common and may occur between classes. For example, patients allergic to penicillins may also be allergic to cephalosporins.

Antibiotic therapy that alters the host normal flora may produce superinfections, diarrhea, flatulence, angular cheilitis, furry tongue, and glossodynia. Superinfections are of concern. Superinfections occur during antibiotic treatment of a primary infection. The organisms involved in superinfections are often resistant to multiple antimicrobials. Species often involved in superinfections include *Clostridium*, *Proteus*, *Pseudomonas*, *Staphylococci*, and *Candida*. Superinfections most commonly occur in the young and the old, patients with pulmonary disease, during prolonged treatments, and with broad-spectrum antibiotics. A serious superinfection is "antibiotic-associated colitis" also known as pseudomembranous colitis. This is caused by an overgrowth of *Clostridium difficile.* The suppression of the normal GI flora as a result of oral antibiotic therapy allows the proliferation of *C. difficile* with production of toxins that produce ulceration, abdominal cramps, fever, diarrhea, and blood/mucus in the stool. Antibiotics often associated with antibiotic-associated colitis are cephalosporins, ampicillin, and clindamycin. Oral vancomycin or metronidazole has been used to treat antibiotic-associated colitis. In addition, penicillin G, penicillin V, all extended-spectrum penicillins, chloramphenicol, and tetracyclines may cause superinfections.

## 12.7 Pharmacology of antibiotics (chemistry and mechanisms of action)

In this section we limit our discussion to those classes of antibiotics that are most efficacious in dealing with endodontic infections, which, for the most part, are also those that are most useful in dealing with general orofacial infections.

As indicated earlier, antibiotics can be succinctly categorized by their mechanisms of action. The most useful antibiotics in endodontics are those agents that interfere with cell wall synthesis. Of this group, the penicillins are the most important. Benzylpenicillin (penicillin G) is the most active of the naturally occurring products of *Penicillium* mold and has a basic structure that includes a β-lactam ring, a thiazolidine ring, and an acyl side chain (Hardman et al. 2005) (Fig. 12.1). When it was discovered that

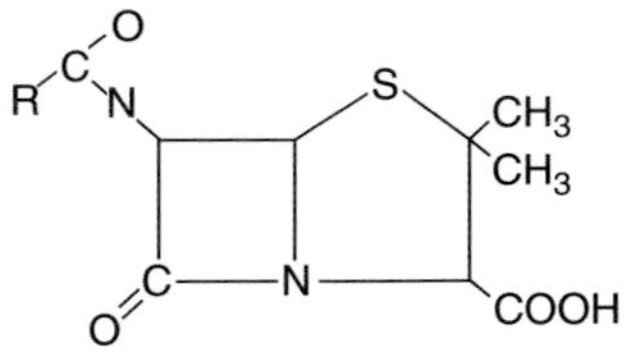

**Fig. 12.1**   Penicillin.

an amidase produced by *P. chrysogenum* could split the peptide linkage between the side chain and 6-aminopenicillanic acid, this opened the way to the development of semisynthetic penicillins with enhanced antimicrobial activity and pharmacokinetic properties through the addition of various side chains onto the 6-aminopenicillanic acid (Hardman et al. 2005). Today, the only natural penicillin in use is penicillin G. All other commercial penicillins are semisynthetic agents.

Penicillins owe their antimicrobial activity to their ability to bind to a variety of membrane-bound proteins, collectively referred to as "penicillin-binding proteins," or PBPs (Spratt 1980). PBPs are ubiquitous throughout the microbial kingdom although the density and variety of PBPs vary greatly among microbial species. Gram-positive bacteria are the most sensitive to penicillin primarily due to the importance of a transpeptidase that is also a PBP. This enzyme plays a crucial role in cell wall synthesis, which is particularly significant in Gram-positive bacteria because of their extremely thick cell walls (50–100 molecules vs 1–2 molecules in Gram-negative bacteria) (Hardman et al. 2005). To better understand the mechanism of action of the penicillins, it is useful to review the basics of bacterial cell wall synthesis.

The cell walls of Gram-positive bacteria are sheets of peptidoglycans that are laid down in a linear order and then cross-linked to form a very rigid and stable meshwork much like a lattice. The production of the cell wall is complex and occurs in three stages and requires 30 or more enzymes. The final stage requires the completion of the cross-linking of the glycan chains via transpeptidation, involving the terminal D-alanine and a transpeptidase. The penicillin molecule resembles D-alanine and thus can bind to this transpeptidase and subsequently inactivates it probably by acylation (Waxman et al. 1980) (Fig. 12.2). Although this process is agreed to be an integral part of the bactericidal actions of penicillin, it is still not clear how the inhibition of transpeptidation ultimately leads to cell death.

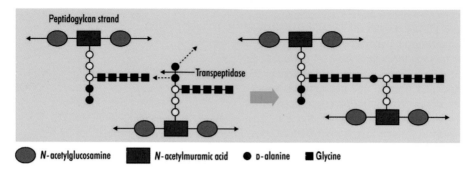

**Fig. 12.2**   This figure represents the cross-linking step during cell wall synthesis in *Staphlococcus aureus*. Transpeptidase cleaves the terminal d-alanine from a pentapeptide chain, which allows the next d-alanine to bind to the terminal glycine from an adjacent peptidoglycan strand. This cross-links the two peptidoglycan strands to provide stability to the cell wall. Penicillins and cephalosporins are structurally similar to the two terminal d-alanines and serve as substrates for transpeptidase. These agents, therefore, will compete with the peptidoglycan d-alanines for this enzyme and thus limit the cross-linking within the cell walls.

The lysis of bacterial cells that result from penicillin action appears to require participation of autolytic enzymes (autolysins or murein hydrolases), but the relationship between cell wall inhibition and release of autolysins is still being actively researched (Tomasz 1986; Ingavale 2003).

The transpeptidase enzyme is just one of many PBPs and the interaction of penicillins with some of these other PBPs does appear to contribute to the antimicrobial actions of penicillins. PBPs have a role in maintaining the shape of certain bacteria, and penicillin, especially at lower concentrations, can induce the formation of spheroplasts as the bacteria lose the integrity of its structural components (Hardman et al. 2005). PBP 4 is another enzyme that appears to play a significant role in the final separation of daughter cells where a shared peptidoglycan is cleaved (Priyadarshini et al. 2006).

Because of the nature of penicillin's actions, it is apparent that it is most effective when the pathogens are in their log phase of growth, that is, when there is rapid cell division occurring that requires considerable cell wall synthesis. Penicillins lose considerable efficacy in long-established infections where the pathogens have gone into the lag growth phase or in situations where a bacteriostatic antibiotic has been used prior to the penicillin (Yagiela et al. 1998).

Of the penicillins commercially available, penicillin V (the phenoxymethyl derivative of penicillin G) is the mainstay for treating endodontic infections. However, amoxicillin does offer advantages in certain situations over penicillin V. Amoxicillin is an aminopenicillin that was developed as an orally effective analog of ampicillin. Although it is typically considered a "broader-spectrum" antibiotic than penicillin V, this expanded spectrum mainly refers to activity against organisms not routinely associated with endodontic infections, such as *Haemophilus influenzae*, *E. coli*, etc. (Johnson 1999). Both penicillin V and amoxicillin are susceptible to inactivation by β-lactamase. To counteract this problem, penicillins have been compounded with agents that will inhibit the actions of β-lactamase. The most convenient product for use in dentistry is the orally effective form, which is a combination of amoxicillin with clavulanate (Augmentin).

There is one other group of antibiotics that, on rare occasions, can be used for endodontic infections, the cephalosporins. The original cephalosporins were isolated from the fungus *Cephalospoirum acremonium*, which was isolated from a sewer outlet off the Sardinian coast. Like the penicillins, these agents contain a β-lactam ring and closely resemble the structure of penicillin (Fig. 12.3). The antimicrobial activity of

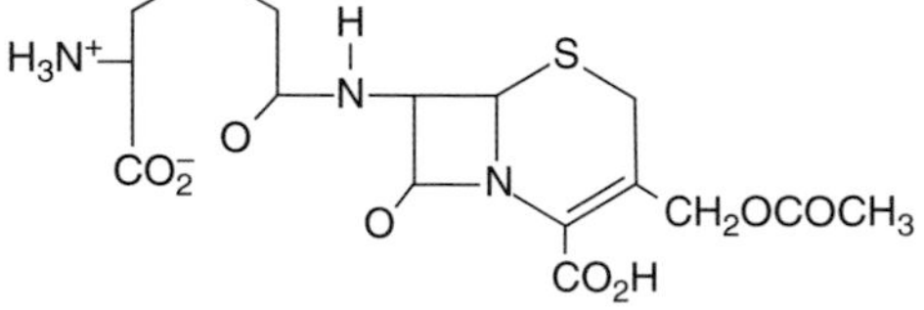

**Fig. 12.3**   Cephalosporin C.

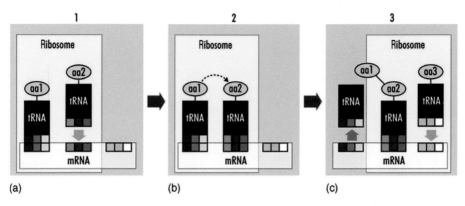

**Fig. 12.4** This figure illustrates the three steps in protein synthesis. (a) Frame 1 represents the initiation step. In this example a small section of a strand of mRNA is depicted with three codons for three different amino acids (aa1, aa2, and aa3). The anticodon for the tRNA carrying aa1 binds to the first codon to initiate protein synthesis. The tetracycline family of antibiotics can inhibit this first step in protein synthesis. (b) Frame 2 represents the elongation step. In this step the adjacent codon has been bound by the complimentary tRNA, which allows transpeptidation to occur, that is, aa1 is freed from its tRNA and bound to the next aa. When this occurs the "donating" tRNA is released from the mRNA. (c) Frame 3 represents the translocation step. In this step the ribosome repositions on the mRNA so that the tRNA carrying the growing peptide chain now occupies the correct location on the ribosome to pass the peptide chain on to the next amino acid. In this illustration, once the tRNA carrying aa3 attaches to the mRNA, the stage will be set for a repeat of the activity portrayed in Frame 2. Macrolide and lincosamide antibiotics will interfere with this third step in protein synthesis.

these agents is enhanced and modified by side-chain additions to the 7-aminocephalosporanic acid. The mechanism of action is similar to that of penicillin, that is, inhibition of cell wall synthesis via inactivation of transpeptidases and other bacterial enzymes (Abraham 1962; Johnson 1999; Hardman et al. 2005). Because the mechanisms of action of these two groups of antibiotics parallel each other, when organisms develop a resistance to one group, it is not uncommon for this resistance to extend to the other group (Jaffe et al. 1982; Davies 1994). Likewise, because the basic structures of the penicillins and cephalosporins are so similar, cross-allergenicity occurs (Kalant and Roschlau 1998; Romano et al. 2004).

Drugs of second choice in treating endodontic infections can be found among those agents that owe their antimicrobial activity to an ability to inhibit microbial protein synthesis. Because of the importance of proteins as receptors, enzymes, transport channels, etc., these agents can interfere with routine functions within the microorganisms. Generally speaking, these agents are considered bacteriostatic; however, at higher concentrations and with specific organisms they can be bactericidal (Johnson 1999). To understand the mechanisms of action of these agents, it is worthwhile to briefly review the process of bacterial protein synthesis.

Bacterial protein synthesis proceeds in three stages: initiation, elongation, and termination (Kalant and Roschlau 1998; Hardman et al. 2005). Initiation involves the formation of the 70S ribosomal subunit. To start the process mRNA must bind to a 30S ribosomal subunit. Once this union has been established, the appropriate tRNA, with its accompanying amino acid, can bind to the mRNA, and this complex now can bind to a 50S ribosomal subunit to complete the formation of the 70S ribosome. Once the tRNA is bound to the "A" (acyl) site on the ribosome, elongation of the peptide chain can occur. This entails the translocation of the tRNA-aa to the "P" (peptidyl) site where the amino acid is added to the growing peptide chain. The final step is termination, which is initiated by a termination codon on the mRNA, and causes the elongation process to cease and the peptide chain to be released from the ribosome (Fig. 12.4). Although protein synthesis in mammalian cells proceeds in an almost identical manner as described previously, we are fortunate that the mammalian ribosome is an 80S molecule that is not easily split into subunits. Therefore, the antibiotics that inhibit protein synthesis in

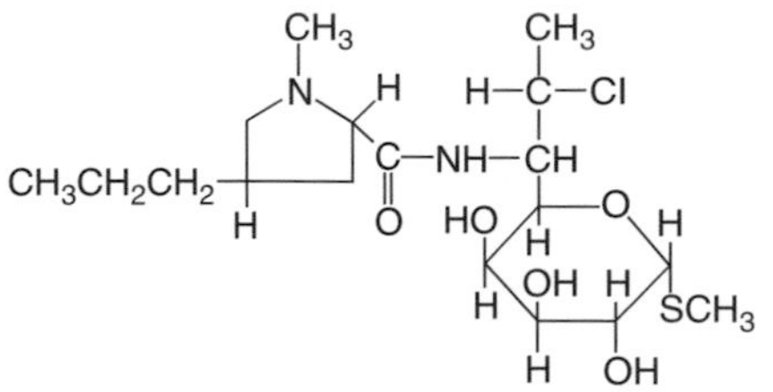

**Fig. 12.5**   Clindamycin.

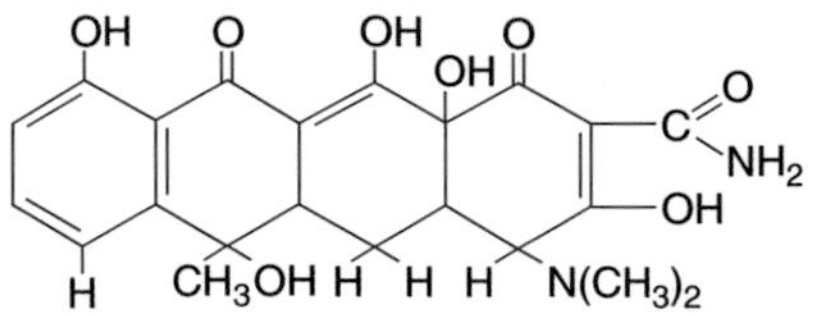

**Fig. 12.7**   Tetracycline.

microbes have little influence on this process in mammals. Most of the adverse side effects associated with the use of these antibiotics can be traced back to overgrowth of nonsusceptible bacteria or direct cellular actions not related to protein synthesis.

The three families of antibiotics used in endodontics that inhibit protein synthesis are the lincosamides, the macrolides, and the tetracyclines. Of these the most useful are the lincosamides and specifically clindamycin. Clindamycin is a derivative of lincomycin, which was named in honor of Lincoln, Nebraska, where the source organism was first isolated (Fig. 12.5). Clindamycin is able to inhibit bacterial protein synthesis by binding to the 50S subunit, which subsequently inhibits translocation and peptide chain extension (Kalant and Roschlau 1998).

The macrolide family of antibiotics (the erythromycins) was so named because of its large molecular structure (Fig. 12.6). Although they show little

resemblance structurally to the lincosamides, their mechanisms of action are identical (Brisson-Noel et al. 1998; Kalant and Roschlau 1998). As a result when organisms develop a resistance to the macrolides, they will also show cross-resistance to clindamycin. Although the usefulness of all the macrolides is limited in endodontics, azithromycin and clarithromycin are preferred over erythromycin because of their wider spectrum of activity and their longer duration of action (Hardman et al. 2005). In addition, these two macrolides have little effect on gastrointestinal motility, which is a major drawback to the use of erythromycin for many patients (Katagiri et al. 2005). The use of the macrolides also introduces additional concerns in medically compromised patients who are taking drugs at the time of their dental visit. The macrolides are notorious for their ability to inhibit various cytochrome $P_{450}$ enzymes, especially CYP3A4, a crucial enzyme in the metabolism of numerous drugs (von Rosenstiel and Adam 1995). As a result there is an increased risk for significant drug interactions when this family of antibiotics is used.

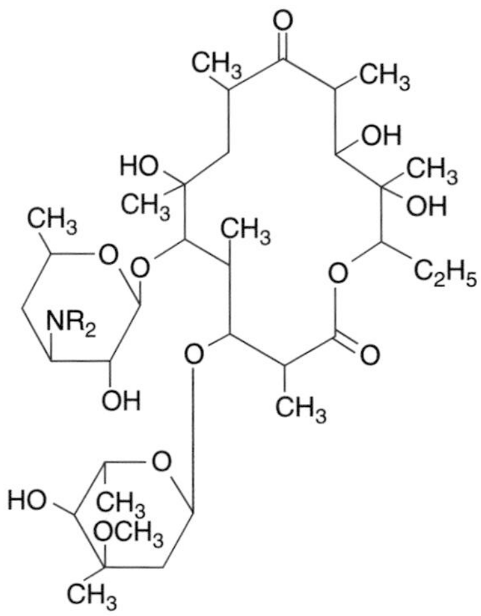

In the unlikely situation where a tetracycline antibiotic would be used in endodontic treatment, doxycycline or minocycline would be the drug of choice (Fig. 12.7). These agents inhibit protein synthesis but at a different site than seen with the macrolides and lincosamides. They will bind to the 30S subunit and prevent the attachment of tRNA, which will block the initiation step in protein synthesis (Kalant and Roschlau 1998). They also seem to inhibit bacterial function by chelating cations (especially magnesium) within the cytoplasm of the microbe. Unlike the previous antibiotics discussed, these agents can inhibit mammalian protein synthesis, but certain selectivity prevails in treatment because bacteria will concentrate the tetracyclines via active transport (Kalant and Roschlau 1998; Hardman et al. 2005). As a result, antibacterial activity can normally be achieved by blood levels of tetracyclines that have minimal influence on mammalian protein synthesis.

**Fig. 12.6**   Erythromycin.

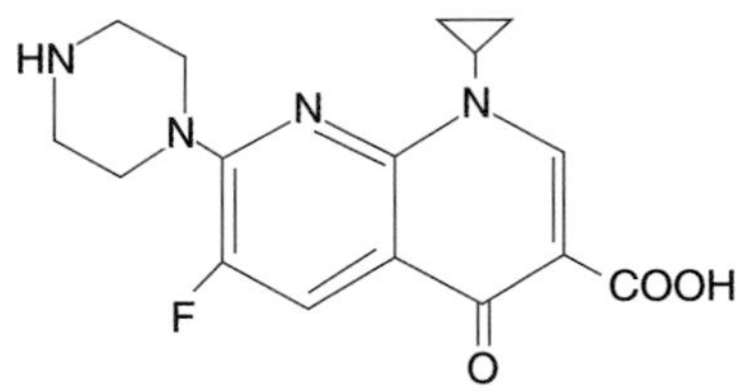

**Fig. 12.8**  Ciprofloxacin.

There are two other families of antibiotics occasionally used to treat endodontic infections that owe their antimicrobial activity to their abilities to interfere with DNA function in bacteria. The quinolone family of antibiotics is fluorinated derivatives of nalidixic acid (Fig. 12.8). Ciprofloxacin is the most well-known of these agents, but none of them is used routinely for endodontic infections. On entering a susceptible bacteria these agents will bind to and inactivate the enzyme DNA gyrase (Cozzarelli 1980; Mandell et al. 1995). This enzyme is active when DNA is replicating or during transcription. As the DNA helix separates, a "supercoiling" is generated in front of the area of separation. In order to rectify this supercoiling, a negative coiling has to be introduced that can be accomplished by cutting a DNA strand to allow a reconfiguration that is then followed by a resealing of the loose ends of the DNA strand (Hardman et al. 2005). DNA gyrase catalyzes the cutting and resplicing of the DNA strand. By preventing this process the quinolones allow for multiple inclusions of supercoils in the DNA, which ultimately leads to the loss of useful DNA activity. Although eucaryotic cells do not contain DNA gyrase, they do use a related enzyme to eliminate supercoiling during transcription. Luckily, this enzyme has far less affinity for the quinolone structure and will only be blocked at much higher blood levels than are needed to provide antimicrobial activity.

The other family of antibiotics that attack bacterial DNA is the nitroimidazole group. The most useful of these in endodontics is metronidazole (Fig. 12.9). This family of antibiotics was developed primarily for their

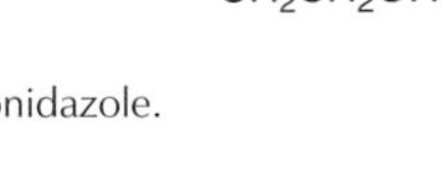

**Fig. 12.9**  Metronidazole.

antiprotozoal activity and is very useful in treating trichomoniasis, amebiasis, and giardiasis. It has also been found to be very effective against certain anaerobes found in endodontic infections. Metronidazole is actually a "prodrug" that acquires its antimicrobial potential only after it is reduced by ferrodoxins or other electron donors in the bacterial cytoplasm. The reductions generate nitro radical anions, which are highly reactive and attack DNA, proteins, and membrane elements (Docampo 1990; Johnson 1993). It is presumed that DNA damage is the major insult to the bacterium. Metronidazole is bacteriocidal and usually prescribed with penicillin.

## 12.8 Management of endodontic infections

The correct diagnosis of an endodontic infection and removal of the source of the infection are of primary importance for a successful outcome. Effective treatment of an endodontic infection includes either chemomechanical debridement of the root canal system or tooth extraction. Cleaning and shaping of the root canal system should always be done with rubber dam isolation and disinfection of the field to prevent further microbial contamination of the pulp cavity. Removal of the reservoir of infection establishes an environment favorable to healing. An antimicrobial dressing (e.g., calcium hydroxide) should be placed in the canals between appointments. Surgical drainage from a periradicular swelling associated with an abscess or cellulitis is also indicated. Drainage from an abscess or cellulitis decreases the amount of irritants (bacteria, bacterial by-products) and inflammatory mediators. By draining an abscess/cellulitis, circulation to the area is improved for better delivery of systemic antibiotics.

A cellulitis is inflammation of connective tissue through which transudate and exudate spread via interstitial and tissue spaces (Baumgartner et al. 2006). With accumulation of inflammatory cells, an abscess forms. An abscess is a localized collection of pus in a cavity formed by disintegration of surrounding tissues. Pus is an exudate consisting of dead cells, cellular debris, bacteria, and bacterial by-products. Virulence factors produced by bacteria stimulate the production of inflammatory mediators (Murakami et al. 2001). In addition, enzymes released from degenerating neutrophils will damage surrounding tissues. Clinicians should consider cellulitis and abscess formation as contiguous events. Needle aspiration of an area of cellulitis will often detect pockets of purulence before development of fluctuant tissue characteristic of a localized abscess. Drainage of accumulated transudate/exudate from an abscess or cellulitis improves the possibility of antibiotics reaching an MIC. An accumulation of purulence may not allow antibiotics to diffuse in a high enough concentration to be effective. The importance of drainage and removing the source of the infection cannot be overemphasized. Incision for drainage improves circulation and increases the concentration of the antibiotic to the infected tissues. It may also increase growth rates, making the bacteria more susceptible to antibiotics.

The bacteria of greatest concern in a polymicrobial infection are those in greatest numbers with virulence factors and resistance to antibiotics. Bacteria most associated with endodontic abscesses/cellulitis are anaerobes and facultative organisms. Superoxide dismutase is produced by some anaerobes to allow them to be tolerant of oxygen levels in periapical tissues (Murdoch 1998). In addition, facultative bacteria may consume oxygen to allow anaerobic bacteria to be established in the ecosystem (Murdoch 1998). If untreated, periapical abscesses may penetrate through the bone and involve the overlying fascial spaces (Baumgartner and Hutter 2001). Usually three or more organisms can be cultivated from periapical abscesses (Von Konow et al. 1981; Lewis et al. 1986; Brook et al. 1991; Fisher and Russell 1993; Baumgartner and Watkins 1994; Sakamoto et al. 1998; Baumgartner et al. 1999; Kuriyama et al. 2000; Siqueira et al. 2001). Similar to an infected root canal, bacterial isolates from these polymicrobial infections are mostly anaerobes. However, facultative streptococci of the milleri group are often isolated from periapical abscesses (Von Konow et al. 1981; Lewis et al. 1986; Fisher and Russell 1993). Some studies have shown that the streptococci milleri group is predominant or sole isolates in the early stages of odontogenic infections (Lewis et al. 1986; Fisher and Russell 1993). This suggests that streptococci may be associated with the spread of infection because they produce enzymes that break down connective tissue. The streptococci milleri group, by removing oxygen, decreases the redox potential in the surrounding tissues, allowing anaerobic bacteria to invade the tissues. Because some anaerobes have aerotolerance especially those with superoxide dismutase, progressive cellulitis may not require

facultative anaerobes or aerobes (Heimdahl and Nord 1985; Heimdahl et al. 1985; Gossling 1988; Baumgartner and Cuenin 1992). More research to understand the mechanisms associated with the spread of cellulitis is needed.

Bacteria often associated with the most severe infections include dark-pigmented bacteria, fusobacteria, Gram-positive anaerobic cocci, and the streptococci of the milleri group. Synergistic relationships have been demonstrated among these organisms (Sundqvist et al. 1979; Lewis et al. 1988; Baumgartner et al. 1992; Gomes et al. 1994; Drucker et al. 1997; Takemoto et al. 1997; Siqueira et al. 1998). *Fusobacterium nucleatum* has been regarded as a pathogen even without synergistic relationships (Heimdahl and Nord 1985; Heimdahl et al. 1985; Lewis et al. 1988; Baumgartner et al. 1992; Siqueira et al. 1998; Murakami et al. 2001). In some way it seems to inhibit the production of antibodies by plasma cells in periapical tissues that specifically target this organism (Baumgartner and Falkler 1991a, b). *F. nucleatum* has been shown to increase the abscessogenic ability of *Streptococcus constellatus* and decrease its killing by polymorphonuclear leukocytes (Nagashima et al. 1999). Other synergistic relationships among bacteria associated with periapical infections include *Fusobacterium* spp.–dark-pigmented bacteria, *Fusobacterium* spp.–*Eubacterium* spp., *Prevotella* spp.–*Peptostreptococcus micros*, and *P. micros*–*Bacteroides forsythus* (Sundqvist et al. 1979; Lewis et al. 1988; Kinder and Holt 1989; Baumgartner et al. 1992; Gomes et al. 1994; Drucker et al. 1997; Takemoto et al. 1997; Siqueira et al. 1998).

The geographic diversity of oral organisms has been investigated (Baumgartner et al. 2002; Haffajee et al. 2004; Haffajee et al. 2005; Siqueira et al. 2005; Rôças et al. 2006). It is now clear that oral microbial populations do vary with geographic population. Differences include the presence of species and the percentage of the species in a polymicrobial population. The prevalence of various species has been shown to vary significantly in endodontic abscesses from different geographical locations (Baumgartner et al. 2004; Siqueira et al. 2005; Rôças et al. 2006). When molecular methods were used to detect all species of bacteria in an endodontic infection, it was found that only 40–55% of the species were cultivable (Munson et al. 2002; Sakamoto et al. 2006). When genetic fingerprinting was accomplished using denaturing gradient gel electrophoresis to evaluate bacteria in endodontic abscesses, there were not only significant differences in species between geographic location but also between patients in the same location (Machado de Oliveira et al. 2006). Of 99 distinct bands (species) in the gel, 27 were unique to samples from the United States and 14 were unique to samples from Brazil (Machado de Oliveira et al. 2006). About 40% of the species were not shared by the other location. Of significance is the fact that in no case the fingerprint of any endodontic infection was ever identical to another. The polymicrobial makeup of all the infections was different from all others (Machado de Oliveira et al. 2006). The all 35 abscesses in this study were unique in terms of diversity of bacteria (Machado de Oliveira et al. 2006). An obvious question to be researched is whether microbial diversity among patients influences clinical outcome. In the future, novel chairside technologies will allow the prescription of antibiotics targeted for the specific microorganisms causing the infection.

Considering that there are hundreds of species of microbes in the human mouth available to produce infections, a relative restricted group of microbes have been associated with infections of the root canal system and periradicular tissues. As was discussed in Chapters 3 and 4, organisms most associated with endodontic infections include the dark-pigmented bacteria, fusobacteria, Gram-positive anaerobic cocci/bacilli, and microaerophilic streptococci (Murdoch 1998). Because of the polymicrobial nature of endodontic infections, it seems unlikely that any antibiotic can be effective against all the organisms. Indeed, although there have been numerous studies evaluating antibiotic susceptibility, many organisms have not been isolated and evaluated for susceptibility to antimicrobials. It is probable that many as yet uncultivable organisms are involved in the polymicrobial infections. There are no data about pathogenicity or antibiotic susceptibility of these organisms. Because many cultivable bacteria are slow growing, it often takes several days to isolate them in pure culture for susceptibility testing. Thus, when indicated, antibiotics are usually prescribed based on available susceptibility testing and clinical experience.

## 12.9 Indications for adjunctive antibiotic therapy

Indications for adjunctive antibiotic therapy include systemic signs associated with the infection, progressive swellings, and cellulitis. Signs and symptoms include fever above 100°F, malaise, cellulitis, unexplained trismus, and swelling beyond a simple localized mucosal enlargement. It is important to provide a pathway of drainage for purulent material consisting of bacteria, bacterial by-products, disintegrated inflammatory cells, enzymes (spreading factors), and other inflammatory mediators. In addition, incision for drainage by decompressing pockets of purulence improves circulation and delivery of the antibiotic to the area. An MIC of antimicrobial is needed to be effective. Without drainage it is difficult for a diffusion gradient of antibiotic in a bolus of purulence to reach an MIC. It is unlikely that any antibiotic will be effective against all the strains of infecting bacteria in a polymicrobial infection. However, it hopefully will disrupt the microbial ecosystem by inhibiting some species that may be producing by-products that are used as nutrients by other organisms. Following removal of the reservoir of infection and a prescription for an antibiotic, clinical signs and symptoms will often diminish in 2–4 days. The antibiotic should be continued for additional 2–3 days to prevent rebound of the infections. Thus a 7-day prescription is usually adequate. A sample from the area of infection for identification and susceptibility may be indicated if the patient is immunocompromised. Recent culture and susceptibility tests for organisms isolated from endodontic abscesses and cellulitis mainly reveal the presence of strict anaerobic bacteria and facultative anaerobic bacteria (Khemaleelakul et al. 2002; Baumgartner and Xia 2003). If the patient is medically compromised or the patient's condition deteriorates, consultation with a specialist and referral should be considered.

*Penicillin VK* remains an effective antibiotic for infections of endodontic origin. Recent studies evaluating antibiotic susceptibility show that it is still effective against most facultative and anaerobic microorganisms associated with endodontic infections (Khemaleelakul et al. 2002; Baumgartner and Xia 2003). In addition, penicillin VK has also low toxicity and low cost, but about 10% of the population will give a history of allergic reactions to penicillin. Penicillin VK should be administered every 4–6 h to achieve a steady serum level (Pallasch 1996). A loading dose of 1,000 mg of penicillin VK should be orally administered, followed by 500 mg every 4–6 h for 5–7 days. If significant improvement of the infection is not seen in 48–72 h, consider adding metronidazole or switching to another antibiotic (e.g., clindamycin). Of course, reviewing the diagnosis and further clinical treatment may also be indicated.

*Amoxicillin* is an analog of penicillin that is rapidly absorbed, has a longer half-life, broader spectrum of activity, and produces a more sustained serum level than penicillin VK. Food does not interfere with the absorption of amoxicillin. Because of these characteristics, amoxicillin is often used for antibiotic prophylaxis of patients who are medically compromised (Dajani 1997; ADA 2003; Wilson et al. 2007). When used to treat endodontic infections, its extended spectrum may select for additional resistant strains of bacteria. The usual oral dosage for amoxicillin is 1,000 mg loading dose followed by 500 mg every 8 h for 5–7 days. An alternate dosage is 875 mg every 12 h. In recent antibiotic susceptibility studies, the combination of amoxicillin with clavulanate (Augmentin) was the most effective antibiotic combination (Le Goff 1997; Khemaleelakul et al. 2002; Baumgartner and Xia 2003; Jacinto et al. 2003; Bresco-Salinas et al. 2006). The usual oral dosage for amoxicillin with clavulanate is the same as for amoxicillin by itself.

*Clindamycin* exhibits bactericidal action against a number of susceptible bacteria at concentrations readily achieved in vivo. It is active against most Gram-positive aerobes and both Gram-positive and Gram-negative anaerobes. Clindamycin is a good choice if a patient is allergic to penicillin or if a change in antibiotic is indicated. Both penicillin and clindamycin have been shown to produce good results in treating odontogenic infections (Gilmore et al. 1988; Khemaleelakul et al. 2002; Baumgartner and Xia 2003). Clindamycin is rapidly absorbed even in the presence of food in the stomach (Hardman et al. 2005). Clindamycin is well distributed throughout most body tissues and reaches a concentration in bone approximating that of plasma. The oral adult dosage for serious endodontic infections is a 600 mg loading dose followed by 300 mg every 6 h for 5–7 days.

*Metronidazole* is a nitroimidazole that is active against protozoal parasites and anaerobic bacteria (Sobottka et al. 2002). It is not effective for aerobic and facultative bacteria but may be used in combination with penicillin. Susceptibility studies have shown that many bacteria are resistant to metronidazole (Khemaleelakul et al. 2002; Baumgartner and Xia 2003). If a patient's symptoms do not improve in 48–72 h after initial treatment and therapy with penicillin, metronidazole may be added to the original prescription of penicillin. Of course, the diagnosis should be con-

firmed and any needed additional treatment to remove the reservoir of infection should be undertaken. It is important that the patient continues to take penicillin that is effective against the facultative bacteria and those anaerobes resistant to metronidazole. The usual oral dosage for metronidazole is a 1,000 mg loading dose followed by 500 mg every 6 h for 5 days. When patients fail to respond to treatment, consultation with a specialist is recommended.

*Clarithromycin* and *azithromycin* are macrolides that include a spectrum of activities for many microbes involved in endodontic infections. Erythromycin is the older macrolide traditionally prescribed for patients allergic to penicillin; however, it is not effective against anaerobic bacteria. Clarithromycin and azithromycin have better efficacy against microbes often found in endodontic infections and offer improved pharmacokinetics (Topazian and Goldberg 2002). Clarithromycin is rapidly absorbed from the gastrointestinal tract. Food delays absorption but does not alter the ultimate blood levels of clarithromycin. Azithromycin is absorbed rapidly but concomitant administration of aluminum and magnesium antacids will decrease the peak serum concentration. The oral dosage for clarithromycin is a 500 mg loading dose followed by 250 mg every 12 h for 5–7 days. A larger dose of 500 mg every 12 h may be used for severe infections. The oral dosage for azithromycin is a 500 mg loading dose followed by 250 mg once a day for 5–7 days.

*Cephalosporins* are usually not indicated for the treatment of endodontic infections. First-generation cephalosporins do not have activity against the anaerobes usually involved in endodontic infections. Second- and third-generation cephalosporins have some efficacy for anaerobes; however, there is a possibility of cross-reactivity in as many as 20% of patients that are allergic to penicillin (Hardman et al. 2005). All except cefuroxime (Ceftin) require parenteral administration but may be desired for hospitalized patients (Hardman et al. 2005). Parenteral forms of cephalosporins with a spectrum including odontogenic anaerobes include cefoxitin, cefmetazole, cefoperazone, and cefotaxime (Murdoch 1998; Kuriyama et al. 2000).

*Doxycycline* or *minocycline* (Kuriyama et al. 2000) occasionally may be indicated when the above antibiotics are contraindicated. However, many strains of bacteria have become resistant to the tetracyclines.

*Ciprofloxacin* is a quinolone antibiotic that is not effective against anaerobic bacteria usually found in endodontic infections (Hardman et al. 2005). With a persistent infection it may be indicated if culture and sensitivity tests demonstrate the presence of susceptible organisms. A fluoroquinolone (moxifloxacin) has been shown to have efficacy for anaerobes in odontogenic infections (Sobottka et al. 2002). However, significant emergence of resistance to fluoroquinolones (trovafloxacin and moxifloxacin) has been detected (Golan et al. 2003).

## 12.10 Microbial sampling of abscesses/cellulitis

Medically compromised patients, especially those that are immunocompetent, are at high risk for infections, and a culture for identification and antibiotic susceptibility testing may be indicated. Unfortunately, identification and the results of susceptibility tests may take several days to a couple of weeks, depending on the microorganisms involved in an infection. In the meantime empiric prescription of an antimicrobial is required.

It is important to have good communication with a laboratory to ensure that the sample is properly collected, transported, cultured, and identified. Request a Gram stain of the sample to determine what types of microorganisms are predominant. The laboratory personnel should be made aware that normal oral flora can become opportunistic pathogens and should not be dismissed as normal oral flora.

The microbial sample must be collected aseptically. If the sample is from a tooth, it must be isolated with a rubber dam and the field disinfected with sodium hypochlorite or another disinfectant. Sterile burs and instruments are used to gain access to the pulp cavity. If exudate is present in the tooth, it may be sampled with a sterile paper point and immediately placed in a prereduced transport medium provided by a laboratory. If there is copious drainage, it may be aspirated into a sterile syringe with a 16- to 18-gauge needle. The aspirate is then injected into a prereduced transport medium. Some laboratories will accept the sample in the syringe if it can be transported to the laboratory within just a few minutes.

If the contents of a mucosal swelling are to be sampled, a needle aspirate is preferred to prevent contamination from saliva and mucosa bacteria. The patient should rinse the area with a chlorhexidine mouthwash and the mucosal surface disinfected with a Povidone-Iodine Prep Pad (PDI, Orangeburg, NY). A 16- to 18-gauge needle and syringe should be used to aspirate the exudate. The exudate is injected into a prereduced transport container such as the Port-A-Cul Vial (Becton Dickinson, Sparks, MD). If a sample cannot be aspirated, a sample can be collected on a swab after the incision for drainage is made, but it may be contaminated with normal oral flora. The specimen should be promptly transported to the laboratory.

The cultivated bacteria are identified and antibiotic susceptibility testing is undertaken to establish the MIC of the antibiotics available. Antibiotics can usually be chosen to treat anaerobic infections based on the identification of the isolates and without susceptibility testing. The laboratory must have the capability to do susceptibility testing for strict anaerobes. Many bacteria are not cultivable and the results are only for those organisms that will grow on the media used in the laboratory.

## 12.11 Indications for prophylactic antibiotic therapy

Antibiotic prophylaxis is the administration of an antibiotic for the purpose of preventing bacterial colonization and to reduce the potential for post-treatment complications. Prospective, double-blind, placebo-controlled studies have shown that prophylactic antibiotics are not effective in managing patients with irreversible pulpitis, a localized abscess without systemic signs or symptoms, or to prevent flare-ups (Walton and Chiappinelli 1993; Fouad et al. 1996; Nagle et al. 2000; Henry et al. 2001; Pickenpaugh et al. 2001). In addition, a recent prospective, double-blind study compared clindamycin prophylaxis with a placebo and found no difference in the incidence of postoperative infections following periradicular endodontic surgery (Lindeboom et al. 2005).

For medically compromised patients with cardiac risks, the American Heart Association recommendations for prophylactic antimicrobial therapy should be followed. Of special concern are immunocompromised patients and the possibility of infection with organisms not normally found in infections of endodontic origin. Culture and antibiotic susceptibility testing is indicated for patients that are immunocompromised with rapidly spreading infections or if initial empirical antibiotic therapy is not effective. Although the incidence of bacteremia during nonsurgical endodontic treatment is low, studies have demonstrated transient bacteremias (Bender et al. 1960; Baumgartner et al. 1976, 1977; Debelian et al. 1995; Savarrio et al. 2005). Medically compromised patients at increased risk of infection must receive prophylactic antibiotics that follow the recommendations of the American Heart Association (Wilson et al. 2007). Patients recommended for prophylactic antibiotic therapy include prosthetic cardiac valves; pervious bacterial endocarditis; unrepaired cyanotic congenital heart disease (CHD), including palliative shunts and conduits; completely repaired congenital heart defect with prosthetic material or device (whether placed by surgery or by catheter intervention), during the first 6 months after the pro-

cedure; repaired CHD with residual defects at the site or adjacent to the site of a prosthetic patch or prosthetic device (which inhibits endothelialization); and cardiac transplant recipients who develop cardiac valvulopathy (Wilson et al. 2007). The standard general prophylaxis is amoxicillin 2 g 1 h before the dental procedure. See Table 12.1 for alternative regimens.

The American Dental Association and American Academy of Orthopedic Surgeons have also issued an advisory statement on the use of antibiotic prophylaxis for dental patients with total joint replacement (ADA 2003). Patients at potential increased risk of hematogenous total joint infection include immunocompromised or immunosuppressed patients

**Table 12.1** Prophylactic regimens for dental procedures.[a]

| Situation | Agent | Regimen: single dose 30–60 min before procedure |
|---|---|---|
| Standard general prophylaxis—oral | Amoxicillin | Adults: 2.0 g |
| | | Children: 50 mg/kg |
| Unable to take oral medications | Ampicillin | Adults: 2.0 g intramuscularly (IM) or intravenously (IV) |
| | | Children: 50 mg/kg IM or IV |
| | Or | |
| | Cefazolin[b,c] or ceftriaxone | Adults: 2 g IM or IV |
| | | Adults: 1 g IM or IV |
| | | Children: 50 mg/kg IM or IV |
| Allergic to penicillins | Clindamycin | Adults: 600 mg |
| | | Children: 20 mg/kg |
| | Or | |
| | Cephalexin[b,c] | Adults: 2.0 g |
| | | Children: 50 mg/kg |
| | Azithromycin or clarithromycin | Adults: 500 mg |
| | | Children: 15 mg/kg |
| Allergic to penicillin and unable to take oral medications | Clindamycin | Adults: 600 mg IM or IV |
| | | Children: 20 mg/kg IM or IV |
| | Or | |
| | Cefazolin[b,c] or ceftriaxone | Adults: 1.0 g IM or IV |
| | | Children: 50 mg/kg IM or IV |

[a] From Wilson et al. Prevention of infective endocarditis: Guidelines from the American Heart Association: A guideline from the American Heart Association Rheumatic Fever, Endocarditis and Kawasaki Disease Committee, Council on Cardiovascular Disease in the Young, and the Council on Clinical Cardiology, Council on Cardiovascular Surgery and Anesthesia, and the Quality of Care and Outcomes Research Interdisciplinary Working Group. JADA 2008; 139 : Table 2 on page 20 S. Copyright © 2008 American Dental Association. All rights reserved. Reprinted by permission.

[b] Cephalosporins should not be used in patients with a history of anaphylaxis, angioedema, or urticaria with penicillins or ampicillin.

[c] Or other first- or second-generation oral cephalosporin in equivalent adult to pediatric dosage.

with inflammatory arthropathies such as rheumatoid arthritis and systemic lupus erythematosus, or disease/drug/radiation-induced immunosuppression. Other patients that may be considered for prophylactic antibiotics include those with insulin-dependent (type 1) diabetes, the first 2 years following joint replacement, previous prosthetic joint infections, malnourishment, and hemophilia. The standard regimen for antibiotic prophylaxis is the administration of amoxicillin (2 g) 1 h before the dental procedure (ADA 2003).

The use of improved culturing and molecular methods now detect the presence of many more organisms in endodontic infections than previously determined. It is important that the clinicians understand the nature of polymicrobial endodontic infections and realize the importance of removing the reservoir of infection by endodontic treatment or tooth extraction. The prescription of antibiotics should be considered adjunctive to the clinical treatment of the patient. Antibiotics should not be substituted for root canal debridement and drainage of purulence from a periradicular swelling.

## 12.12 References

Abraham EP. 1962. The cephalosporins. *Pharm Rev* 14: 473–500.

ADA. 2003. Antibiotic prophylaxis for dental patients with total joint replacements. *JADA* 134: 895–99.

Baumgartner JC and Cuenin PR. 1992. Efficacy of several concentrations of sodium hypochlorite for root canal irrigation. *J Endod* 18(12): 605–12.

Baumgartner JC, Falkler WA, and Beckerman T. 1992. Experimentally induced infection by oral anaerobic microorganisms in a mouse model. *Oral Microbiol Immunol* 7: 253–56.

Baumgartner JC and Falkler WA, Jr. 1991a. Biosynthesis of IgG in periapical lesion explant cultures. *J Endod* 17(4): 143–46.

Baumgartner JC and Falkler WA, Jr. 1991b. Reactivity of IgG from explant cultures of periapical lesions with implicated microorganisms. *J Endod* 17(5): 207–12.

Baumgartner JC, Heggers J, and Harrison J. 1976. The incidence of bacteremias related to endodontic procedures. I. Nonsurgical endodontics. *J Endod* 2: 135–40.

Baumgartner JC, Heggers JP, and Harrison JW. 1977. Incidence of bacteremias related to endodontic procedures. II. Surgical endodontics. *J Endod* 3(10): 399–404.

Baumgartner JC and Hutter JW. 2001. Endodontic microbiology and treatment of infections. In: Cohen S and Burns R (eds), *Pathways of the Pulp*. St Louis, MO: C.V. Mosby.

Baumgartner JC, Hutter JW, and Siquiera JJF. 2006. Endodontic microbiology and treatment of infections. In:

Cohen S and Hargreaves KM (eds), *Pathways of the Pulp*. St Louis, MO: Mosby, 580–609.

Baumgartner JC, Sequeira JF, Jr, Xia T, and Rocas IN. 2002. Geographical differences in bacteria detected in endodontic infections using PCR. *J Endod* 28: 238.

Baumgartner JC, Siqueira JF, Xia T, and Rocas IN. 2004. Geographical differences in bacteria detected in endodontic infections using polymerase chain reaction. *J Endod* 30(3): 141–44.

Baumgartner JC and Watkins BJ. 1994. Prevalence of black-pigmented bacteria associated with root canal infections. *J Endod* 20(4): 191.

Baumgartner JC, Watkins JB, Bae KS, and Xia T. 1999. Association of black-pigmented bacteria with endodontic infections. *J Endod* 25(6): 413–15.

Baumgartner JC and Xia T. 2003. Antibiotic susceptibility of bacteria associated with endodontic abscesses. *J Endod* 29(1): 44–47.

Bender IB, Seltzer S, and Yermish M. 1960. The incidence of bacteremia in endodontic manipulation. *Oral Surg Oral Med Oral Pathol Oral Radiol Endod* 13(3): 353–60.

Bresco-Salinas M, Costa-Riu N, Berini-Aytes L, and Gay-Escoda C. 2006. Antibiotic susceptibility of the bacteria causing odontogenic infection. *Med Oral Pathol Oral Cir Bucal* 11(1): 70–75.

Brisson-Noel A, Trieu-Cuot P, and Courvalis P. 1998. Mechanism of action of spiramycin and other macrolides. *J Antimicrob Chemother* 22(Suppl. B): 13–23.

Brook I, Frazier EH, and Gher ME. 1991. Aerobic and anaerobic microbiology of periapical abscess. *Oral Microbiol Immunol* 6: 123–25.

Cozzarelli NR. 1980. DNA gyrase and the supercoiling of DNA. *Science* 207: 957.

Dajani AS, Taubert K, Wilson W, Bolger AF, Bayer A, Ferrieri P, Gewitz MH, Shulman ST, Soraya N, Newburger JW, Hutto C, Pallasch TJ, Gage TW, Levison M, Peter G, and Zuccaro G. 1997. Prevention of bacterial endocarditis: Recommendations by the American Heart Association. *JAMA* 277(22): 1794–1801.

Davies J. 1994. Inactivation of antibiotics and the dissemination of resistance genes. *Science* 264: 375–82.

Debelian GF, Olsen I, and Tronstad L. 1995. Bacteremia in conjunction with endodontic therapy. *Endod Dent Traumatol* 11(3): 142–49.

Docampo R. 1990. Sensitivity of parasites to free radical damage by antiparasite drugs. *Chem Biol Interact* 73: 1–27.

Drucker DB, Gomes B, and Lilley JD. 1997. Role of anaerobic species in endodontic infection. *Clin Infect Dis* 25(Suppl. 2): S220-S21.

Fisher LE and Russell RRB. 1993. The isolation and characterization of *Milleri* group streptococci from dental periapical abscesses. *J Dent Res* 72: 1191–93.

Fouad AF, Rivera EM, and Walton RE. 1996. Penicillin as a supplement in resolving the localized acute apical abscess. *Oral Surg Oral Med Oral Pathol Oral Radiol Endod* 81(5): 590–95.

Gilmore WC, Jacobus NV, Gorbach SL, and Doku HC. 1988. A prospective double-blind evaluation of penicillin versus clindamycin in the treatment of odontogenic infections. *J Oral Maxillofac Surg* 46: 1065–70.

Golan Y, McDermott L, Jacobus N, and Goldstein E. 2003. Emergence of fluoroquinolone resistance among *Bacteroides* species. *J Antimicrob Chemother* 52: 208–13.

Gomes BPFA, Drucker DB, and Lilley JD. 1994. Association of specific bacteria with some endodontic signs and symptoms. *Int Endod J* 27(6): 291–98.

Gossling J. 1988. Occurrence and pathogenicity of the *Streptococcus milleri* group. *Rev Infect Dis* 10: 257–85.

Haffajee AD, Bogren A, Hasturk H, Geres M, Lopez NJ, and Socransky SS. 2004. Subgingival microbiota of chronic periodontitis subjects from different geographic locations. *J Clin Periodontol* 31: 996–1002.

Haffajee AD, Japlit M, Bogren A, Kent RL, and Goodson JM. 2005. Differences in the subgingival microbiota of Swedish and USA subjects who were periodontally healthy or exhibited minimal periodontal disease. *J Clin Periodontol* 32: 33–39.

Hardman JG, Limbird LE, Molinoff PB, Ruddon RW, and Gilman AG, Eds. 2005. *Goodman and Gilman's the Pharmacological Basis of Therapeutics*. New York, McGraw-Hill.

Heimdahl A and Nord CE. 1985. Treatment of orofacial infections of odontogenic origin. *Scand J Infect Dis* 46: 101–5.

Heimdahl A, Von Konow L, Satoh T, and Nord CE. 1985. Clinical appearance of orofacial infections of odontogenic origin in relation to microbiological findings. *J Clin Microbiol* 22: 299–302.

Henry M, Reader A, and Beck M. 2001. Effect of penicillin on postoperative endodontic pain and swelling in symptomatic necrotic teeth. *J Endod* 27(2): 117–23.

Ingavale SS, Van Wamel W, and Cheung A. 2003. Characterization of RAT, an autolysis regulator in *Staphylococcus aureus*. *Mol Microbiol* 48(6): 1451–66.

Jacinto RC, Gomes BP, Ferraz CC, Zaia AA, and Filho FJ. 2003. Microbiological analysis of infected root canals. *Oral Microbiol Immunol* 18(5): 285–92.

Jaffe A, Chabbert YA, and Semonin O. 1982. Role of porin proteins OmpF and OmpC in the permeation of beta-lactams. *Antimicrob Agents Chemother* 22: 942–48.

Johnson BS. 1999. Principles and practice of antibiotic therapy. *Infect Dis Clin N Am* 13: 851–70.

Johnson PJ. 1993. Metronidazole and drug resistance. *Parasitol Today* 9: 183–86.

Kalant H and Roschlau WHE, Eds. 1998. *Principles of Medical Pharmacology*. New York: Oxford University Press.

Katagiri F, Itoh H, and Takeyama M. 2005. Effects of erythromycin on gastrin, somatostatin, and motilin levels in healthy volunteers and postoperative cancer patients. *Biol Pharm Bull* 28(7): 1307–10.

Khemaleelakul S, Baumgartner JC, and Pruksakorn S. 2002. Identification of bacteria in acute endodontic infections and their antimicrobial susceptibility. *Oral Surg Oral Med Oral Pathol Oral Radiol Endod* 94(6): 746–55.

Kinder SA and Holt SC. 1989. Characterization of coaggregation between *Bacteroides gingivalis* T22 and *Fusobacterium nucleatum* T18. *Infect Immun* 57: 3425–33.

Kuriyama T, Karasawa T, Nakagawa K, Saiki Y, Yamamoto E, and Nakamura S. 2000. Bacteriologic features and antimicrobial susceptibility in isolates from orofacial odontogenic infections. *Oral Surg Oral Med Oral Pathol Oral Radiol Endod* 90: 600–608.

Le Goff A, Bunetel L, Mouton C, and Bonnaure-Mallet M. 1997. Evaluation of root canal bacteria and their antimicrobial susceptibility in teeth with necrotic pulp. *Oral Microbiol Immunol* 12: 318–22.

Lewis MAO, MacFarlane TW, and McGowan DA. 1986. Quantitative bacteriology of acute dento-alveolar abscesses. *J Med Microbiol* 21: 101–4.

Lewis MAO, MacFarlane TW, McGowan DA, and MacDonald DG. 1988. Assessment of the pathogenicity of bacterial species isolated from acute dentoalveolar abscesses. *J Med Microbiol* 27: 109–16.

Lindeboom JAH, Frenken JWFH, Valkenburg P, and Van Den Akker HP. 2005. The role of preoperative prophylactic antibiotic administration in periapical endodontic surgery: A randomized, prospective double-blind placebo-controlled study. *Int Endo J* 38: 877–81.

Machado de Oliveira JC, Siqueira JF, Rôças IN, Baumgartner JC, Xia T, Peixoto RS, and Rosado AS. 2006. Bacterial community profiles of endodontic abscesses from Brazilian and US subjects as compared by denaturing gradient gel electrophoresis analysis. *Oral Microbiol Immunol* 22: 14–18.

Mandell GL, Bennett JE, and Dolin R, Eds. 1995. *Mandell, Douglas, and Bennett's Principles and Practice of Infectious Diseases*. New York: Churchill Livingston Inc.

Munson MA, Pitt-Ford T, Chong B, Weightman A, and Wade WG. 2002. Molecular and cultural analysis of the microflora associated with endodontic infections. *J Dent Res* 81: 761–66.

Murakami Y, Hanazawa S, Iwahashi H, Yamamoto A, and Fujisawa S. 2001. A possible mechanism of maxillofacial abscess formation: Involvement of *Porphyromonas endodontalis* liptpolysaccharide via the expression of inflammatory cytokines. *Oral Microbiol Immunol* 16: 321–25.

Murdoch DA. 1998. Gram-positive anaerobic cocci. *Clin Microbiol Rev* 11: 81–120.

Nagashima H, Takao A, and Maeda N. 1999. Abscess forming ability of *Streptococcus milleri* group: Synergistic effect with *Fusobacterium nucleatum*. *Microbiol Immunol* 43: 207–16.

Nagle D, Reader A, Beck M, and Weaver J. 2000. Effect of systemic penicillin on pain in untreated irreversible pulpitis. *Oral Surg Oral Med Oral Pathol Oral Radiol Endod* 90: 636–40.

Pallasch TJ. 1996. Pharmacokinetic principles of antimicrobial therapy. *Periodontol 2000* 10: 5–11.

Pickenpaugh L, Reader A, Beck M, Meyers WJ, and Peterson LJ. 2001. Effect of prophylactic amoxicillin on endodontic flare-up in asymptomatic, necrotic teeth. *J Endod* 27(1): 53–56.

Priyadarshini R, Popham DL, and Young KD. 2006. Daughter cell separation by penicillin-binding proteins and peptiglycan amidasis in *Escherichia coli*. *J Bact* 188(15): 5345–55.

Rôças IN, Baumgartner JC, Xia T, and Siqueira JF, Jr. 2006. Prevalence of selected bacterial named species and uncultivated phylotypes in endodontic abscesses from two geographic locations. *J Endod* 32(12): 1135–38.

Romano A, Gueant-Rodriguez R, Viola M, Petinato R, and Gueant J. 2004. Cross-reactivity and tolerability of cephalosporins in patients with immediate-hypersenitivity to penicillins. *Ann Int Med* 14(1): 16–22.

Sakamoto H, Kato H, Sato T, and Sasaki J. 1998. Semiquantitative bacteriology of closed odontogenic abscesses. *Bull Tokyo Dent Coll* 39: 103–7.

Sakamoto M, Rôças IN, Siqueira JF, and Benno Y. 2006. Molecular analysis of bacteria in asymptomatic and symptomatic endodontic infections. *Oral Microbiol Immunol* 21: 112–22.

Savarrio L, Mackenzie D, Riggio M, Saunders WP, and Bagg J. 2005. Detection of bacteraemias during non-surgical root canal treatment. *J Dent* 33: 293–303.

Siqueira JF, Jr, Magalhaes FAC, Lima KC, and de Uzeda M. 1998. Pathogenicity of facultative and obligate anaerobic bacteria in monoculture and combined with either *Prevotella intermedia* or *Prevotella nigrescens*. *Oral Microbiol Immunol* 13: 368–72.

Siqueira JF, Jr, Rôças IN, Souto R, de Uzeda M, and Colombo AP. 2001. Microbiological evaluation of acute periradicular abscesses by DNA-DNA hybridization. *Oral Surg Oral Med Oral Pathol Oral Radio Endod* 92: 451–57.

Siqueira JF, Jung IY, Rôças IN, and Lee CY. 2005. Differences in prevalence of selected bacterial species in primary endodontic infections from two distinct geographic locations. *Oral Surg Oral Med Oral Pathol Oral Radiol Endod* 99: 641–47.

Sobottka I, Cachovan G, Stürenburg E, Ahlers MO, Laufs R, and Platzer U. 2002. In vitro activity of moxifloxacin aginst bacteria isolated from odontogenic infections. *Antimicrob Agents Chemother* 49: 4019–21.

Spratt BG. 1980. Biochemical and genetic approaches to the mechanism of action of penicillin. *Philos Trans R Soc Lond [Biol]* 289: 273–83.

Sundqvist GK, Eckerbom MI, Larsson ÅP, and Sjögren UT. 1979. Capacity of anaerobic bacteria from necrotic dental pulps to induce purulent infections. *Infect Immun* 25: 685–93.

Takemoto T, Kurihara H, and Dahlén G. 1997. Characterization of *Bacteroides forsythus* isolates. *J Clin Microbiol* 35: 1378–81.

Tomasz A. 1986. Penicillin-binding proteins and the antibacterial effectiveness of β-lactam antibiotics. *Rev Infect Dis* 8: 5270–78.

Topazian RG and Goldberg MH. 2002. *Oral and Maxillofacial Infections*. Philadelphia, WB Saunders.

Von Konow L, Nord CE, and Nordenram Å. 1981. Anaerobic bacteria in dentoalveolar infections. *Int J Oral Surg* 10: 313–22.

von Rosenstiel NA and Adam D. 1995. Macrolide antibacterials. Drug interactions of clinical significance. *Drug Saf* 13: 105–22.

Walton RE and Chiappinelli J. 1993. Prophylactic penicillin: Effect on posttreatment symptoms following root canal treatment of asymptomatic periapical pathosis. *J Endod* 19(9): 466–70.

Waxman DJ, Yocum RR, and Strominger JL. 1980. Penicillins and cephalsporins are active site-directed acylating agents: Evidence in support of the substrate analogue hypothesis. *Philos Trans R Soc Lond [Biol]* 289: 257–71.

Wilson W, Traubert KA, and Gewitz M. 2007. Prevention of infective endocarditis: Guidelines from the American Heart Association. *J Am Dent Assoc* 138: 739–60.

Yagiela JA, Neidle EA, and Dowd FJ. 1998. *Pharmacology and Therapeutics for Dentistry*. St Louis, MO: Mosby.

# Chapter 13

# Topical Antimicrobials in Endodontic Therapy

*Tuomas Waltimo and Matthias Zehnder*

## 13.1 Endodontic infection

The dental pulp is a soft tissue organ capable to react against microbial challenges, which are usually caused by caries but may also be associated with other dental hard tissue defects or injuries such as attrition, abrasion, erosion, trauma, or fractures. The pulp in its sterile hard tissue compartment does not have a protective epithelial lining but is surrounded by a layer of odontoblasts and their processes in the dentinal tubules. Whenever exposed, dentin as a porous hard tissue does not prevent diffusion into the pulp (Gerzina and Hume 1995). Through dentinal tubules enzymes, toxins, and antigens of microbial origin trigger an inflammatory reaction in the pulp tissue. The pulp–dentin complex reacts with outward flow of dentinal fluid, sclerosis of dentinal tubules, and reactionary/reparative dentin production (Langeland 1987; Ekstrand et al. 1991). These responses can be considered as barrier functions against microbial or any other irritation. A prolonged and intensive microbial challenge such as dentin caries ultimately results in the invasion of the pulp tissue by microorganisms. Without elimination of the hard tissue infection by the dentist, the acute inflammatory pulp response does not suffice to fend off microbial invaders. Neutrophils infiltrate the tissue in large numbers, their proteolytic enzymes dissolve the pulp matrix (Gusman et al. 2002), and microabscesses develop (Langeland 1987). Hence, the soft tissue in the pulp chamber and later the entire root canal system become necrotic and infected.

The presence of microorganisms in the pulpless root canal can be classified into *primary*, *persistent*, and *secondary* infections. Acute or chronic root canal infections before root canal therapy are termed primary. Persistent infections are those that do not respond favorably to conservative therapy due to various reasons, such as difficult anatomical conditions or microorganisms resistant to the medicaments used. Secondary infections appear after the treatment has been successfully completed, but the coronal and the root canal seal are breached. A pulpless tooth after root canal treatment is essentially defenseless against the invasion of oral microbiota. Hence, only a bacteria-tight root filling and/or tight coronal seal can defend against a secondary infection of the root canal system.

Regardless of the type of infection, the microorganisms contained in the pulp space may gain access to the surrounding periodontium via anatomical pathways (Zehnder et al. 2002a). These are the openings of the main canal and its anatomical ramifications, which contain the nerve-vessel bundles to the pulp in a healthy state (Russell and Kramer 1956). Microorganisms in the necrotic root canal space cannot be reached or eliminated by the host defense system, and consequently, the host response aims at confining the root canal infection to the canal space (Nair 1997), thus preventing microbial invasion of the bone (Roane and Marshall 1972). The inflammation in the bone surrounding the canal openings apparent as a radiolucent area (Barthel et al. 2004) is termed apical periodontitis. In an acute stage, this is typically an abscess; if the disease process is chronic, the inflammatory lesion may be a granuloma or a cyst (Simon 1980). Based on currently available diagnostic tools, a diagnosis of the histological type of apical lesion is not possible (Trope et al. 1989). Consequently, according to the World Health Organization (WHO), there are only two possible clinical diagnoses for a nonvital tooth with an apical inflammation: acute or chronic apical periodontitis (ICD Version 2006, K04.4 and K04.5, respectively).

## 13.2 Microorganisms in the root canal system

Microbiology of root canal infection has been described in detail in the previous chapters. Therefore, this chapter provides only a brief summary of some general characteristics of the microbiology, and lists some factors, which may affect the response of microorganisms to topical antimicrobials used in endodontics.

Root canal infections are typically polymicrobial, with 3–10 different culturable taxa being present (Sundqvist 1994; Gomes et al. 2004). Not-yet-cultivated microorganisms are also found and contribute a remarkable part to endodontic infections (Munson et al. 2002; Siqueira and Rocas 2005). Microorganisms in long-standing endodontic infections are most frequently found close to the portals of exit of the root canal system, tentatively because there they have access to nutrients from the host.

Primary root canal infections are dominated by strict anaerobes (Sundqvist 1994; Gomes et al. 2004). In contrast, the majority of culturable taxa in secondary and persistent infections are facultative anaerobes (Siren et al. 1997; Molander et al. 1998; Gomes et al. 2004). The difference between primary and secondary/persistent infections appears to be largely due to the different ecological conditions in the root canal system before and after the treatment. Adaptation to the harsh environmental changes and often poor nutrient conditions as well as to the microecological pressure is the key characteristic required for survival. This explains the remarkable change in the microflora after the initiation of the root canal treatment (Chavez De Paz et al. 2003). Surviving microbial species may be present only in low proportions in the initial microflora and gain their predominance within the environmental changes induced by root canal treatment. Some species may reside in the parts of the root canal system, such as in ramifications in the apical delta, or in dentinal tubules, where no significant changes in the ecological conditions can be affected by current treatment protocols. Sufficient concentrations of antimicrobial agents in these areas may be difficult to reach. Alternatively or in addition, taxa not initially present may also invade into the root canal as contaminants during the therapy if strict asepsis has been neglected (Siren et al. 1997).

Interestingly, as in any microbial community there are positive as well as negative interactions between microbial species also occurring in infected root canals (Sundqvist 1992; Gomes et al. 2004). Positive interactions may be reflected in the microscopically visible plaque-like aggregates or biofilms found inside the infected root canal system (Nair 1997). However, it is to be mentioned that the term "biofilm" is often misused, and it is hitherto not clear whether the polymicrobial root canal aggregates are true biofilms with an expression of specific genes and extracellular proteins, exopolysaccharides, interspecies protein exchange, organized nutrition channels, and a tight adherence to the root canal walls (Costerton et al. 1995; Lasa 2006). True biofilms are often found in dynamic systems with a constant flow of liquid such as on stones in rivers, in dental unit waterlines, or on teeth surfaces exposed to the oral cavity (supragingival plaque). In contrast, in the root canal, there is not necessarily an ecological pressure to adhere and form biofilms in the true sense of the expression. Whether or not such true biofilms fulfilling the strictest criteria are present in the root canal, microbial communities of multiple species are of great interest in terms of dentin and root canal disinfection.

## 13.3 General treatment considerations

The microbial etiology and pathogenesis of pulpal and periapical disease dictates the treatment. Treatment goals and their expected success differ based on the diagnosis. Strategies applied to the treatment of an exposed vital dentin–pulp complex after a traumatic injury are different from those applied to the treatment of a tooth associated with an apical periodontitis. Root canal treatment of a vital tooth with a largely noninfected canal system has a roughly 10–15% higher chance for long-term periapical health compared to necrotic teeth with radiographic signs of an apical periodontitis (Basmadjian-Charles et al. 2002). Consequently, there appears to be a difference in expected treatment outcome between prevention and treatment of the root canal infection. Independent on the initial diagnosis, asepsis and disinfection are cornerstones of any endodontic therapy. From a microbiological standpoint, however, *antisepsis* is the biggest challenge when treating a nonvital tooth with apical periodontitis and an established root canal infection. Consequently, the following text concentrates on means to disinfect teeth with apical periodontitis. Disinfection of the entire root canal system is the primary treatment goal. This concept is widely accepted and has been recently confirmed by studies in monkeys with induced apical periodontitis that showed that the great majority of apical lesions heal if the root canal infection is sufficiently reduced or eliminated (Fabricius et al. 2006).

## 13.4 The chemomechanical disinfection approach

Regarding root canal disinfection, it should be realized that the dead and infected pulp tissue in the root canal space represents a unique environment in the human body. Apart from a maximal reduction of microbiota, necrotic tissue remnants should be dissolved, as proteins and peptides can potentially act as a nutrient source for microorganisms left behind after the treatment (Shah and Gharbia 1995). Moreover, the outer membrane of Gram-negative bacteria contains endotoxin, which is present in all necrotic teeth with periapical lesions (Dahlen and Bergenholtz 1980), and is able to trigger an inflammatory response even in the absence of viable bacteria (Dwyer and Torabinejad 1981; Hong et al. 2004). Moreover, the levels of endotoxin in necrotic root canals are positively correlated to clinical symptoms such as spontaneous pain and tenderness to percussion (Jacinto et al. 2005).

Based on the above facts, it appears pivotal to not just disinfect but to *cleanse* the necrotic root canal system to prevent the root canal from maintaining periapical inflammation. To this end, methods unique to the specific root canal environment can be employed. Instrumentation of the root canal system to mechanically remove infected tissue and debris is a major part of this procedure. However, mechanical treatment does not suffice to render the root canal system free of culturable microorganisms (Cvek et al. 1976; Bystrom and Sundqvist 1981). Current evidence dictates that mechanical preparation of the root canals is combined with the application of chemically active substances in order to obtain sufficient root canal cleanliness. As of yet, methods relying solely on chemical canal disinfection have not proven to be clinically successful (Attin et al. 2002). Therefore, the chemomechanical approach is recommended.

During the past two decades, the introduction of rotary nickel–titanium instruments has simplified mechanical root canal preparation. However, as shown in a randomized clinical trial, rotary instrumentation does not necessarily reduce bacterial counts in infected root canals to a greater extent than conventional hand instrumentation does (Dalton et al. 1998). Nevertheless, rotary instrumentation using nickel–titanium files has significantly improved the technical quality of root canal treatments performed by general practitioners (Molander et al. 2007). The central factor of treatment success, however, is still the dentist treating the patient. Even the most modern technical gadgets cannot compensate for treatment errors such as missing a canal, which will always result in a treatment failure in a case of apical periodontitis (Fig. 13.1). The importance of the technical preparation quality is essential for the application of topical antimicrobials and should not be underestimated. Poor success rate of endodontic treatment performed by inexperienced practitioners is linked to technical inadequacy (Imfeld 1991). Moreover, a clinical study on endodontic retreatments has shown that success rates dropped dramatically when the anatomy was violated in the initial treatment, that is, when it was impossible to mechanically reach the infected root canal area during the retreatment procedure (Gorni and Gagliani 2004).

Even with contemporary instrumentation techniques that maintain the main canal anatomy during

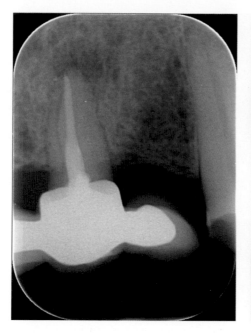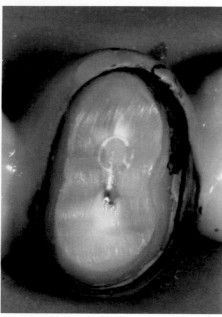

**Fig. 13.1**  Maxillary second premolar that was retreated 1 year previously. Patient appeared as an emergency, and the tooth was extracted at the oral surgery department. Note the missed palatal canal that was filled with pus and debris.

preparation, a large canal wall area is left untouched by the instruments (Peters 2004). Fins and ramifications of the main canals cannot be instrumented and contain ample space and necrotic tissue remnants. These locations in the necrotic root canal space may not be reached by sufficient concentrations of the topical antiseptics applied to the instrumented part of the canal, thus allowing microorganisms to persist (Fig. 13.2).

Apart from the inability to clean sufficiently, mechanical treatment has two side effects. First, it produces a smear layer consisting of dentinal filings and

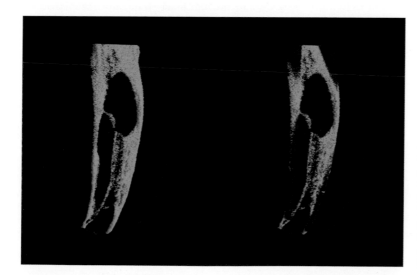

**Fig. 13.2**  Microcomputer tomography of a mandibular first molar prior to (left) and after (right) instrumentation using a contemporary rotary nickel–titanium instrument system. On the right panel, note the green areas, which reflect the so-called static voxels or areas that were not touched by the instrument.

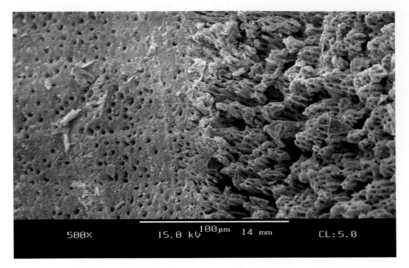

**Fig. 13.3** Canal wall of a mid-root aspect of a mandibular second premolar that was instrumented using a contemporary rotary nickel–titanium instrument system and copiously rinsed with a 1% NaOCl solution. Note the denuded calcospherites on the right aspect of this micrograph, which reflect an uninstrumented area with dissolved predentin. In the area touched by the instrument (left aspect), the calcospherites were scraped off by the rotary instrument, and a smear layer was created.

pulp tissue remnants on the canal wall (Fig. 13.3), and smear and debris are pressed into the fins and ramifications of the canal space. It has been shown in laboratory models that the endodontic smear layer may offer protection to microbiota adhering to root canal walls (Sen et al. 1999). The smear layer also impedes adaptation of root canal sealers to the canal wall, and thus promotes leakage in filled root canals (Shahravan et al. 2007).

The second untoward effect of mechanical root canal preparation is that dentin is removed during the procedure. Depending on the amount of hard tissue loss, this may compromise the tooth in withstanding loading stress during mastication (Lang et al. 2006). Therefore, the amount of dentin to be removed should be kept minimal. Ideally, the mechanical preparation follows the anatomical configuration of the root canals, allowing efficient application of topical antimicrobials.

## 13.5 Irrigants and dressings

Topical antimicrobials for endodontic usage can be divided into two groups: *irrigants* and *dressings*. Irrigants are solutions used to rinse the root canal system during and after preparation. Dressings are applied between two visits. As delineated in the previous paragraphs, the goal of the chemomechanical root canal procedure is to get the necrotic root canal system as clean as possible. Consequently, the chemical aspect

of this procedure should reduce the microbial level in the root canal system to a minimum, dissolve necrotic tissue remnants, prevent or dissolve the smear layer and accumulated debris, and inactivate endotoxin. On the other hand, the systemic and local side effects of these chemicals must be minimal. However, the root canal system is a semi-isolated space, which allows a limited communication with the surrounding tissues. Therefore, within certain limitations and with an awareness of risk factors, substances commonly classified as disinfectants are useful in conservative endodontic therapy. Based on the above requirements for topical antimicrobial substances (Table 13.1), sodium hypochlorite solutions and calcium hydroxide suspensions are the most frequently recommended irrigants and intervisit dressings in endodontics, respectively.

*Sodium hypochlorite* (NaOCl or NaClO) is industrially produced via electrolysis from sodium chloride, and is used for the chlorination of swimming pools, laundry bleaching, and multiple other purposes. It is cheap, available from many sources, and therefore also popular among clinicians (Frais et al. 2001). Sodium hypochlorite dissociates in water to $Na^+$ and $OCl^-$ (hypochlorite). $OCl^-$ is in a pH-dependent equilibrium with HOCl (hypochlorous acid). Pure aqueous sodium hypochlorite solutions, depending on their concentration, have a pH of 11–12. They are fairly stable if stored correctly, that is, in dark, airtight containers (Frais et al. 2001). However, in low concentrations as used for pool water disinfection, hypochlorous

**Table 13.1** Overview on the features of topical antimicrobials used in endodontics.

| Compound | Type | Action on biofilm | Tissue dissolution capacity | Endotoxin inactivation | Action on smear layer | Caustic potential | Allergic potential |
|---|---|---|---|---|---|---|---|
| Irrigants | | | | | | | |
| Hydrogen peroxide | Peroxygen | ++ | – | – | – | † | – |
| Sodium hypochlorite | Halogen-releasing agent | +++ | +++ | + | ++[a] | † | + |
| Iodine potassium iodide | Halogen-releasing agent | ++ | – | † | – | – | ++ |
| Chlorhexidine | Biguanide | ++ | – | + | – | † | + |
| EDTA | Polyprotic acid | †† | – | – | ++[b] | – | – |
| Citric acid | Organic acid | †† | – | – | +++[b] | – | – |
| Dressings | | | | | | | |
| Calcium hydroxide | Alkaline earth hydroxide | ++ | ++ | +++ | +[a] | – | – |
| CMCP | Phenol | †† | – | – | – | ++ | ++ |
| Iodoform | Triiodomethane | †† | – | †† | – | – | ++ |

EDTA, ethylenediamine tetraacetic acid; CMCP, camphorated mentholated chlorophenol. Effects: –, absent or minor; +, reported; ++, definitely present; +++, strong;+†, depending on concentration; ††, no information available.
[a]On organic moieties.
[b]On inorganic moieties.

acid (HOCl) is the more effective species than OCl⁻
(Bloomfield and Miles 1979). Chemistry dictates that
HOCl can only be present at physiological pH values or in an acidic environment ($pK_a$ of hypochlorous acid is 7.46). Henry Dakin originally proposed
a bisphosphonate-buffered solution for the cleansing
of open wounds, under the assumption that a solution of physiological pH would be less caustic (Dakin
1915). However, NaOCl solutions with an adjusted
pH are highly unstable, even autoreactive, and the
difference in antimicrobial efficacy and caustic effect between pure NaOCl solutions and counterparts
with adjusted pH is most likely clinically irrelevant at
the concentrations applied to the root canal system
(Cotter et al. 1985; Zehnder et al. 2002b). Both OCl⁻
and HOCl are extremely reactive oxidizing agents
(Dychdala 1991); they destroy the cellular activity of
proteins (McDonnell and Russell 1999).

After its introduction as a wound-cleansing agent
during World War I by Henry Dakin and Alexis Carrel,
hypochlorite was soon adopted as an endodontic irrigating solution (Coolidge 1919). Ever since, sodium
hypochlorite has maintained its status as the ideal root
canal irrigant. Hypochlorite has the unique feature
that it can clean the canal system from tissue remnants
and organic smear layer components (Grossman and
Meiman 1941; Baumgartner and Cuenin 1992; Naenni
et al. 2004). However, additional cleansing agents that
dissolve the inorganic dentin components are recommended to render the canal system clean prior to filling
it (Torabinejad et al. 2002).

Aqueous solutions of chelators such as *ethylenediamine tetraacetic acid (EDTA)* (Nygaard Ostby 1957)
or *citric acid* (Loel 1975) are the most frequently advocated agents for this purpose (Hulsmann et al. 2003).
Both are fairly efficient in dissolving the smear layer,
but interfere with the hypochlorite action and thus
should not be used concomitantly (Zehnder et al.
2005). In addition to their effect on the smear layer,
chelating agents also appear to affect a microorganisms organized in biofilms (see below).

*Hydrogen peroxide ($H_2O_2$)* solutions have also been
used traditionally together with sodium hypochlorite
as root canal irrigants (Grossman 1943). Peroxygens
are oxidants, which act via the liberation of hydroxyl-
free radicals. These attack essential cell components,
including lipids, proteins, and DNA (Block 1991). The
so-called effervescence or bubbling effect when $H_2O_2$
was used concomitantly with sodium hypochlorite was
thought to increase the cleansing ability of hypochlo-

rite. However, hydrogen peroxide is a fairly weaker
antiseptic than sodium hypochlorite (McDonnell and
Russell 1999), and, contrary to what has traditionally been believed without any hard data to prove the
concept (Grossman 1943), peroxide actually interferes
with the hypochlorite in solution, rendering the latter
agent ineffective. The bubbling effect is simply oxygen
gas that results from the reduction of hypochlorite by
peroxide in water (Baumgartner and Ibay 1987).

*Chlorhexidine* is a cationic biguanide with a broad-
spectrum, covering both Gram-positive taxa, oral
yeasts, and, to a lesser extent, Gram-negative bacteria
(Davies et al. 1954; Law and Messer 2004). Chlorhexidine is a base and most stable in form of its salts.
Chlorhexidine digluconate is the commonly available
form. The adsorption of chlorhexidine in low concentrations to the cell wall causes leakage of cytoplasmic
substances such as potassium and phosphorus, resulting in growth inhibition. Higher concentrations kill
microorganisms directly by coagulation and precipitation of the cytoplasm (Fardal and Turnbull 1986).
Chlorhexidine retention and slow release from exposed surfaces with prolonged antimicrobial activity
have revealed various applications in dentistry (Rolla
et al. 1971). However, chlorhexidine is relatively ineffective against various Gram-negative rods (Hennessey 1973; Slots et al. 1991), which represent the
major part of bacteria in primary root canal infections
(Sundqvist 1994).

*Iodine potassium iodide* is a widely used antiseptic and disinfectant in medicine. Iodine solutions are
unstable. Potassium iodide or, in more recent products, iodine carriers such as povidone or poloxamer
are added to the solution to keep the antimicrobially
active moiety, which is the molecular iodine ($I_2$), in
solution (McDonnell and Russell 1999). As a halogen-
releasing agent, iodine has a similar mode of action as
sodium hypochlorite (McDonnell and Russell 1999).
However, it lacks the ability to dissolve necrotic tissue (Naenni et al. 2004). Although less reactive than
chlorine, iodine is rapidly bactericidal, fungicidal, viru-
cidal, and sporicidal. Due to its antimicrobial potential and broad-spectrum, it has been considered as a
possible antiseptic in endodontic therapy (Engstrom
and Spangberg 1969; Safavi et al. 1985). Lower toxicity of iodine in comparison to commonly used irrigants could also indicate its possible usefulness as
a root canal irrigant (Engstrom and Spangberg 1967,
1969). However, its allergenic potential and possible
tooth-staining effect in long-term use may have had

a negative influence on the popularity of this potent antimicrobial among clinicians.

When it comes to root canal dressings, *calcium hydroxide* is the most frequently recommended agent. Calcium hydroxide or slaked lime is an alkaline earth hydroxide. A white, solid $Ca(OH)_2$ dissolves sparingly in water. Calcium hydroxide is an old remedy against tooth infections; it was recommended for the treatment of "fistula dentalis" as early as 1838 (Nygren 1838). Calcium hydroxide suspensions gained wide acceptance as root canal dressings after the landmark study by Bernhard Hermann (Hermann 1920). Because it disassociates into calcium and hydroxyl ions slowly, calcium hydroxide exerts a depot effect in the wet root canal system, that is, hydroxide ions continuously go in solution and interact with their environment (Nerwich et al. 1993). This alkaline capacity creates a high-pH environment, which interferes with microbial viability (Proell 1949). Furthermore, soft tissue is lysed (Hasselgren et al. 1988), and endotoxin is inactivated (Safavi and Nichols 1993), two features of calcium hydroxide not found with other commonly used root canal dressings.

Other frequently recommended "classical" root canal dressings are based on *camphorated and/or mentholated chlorophenol (CMCP)*. Phenolic compounds have long been used for their antiseptic, disinfectant, and preservative properties (McDonnell and Russell 1999). While being biocides with little specific action on microbiota, they do have membrane-active properties, which contribute to their overall activity (Denyer 1995). However, CMCPs are highly cytotoxic (Soekanto et al. 1996), allergenic (Pascher 1978), and fairly irritating to periapical tissues (Cruz and Barbosa 2005).

Finally, *iodoform* is a compound that has traditionally been used as an intervisit dressing or root filling, especially in primary teeth (Ranly and Garcia-Godoy 2000). The formula for iodoform ($CHI_3$) shows that this compound is related to chloroform ($CHCl_3$). It is synthesized by the reaction of iodine and sodium hydroxide with organic compounds. It was used in medicine as a healing and antiseptic dressing for wounds and sores around the beginning of the twentieth century, but has since been replaced by more potent antiseptics. Nevertheless, based on their biocompatibility, resorbability, and long-lasting antimicrobial effect, iodoform pastes are still successfully used after pulpectomies in primary teeth (Ranly and Garcia-Godoy 2000). In "adult" endodontics, iodoform was also applied in combination with CMCP, a root canal dressing known as Walkhoff's paste (Walkhoff 1882). However, the use of this combination is not recommended based on the relatively high toxicity compared to the antimicrobial effect and the excessive staining of dentin promoted by this dressing (Gutierrez and Guzman 1968).

## 13.6 Common in vitro methodologies to test antiseptic effects

There are hundreds of published articles in international peer-reviewed journals describing root canal and dentin disinfection. The great majority of these reports are of an in vitro nature reflecting the difficulty to obtain well-controlled clinical data about the efficacy and other characteristics of agents used in endodontics. As of yet, apart from studies comparing the one-visit treatment to the two-visit approach using a root canal dressing for the interim, no clinical investigation has related the use of antimicrobial substances to the clinical outcome of endodontic cases (Trope et al. 1999; Torabinejad et al. 2005; Waltimo et al. 2005). Because the in vitro papers involve a number of more or less suitable methodologies with respect to the clinical context, a critical approach in the interpretation of the results should be applied. Moreover, drawing conclusions based on experimental study designs must be executed with great care.

For better understanding of the strengths and weaknesses of the contemporary knowledge on the field of antimicrobials in endodontics, a critical approach to the methodology is essential. Therefore, some of the key determinants and general methodologies are described in the following paragraphs.

Antimicrobial agents may have three types of action that can be determined in a closed experimental system as follows: *Bacteriostatic* action refers to the inhibition of microbial growth; that is, both the total and the viable cell counts remain stable. Bacteriostatic agents are typically antibiotics that commonly inhibit protein synthesis by reversible concentration-dependent binding to ribosomes. *Bactericidal* action reduces viable cell counts whereas the total cell count is not affected. Thus, such agents bind irreversibly to their cellular targets, and kill the cells but do not cause cell rupture. *Bacteriolytic* action results in the reduction of both total and viable cell counts. It involves killing and destruction of the cell integrity by inhibition of cell wall synthesis such as penicillin does, or by the damage of the cell membrane and leakage of the cell contents as is the case with several antiseptics or disinfectants. As mentioned above, the action of disinfectants is usually based on protein precipitation, oxidization or iodization, alkalization, or on various interactions with cell wall phospholipids (McDonnell and Russell 1999).

*Agar diffusion* methods for determination of the bacteriostatic action use an agar medium evenly inoculated with the test organisms. Antimicrobial agents of known amounts are placed with paper disks or inside wells in the agar. The diffusion of the agent into the agar forms a concentration gradient; the highest concentration is close to the disk or well and gradually decreases toward the periphery. A zone of inhibition of the microbial growth can then be measured. This method and its several commercially available variants are simple, effective, and commonly used in clinical laboratories for antibiotic susceptibility testing. However, any agent with an uneven diffusion into the medium or neutralization by contents of the agar is not suitable to be tested with this methodology. Comparing disinfectants of a different chemical nature by agar diffusion should thus be considered questionable (Estrela et al. 2001; Haenni et al. 2003; Torabinejad et al. 2003b).

*Broth dilution* is a method comparable to agar diffusion with the difference that the concentration range is obtained by serial dilutions of the tested agent in a nutrient broth. The minimum inhibitory concentration or MIC of the agent under investigation is determined as the weakest concentration without visible growth. Broth dilution test can be extended by plating the broth of MIC and the concentrations above in order to detect minimum bactericidal (or bacteriolytic) concentration (MBC).

*Direct exposure* of a known amount of microbial colony-forming units (CFUs) of planktonic, adherent, or biofilm-associated microbes to antimicrobial solutions is a further commonly used test to determine bactericidal or bacteriolytic action. As such it is actually a modification of MBC determination allowing desired incubation periods and growth phases of the cultures, as well as adjusting other conditions without possibly confounding effects of nutrient broths (Portenier et al. 2005). After chosen exposure periods the antimicrobial agent is either neutralized or diluted, the microorganisms are harvested and resuspended in a buffer solution. Subsequently, the number of CFU is determined by cultivation on appropriate agar media. Of particular importance in this context is the theoretical detection limit, that is, the smallest proportion of the microorganisms, which can be detected. Direct exposure is suitable for irrigants such as sodium hypochlorite or chlorhexidine, whereas suspensions such as calcium hydroxide may be problematic to test in their clinically applied form due to their two-phase characteristics.

## 13.7 Contemporary knowledge based on the common in vitro methodologies

Our contemporary understanding on the relative susceptibility of microorganisms associated with root canal infections is largely based on investigations using the common in vitro methodologies described above. Among the latter, direct exposure tests can clinically provide the most useful data. Using a direct exposure test by screening the time required to kill 99.9% or more of the cells, difference between bacterial species in their in vitro susceptibility against a saturated aqueous calcium hydroxide solution, pH 12.5, was demonstrated (Bystrom et al. 1985). *Enterococcus faecalis* was found to be the most resistant species with a survival for time periods exceeding 6 min. In contrast, the majority of the species tested, such as oral streptococci and anaerobic Gram-negative rods, were killed within 1 min. Anaerobic Gram-positive rods such as *Actinomyces* spp., *Propionibacterium* spp., and *Eubacterium* spp. survived up to 6 min. Interestingly, solutions adjusted to pH 11.5 were not efficient against *E. faecalis*. The resistance of *E. faecalis* against high alkalinity has been confirmed in a number of studies using varying techniques (Safavi et al. 1985, 1990; Haapasalo and Orstavik 1987, Heling et al. 1992; Gomes et al. 2003). Resistance against calcium hydroxide comparable to that of *E. faecalis*, or even higher, has been shown within oral *Candida* spp. (Waltimo et al. 1999).

Commonly used irrigants are in general experimentally very potent in killing microorganisms isolated from root canal infections at concentrations fairly below the ones applied clinically. Sodium hypochlorite has been investigated and found effective against relevant microbiota in various studies using common in vitro methodologies (Spangberg et al. 1973; Waltimo et al. 1999). Contrary to other bacteria tested but also to earlier investigations, survival of *E. faecalis* in a direct exposure test using 0.5% sodium hypochlorite for periods up to 30 min has been reported (Gomes et al. 2001). A very low detection limit applied in the study combined with the used high bacteria/medicament ratio may explain this interesting finding.

Direct exposure tests have revealed in vitro killing efficacy of chlorhexidine in clinically appropriate concentration ranges of 0.05–2.0% against facultative Gram-positive bacteria and yeasts (Spangberg et al. 1973; Waltimo et al. 1999; Estrela et al. 2003). Several root canal isolates of *Candida albicans* were killed by a 0.05% solution within 1 h of incubation (Waltimo et al. 1999). In the same series of experiments on *C. albicans*, iodine-potassium iodide demonstrated systematically stronger fungicidal efficacy than chlorhexidine acetate at clinically applicable concentrations (Waltimo et al. 1999). Susceptibility of *E. faecalis* comparable to that of *C. albicans* has been reported (Portenier et al. 2002). However, chlorhexidine showed a markedly lesser activity against Gram-negative bacteria such as *Pseudomonas aeruginosa* and Gram-positive obligate anaerobe *Bacillus subtilis* than sodium hypochlorite in a comparative direct exposure experiment (Estrela et al. 2003).

The in vitro efficacy documented for sodium hypochlorite, chlorhexidine, and iodine has initiated investigations on possibly beneficial synergism between the modes of action of these agents and calcium hydroxide. Direct exposure tests have indicated that antimicrobial efficacy and spectrum of calcium hydroxide can be improved, although there is also evidence that the high alkalinity or other characteristics of calcium hydroxide may hinder the efficacy of other disinfectants, in particular that of iodine and chlorhexidine (Waltimo et al. 1999; Haenni et al. 2003). In contrast, in terms of chemical compatibility, hypochlorite and calcium hydroxide are a perfect match, as the calcium hydroxide in suspension does not alter the available chlorine (Zehnder et al. 2003), and the hypochlorite does not affect the calcium hydroxide pH effect (Haenni et al. 2003).

## 13.8 Modified in vitro tests and their outcome related to endodontics

The common methods described above are suitable for the evaluation of the relative efficacy of antimicrobial agents and their combinations, but the interpretation and/or extrapolation of the data to clinical endodontics is difficult. Even the determination of bactericidal or bacteriolytic efficacy with direct exposure tests may reveal clinically useless or misleading information due to the complexity of the clinical reality. In order to obtain more relevant data and to gain understanding on the factors that might affect the clinical efficacy of topical antimicrobials, a number of modifications of commonly used tests have been developed. These methods consider on one hand the inhibitory effects of microenvironmental factors, such as the presence of organic and inorganic dentin compounds in the root canal system, and on the other hand the possibly increased resistance of adherent and/or mixed microbial populations against antimicrobial agents.

Chronologically seen, the bovine dentin block model may have been the first serious attempt with a wide acceptance in the endodontic community to mimic the clinical reality in the context of topical antimicrobial efficacy in the root canal system (Haapasalo and Orstavik 1987). The model involves the preparation of blocks obtained from bovine tooth roots to a standardized inner diameter. These blocks are then infected with an *E. faecalis* strain. After application of topical antiseptics for the desired period of time, the blocks are washed and dentin filings from different dentin depths are harvested using sterile burs of increasing size. Although this in vitro model has its own weaknesses, its contribution to our knowledge as a pioneering study is beyond doubt. Some advantages of the dentin block model over commonly used laboratory tests are (i) the colonization of root canal walls and dentinal tubules; (ii) possible interactions between the agents tested and dentin; and (iii) the periods of dressing times clinically relevant. On the other hand, the list of disadvantages also contains some remarkable factors such as (i) bovine dentin structure differs from that of human dentin (Schilke et al. 2000) and, therefore, the infection as well as penetration of antimicrobial agents into dentinal tubules may not reflect the clinical reality (see Fig. 13.4); (ii) the large size of the canal lumen and subsequent medicament volume in relation to the volume of infected dentin are disproportional; and (iii) a monospecies infection rarely reflects the clinical situation. However, keeping these limitations in mind, this model has, amongst other observations, indicated that adherence on dentin and penetration into dentinal tubules by microbiota inhibits the killing efficacy of antiseptics. These and other results obtained with this model can provide useful preclinical information.

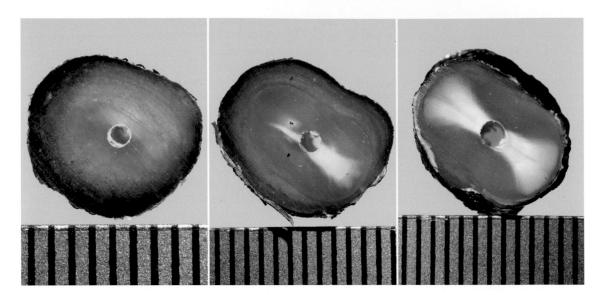

**Fig. 13.4**   Human mandibular premolar that was irrigated with a 2% patent blue solution. Cross sections (from left to right) 3, 6, and 9 mm from the apex. The blue color can only diffuse into areas with open tubules. The figure illustrates the typical sclerosis pattern of adult teeth, that is, there are no open tubules in the apical section. Mesiodistal aspects of the entire root become sclerotic with increasing age, followed by bucco-oral aspects (Vasiliadis et al. 1983; Paque et al. 2006).

In order to study the interactions between dentin and antimicrobial agents further, a modification of commonly used direct exposure tests has been developed. In this setup, the agents of interest are allowed to interact with powdered dentin or its components, and then the antimicrobial efficacy against chosen microorganisms is tested (Haapasalo et al. 2000). It seems that dentin efficiently inhibits the antimicrobial efficacy of calcium hydroxide and iodine, whereas chlorhexidine is less affected. The inhibition of calcium hydroxide is based on the buffering capacity of dentin. Both hydroxylapatite and albumin are also capable to inhibit the efficacy of calcium hydroxide but not that of iodine. Binding to organic material inhibits chlorhexidine, but hydroxylapatite alone does not seem to hamper its killing efficacy (Haapasalo et al. 2000; Portenier et al. 2001)

One further method to study topical antimicrobials in endodontics is their application in human teeth ex vivo. This approach bears the advantage that it is controllable and repeatable under laboratory conditions. Moreover, in comparison to the bovine dentin block model, the dentin structure and medicament to infected dentin ratio reflect the clinical reality. Anatomically (radiologically) matched pairs of teeth,

or even better, contralaterally extracted premolars allow well-controlled comparisons between the agents tested (Zehnder et al. 2006a), as these matched pairs have a close to identical macroscopic and tubular anatomy (Wood and Green 1969; Vasiliadis et al. 1983). Up to date, such ex vivo studies using human teeth have applied *E. faecalis* single-species infection protocols comparable to the dentin block model of Haapasalo and Orstavik (1987). However, because monoinfections with Gram-positive bacteria are rare, the development of a multi-species ex vivo model would be rational.

The latest development in the testing of topical antimicrobials is the use of multispecies biofilms (Guggenheim et al. 2001; Chavez de Paz et al. 2007). As our understanding of heterogenic microbial communities has improved, it has also become evident that biofilms can also be approached as such instead of the traditional separate species-based approach, determining on one hand, their expression of virulence factors, and on the other hand, their collective susceptibility to antimicrobials (Chavez de Paz et al. 2007). This approach will certainly clarify some of the key questions in contemporary endodontology. An interesting recent finding in this context is that chlorhexidine, in contrast to hypochlorite, has only limited efficacy against a multispecies biofilm of anaerobic oral microbiota (Muller et al. 2007). This in vitro observation correlates well with the results of a recently published randomized clinical trial. In human teeth with apical periodontitis, 2.5% NaOCl reduced microbial loads significantly better than 2% chlorhexidine in terms of both viable counts and PCR amplification of ribosomal RNA genes (Vianna et al. 2006). One more example underlining the usefulness of biofilm models to study the effectiveness of topical antimicrobials in endodontics is EDTA. EDTA has only a limited antimicrobial effect against planktonic bacteria (Zehnder et al. 2005), but is a powerful biofilm disruptor (Banin et al. 2006). Calcium plays an important role in the cohesiveness of biofilms (Ahimou et al. 2007). This might explain why EDTA has a strong effect in reducing bacterial loads in root canals clinically (Yoshida et al. 1995).

## 13.9 Animal models

Animal models bear the advantage that the effect of topical antimicrobials on healing of (previously induced) apical periodontitis can be monitored histologically. Human studies on the histological assessment of periapical healing, as has been done using block sections of the root apex and surrounding tissues, would be impossible to get through ethics committees these days (Seltzer et al. 1968). However, animal tests are also questionable from an ethical point of view. It may be argued that the biocompatibility of antimicrobials has usually already been tested in animals for other purposes. Furthermore, animal teeth other than those of certain monkey species are not very similar to human teeth anatomically.

Studies in dog's teeth with induced apical periodontitis consistently showed a positive effect of a calcium hydroxide dressing on bone healing (Katebzadeh et al. 1999; Holland et al. 2003; Leonardo et al. 2006). Whether this was via a greater bacterial reduction, the effect of calcium hydroxide on endotoxin (Silva et al. 2002), or an effect of the root canal dressing on periapical tissues remains unknown. In a similar dog model, irrigation with 2% chlorhexidine resulted in a significantly better periapical repair than irrigation with 5.25% sodium hypochlorite (Tanomaru Filho et al. 2002). The concentrated hypochlorite solution, on the other hand, was better in promoting healing

than a 0.5% solution (Leonardo et al. 1995). However, as indicated above, these findings do not necessarily correlate with clinical observations, and results have to be interpreted with care.

## 13.10 Randomized clinical trials

The ultimate outcome of an endodontic treatment is the retention, function, and/or periapical health of the teeth that were treated. To demonstrate causality during a treatment procedure such as the use of a topical antimicrobial agent on outcome, this step has to be exercised randomly versus a similar treatment using a control agent, while all other treatment steps should be identical. However, it can take years for apical periodontitis to heal radiographically (Orstavik 1996), and consequently, few randomized trials on the effect of topical antimicrobials on endodontic treatment outcome have been performed (Trope et al. 1999; Weiger et al. 2000; Peters and Wesselink 2002; Waltimo et al. 2005). These studies found no difference in healing rates in teeth that were irrigated with sodium hypochlorite and subsequently dressed with calcium hydroxide between visits compared to counterparts that were left empty between visits or filled directly after the first visit. It may be argued that based on the high efficacy of sodium hypochlorite expected differences were small and case numbers too low to address the experimental question.

One possibility to avoid issues related to outcome studies is to define the so-called surrogate outcomes, which are outcome variables that are easier to attain, but are linked to the true treatment outcome. In the case of endodontic antimicrobials, this surrogate outcome is microbial load after treatment, which has been shown in humans as well as in monkeys to be related to healing (Sjogren et al. 1997; Waltimo et al. 2005; Fabricius et al. 2006). However, as has been outlined that microbial sampling from infected root canals is not free of possible methodological pitfalls (Reit et al. 1999). Unfortunately, few studies of an acceptable quality have assessed the effectiveness of topical antimicrobials other than calcium hydroxide (Law and Messer 2004). A systematic review and meta-analysis on the antibacterial efficacy of calcium hydroxide revealed that this dressing material (used after sodium hypochlorite irrigation) has limited effectiveness in eliminating bacteria from human root canals when assessed by culture techniques (Sathorn et al. 2007). It was also shown that leaving a 5% iodine potassium iodide solution in the canal system for 10 min after chemomechanical debridement with 0.5% NaOCl during a one-visit treatment approach results in a similar microbial reduction as the two-visit approach with an intervisit calcium hydroxide dressing (Kvist et al. 2004). As described above, in vitro studies have suggested that calcium hydroxide powder could be mixed with an antiseptic irrigant rather than an inert aqueous solution to improve its effectiveness. However, the combination of calcium hydroxide powder with a 2% chlorhexidine solution was not more effective in reducing intracanal microbial loads than a calcium hydroxide suspension in an inert solution (Zerella et al. 2005). As of yet, no clinical trial has investigated the effectiveness of a calcium hydroxide/sodium hypochlorite suspension.

Studies on different irrigating protocols have shown that a final rinse with 2% chlorhexidine resulted in significantly less growth-positive root canals than a control treatment with saline solution after irrigation with 1% NaOCl during instrumentation (Zamany et al. 2003). However, a final chlorhexidine flush was not compared to a corresponding treatment with sodium hypochlorite. When canals of single-rooted teeth were randomly irrigated with 2.5% NaOCl or 0.2% chlorhexidine throughout the treatment and left empty between treatments, significantly more anaerobic microbiota were recovered in the chlorhexidine group compared to the NaOCl group, while no difference was found regarding the recovery of facultative taxa. Furthermore, significantly more culture reversals from negative to positive occurred in the chlorhexidine compared to the hypochlorite group (Ringel et al. 1982). These results are in line with the above-mentioned fact that chlorhexidine is more effective against facultatives than strict anaerobes. Furthermore, as indicated above, chlorhexidine does not remove the organic tissue remnants from the canal system, which could explain the higher culture reversal rate in comparison to NaOCl. Another randomized clinical trial revealed that 2% chlorhexidine not only caused a lesser reduction in culturable microorganisms from infected root canals than 2.5% NaOCl, but also resulted in less reduction of bacteria-specific 16S rRNA genes (Vianna et al. 2006). Hence, when used as the main irrigant, hypochlorite appears to be more efficient than chlorhexidine in reducing both culturable and nonculturable taxa.

## 13.11 Ideal application modes of topical antimicrobials

To maximize the success of chemomechanical root canal treatment, there are some key points to be followed: first, the technical aspects that are the mechanical part of the treatment should be exercised with great care. There are no biologically acceptable chemicals that could compensate for operator failures such as large uninstrumented areas or missed canals in teeth with apical periodontitis. Second, the kinetics and interactions of the chemicals that are applied during and after mechanical treatment should be understood. The first issue is not the topic of this chapter, so this subsection focuses on chemical aspects. The most important notion in this context is that chemical disinfection is a *function of time and concentration*. Understanding the kinetics of the chemical reactions in the root canal system is pivotal to time the individual treatment steps. Unfortunately, this exact issue is the one point that has received far too little attention in published investigations, and so we have to relate to deductive considerations based on basic research information.

As delineated above, sodium hypochlorite should be the main, or, in fact, the only irrigant used during instrumentation. This increases the NaOCl reaction time on root canal microbiota. In addition, keeping an aqueous hypochlorite solution in the root canal during debridement reduces stress on rotary instruments (Boessler et al. 2007). As hypochlorite is extremely reactive, the active moieties in the canal are probably used up fairly quickly, although there are no hard data on this issue. To increase active $OCl^-$ levels in the canal, one could use a concentrated (5–6%) NaOCl solution, as is commercially available (household bleach). The antimicrobial efficacy of hypochlorite is concentration dependent (Radcliffe et al. 2004). However, because hypochlorite is a nonspecific oxidizing agent, it has also a dose-dependent caustic effect if inadvertently pressed over the apex during irrigation (Hulsmann and Hahn 2000). Furthermore, sodium hypochlorite not only dissolves necrotic, and, to a lesser extent, vital pulp tissue, but also attacks the organic dentin matrix (Driscoll et al. 2002). In an ex vivo setup, repeated flushing with 5.25% NaOCl over 30 min significantly increased tooth strain under standard cyclic nondestructive loading compared to a control treatment with 0.5% NaOCl (Sim et al. 2001). These data suggest that the excessive use of concentrated

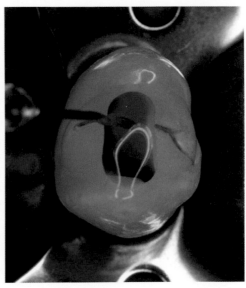

**Fig. 13.5**    Irrigating needle (30-gauge) with a safety tip that can be introduced to full working length if the canal is prepared to an ISO size 40 (Zehnder 2006).

hypochlorite might render a tooth more prone to fracture. Based on these concerns and the fact that neither clinical nor ex vivo studies on extracted human teeth have found an increased efficacy of 5% in comparison to 0.5 or 1% NaOCl solutions (Bystrom and Sundqvist 1983, 1985; Siqueira et al. 2000), hypochlorite should not be used at concentrations over 1–3% to irrigate root canals. To compensate for the lower concentration, higher volumes of the NaOCl irrigant should be used (Siqueira et al. 2000; Fouad and Barry 2005). This will deliver new active moieties to the target area, which is the zone of infection. This zone is close to the portals of exit of the root canal system. Penetration of an irrigant into the instrumented root canal, on the other hand, is related to the depth that the irrigating needle can be introduced into the canal (Ram 1977). Consequently, it appears important to use a slim irrigating needle with a safety tip that reaches the apical area (Fig. 13.5).

After the mechanical treatment is completed, canals should be flushed for roughly 1 min with a 17% EDTA or 10% citric acid solution. This will remove the smear layer that was created during instrumentation. However, no information is available as to how long exactly the irrigation with chelators should be exercised. Ex vivo experiments suggest that 30 s contact to 17% EDTA is enough to remove the smear from root dentin; however, that only goes for exposed areas and may be different in the complex anatomy of root canals (De-Deus et al. 2007). In addition to improved canal cleanliness, the microbial load is most likely further reduced by this treatment step (Bystrom and Sundqvist 1985), and the seal of a subsequent root filling is improved (Shahravan et al. 2007). The next step depends on what is planned with the specific case. If the canal system appears clean and it is decided to fill the root canal system, chlorhexidine can be used as the final irrigant, in particular if specifically targeted against Gram-positive bacteria of persistent infections.

If calcium hydroxide is used as the interim dressing, the canals should be flushed with sodium hypochlorite. Calcium hydroxide powder can then be mixed with sodium hypochlorite to prolong the hypochlorite action (Zehnder et al. 2003). In contrast to the chemical reactions promoted by sodium hypochlorite, the effect of the calcium hydroxide in the suspension has a slow onset but is long lasting (Zehnder et al. 2003). However, similar to sodium hypochlorite, the lytic properties of calcium hydroxide are not only limited to pulp tissue, but also break down the dentin matrix. Consequently, calcium hydroxide can make teeth more brittle (Cvek 1992; Rosenberg et al. 2007). Hence, canals should be dressed for at least 1 week (Sjogren et al. 1991), but probably not for periods longer than

1 month. The suspension should be relatively thin to improve the antimicrobial effect of the calcium hydroxide (Behnen et al. 2001). The calcium hydroxide suspension should be placed using a lentulo spiral to working length (Peters et al. 2005). To facilitate the removal of the calcium hydroxide at the subsequent appointment, a chelating solution such as EDTA or citric acid can be employed (Nandini et al. 2006).

## 13.12 Considerations on future research and possibilities

More efficient and simplified treatment regimens fulfilling the requirements of cleansing and disinfection of the entire root canal system are of interest. A recently introduced approach combines agents with different effects—a tetracycline isomer, an acid, and a detergent (MTAD)—in order to provide an all-in-one irrigant (Torabinejad et al. 2003a). Due to methodological pitfalls revealing questionable data using agar diffusion to compare agents of different nature, and on the other hand, due to contradictory results on the antimicrobial efficacy of MTAD in root canal and dentin disinfection, the advantages of this irrigant are hitherto not clear (Torabinejad et al. 2003b; Kho and Baumgartner 2006; Ruff et al. 2006; Krause et al. 2007). It appears that MTAD cannot replace sodium hypochlorite, but may be used as an alternative to (pure) citric acid or EDTA after instrumentation/NaOCl irrigation. Alternatively, the use of a weak chelator compatible with sodium hypochlorite has been proposed (Zehnder et al. 2005). Etidronic acid (chemical name 1-hydroxyethylidene-1,1-bisphosphonate or HEBP) could be mixed with sodium hypochlorite to obtain an all-in-one irrigant that would prevent the formation of a smear layer during instrumentation, rather than removing it after instrumenting as is done currently. However, further research is necessary to evaluate this concept.

Activation of a final hypochlorite irrigant in the instrumented canal system via passive ultrasonic activation (Martin 1976) or heat (Sirtes et al. 2005) is further interesting approaches that should be evaluated in clinical trials. However, as indicated above, desired and untoward hypochlorite effects should be weighed against each other, or, in other words, the question of "how much hypochlorite is good enough?" should be carefully addressed. Tentatively, the amount and activity of hypochlorite needed to render a canal system free of microbiota and tissue remnants depend on the complexity of the anatomy.

Biocompatible alternatives to calcium hydroxide are also of interest. One group of materials that could possibly replace calcium hydroxide is bioactive glasses of the $SiO_2$–$Na_2O$–$CaO$–$P_2O_5$ system (Zehnder et al. 2004). Similar to calcium hydroxide, these create a high-pH environment (Zehnder et al. 2006b), but bear the advantage that they mineralize dentin rather than weaken it (Forsback et al. 2004). However, with the currently available materials, disinfection efficacy is less than that observed with calcium hydroxide (Zehnder et al. 2006a). Latest developments using nanometric glass particulates have improved the antimicrobial effect of bioactive glass (Waltimo et al. 2007). This approach, however, needs further assessment before being considered clinically.

Technical developments in medicine may allow a wide variety of new approaches in the control of infectious and other diseases. For example, there is currently a rising interest in photodynamic therapy in various fields of medicine. This may also provide an efficient strategy for the infection control in endodontics. However, this remains to be shown in well-controlled in vitro, ex vivo, and clinical studies.

## 13.13 References

Ahimou F, Semmens MJ, Novak P, and Haugstad G. 2007. Biofilm cohesiveness measurement using a novel AFM methodology. *Appl Environ Microbiol* 73: 2897–2904.

Attin T, Buchalla W, Zirkel C, and Lussi A. 2002. Clinical evaluation of the cleansing properties of the noninstrumental technique for cleaning root canals. *Int Endod J* 35: 929–33.

Banin E, Brady K, and Greenberg E. 2006. Chelator-induced dispersal and killing of *Pseudomonas aeruginosa* cells in a biofilm. *Appl Environ Microbiol* 72: 2064–69.

Barthel C, Zimmer S, and Trope M. 2004. Relationship of radiologic and histologic signs of inflammation in human root-filled teeth. *J Endod* 30: 75–79.

Basmadjian-Charles CL, Farge P, Bourgeois D, and Lebrun T. 2002. Factors influencing the long-term results of endodontic treatment: A review of the literature. *Int Dent J* 52: 81–86.

Baumgartner J and Cuenin P. 1992. Efficacy of several concentrations of sodium hypochlorite for root canal irrigation. *J Endod* 18: 605–12.

Baumgartner J and Ibay A. 1987. The chemical reactions of irrigants used for root canal debridement. *J Endod* 13: 47–51.

Behnen M, West L, Liewehr F, Buxton T, and Mcpherson JR. 2001. Antimicrobial activity of several calcium hydroxide preparations in root canal dentin. *J Endod* 27: 765–67.

Block SS. 1991. Peroxygen compounds. In: Block SS (ed), *Disinfection, Sterilization, and Preservation*. Philadelphia: Lea and Febinger, pp. 167–81.

Bloomfield S and Miles G. 1979. The antibacterial properties of sodium dichloroisocyanurate and sodium hypochlorite formulations. *J Appl Bacteriol* 46: 65–73.

Boessler C, Peters OA, and Zehnder M. 2007. Impact of lubricant parameters on rotary instrument torque and force. *J Endod* 33: 280–83.

Bystrom A, Claesson R, and Sundqvist G. 1985. The antibacterial effect of camphorated paramonochlorophenol, camphorated phenol and calcium hydroxide in the treatment of infected root canals. *Endod Dent Traumatol* 1: 170–75.

Bystrom A and Sundqvist G. 1981. Bacteriologic evaluation of the efficacy of mechanical root canal instrumentation in endodontic therapy. *Scand J Dent Res* 89: 321–28.

Bystrom A and Sundqvist G. 1983. Bacteriologic evaluation of the effect of 0.5 percent sodium hypochlorite in endodontic therapy. *Oral Surg Oral Med Oral Pathol* 55: 307–12.

Bystrom A and Sundqvist G. 1985. The antibacterial action of sodium hypochlorite and EDTA in 60 cases of endodontic therapy. *Int Endod J* 18: 35–40.

Chavez De Paz LE, Bergenholtz G, Dahlen G, and Svensater G. 2007. Response to alkaline stress by root canal bacteria in biofilms. *Int Endod J* 40: 344–55.

Chavez De Paz LE, Dahlen G, Molander A, Moller A, and Bergenholtz G. 2003. Bacteria recovered from teeth with apical periodontitis after antimicrobial endodontic treatment. *Int Endod J* 36: 500–508.

Coolidge E. 1919. The diagnosis and treatment of conditions resulting from diseased dental pulps. *J Am Dent Assoc* 6: 337–49.

Costerton J, Lewandowski Z, Caldwell D, Korber D, and Lappin-Scott HM. 1995. Microbial biofilms. *Annu Rev Microbiol* 49: 711–45.

Cotter J, Fader R, Lilley C, and Herndon DN. 1985. Chemical parameters, antimicrobial activities, and tissue toxicity of 0.1 and 0.5% sodium hypochlorite solutions. *Antimicrob Agents Chemother* 28: 118–22.

Cruz R and Barbosa SV. 2005. Histologic evaluation of periradicular tissues in dogs treated with calcium hydroxide in combination with HCT20 and camphorated P-chlorophenol. *Oral Surg Oral Med Oral Pathol Oral Radiol Endod* 100: 507–11.

Cvek M. 1992. Prognosis of luxated non-vital maxillary incisors treated with calcium hydroxide and filled with gutta-percha. A retrospective clinical study. *Endod Dent Traumatol* 8: 45–55.

Cvek M, Nord C, and Hollender L. 1976. Antimicrobial effect of root canal debridement in teeth with immature root. A clinical and microbiologic study. *Odontol Revy* 27: 1–10.

Dahlen G and Bergenholtz G. 1980. Endotoxic activity in teeth with necrotic pulps. *J Dent Res* 59: 1033–40.

Dakin H. 1915. On the use of certain antiseptic substances in treatment of infected wounds. *Br Med J* 2: 318–20.

Dalton B, Orstavik D, Phillips C, Pettiette M, and Trope M. 1998. Bacterial reduction with nickel-titanium rotary instrumentation. *J Endod* 24: 763–67.

Davies G, Feancis J, Martin AR, Rose F, and Swain G. 1954. 1:6-Di-4'-chlorophenyldiguanidohexane (hibitane); laboratory investigation of a new antibacterial agent of high potency. *Br J Pharmacol Chemother* 9: 192–96.

De-Deus G, Reis C, Fidel R, Fidel S, and Paciornik S. 2007. Co-site digital optical microscopy and image analysis: An approach to evaluate the process of dentine demineralization. *Int Endod J* 40: 1–12.

Denyer SP. 1995. Mechanisms of action of antibacterial biocides. *Int Biodeterior Biodegrad* 36: 227–45.

Driscoll C, Dowker S, Anderson P, Wilson R, and Gulabivala K. 2002. Effects of sodium hypochlorite solution on root dentine composition. *J Mater Sci Mater Med* 13: 219–23.

Dwyer T and Torabinejad M. 1981. Radiographic and histologic evaluation of the effect of endotoxin on the periapical tissues of the cat. *J Endod* 7: 31–35.

Dychdala GR. 1991. Chlorine and chlorine compounds. In: Block SS (ed), *Disinfection, Sterilization and Preservation*. Philadelphia: Lea and Febiger, pp. 131–51.

Ekstrand K, Carlsen O, and Thylstrup A. 1991. Morphometric analysis of occlusal groove-fossa-system in mandibular third molar. *Scand J Dent Res* 99: 196–204.

Engstrom B and Spangberg L. 1967. Studies on root canal medicaments. I. Cytotoxic effect of root canal antiseptics. *Acta Odontol Scand* 25: 77–84.

Engstrom B and Spangberg L. 1969. V. Toxic and antimicrobial effects of antiseptics in vitro. *Sven Tandlak Tidskr* 62: 543–49.

Estrela C, Bammann LL, Pimenta FC, and Pecora J. 2001. Control of microorganisms in vitro by calcium hydroxide pastes. *Int Endod J* 34: 341–45.

Estrela C, Estrela C, Reis C, Bammann LL, and Pecora J. 2003. Control of microorganisms in vitro by endodontic irrigants. *Braz Dent J* 14: 187–92.

Fabricius L, Dahlen G, Sundqvist G, Happonen R, and Moller A. 2006. Influence of residual bacteria on periapical tissue healing after chemomechanical treatment and root filling of experimentally infected monkey teeth. *Eur J Oral Sci* 114: 278–85.

Fardal O and Turnbull R. 1986. A review of the literature on use of chlorhexidine in dentistry. *J Am Dent Assoc* 112: 863–69.

Forsback A, Areva S, and Salonen JI. 2004. Mineralization of dentin induced by treatment with bioactive glass S53P4 in vitro. *Acta Odontol Scand* 62: 14–20.

Fouad AF and Barry J. 2005. The effect of antibiotics and endodontic antimicrobials on the polymerase chain reaction. *J Endod* 31: 510–13.

Frais S, Ng Y, and Gulabivala K. 2001. Some factors affecting the concentration of available chlorine in commercial sources of sodium hypochlorite. *Int Endod J* 34: 206–15.

Gerzina TM and Hume W. 1995. Effect of hydrostatic pressure on the diffusion of monomers through dentin in vitro. *J Dent Res* 74: 369–73.

Gomes B, Ferraz CC, Vianna M, Berber V, Teixeira F, and Souza-Filho FJ. 2001. In vitro antimicrobial activity of several concentrations of sodium hypochlorite and chlorhexidine gluconate in the elimination of *Enterococcus faecalis*. *Int Endod J* 34: 424–28.

Gomes B, Pinheiro E, Gade-Neto CR, Sousa E, Ferraz C, Zaia A, Teixeira F, and Souza-Filho FJ. 2004. Microbiological examination of infected dental root canals. *Oral Microbiol Immunol* 19: 71–76.

Gomes B, Souza S, Ferraz C, Teixeira F, Zaia A, Valdrighi L, and Souza-Filho FJ. 2003. Effectiveness of 2% chlorhexidine gel and calcium hydroxide against *Enterococcus faecalis* in bovine root dentine in vitro. *Int Endod J* 36: 267–75.

Gorni F and Gagliani M. 2004. The outcome of endodontic retreatment: A 2-yr follow-up. *J Endod* 30: 1–4.

Grossman L. 1943. Irrigation of root canals. *J Am Dent Assoc* 30: 1915–17.

Grossman L and Meiman B. 1941. Solution of pulp tissue by chemical agents. *J Am Dent Assoc* 28: 223–25.

Guggenheim M, Shapiro S, Gmur R, and Guggenheim B. 2001. Spatial arrangements and associative behavior of species in an in vitro oral biofilm model. *Appl Environ Microbiol* 67: 1343–50.

Gusman H, Santana RB, and Zehnder M. 2002. Matrix metalloproteinase levels and gelatinolytic activity in clinically healthy and inflamed human dental pulps. *Eur J Oral Sci* 110: 353–57.

Gutierrez J and Guzman M. 1968. Tooth discoloration in endodontic procedures. *Oral Surg Oral Med Oral Pathol* 26: 706–11.

Haapasalo H, Siren E, Waltimo TM, Orstavik D, and Haapasalo M. 2000. Inactivation of local root canal medicaments by dentine: An in vitro study. *Int Endod J* 33: 126–31.

Haapasalo M and Orstavik D. 1987. In vitro infection and disinfection of dentinal tubules. *J Dent Res* 66: 1375–79.

Haenni S, Schmidlin P, Mueller B, Sener B, and Zehnder M. 2003. Chemical and antimicrobial properties of calcium hydroxide mixed with irrigating solutions. *Int Endod J* 36: 100–105.

Hasselgren G, Olsson B, and Cvek M. 1988. Effects of calcium hydroxide and sodium hypochlorite on the dissolution of necrotic porcine muscle tissue. *J Endod* 14: 125–27.

Heling I, Steinberg D, Kenig S, Gavrilovich I, Sela M, and Friedman M. 1992. Efficacy of a sustained-release device containing chlorhexidine and Ca(OH)$_2$ in preventing secondary infection of dentinal tubules. *Int Endod J* 25: 20–24.

Hennessey T. 1973. Some antibacterial properties of chlorhexidine. *J Periodontal Res Suppl* 12: 61–67.

Hermann BW. 1920. Calciumhydroxid Als Mittel Zum Behandeln Und Füllen Von Zahnwurzelkanälen. Med Diss. Würzburg.

Holland R, Otoboni Filho JA, De Souza V, Nery M, Bernabe PF, and Dezan EJ. 2003. A comparison of one versus two appointment endodontic therapy in dogs' teeth with apical periodontitis. *J Endod* 29: 121–24.

Hong C, Lin S, Kok S, Cheng S, Lee M, Wang T, Chen C, Lin L, and Wang J. 2004. The role of lipopolysaccharide in infectious bone resorption of periapical lesion. *J Oral Pathol Med* 33: 162–69.

Hulsmann M and Hahn W. 2000. Complications during root canal irrigation—literature review and case reports. *Int Endod J* 33: 186–93.

Hulsmann M, Heckendorff M, and Lennon A. 2003. Chelating agents in root canal treatment: Mode of action and indications for their use. *Int Endod J* 36: 810–30.

Imfeld T. 1991. Prevalence and quality of endodontic treatment in an elderly urban population of Switzerland. *J Endod* 17: 604–7.

Jacinto R, Gomes B, Shah H, Ferraz C, Zaia A, and Souza-Filho FJ. 2005. Quantification of endotoxins in necrotic root canals from symptomatic and asymptomatic teeth. *J Med Microbiol* 54: 777–83.

Katebzadeh N, Hupp J, and Trope M. 1999. Histological periapical repair after obturation of infected root canals in dogs. *J Endod* 25: 364–68.

Kho P and Baumgartner J. 2006. A comparison of the antimicrobial efficacy of NaOCl/Biopure MTAD versus NaOCl/EDTA against *Enterococcus faecalis*. *J Endod* 32: 652–55.

Krause TA, Liewehr FR, and Hahn C. 2007. The antimicrobial effect of MTAD, sodium hypochlorite, doxycycline, and citric acid on *Enterococcus faecalis*. *J Endod* 33: 28–30.

Kvist T, Molander A, Dahlen G, and Reit C. 2004. Microbiological evaluation of one- and two-visit endodontic treatment of teeth with apical periodontitis: A randomized, clinical trial. *J Endod* 30: 572–76.

Lang H, Korkmaz Y, Schneider K, and Raab W. 2006. Impact of endodontic treatments on the rigidity of the root. *J Dent Res* 85: 364–68.

Langeland K. 1987. Tissue response to dental caries. *Endod Dent Traumatol* 3: 149–71.

Lasa I. 2006. Towards the identification of the common features of bacterial biofilm development. *Int Microbiol* 9: 21–28.

Law A and Messer H. 2004. An evidence-based analysis of the antibacterial effectiveness of intracanal medicaments. *J Endod* 30: 689–94.

Leonardo M, Almeida W, Da Silva LA, and Utrilla L. 1995. Histopathological observations of periapical repair in teeth with radiolucent areas submitted to two different methods of root canal treatment. *J Endod* 21: 137–41.

Leonardo M, Hernandez M, Silva L, and Tanomaru-Filho M. 2006. Effect of a calcium hydroxide-based root canal dressing on periapical repair in dogs: A histological study. *Oral Surg Oral Med Oral Pathol Oral Radiol Endod* 102: 680–85.

Loel D. 1975. Use of acid cleanser in endodontic therapy. *J Am Dent Assoc* 90: 148–51.

Martin H. 1976. Ultrasonic disinfection of the root canal. *Oral Surg Oral Med Oral Pathol* 42: 92–99.

Mcdonnell G and Russell AD. 1999. Antiseptics and disinfectants: Activity, action, and resistance. *Clin Microbiol Rev* 12: 147–79.

Molander A, Caplan D, Bergenholtz G, and Reit C. 2007. Improved quality of root fillings provided by general dental practitioners educated in nickel-titanium rotary instrumentation. *Int Endod J* 40: 254–60.

Molander A, Reit C, Dahlen G, and Kvist T. 1998. Microbiological status of root-filled teeth with apical periodontitis. *Int Endod J* 31: 1–7.

Muller P, Guggenheim B, and Schmidlin P. 2007. Efficacy of gasiform ozone and photodynamic therapy on a multispecies oral biofilm in vitro. *Eur J Oral Sci* 115: 77–80.

Munson M, Pitt-Ford T, Chong B, Weightman A, and Wade W. 2002. Molecular and cultural analysis of the microflora associated with endodontic infections. *J Dent Res* 81: 761–66.

Naenni N, Thoma K, and Zehnder M. 2004. Soft tissue dissolution capacity of currently used and potential endodontic irrigants. *J Endod* 30: 785–87.

Nair P. 1997. Apical periodontitis: A dynamic encounter between root canal infection and host response. *Periodontol 2000* 13: 121–48.

Nandini S, Velmurugan N, and Kandaswamy D. 2006. Removal efficiency of calcium hydroxide intracanal medicament with two calcium chelators: Volumetric analysis using spiral CT, an in vitro study. *J Endod* 32: 1097–1101.

Nerwich A, Figdor D, and Messer H. 1993. pH changes in root dentin over a 4-week period following root canal dressing with calcium hydroxide. *J Endod* 19: 302–6.

Nygaard Ostby B. 1957. Chelation in root canal therapy. *Odontol Tidskr* 65: 3–11.

Nygren JA. 1838. *Rådgivare Angående Bästa Sättet Att Vårda Och Bevara Tändernasfrisket*. Stockholm.

Orstavik D. 1996. Time-course and risk analyses of the development and healing of chronic apical periodontitis in man. *Int Endod J* 29: 150–55.

Paque F, Luder H, Sener B, and Zehnder M. 2006. Tubular sclerosis rather than the smear layer impedes dye penetration into the dentine of endodontically instrumented root canals. *Int Endod J* 39: 18–25.

Pascher F. 1978. Systemic reactions to topically applied drugs. *Int J Dermatol* 17: 768–75.

Peters C, Koka R, Highsmith S, and Peters OA. 2005. Calcium hydroxide dressings using different preparation and application modes: Density and dissolution by simulated tissue pressure. *Int Endod J* 38: 889–95.

Peters L and Wesselink P. 2002. Periapical healing of endodontically treated teeth in one and two visits obturated in the presence or absence of detectable microorganisms. *Int Endod J* 35: 660–67.

Peters OA. 2004. Current challenges and concepts in the preparation of root canal systems: A review. *J Endod* 30: 559–67.

Portenier I, Haapasalo H, Orstavik D, Yamauchi M, and Haapasalo M. 2002. Inactivation of the antibacterial activity of iodine potassium iodide and chlorhexidine digluconate against *Enterococcus faecalis* by dentin, dentin matrix, type-I collagen, and heat-killed microbial whole cells. *J Endod* 28: 634–37.

Portenier I, Haapasalo H, Rye A, Waltimo T, Orstavik D, and Haapasalo M. 2001. Inactivation of root canal medicaments by dentine, hydroxylapatite and bovine serum albumin. *Int Endod J* 34: 184–88.

Portenier I, Waltimo T, Orstavik D, and Haapasalo M. 2005. The susceptibility of starved, stationary phase, and growing cells of *Enterococcus faecalis* to endodontic medicaments. *J Endod* 31: 380–86.

Proell F. 1949. Über die Eigenschaften des Calxyls und seine Vorzüge vor anderen in der zahnärtzlichen Praxis angewandten Medikamenten. *Zahnärztl Rdsch* 14: 255–59.

Radcliffe C, Potouridou L, Qureshi R, Habahbeh N, Qualtrough A, Worthington H, and Drucker D. 2004. Antimicrobial activity of varying concentrations of sodium hypochlorite on the endodontic microorganisms *Actinomyces israelii*, *A. naeslundii*, *Candida albicans* and *Enterococcus faecalis*. *Int Endod J* 37: 438–46.

Ram Z. 1977. Effectiveness of root canal irrigation. *Oral Surg Oral Med Oral Pathol* 44: 306–12.

Ranly D and Garcia-Godoy F. 2000. Current and potential pulp therapies for primary and young permanent teeth. *J Dent* 28: 153–61.

Reit C, Molander A, and Dahlen G. 1999. The diagnostic accuracy of microbiologic root canal sampling and the influence of antimicrobial dressings. *Endod Dent Traumatol* 15: 278–83.

Ringel AM, Patterson S, Newton C, Miller C, and Mulhern J. 1982. In vivo evaluation of chlorhexidine gluconate solution and sodium hypochlorite solution as root canal irrigants. *J Endod* 8: 200–204.

Roane J and Marshall F. 1972. Osteomyelitis: A complication of pulpless teeth. Report of a case. *Oral Surg Oral Med Oral Pathol* 34: 257–61.

Rolla G, Loe H, and Schiott C. 1971. Retention of chlorhexidine in the human oral cavity. *Arch Oral Biol* 16: 1109–16.

Rosenberg B, Murray PE, and Namerow K. 2007. The effect of calcium hydroxide root filling on dentin fracture strength. *Dent Traumatol* 23: 26–29.

Ruff M, Mcclanahan S, and Babel B. 2006. In vitro antifungal efficacy of four irrigants as a final rinse. *J Endod* 32: 331–33.

Russell L and Kramer IRH. 1956. Observations on the vascular architecture of the dental pulp. *J Dent Res* 35: 957.

Safavi K and Nichols FC. 1993. Effect of calcium hydroxide on bacterial lipopolysaccharide. *J Endod* 19: 76–78.

Safavi K, Spangberg L, and Langeland K. 1990. Root canal dentinal tubule disinfection. *J Endod* 16: 207–10.

Safavi KE, Dowden WE, Introcaso J, and Langeland K. 1985. A comparison of antimicrobial effects of calcium hydroxide and iodine-potassium iodide. *J Endod* 11: 454–56.

Sathorn C, Parashos P, and Messer H. 2007. Antibacterial efficacy of calcium hydroxide intracanal dressing: A systematic review and meta-analysis. *Int Endod J* 40: 2–10.

Schilke R, Lisson J, Bauss O, and Geurtsen W. 2000. Comparison of the number and diameter of dentinal tubules in human and bovine dentine by scanning electron microscopic investigation. *Arch Oral Biol* 45: 355–61.

Seltzer S, Soltanoff W, Sinai I, Goldenberg A, and Bender I. 1968. Biologic aspects of endodontics. 3. Periapical tissue reactions to root canal instrumentation. *Oral Surg Oral Med Oral Pathol* 26: 694–705.

Sen B, Safavi K, and Spangberg L. 1999. Antifungal effects of sodium hypochlorite and chlorhexidine in root canals. *J Endod* 25: 235–38.

Shah H and Gharbia SE. 1995. The biochemical milieu of the host in the selection of anaerobic species in the oral cavity. *Clin Infect Dis* 20(Suppl 2): S291–300.

Shahravan A, Haghdoost A, Adl A, Rahimi H, and Shadifar F. 2007. Effect of smear layer on sealing ability of canal obturation: A systematic review and meta-analysis. *J Endod* 33: 96–105.

Silva L, Nelson-Filho P, Leonardo M, Rossi M, and Pansani CA. 2002. Effect of calcium hydroxide on bacterial endotoxin in vivo. *J Endod* 28: 94–98.

Sim T, Knowles JC, Ng Y, Shelton J, and Gulabivala K. 2001. Effect of sodium hypochlorite on mechanical properties of dentine and tooth surface strain. *Int Endod J* 34: 120–32.

Simon J. 1980. Incidence of periapical cysts in relation to the root canal. *J Endod* 6: 845–48.

Siqueira JJ and Rocas I. 2005. Exploiting molecular methods to explore endodontic infections: Part 2—Redefining the endodontic microbiota. *J Endod* 31: 488–98.

Siqueira JJ, Rocas I, Favieri A, and Lima K. 2000. Chemomechanical reduction of the bacterial population in the root canal after instrumentation and irrigation with 1%, 2.5%, and 5.25% sodium hypochlorite. *J Endod* 26: 331–34.

Siren E, Haapasalo M, Ranta K, Salmi P, and Kerosuo E. 1997. Microbiological findings and clinical treatment procedures in endodontic cases selected for microbiological investigation. *Int Endod J* 30: 91–95.

Sirtes G, Waltimo T, Schaetzle M, and Zehnder M. 2005. The effects of temperature on sodium hypochlorite short-term stability, pulp dissolution capacity, and antimicrobial efficacy. *J Endod* 31: 669–71.

Sjogren U, Figdor D, Persson S, and Sundqvist G. 1997. Influence of infection at the time of root filling on the outcome of endodontic treatment of teeth with apical periodontitis. *Int Endod J* 30: 297–306.

Sjogren U, Figdor D, Spangberg L, and Sundqvist G. 1991. The antimicrobial effect of calcium hydroxide as a short-term intracanal dressing. *Int Endod J* 24: 119–25.

Slots J, Rams T, and Schonfeld S. 1991. In vitro activity of chlorhexidine against enteric rods, pseudomonads and acinetobacter from human periodontitis. *Oral Microbiol Immunol* 6: 62–64.

Soekanto A, Kasugai S, Mataki S, Ohya K, and Ogura H. 1996. Toxicity of camphorated phenol and camphorated parachlorophenol in dental pulp cell culture. *J Endod* 22: 284–89.

Spangberg L, Engstrom B, and Langeland K. 1973. Biologic effects of dental materials. 3. Toxicity and antimicrobial effect of endodontic antiseptics in vitro. *Oral Surg Oral Med Oral Pathol* 36: 856–71.

Sundqvist G. 1992. Associations between microbial species in dental root canal infections. *Oral Microbiol Immunol* 7: 257–62.

Sundqvist G. 1994. Taxonomy, ecology, and pathogenicity of the root canal flora. *Oral Surg Oral Med Oral Pathol* 78: 522–30.

Tanomaru Filho M, Leonardo M, and Da Silva LA. 2002. Effect of irrigating solution and calcium hydroxide root canal dressing on the repair of apical and periapical tissues of teeth with periapical lesion. *J Endod* 28: 295–99.

Torabinejad M, Handysides R, Khademi A, and Bakland L. 2002. Clinical implications of the smear layer in endodontics: A review. *Oral Surg Oral Med Oral Pathol Oral Radiol Endod* 94: 658–66.

Torabinejad M, Khademi AA, Babagoli J, Cho Y, Johnson WB, Bozhilov K, Kim J, and Shabahang S. 2003a. A new solution for the removal of the smear layer. *J Endod* 29: 170–75.

Torabinejad M, Kutsenko D, Machnick T, Ismail A, and Newton CW. 2005. Levels of evidence for the outcome of nonsurgical endodontic treatment. *J Endod* 31: 637–46.

Torabinejad M, Shabahang S, Aprecio RM, and Kettering J. 2003b. The antimicrobial effect of MTAD: An in vitro investigation. *J Endod* 29: 400–403.

Trope M, Delano EO, and Orstavik D. 1999. Endodontic treatment of teeth with apical periodontitis: Single vs. multivisit treatment. *J Endod* 25: 345–50.

Trope M, Pettigrew J, Petras J, Barnett F, and Tronstad L. 1989. Differentiation of radicular cyst and granulomas using computerized tomography. *Endod Dent Traumatol* 5: 69–72.

Vasiliadis L, Darling AI, and Levers B. 1983. The amount and distribution of sclerotic human root dentine. *Arch Oral Biol* 28: 645–49.

Vianna M, Horz H, Gomes B, and Conrads G. 2006. In vivo evaluation of microbial reduction after chemo-mechanical

preparation of human root canals containing necrotic pulp tissue. *Int Endod J* 39: 484–92.

Walkhoff O. 1882. Vereinfachte Behandlung der Pulpakrankheiten mittels Jodoformknorpel und Chlorphenol. *Dtsch Monatsschr Zahnheilkd* 1: 192–201.

Waltimo T, Brunner TJ, Vollenweider M, Stark W, and Zehnder M. 2007. Antimicrobial effect of nanometric bioactive glass 45S5. *J Dent Res* 86(8): 754–57.

Waltimo T, Trope M, Haapasalo M, and Orstavik D. 2005. Clinical efficacy of treatment procedures in endodontic infection control and one year follow-up of periapical healing. *J Endod* 31: 863–66.

Waltimo TM, Orstavik D, Siren E, and Haapasalo M. 1999. In vitro susceptibility of *Candida albicans* to four disinfectants and their combinations. *Int Endod J* 32: 421–29.

Weiger R, Rosendahl R, and Lost C. 2000. Influence of calcium hydroxide intracanal dressings on the prognosis of teeth with endodontically induced periapical lesions. *Int Endod J* 33: 219–26.

Wood B and Green L. 1969. Second premolar morphologic trait similarities in twins. *J Dent Res* 48: 74–78.

Yoshida T, Shibata T, Shinohara T, Gomyo S, and Sekine I. 1995. Clinical evaluation of the efficacy of EDTA solution as an endodontic irrigant. *J Endod* 21: 592–93.

Zamany A, Safavi K, and Spangberg L. 2003. The effect of chlorhexidine as an endodontic disinfectant. *Oral Surg Oral Med Oral Pathol Oral Radiol Endod* 96: 578–81.

Zehnder M. 2006. Root canal irrigants. *J Endod* 32: 389–98.

Zehnder M, Gold S, and Hasselgren G. 2002a. Pathologic interactions in pulpal and periodontal tissues. *J Clin Periodontol* 29: 663–71.

Zehnder M, Grawehr M, Hasselgren G, and Waltimo T. 2003. Tissue-dissolution capacity and dentin-disinfecting potential of calcium hydroxide mixed with irrigating solutions. *Oral Surg Oral Med Oral Pathol Oral Radiol Endod* 96: 608–13.

Zehnder M, Kosicki D, Luder H, Sener B, and Waltimo T. 2002b. Tissue-dissolving capacity and antibacterial effect of buffered and unbuffered hypochlorite solutions. *Oral Surg Oral Med Oral Pathol Oral Radiol Endod* 94: 756–62.

Zehnder M, Luder H, Schatzle M, Kerosuo E, and Waltimo T. 2006a. A comparative study on the disinfection potentials of bioactive glass S53P4 and calcium hydroxide in contra-lateral human premolars ex vivo. *Int Endod J* 39: 952–58.

Zehnder M, Schmidlin P, Sener B, and Waltimo T. 2005. Chelation in root canal therapy reconsidered. *J Endod* 31: 817–20.

Zehnder M, Soderling E, Salonen J, and Waltimo T. 2004. Preliminary evaluation of bioactive glass S53P4 as an endodontic medication in vitro. *J Endod* 30: 220–24.

Zehnder M, Waltimo T, Sener B, and Soderling E. 2006b. Dentin enhances the effectiveness of bioactive glass S53P4 against a strain of *Enterococcus faecalis*. *Oral Surg Oral Med Oral Pathol Oral Radiol Endod* 101: 530–35.

Zerella JA, Fouad A, and Spangberg L. 2005. Effectiveness of a calcium hydroxide and chlorhexidine digluconate mixture as disinfectant during retreatment of failed endodontic cases. *Oral Surg Oral Med Oral Pathol Oral Radiol Endod* 100: 756–61.

# Chapter 14

# Endodontic Infections in Incompletely Developed Teeth

*George T.-J. Huang*

## 14.1 Introduction

Two specific aspects are encountered when performing endodontic treatment on immature teeth. One is large canal space and the other is open apex resulting from incomplete root formation. These aspects necessitate special procedures during treatment and affect the outcome of treatment. Conservative approach is the preferred treatment option as long as there is a possibility to allow complete root maturation. When pulp is totally necrotic and infected, disinfection of the wide-open canal is a challenge. Routine practice calls for providing an appropriate environment to form an apical calcific barrier for these teeth to ensure effective instrumentation and filling of the canal space.

Generally, the clinical diagnosis for immature teeth follows the same criteria as mature teeth. Vital pulp therapy is an important treatment of choice for immature teeth with a vital pulp. Basically, when diseased pulp tissue is partially removed by pulpotomy, continued root formation or apexogenesis is initiated. Whereas when teeth are diagnosed with nonvital pulp, the entire canal system is debrided, cleansed, and

apexification is contemplated. However, the clinical definition of pulp vitality is determined by diagnostic tools that show sensitivity, rather than vitality, and the histological condition of the pulp cannot be clinically determined. There have been a number of sporadic case reports in the literature that suggest the potential of immature teeth to complete the root formation even after being clinically diagnosed with nonvital pulps.

Recently, more convincing and well-presented clinical cases further demonstrate that despite the formation of periapical abscesses with extensive periradicular bone resorption as the result of root canal infection in immature teeth, conservative treatment that involves disinfection of the canal space may allow root development to their maturation (Iwaya et al. 2001; Banchs and Trope 2004; Chueh and Huang 2006). The interest generated from this clinical observation appears to be underscored by the recent discovery of various dental stem cells, including dental pulp stem cells (DPSCs) (Gronthos et al. 2000), which may shed light on the understanding of the repair potential of pulp–dentin complex. Furthermore, a recent report on the discovery of dental stem cells from apical papilla

(SCAP) (Sonoyama et al. 2006, 2008) may explain the mechanisms underlying the continued maturation of roots after endodontic treatment of immature teeth.

This chapter emphasizes this new perspective of the management of immature teeth in addition to providing an overview of the traditional protocols. Furthermore, in order to manage clinical cases from an indepth biological point of view, it is important to review and understand the recent advancement in stem cell biology, especially dental tissue–derived stem cells. This new understanding may reshape our traditional way of decision making in the management of immature teeth. In light of the fact that the discovery and characterization of dental stem cells extend not just to facilitate our clinical endodontic treatment planning, this chapter also introduces the potential applications of these stem cells for regenerative endodontics, for example, dentin/pulp regeneration and generation of bio-roots.

The immature permanent tooth has a higher prognosis for healing compared to teeth with completely formed apex. This is observed clinically when comparing the pulpal and periodontal healing of teeth with open apex and fully mature teeth in response to caries, luxation and avulsion injuries, transplantation, or simple operative procedures. This is also observed in the remarkable potential for pulpal regeneration that has been described in classical studies, but has more recently been documented and explored clinically. This regenerative ability may be due to the higher prevalence of progenitor or stem cells in the young pulp, greater primary and collateral circulation due to the open apex, or the stronger ability of the inflamed pulp to cope with fluctuations in pulpal blood flow or neuropeptide content given its large communication with apical soft tissues. Therefore, the pathogenesis of endodontic infections tend to take a different route in these teeth, despite the lack of any evidence of a difference in microbial or other host-related differences in endodontic infections in these patients. Clearly, if untreated these infections can take a dramatic route and lead to significant morbidity. However, the clinical presentation generally tends to show greater healing potentials in teeth with immature apex than in mature teeth. In this chapter a description of the unique presentation of endodontic infections in immature teeth and their treatment, as well as the recent interest in pulpal regeneration in

these teeth, is presented. The immature tooth may be more biologically suited to deal with infection, but it tends to be structurally weaker, and its development is necessary for jawbone development. Thus, there is a significant clinical interest in reversing the course of endodontic infections in immature teeth, in order to allow them to fully maturate.

## 14.2 Review of tooth development as it relates to endodontic pathosis

Tooth is a unique and complex structure developed via interactions between dental epithelium and mesoderm. From the formation of dental lamina in the epithelium, teeth develop in sequential stages including lamina, bud, cap, early bell, and late bell stages (Bhasker 1991; D'Souza 2002). The dental lamina induces underlying ectomesenchyme to form the dental papilla. At the cap stage, the entire tooth organ is surrounded by dental follicle made of mesenchymal cells. At the early bell stage, the dental epithelium is stratified and the junction where the internal and external dental epithelial layers meet is termed cervical loop, which plays an important role in the root development. Odontoblasts differentiated from the mesenchymal cells in the dental papilla produce dentin in the late bell stage. As the coronal dentin encases the dental papilla, it evolves to dental pulp. During this stage, the crown is developed followed by root formation as the epithelial cells from the cervical loop proliferate apically and influence the differentiation of odontoblasts from the dental papilla and cementoblasts from follicle mesenchyme. This apically extending two-layered epithelial wall (merging of the inner and outer enamel epithelium) forms the Hertwig's epithelial root sheath (HERS) that is responsible for determining the shape of the root(s). The epithelial diaphragm surrounds the apical opening to the pulp and eventually becomes the apical foramen. When the first layer of dentin has been laid down, HERS begins to disintegrate leaving behind discontinued epithelial cell rests of Malassez in the periodontal ligament.

The apical portion of the dental papilla during the stage of root development has not been well described in the literature and was only characterized quite recently. The physical and histological characteristics of the dental papilla located at the apex of developing human permanent teeth have been defined and this

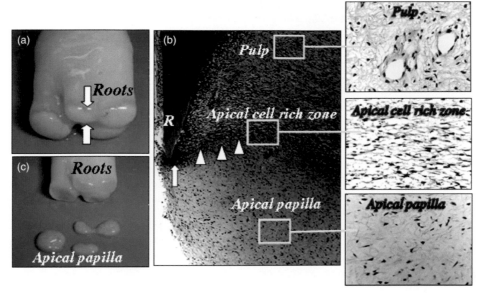

**Fig. 14.1**   Anatomy of apical papilla. (a) An extracted human third molar depicting root attached to the root apical papilla (open arrows) at developmental stage. (b) Hematoxylin and eosin (H&E) staining of human developing root (*R*) depicting epithelial diaphragm (open arrows) and apical cell–rich zone (open arrowheads). (c) Harvested root apical papilla for stem cell isolation. (Adapted from Sonoyama et al. 2008 with permission.)

tissue termed apical papilla (Sonoyama et al. 2008; Huang et al. 2008). The tissue is loosely attached to the apex of the developing root and can be easily detached with a pair of tweezers (Fig. 14.1). Apical papilla is apical to the epithelial diaphragm and there is an apical cell–rich zone lying between the apical papilla and the pulp. Importantly, there are stem/progenitor cells located in both dental pulp and apical papilla, but they have somewhat different characteristics (Sonoyama et al. 2006, 2008). Because of the apical location of the apical papilla, this tissue may be benefited by its collateral circulation that enables it to survive during the process of pulp necrosis.

Generally the roots are halfway to two-third formed at the eruption with wide-open apices and the pulp volume is still large with thin dentin walls. Three more years after tooth eruption are normally needed for the further deposition of dentin and maturation of the apex. However, as the eruption takes place, the tooth is susceptible to trauma and caries invasion, both of which may threaten the viability and functionality of the cells involved in the root development. Damage of the pulp, apical papilla, and/or the HERS

resulting in cell death impedes the root formation. Loss of HERS ceases the continued root apical extension because odontoblasts can no longer be differentiated from the apical papilla. However, hard tissue can still be formed by cementoblasts that are normally present in the apical region and by mesenchymal cells in the dental follicle that may differentiate after the injury into hard tissue-producing cells (Torneck 1982). Cooke and Robotham (1988) advised avoidance of trauma to the tissue around the apex. They speculated that the remnants of the survived HERS at the apices of immature teeth, after disinfection, may organize the apical mesodermal tissue into root components. This mesodermal tissue is now considered to be the apical papilla as mentioned above.

The fate of human pulp tissue after dental trauma has been observed in clinical radiographs. Excellent radiographic images have been reported demonstrating the ingrowth of bone and periodontal ligament (PDL) (next to the inner dentinal wall) into the canal space with arrested root formation after replantation of avulsed maxillary incisors (Kling et al. 1986; Andreasen et al. 1995a, b), suggesting a complete loss of the viability of pulp, apical papilla, and HERS. Some cases demonstrated partial formation of root accompanied with ingrowth of bone and PDL into the canal space, and in some cases the roots continued to develop to their completion, suggesting at least partial pulp or apical papilla survival after the replantation. It is noted, however, that severe narrowing of canal space is usually associated with survived pulp.

## 14.3 Etiology, prevalence, and pathogenesis of pulp disease in incompletely developed teeth

Children and young adults aged 6–18 years have sequential permanent tooth eruption and maturation of the roots. Any factors that impinge on the vitality of the pulp may interfere with the completion of root development. Traumatic injuries to young permanent teeth affect 30% of children (Andreasen and Ravn 1972; Andreasen and Andreasen 1994). Anterior teeth are particularly susceptible to this type of damage. First molars erupt around 6–7 years of age when oral hygiene is more difficult to maintain than it is for older children. Caries invasion is a common cause of pulp necrosis. Losing immature teeth is difficult to manage as the jaws are still in development, and any restorative procedure is likely to be a temporary measure. Because of the important role of HERS in continued root development, after pulpal injury, every effort should be made to maintain its viability (Rafter 2005).

Two distinct clinical situations should be well defined before a clinical treatment approach can be determined, because their clinical outcomes are dramatically different. The first is the immature tooth with vital pulp. A procedure called *apexogenesis* is defined as clinical treatment of immature teeth, with vital pulp to preserve remaining normal vital tissue and to allow completion of root formation and apical maturation. The second is immature tooth with totally necrotic, and frequently infected, pulp. Apexification is then undertaken and is defined as clinical treatment of immature teeth with nonvital pulp by inducing a calcific barrier at the open apex (Goldstein et al. 1999; Rafter 2005). Teeth after apexogenesis with vital pulp therapy develop a normal thickness of dentin, root length, and apical morphology. Teeth receiving apexification

normally gain only an apical hard tissue bridge without a further development of root.

## 14.4 Management of immature teeth

### 14.4.1 Diagnosis

Advanced technologies for endodontic diagnostic methods such as laser Doppler flowmetry tests were shown to be promising (Yanpiset et al. 2001; Roeykens et al. 2002); however, traditional sensitivity tests by cold, heat, and electrical pulp testing remain to be popular diagnostic tools to differentiate vital versus nonvital pulps. The limitation and accuracy of these tests were found to be 86% for the cold test, 71% for the heat, and 81% for the electrical test (Petersson et al. 1999). Whether a negative response means a totally or partially necrotic pulp, or simply a severely calcified pulp, cannot be distinguished. Inflammation and necrosis of pulp is a progressive process, and it has been long established that there is no close correlation between the results of these clinical tests and the histological diagnosis (Seltzer et al. 1963). Teeth that respond normally to cold and electrical pulp test compared to their control teeth are clinically considered having a normal pulp, whereas those respond to these tests with a brief hypersensitive reaction are considered having a reversible pulpitis. Teeth with spontaneous pain or lingering pain following stimulation are considered having an irreversible pulpitis.

The reliability of the diagnosis for immature teeth is complicated by the fact that the sensory plexus of nerves in the sub-odontoblastic region is not well developed and that not all pulpal nerves end among the odontoblasts, or into predentin or dentin, as in fully developed teeth in occlusion (Bernick 1964; Fulling and Andreasen 1976; Klein 1978; Fuss et al. 1986). History of symptoms, radiographic examination, and information on the duration and characteristics of the pain are of critical importance. Other clinical tests including visual examination for color change or cracks, and percussion testing should be conducted. The presence of a swelling, sinus tract, and extensive radiolucency associated with the tooth is more indicative of pulp necrosis and periradicular infection for mature teeth than is for the case of immature teeth (Iwaya et al. 2001; Chueh and Huang 2006). Therefore, it may be helpful to make the final diagnosis after the tooth is accessed and the pulp space directly observed.

## 14.4.2 Clinical management

### 14.4.2.1 Apexogenesis

Apexogenesis treatment procedure falls in the category of vital pulp therapy for cases with open apices. Vital pulp therapy includes indirect and direct pulp capping and pulpotomy treatment. The common clinical indications are cases demonstrating signs of typical reversible pulpitis, or if the pulp is simply exposed by fracture due to trauma or by caries removal. Direct pulp capping may be carried out as needed. A shallow pulpotomy (Cvek pulpotomy) is performed when the pulp is minimally exposed by mechanical reason or trauma. In principle, immature teeth should be treated as conservatively as practical to allow apexogenesis to occur (Weisleder and Benitez 2003). Therefore, pulp capping is the first treatment option if the pulp is considered largely normal. For cases having more extensive trauma or caries, pulpotomy is the next alternative approach.

*14.2.2.1.1 Microbiology and infection control*

*Indirect pulp capping.* Teeth undergoing operative procedures should be isolated with rubber dam. Indirect pulp capping is to remove infected dentin leaving behind partially decalcified but not infected dentin (Leksell et al. 1996; Bjornal and Kidd 2005) and place a layer of calcium hydroxide cement without exposing the pulp. The problem is the difficulty to determine clinically this decalcified but not infected dentin (Alacam 1985), and there has been a lack of studies using caries staining agent for indirect pulp capping. The goal, however, is to provide the pulp an opportunity to lay down new dentin underneath the caries removal site in order to prevent pulp exposure. Approximately 6–8 weeks later, the tooth will be assessed and examined for the formation of new dentin. Recent reports indicated success rates of 83–96% at 2–4 years of follow-ups in primary teeth that received indirect pulp capping treated with calcium hydroxide or adhesive resin system (Falster et al. 2002; Marchi et al. 2006). Although there lacks systematic clinical studies and reports on the outcome of indirect pulp capping of immature permanent teeth, it may be considered as a clinical option for treating immature permanent teeth. Alternatively, more aggressive removal of caries at the cost of pulp exposure and treated with direct

pulp capping may be undertaken. Direct pulp capping after carious exposure with mineral trioxide aggregate (MTA) has shown a high success rate for permanent teeth in recent clinical observations discussed below.

*Direct pulp capping.* Pulp is able to repair after treatment if contaminated by oral flora 1 h after mechanical exposure. Longer exposure to oral cavity (1–7 days) mounts pronounced inflammatory infiltrates based on the studies in monkeys (Cox et al. 1982). Clinically, when pulp is exposed either as a result of mechanical nature or due to caries removal, the exposed site should be kept clean and irrigated with 5.25 or 6% NaOCl or on a soaked cotton pellet to achieve hemostasis. After which, the cavity is carefully rinsed with normal saline and dried with wetted sterile cotton pellets or paper points. Mixed MTA cement is placed over the exposed spot followed by overlaying a wetted cotton pellet, and the tooth is provisionally restored with unbonded resin-based composite. A final restoration is then placed within 5–10 days if the tooth is asymptomatic and responds normally to cold (Bogen et al. 2008). It was reported that hard tissue bridge formation is more complete after MTA treatment than BMP-7 treatment of pulp exposure in rats (Andelin et al. 2003). The diameter of the exposure site is normally in the range of 1 mm. Larger sizes of exposure theoretically should not affect the outcome so long as the management is properly performed. It is suggested that there should be good hemorrhage control and prevention of blood clot formation before the capping (Stanley 1989). The blood clot may subject the exposed site to secondary infection. A recent study has shown that the success rate of direct pulp capping is 92.9% with mechanical exposure and 33.3% with carious exposure based on at least 3 years of follow-up (Al-Hiyasat et al. 2006). Another study revealed 44.5% failures in 5 years and 79.7% failures in 10-year follow-ups for capping of carious exposure (Barthel et al. 2000). The above two studies did not specify mature versus immature teeth. Farsi and coworkers found a success rate of 93% with evidence of continued root growth after using MTA as direct pulp capping to treat carious exposure in young permanent teeth (Farsi et al. 2006). Bogen et al. (2008) found a 97.96% success rate of direct pulp capping after removal of deep caries with an observation period of 9 years. Among 49 teeth they treated with follow-ups, 9 teeth were presumably immature teeth based on the documentation of patients' age and tooth type.

*Pulpotomy.* Pulpotomy requires aggressive removal of the inflamed pulp tissue, although clinically it is difficult to determine which part of the pulp is actually inflamed. Cvek reported a 96% success of mature and immature teeth treated with partial pulpotomy and capping with calcium hydroxide for pulp exposure resulting from crown fracture (Cvek 1978). The interval between accident and treatment varied from 1 h to 90 days. A 0.5% chlorhexidine solution was used to clean the teeth, and the partial pulpotomy removed approximately 2 mm in depth into the pulp tissue at the exposed site. Using monkey as a study model, it was found that pulp inflammation extends only approximately 2 mm even up to 7 days after pulp exposure by fracture and the teeth left open to oral environment (Cvek et al. 1982). Fong and Davis reviewed the success of partial pulpotomy for immature permanent teeth that had trauma, carious exposure, or coronal pulpitis. The success rates ranged from 83 to 100% (Fong and Davis 2002). Trope's group reported that, in a canine study model, partial pulpotomy with calcium hydroxide capping with a bacteria-tight coronal restoration is a viable approach to treat inflamed pulp (Trope et al. 2002).

The complete pulpotomy removes the entire pulp tissue in the pulp chamber down to approximately the orifice levels. Hemorrhage control is accomplished by pressure on the canal orifices, with a hemostatic agent such as ferric sulfate, the use of a cotton pellet with sodium hypochlorite, or the use of calcium hydroxide powder. This is followed by placing either calcium hydroxide or MTA in the same way as for direct pulp capping.

The understanding of the effects of certain growth factors such as transforming growth factor-$\beta$ (TGF-$\beta$) and bone morphogenic proteins (BMPs) on dentinogenesis provides a potentially potent vital pulp therapy approach via biologically directed tissue repair (Rutherford and Gu 2000; Tziafas et al. 2000). The delivery of growth factors such as BMP-7 and growth/differentiation factor 11 by ex vivo transduced cells or stem cells implanted into exposed pulp to induce reparative dentin formation has also been tested in animal models (Rutherford 2001; Iohara et al. 2004; Nakashima et al. 2004). Cell-based therapy with the combination of growth factors may become future mode of vital pulp treatment approaches.

### 14.4.2.2 Apexification

*14.2.2.2.1 Microbiology and infection control*

Disinfection of nonvital immature teeth is generally similar to that of mature teeth, except that the nature of canal size and the divergent canal shape toward apex render the shaping procedure somewhat different. Circumferential filing with hand instruments may be more effective than with rotary instrumentation. To reach the blunderbuss apical canal, precurved stainless steel files may be used. Depending on the level of the tooth maturation, the less tooth structure there is, the less aggressive filing should be undertaken to preserve as much dentin as possible. Therefore, chemical cleansing will play a more important role to disinfect the canal. Cvek and coworkers found that antimicrobial effect by mechanical cleansing is very low for mature and even lower for immature teeth (Cvek et al. 1976). Irrigation with up to 5% NaOCl enhances the antimicrobial effect but is still inadequate. Current chemical agents used for canal irrigation besides NaOCl include chlorhexidine, ethylene diamine tetraacetic acid (EDTA), IKI (iodine-potassium iodide) (Safavi et al. 1985), and triple antibiotic paste consisting of metronidazole, ciprofloxacin, and minocycline. Sato's group first described the use of the mixture of these three antibiotics against *Escherichia coli*–infected dentin in vitro (Sato et al. 1996). The same group also tested their bactericidal efficacy against microbes from carious dentin and infected pulp and found them to be sufficiently potent to eradicate the bacteria. The respective antibiotic alone substantially decreases bacterial recovery, but cannot kill all the bacteria as do the mixture of antibiotics (Hoshino et al. 1996). It was found that this triple antibiotic paste is more effective than 1.25% NaOCl in disinfecting immature teeth in vivo in a dog study model (Windley et al. 2005). The triple antibiotic paste was mixed at a concentration of 20 mg of each antibiotic per milliliter, inserted into canal with a sterile Lentulo spiral filler, and left in the canal for a period of 2 weeks. It is generally difficult to test the antimicrobial effectiveness of antibiotics in endodontic models, because of the absence of an inactivator to prevent the carryover effect. Furthermore, the effectiveness of bactericidal and bacteriostatic antibiotic combinations and the discoloring effects of tetracycline analogs, such as minocycline, raise some concerns about whether this is the most suitable antibiotic combination for this purpose.

To date, the most commonly used medication associated with apexification is calcium hydroxide paste (reviewed by Rafter 2005). The purpose of its use is twofold: disinfection and induction of apical calcific barrier. When first introduced (Kaiser 1964), it was proposed to mix calcium hydroxide with camphorated parachlorophenol (CMCP). To avoid the potential cytotoxicity of CMCP, calcium hydroxide was later tested for its efficacy by mixing with just saline or sterile water and found to have similar clinical success (Michanowicz and Michanowicz 1967; Binnie and Rowe 1973). Calcium hydroxide is antimicrobial due to its release of hydroxyl ions that can cause damage to the bacterial cellular components. The best example is the demonstration of its effect on lipopolysaccharide (LPS). Calcium hydroxide chemically alters LPS that affects its various biological properties (Safavi and Nichols 1993, 1994; Barthel et al. 1997; Nelson-Filho et al. 2002; Jiang et al. 2003).

It has been unclear as to whether calcium hydroxide per se possesses the ability to induce apical barrier or it is its antimicrobial activity that provides a microbial-free environment such that calcific tissue is able to develop. Early in vivo studies have suggested the ectopic induction of bone by calcium hydroxide (Mitchell and Shankwalker 1958). It was considered that the layer of firm necrosis created by the contact of calcium hydroxide generates a low-grade irritation of the underlying tissue sufficient to produce a matrix that mineralizes (Schroder and Granath 1971; Holland et al. 1977). Calcium is attracted to the area and mineralization of newly formed collagenous matrix is initiated from the calcified foci. High pH was considered an important factor in its ability to induce hard tissue formation (Javelet et al. 1985). The time required for apical barrier formation in apexification using calcium hydroxide varies from 3 to 20 months (Yates 1988; Sheehy and Roberts 1997; Finucane and Kinirons 1999). Other conditions such as age and presence of symptoms or periradicular radiolucencies may play a role in the extent of time needed to form apical barrier (Cvek 1972; Ghose et al. 1987; Mackie et al. 1988). Refreshing the calcium hydroxide paste usually takes place every 3 months.

Obturation of the root canal takes place normally when the apical calcific barrier is formed (Fig. 14.2). Without the barrier, there is nothing against which the traditional gutta-percha filling material can be condensed. The barrier may be verified radiographically and by detecting the apical stop with endodontic files. In recent years, the use of MTA to fill the apical end without the need for calcific barrier formation has been reported with success. MTA was first introduced in 1993 by Torabiajed and his coworkers (Lee et al. 1993; Torabinejad et al. 1993) and approved by the Food and Drug Administration (FDA) in 1998. This material has demonstrated good sealability and biocompatibility. It has a pH of 12.5 after setting, which is similar to that of calcium hydroxide (Torabinejad et al. 1995a). The first popular use of MTA was perforation repairs in roots (Lee et al. 1993) or furcations (Ford et al. 1995; Arens and Torabinejad 1996) for its excellent tissue compatibility. The use then extended to direct pulp capping, root-end retrograde filling, and apical plug in apexification (Ford et al. 1996; Torabinejad et al. 1997; Shabahang and Torabinejad 2000). Additionally, MTA appears to exhibit antimicrobial properties (Torabinejad et al. 1995b). In comparison to calcium hydroxide on hard tissue induction, MTA appears to have a greater consistency based on in vivo studies in dogs (Shabahang et al. 1999).

Some reports have shown that using MTA for apexification may shorten the treatment period with more favorable results. After 1-week calcium hydroxide medication in the canal, MTA is condensed down to the apical canal by plugging motion with pluggers or with the blunt end of gutta-percha points, after a biocompatible collagen matrix, such as CollaCote, is placed in the apical region. The sealing and apical extension is carefully controlled and monitored radiographically against the matrix material. Favorable results have been reported (Maroto et al. 2003; El-Meligy and Avery 2006; Pace et al. 2007). One-visit apexification using MTA has also been reported (Witherspoon and Ham 2001). Some authors considered that lengthy treatment protocols are inconvenient and may lead to ultimate failure simply because patients cannot return for the numerous visits each month apart. They proposed a one-visit apexification protocol with MTA as an alternative to the traditional treatment practices with calcium hydroxide (Steinig et al. 2003). This expedient cleaning and shaping of the root canal system followed by its apical seal with MTA makes an immediate placement with a bonded core within the root canal in one visit possible, which may prevent potential fractures of immature teeth. One-visit or two-visit approach contradicts the traditional concept that an apical barrier must be

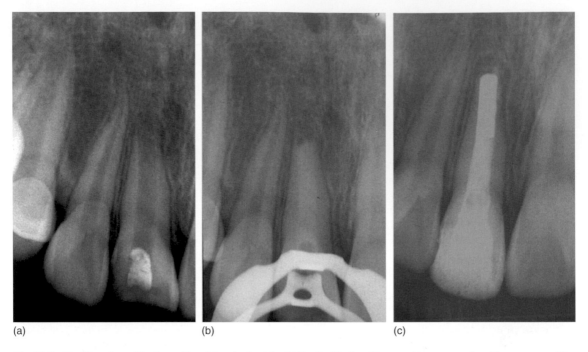

(a)          (b)          (c)

**Fig. 14.2**    Traditional apexification with calcium hydroxide. (a) Tooth #8 of an 11-year-old was necrotic with open apex. (b) The canal was cleaned and shaped and filled with calcium hydroxide every 3 months until an apical stop was detected. (c) The apical half of the canal was filled with gutta-percha and the remaining canal with composite. (Adapted from Huang 2008 with permission.)

first formed in order to obtain a good stop. Since the nature of placing the MTA to the apex is quite different from that of gutta-percha, different techniques have been developed to make MTA apical plug possible. The sealing ability is more superior when MTA is used as a root-end filling material and condensed against a physical barrier than as an orthograde apical plug. A thickness of 1–4 mm for the MTA apical plug may provide adequate retention (Hachmeister et al. 2002; de Leimburg et al. 2004). In immature teeth with necrotic pulps and periapical bone resorption, the irregular dentin walls, the divergent apices, and lack of physical barrier make the adaptation and control of MTA apical placement more difficult. Some authors suggested that hand condensation resulted in better adaptation and fewer voids than ultrasonic compaction (AminosHariae et al. 2003). As noted before, the use of resorbable material such as freeze-dried bone, collagen plug, or other biocompatible materials packed into the apical region to serve as a physical barrier or matrix against which MTA may be

condensed has been suggested (reviewed by Rafter 2005). Proper apical placement of MTA into a tooth with open apex should lead to a favorable outcome (Fig. 14.3).

### 14.4.2.3 Outcome of endodontic therapy on young permanent teeth

The success of pulp capping of mechanical or traumatic exposure is high; therefore, this procedure should be the first choice for immature teeth if conditions allow. In fact, because the immature teeth have great potential to heal and regenerate, efforts should be made toward creating a favorable environment for root maturation to complete. Considering the high levels of success for direct pulp capping with MTA (Bogen et al. 2008), pulp capping and pulpotomy for apexogenesis may be a favorable alternative to apexification for young permanent teeth, although there has been a lack of clinical studies on pulp capping success focusing on immature permanent teeth. Failure

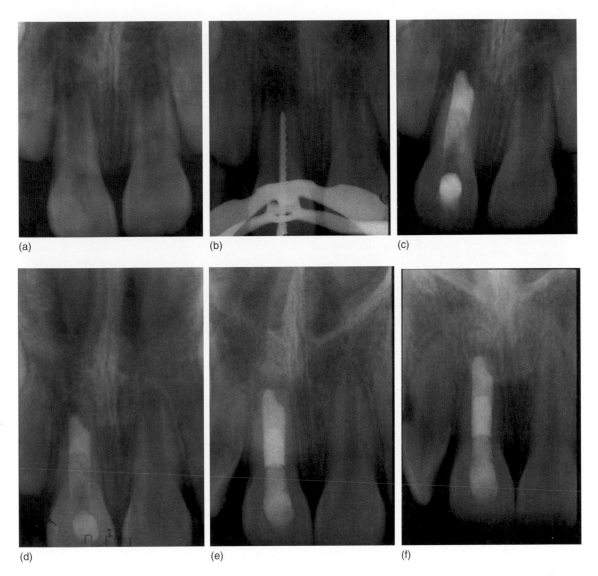

(a)     (b)     (c)

(d)     (e)     (f)

**Fig. 14.3**   An open apex #8 with apex filled with MTA. (a) Before treatment. (b) Canal cleaned and shaped. (c) Apex filled with MTA plug. (d–f) Six-month, 3-year, and 5-year recall, respectively. (Courtesy of Dr G. Bogen.)

of pulp capping cases has been attributed to infection left behind in the pulp that normally should become noticeable early, or attributed to microleakage. It was shown in studies using monkeys that the formation of dentin bridge does not secure the leakage proof as most dentin bridges contain multiple tunnel defects, which remain open to the underlying pulp from the medicament interface. The recurring pulp inflammation observed 1- and 2-year after direct pulp capping

is associated with bacterial contamination (Cox et al. 1985).

With respect to the use of calcium hydroxide for the induction of apical barrier formation and healing in immature permanent teeth, the successful rate is 74–100% in terms of apical barrier formation and is irrespective of the proprietary brand used (reviewed by Sheehy and Roberts 1997). In the traditional apexification treatment, apical closure is accomplished by the formation of cementum-like barrier of various thicknesses. The hard tissue barrier has been described as a cap, bridge, or ingrown wedge that may be composed of cementum, dentin, bone or so-called osteodentin, which can deposit on the inner walls of the canal (Steiner and Van Hassel 1971; Torneck et al. 1973; Lieberman and Trowbridge 1983; Ghose et al. 1987; Walia et al. 2000; Ritter et al. 2004; Rafter 2005). The apical hard tissue bridge is generally described as consisting of cementum, dentin, bone, and osteodentin. Histologically, the distinction of cementum and bone requires careful observation. Cementum formation can proceed from the periphery of the apex toward the center in decreasing concentric circles. Formation of dentin requires the presence of odontoblasts. The remaining soft tissue at the apical region of a nonvital immature tooth cannot be easily identified clinically. If there are viable odontoblasts, formation of some dentin structure is certainly a possibility.

In contrast to apexogenesis, apexification treatment does not generally lead to an additional formation of root dentin, leaving behind a weak root that is susceptible to fracture. Filling the canal from mid-root to coronal third with resin bonding to strengthen the root has been advocated after the completion of apexification (Rabie et al. 1985; Katebzadeh et al. 1998; Pene et al. 2001; Goldberg et al. 2002).

A clinical case was observed and reported of apical bridge formation and the tooth was extracted for orthodontic reason. The histological examinations showed that the apical barrier is basically made of cementum that not only forms a bridge but also extends coronally along the dentinal wall. Sharpey's fibers were not explicitly described by the authors, but appeared to form between the cementum inside the canal and the soft tissue (Lieberman and Trowbridge 1983).

### 14.4.2.4 Paradigm shift

While the decision for teeth undergoing apexogenesis or apexification has been determined by the result of pulp vitality, recent well-presented reports have shown that this straightforward dichotomized approach may sacrifice certain cases that still have the potential to undergo apexogenesis even when clinical condition qualifies for an apexification treatment. These case reports have shown convincingly that immature teeth clinically diagnosed with nonvital pulp and periradicular periodontitis or abscess can undergo apexogenesis (Iwaya et al. 2001; Banchs and Trope 2004; Chueh and Huang 2006; Thibodeau and Trope 2007). In fact, before these recent reports that provoked our traditional treatment decisions, there had been sporadic case reports in the literature documenting this observation (Rule and Winter 1966; Saad 1988; Matusow 1991a, b; Whittle 2000; Selden 2002).

These recent reports stimulated a new perspective as to how we determine the treatment plans for these cases (Huang 2008). A common aspect of the reported cases is that those teeth showing continual maturation of root and apex had developed extensive periradicular lesions with sinus tract formation before the treatment—a condition normally considered a total necrosis and infection of the pulp, and requires apexification (see Figs 14.4–14.6) (Iwaya et al. 2001; Banchs and Trope 2004; Chueh and Huang 2006).

#### 14.4.2.4.1 Clinical diagnosis of partially vital pulp

As mentioned, clinical diagnosis of the vitality of pulp is crude. It was considered in the past that the presence of radiolucent lesions at the apices confirms a total necrosis and infection of the pulp, until Lin et al. biopsied pulp tissue from mature permanent teeth associated with periapical radiolucency and found that vital tissues may be present in some cases, even within the pulp chambers in rare cases. The size of the lesions in that study varied significantly, and some had little change from normal radiographic appearance, while others may be described as moderately extensive (Lin et al. 1984). In the case of immature teeth, association of periapical lesions with remaining vital tissues within the canal could be more common, although there is a lack of histological evidence for this. However, those recent case reports of immature teeth that all have radiolucent lesions, some with quite extensive purulence in the periradicular region, showed remarkable root maturation after conservative treatment (Iwaya et al. 2001; Banchs and Trope 2004; Chueh and Huang 2006). This suggests that vital pulp tissue may have remained in the canals or the ingrowth of nonpulp tissue filling into the canal space. Iwaya et al. (2001) reported that during the treatment of an immature second mandibular premolar with a periradicular abscess, the patient felt the insertion of a smooth broach into the canal before reaching the apex, indicating the partial sensitivity of the pulp. Thirty-five months after the treatment, the root formation was complete and the tooth responded to electrical pulp test. Similarly,

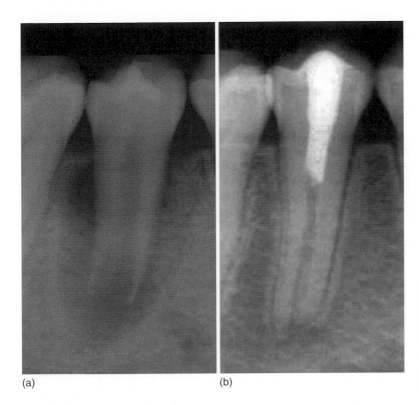

(a)                              (b)

**Fig. 14.4**   (a) Radiograph showing a lower premolar of an 11-year-old patient having an extensive periradicular lesion. (b) Twenty-four-month radiograph after treatment showing complete root development. (Adapted from Banchs and Trope 2004 with permission.)

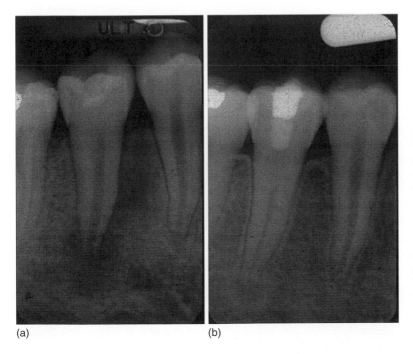

(a)                              (b)

**Fig. 14.5**   Clinical case of a 10-year-old patient. (a) Radiograph showing an immature root of tooth #29 with an open apex and an extensive radiolucency at the periapical and mesial regions of the root. (b) Seven months after the initial treatment showing complete maturation of the root apex, healing of the periradicular bone, a significant increase of the calcified tissue in the root, decrease of root canal space, and the calcified coronal third of the root canal. (Adapted from Chueh and Huang 2006 with permission.)

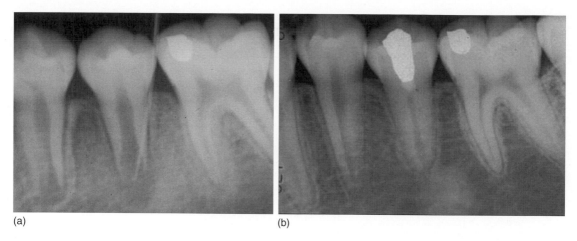

(a)                                    (b)

**Fig. 14.6**    Clinical case of a 10-year-old patient. (a) Radiograph showing a radiolucent lesion at the periapical area of tooth #20 with a wide-open apex (a gutta-percha point into the sinus tract). (b) Thirty-five months after the initial treatment revealing a market reduction of the root canal space and maturation of the root apex. (Adapted from Chueh and Huang 2006 with permission.)

Chueh and Huang (2006) reported four cases of immature teeth with periradicular lesions or abscess. In some of these cases, either the patients felt the entry of instrument into the pulp chamber or vital pulp tissue and hemorrhage were observed.

### 14.4.2.4.2 Disinfection protocol

The management of these immature teeth in the case reports mentioned above has the following common features: (a) minimal or no instrumentation; (b) irrigation with 2.5–5.25% NaOCl, 3% hydrogen peroxide, and/or Peridex; and (c) medication with calcium hydroxide or antimicrobial agents consisting of metronidazole, minocycline, and ciprofloxacin in paste form.

It was noted by some authors that calcium hydroxide is not recommended as intracanal medicament, so as not to damage the remaining pulp tissue and Hertwig epithelial root sheath (Banchs and Trope 2004; Chueh and Huang 2006). Based on these case reports, a clinical protocol to treat immature teeth, after a thorough irrigation with hypochlorite, and possibly with 17% EDTA to remove the smear layer, is summarized as the following:

a.  The antimicrobial paste is to be introduced into the canal carefully to a depth slightly shy of the remaining vital tissue with a lentulo spiral and the accessed cavity sealed with 2 mm cavity as an inner layer and 2 mm durable cement such as glass ionomer as the outer layer.

b.  After 1–2 weeks, the patient should return for evaluation. If asymptomatic and lack of clinical intraoral signs of pathology, the canal will be re-accessed to visually examine the condition under the microscope. If the canal is clean, it will be irrigated again with NaOCl followed by sterile normal saline and then dried. Banchs and Trope (2004) induced hemorrhage by an explore penetrating slightly into the periapical tissue allowing the blood clot to form in the canal and stopped at a level 3 mm below that of the cementoenamel junction (CEJ). MTA was then placed over the blood clot. They considered the blood clot as a scaffold and source of growth factors to facilitate the regeneration and repair of tissues into the canal. Currently, neither there is histological evidence showing that blood clot is required for the formation of repairing tissues, nor there are systematic clinical studies to show that this approach is optimal. However, these cases reports at least provide some guidelines as to the extent of which the healing potential in these immature teeth is capable.

A small piece of resorbable collagenous matrix such as CollacCote (Integra Life Science Corp., Plainsboro, NJ) is placed at the level of the CEJ or slightly apical to it and MTA is placed over it. The accessed cavity will be sealed with glass ionomer cement or composite and the tooth should be followed up periodically to observe the maturation of the root. If after several rounds of intracanal irrigation and medication the clinical symptoms show no sign of improvement, that is, persistent presence of sinus tract, swelling, and/or pain, apexification procedure should then be carried out.

### 14.4.2.4.3 Outcome
Given the right condition, many tissues are programmed and capable of self-regeneration to repair the damaged portion. The literature is replete with the finding that if disinfection can be performed effectively, pulp tissue can heal and new dentin bridge will form. Pulp tissue in immature teeth with open apex has a rich blood supply and therefore it is more potent in defending infection and healing. As an example, pulp polyp is normally seen in immature teeth when the pulp is widely exposed to the oral cavity. The newly formed epithelial layer helps defend the invading microbes and keeps the underlying tissue alive except the tissue is no longer normal and filled with inflammatory infiltrates.

As noted above, teeth inherit a thin and weak root after successful apexification is susceptible to fracture. Shifting apexification to apexogenesis even for nonvital teeth with periradicular periodontitis or abscess is a clinically beneficial approach for patients. These mentioned case reports all show that the partially survived pulp and apical tissue of immature teeth have the potential to regenerate and complete the root formation. The infection has not recurred during the period of follow-up observation.

Whether the thickened root is formed by pulp tissue from the remaining vital pulp tissue at the apical region, or that is formed by PDL tissue that has grown into the root canal from the apical foramen and deposited the cementum onto the inner surface of the root dentin, is not clear. One may speculate that some pulp tissue survived and allowed apexogenesis to occur. DPSCs have been identified to exist in permanent teeth (Gronthos et al. 2000, 2002; Shi and Gronthos 2003). The apical papilla contains stem cells (SCAP) that have been recently described to be more robust stem cells than DPSCs (Sonoyama et al. 2006, 2008). The SCAP may survive the infection to allow root maturation, while the survived DPSCs in the remaining vital pulp rebuild the lost pulp tissue in the canal. Iwaya and colleague (2001) suggested that

the open apex provides a good communication from pulp space to the periapical tissues; therefore, it may be possible for periapical disease to occur while the pulp is only partially necrotic and infected. If this is the case, using revascularization to describe this phenomenon does not encompass the scope of regeneration of pulp tissue in which genuine pulp containing functional odontoblasts is capable of laying down new dentin to complete the root development. Pulp stem cells seeded onto existing dentin may differentiate into odontoblasts and may deposit new dentin. Nör (2006) has shown that by seeding dental stem cells onto synthetic scaffolds seated in pulp chamber, odontoblasts can develop from the stem cells that localize against existing dentin surface. Therefore, the survived pulp tissue fragments may provide source of new odontoblasts that will form dentin, some may be in the center of the pulp and mixed with ingrown cementum, creating a calcified canal space.

Deposition of cementum or bone in the canal may also occur, gradually narrowing the space. The studies using rhesus monkey demonstrated that after total pulp tissue removal in immature teeth, either treated with calcium hydroxide or collagen gel, there was cementum tissue formation at the apex and in the canal. PDL-like tissue can also be found in canal treated with collagen gel (Nevins et al. 1977, 1978). PDL and cementum tissues can be verified histologically by the presence of Sharpey's fibers (A. Nevins, unpublished data).

Animal studies focusing on the changes in pulp tissue after replantation showed that various hard tissues including dentin, cementum, and bone may form in pulp space depending on the level of pulp recovery (Skoglund and Tronstad 1981; Kvinnsland and Heyeraas 1989; Ritter et al. 2004). By tracing the migration of periodontal cells after pulpectomy in immature teeth, it was found that periodontal cells migrate into the apical pulp space during the repair process (Vojinovic and Vojinovic 1993). Therefore, if one assumes the total loss of pulp tissue but remaining in a sterile condition, the outcome is the ingrowth of periodontal tissues. This may explain the increased thickness of the canal wall and the severe shrinkage of canal space. Taken together, if pulp tissue is totally lost, the canal space may be occupied by cementum, PDL, and bone. In this situation, it is difficult to identify clinically via radiographs because the canal space may well be PDL tissue and the thickened root structure be cementum.

## 14.5 Orthodontic considerations in pathologically involved incompletely formed teeth

Orthodontic treatment often initiates at age ranging from 12 to 16 years during that time the dentition is mixed with immature teeth (Anthony 1986). Young individuals in this age range are also more susceptible to dental injuries. Therefore, it is not uncommon to encounter immature teeth that require endodontic therapy while under orthodontic treatment, which can result in root resorption. Continuous forces seem to produce more resorptions than discontinuous forces (Weiland 2006). Orthodontic forces are traumatic to the periodontal tissues and cause the expression of proinflammatory cytokines and release of mediators by periodontal tissue cells, leading to the resorption of bone and cementum. The degree of the induced inflammation depends on various factors including the amount of orthodontic forces and individual variable biological responses or genetic influences (Al-Qawasmi et al. 2006; Filho et al. 2006). Increased angiogenesis and inflammatory response in pulp also occur (Derringer et al. 1996; Vandevska-Radunovic 1999). Endodontically treated teeth under orthodontic treatment do not show more apical root resorption than normal vital teeth (Esteves et al. 2007). Proper orthodontic treatment does not affect the development of normal immature teeth. When immature teeth under orthodontic forces developed pathosis, teeth appear to have a doubled infliction, and therefore the course of healing after endodontic treatment may be compromised. Clinical observations of immature teeth having apexification, however, do not show retardation or inhibition of the deposition of a calcified barrier at the root apex during an active orthodontic movement in children (Anthony 1986) or in adult (Fava 1999). In the case of transplantation of immature teeth followed by orthodontic treatment, studies have shown that orthodontic movement of the transplants appeared to have a tendency to shorten the final root length (Lagerstrom and Kristerson 1986).

## 14.6 Stem cells for pulp and periodontal tissue regeneration

In light of the recent isolation and characterization of adult dental stem cells as well as the progress of regenerative medicine, clinicians should reevaluate current clinical protocols when treatment planning for certain cases. This is especially true for immature teeth as they are still at the growing phase of their lifespan.

Stem cell biology has become an important field for the understanding of tissue regeneration, although much knowledge in this area has been from the in vitro studies. In general, stem cells are defined by having two major properties. First, they are capable of self-renewal. Second, when they divide, some daughter cells give rise to cells that eventually become terminally differentiated cells. Depending on the type of stem cells and their ability and potency to become different tissues, the following categories of stem cells have been established: (i) totipotent stem cells, each cell is capable of developing into an entire organism; (ii) pluripotent stem cells, cells from embryos (embryonic stem cells) when grown in the right environment in vivo are capable of forming all types of tissues; and (iii) multipotent stem cells, postnatal stem cells or commonly called adult stem cells that are capable of giving rise to multiple lineages of cells. Dental stem cells belong to the third category as described below (Robey and Bianco 2006).

Embryonic stem cells are potentially immortal when grown in vitro, whereas multipotent or unipotent stem cells have limited lifespan and become senesced after certain population doublings in culture. Adult stem cells have been isolated from various tissues. Bone marrow is a source of stem cells of multipotency including mainly hematopoietic and mesenchymal stem cells. Mesenchymal stem cells (MSCs) have been identified in many mesenchymal tissues and are capable of becoming many lineages of cells when grown in defined conditions including osteogenic, chondrogenic, adipogenic, myogenic, and neurogenic lineages (Tuan et al. 2003; Baksh et al. 2004).

To date, five types of human dental stem/progenitor cells have been isolated and characterized: (i) dental pulp stem cells (DPSCs) (Gronthos et al. 2000), (ii) stem cells from exfoliated deciduous teeth (SHED) (Miura et al. 2003), (iii) stem cells from apical papilla (SCAP) (Sonoyama et al. 2006), (iv) periodontal ligament stem cells (PDLSCs) (Seo et al. 2004), and follicle precursor cells (DFPCs) (Morsczeck et al. 2005). Among these, all except SHED are from permanent teeth. These dental stem cells are considered mesenchymal-like stem cells and possess different potential for becoming specific tissue-forming cells. DPSCs, SHED, and SCAP are from the developed or developing pulp tissue. These ex vivo expanded cells can differentiate into odontoblast-like cells and produce ectopic dentin-like tissue in vivo. When grown in cultures and induced under specific conditions, DPSCs and SHED can differentiate into neuronal and adipogenic cells in addition to dentinogenic cells (Miura et al. 2003; Zhang et al. 2006). Some reports have shown that DPSCs have also chondrogenic, myogenic, and osteogenic potentials (Laino et al. 2005, 2006; Zhang et al. 2006; d'Aquino et al. 2007). These dental stem cells may potentially be utilized for dental tissue regeneration, that is, pulp/dentin and periodontal ligament (Nakashima and Akamine 2005; Huang et al. 2006, 2008; Ivanovski et al. 2006). More importantly, the identification of these dental stem cells provided us a better understanding on the biology of pulp and periodontal ligament tissues and their regenerative potential after tissue damage. For example, the observation of severely infected pulp in immature teeth able to undergo complete root maturation after proper disinfection procedures may be explained by the possibility that SCAP is relatively resistant to infection.

## 14.7 Recent innovations on the regeneration of tooth form

Dental implants have been a favorable option instead of bridges or removable dentures to restore lost teeth and functions. Recent isolation of various dental stem cells shed light on the possibility of regeneration of tooth structure. Cells isolated from tooth bud can form ectopic teeth in vivo when seeded onto scaffolds (Young et al. 2002, 2005; Duailibi et al. 2004; Komine et al. 2007; Nakao et al. 2007). A few reports have demonstrated the orthotopic regeneration of engineered teeth. Single cells from the dog tooth buds at the bell stage were isolated and directly seeded onto scaffolds and transplanted back to the original tooth sockets. Dentin structure regeneration was achieved, but neither enamel nor root formation (Honda et al. 2006). With a swine model, Chen's research team utilized ex vivo expanded tooth bud cells (from bell stage), cultured onto cylinder-shaped scaffolds and autografted back to the original alveolar sockets. They were able to observe tooth formation with root structures along with periodontium (Kuo et al. 2007). Overall, tooth regeneration still faces many obstacles: (i) lack of formation of normal tooth size, (ii) lack of consistent root formation, and

(iii) no evidence of complete eruption into functional occlusion.

The other approach is to use SCAP and PDLSCs to form a bio-root (Sonoyama et al. 2006). Using a minipig model, autologous SACP and PDLSCs from minipigs were loaded onto HA/TCP (hydroxyapatite/tricalcium phosphate) and Gelfoam scaffolds, respectively, and implanted into sockets of the lower jaw. A post-channel was precreated to leave space for post-insertion. Three months later, the bio-root was exposed and a porcelain crown was inserted (Fig. 14.7). This approach is relatively a fast way of creating a root to which an artificial crown can be installed. The bio-root is different from a natural root in that the root structure is deposited by SCAP in a random manner. Nevertheless, the bio-root is encircled with periodontal ligament tissue and appears to have a natural relationship with the surrounding bone. What remains to be improved is likely the mechanical strength of the bio-root that is approximately two-thirds of a natural tooth.

## 14.8 Conclusion and prospects

Although the healing potential and defense mechanism of pulp have been long recognized, the intensity and the nature of the virulence of infection are still the determining factors for the outcome of pulp recovery. Immature teeth, by having a large and young pulp tissue and an open apex to allow good blood supply, remarkable healing potential has been observed in conditions that would not have been possible for mature teeth. The discovery and understanding of pulp stem cells provide us a better insight into the healing potential of the immature teeth. Along with an improved regimen of canal disinfection, it seems to be the right time to establish a new protocol for a paradigm shift in treating these infected immature teeth.

Currently there is a lack of clinical studies on the success rate of treating immature teeth with periradicular bone resorption. The time lapse between the pulp infection and endodontic intervention in relation to clinical success is a critical issue. Presumably, the longer standing of an infected pulp in immature teeth, the less survived pulp tissue and stem cells are remained. Additionally, the longer the infection, the more likelihood of a deeper penetration of microbial colonies into dentinal tubules, which renders the disinfection more difficult.

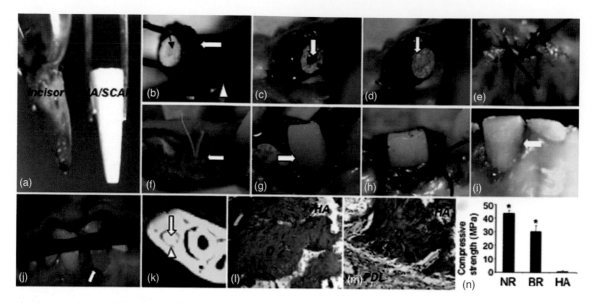

**Fig. 14.7** Swine SCAP/PDLSC-mediated root/periodontal structure as an artificial crown support for the restoration of tooth function in swine. (a) Extracted minipig lower incisor and root-shaped HA/TCP carrier loaded with SCAP. (b) Gelfoam containing $10 \times 10^6$ PDLSCs (open arrow) was used to cover the HA/SCAP (black arrow) and implanted into the lower incisor socket (open triangle). (c) HA/SCAP-Gelfoam/PDLSCs were implanted into a newly extracted incisor socket. A post-channel was precreated inside the root shape HA carrier (arrow). (d) The post-channel was sealed with a temporary filling for affixing a porcelain crown in the next step. (e) The HA/SCAP-Gelfoam/PDLSC implant was sutured for 3 months. (f) The HA/SCAP-Gelfoam/PDLSC implant (arrow) was reexposed and the temporary filling was removed to expose the post-channel. (g) A premade porcelain crown was cemented to the HA/SCAP-Gelfoam/PDLSC structure. (h) The exposed section was sutured. (i and j) Four weeks after fixation, the porcelain crown was retained in the swine after normal tooth use as shown by open arrows. (k) After 3 months implantation, the HA/SCAP-Gelfoam/PDLSC implant had formed a hard root structure (open arrows) in the mandibular incisor area as shown by CT scan image. A clear PDL space was found between the implant and surrounding bony tissue (triangle arrows). (l and m) H&E staining showed that implanted HA/SCAP-Gelfoam/PDLSC contains newly regenerated dentin (d) inside the implant (l) and PDL tissue (PDL) on the outside of the implant (m). (n) Compressive strength measurement showed that newly formed bio-roots have much higher compressive strength than original HA/TCP carrier (*$P =$ 0.0002), but lower than that in natural swine root dentin (*$P = 0.003$) (NR, natural minipig root; BR, newly formed bio-root; HA, original HA carrier). (Reprinted from Sonoyama et al. 2006.)

## 14.9 References

Alacam T. 1985. Evaluation of a tactile hardness test in indirect pulp capping. *Int Endod J* 18(4): 274–76.

Al-Hiyasat AS, Barrieshi-Nusair KM, and Al-Omari MA. 2006. The radiographic outcomes of direct pulp-capping procedures performed by dental students: A retrospective study. *J Am Dent Assoc* 137(12): 1699–1705.

Al-Qawasmi RA, Hartsfield JK, Jr, Everett ET, Weaver MR, Foroud TM, Faust DM, and Roberts WE. 2006. Root resorption associated with orthodontic force in inbred mice: Genetic contributions. *Eur J Orthod* 28(1): 13–19.

Aminoshariae A, Hartwell GR, and Moon PC. 2003. Placement of mineral trioxide aggregate using two different techniques. *J Endod* 29(10): 679–82.

Andelin WE, Shabahang S, Wright K, and Torabinejad M. 2003. Identification of hard tissue after experimental pulp capping using dentin sialoprotein (DSP) as a marker. *J Endod* 29(10): 646–50.

Andreasen JO and Andreasen FM. 1994. *Textbook and Color Atlas of Traumatic Injuries to the Teeth.* Copenhagen, Denmark: Munksgaard.

Andreasen JO, Borum MK, Jacobsen HL, and Andreasen FM. 1995a. Replantation of 400 avulsed permanent incisors. 1. Diagnosis of healing complications. *Endod Dent Traumatol* 11(2): 51–58.

Andreasen JO, Borum MK, Jacobsen HL, and Andreasen FM. 1995b. Replantation of 400 avulsed permanent incisors. 2. Factors related to pulpal healing. *Endod Dent Traumatol* 11(2): 59–68.

Andreasen JO and Ravn JJ. 1972. Epidemiology of traumatic dental injuries to primary and permanent teeth in a Danish population sample. *Int J Oral Surg* 1(5): 235–39.

Anthony DR. 1986. Apexification during active orthodontic movement. *J Endod* 12(9): 419–21.

Arens DE and Torabinejad M. 1996. Repair of furcal perforations with mineral trioxide aggregate: Two case reports. *Oral Surg Oral Med Oral Pathol Oral Radiol Endod* 82(1): 84–88.

Baksh D, Song L, and Tuan RS. 2004. Adult mesenchymal stem cells: Characterization, differentiation, and application in cell and gene therapy. *J Cell Mol Med* 8(3): 301–16.

Banchs F and Trope M. 2004. Revascularization of immature permanent teeth with apical periodontitis: New treatment protocol? *J Endod* 30(4): 196–200.

Barthel CR, Levin LG, Reisner HM, and Trope M. 1997. TNF-alpha release in monocytes after exposure to calcium hydroxide treated *Escherichia coli* LPS. *Int Endod J* 30(3): 155–59.

Barthel CR, Rosenkranz B, Leuenberg A, and Roulet JF. 2000. Pulp capping of carious exposures: Treatment outcome after 5 and 10 years: a retrospective study. *J Endod* 26(9): 525–28.

Bernick S. 1964. Differences in nerve distribution between erupted and non-erupted human teeth. *J Dent Res* 43: 406–11.

Bhasker SN. 1991. *Orban's Oral Histology and Embryology.* St Louis, OM: Mosby-Year Book.

Binnie WH and Rowe AH. 1973. A histological study of the periapical tissues of incompletely formed pulpless teeth filled with calcium hydroxide. *J Dent Res* 52(5): 1110–16.

Bjornal L and Kidd EA. 2005. The treatment of deep dentine caries lesions. *Dent Update* 32(7): 402–4, 407–10, 413.

Bogen G, Kim JS, and Bakland LK. 2008. Direct pulp capping with mineral trioxide aggregate: An observational study. *J Am Dent Assoc* 139(3): 305–15.

Chueh L-H and Huang GTJ. 2006. Immature teeth with periradicular periodontitis or abscess undergoing apexogenesis: A paradigm shift. *J Endod* 32(12): 1205–13.

Cooke C and Rowbotham TC. 1988. The closure of open apices in non-vital immature incisor teeth. *Br Dent J* 165(12): 420–21.

Cox CF, Bergenholtz G, Fitzgerald M, Heys DR, Heys RJ, Avery JK, and Baker JA. 1982. Capping of the dental pulp mechanically exposed to the oral microflora—a 5 week observation of wound healing in the monkey. *J Oral Pathol* 11(4): 327–39.

Cox CF, Bergenholtz G, Heys DR, Syed SA, Fitzgerald M, and Heys RJ. 1985. Pulp capping of dental pulp mechanically exposed to oral microflora: A 1–2 year observation of wound healing in the monkey. *J Oral Pathol* 14(2): 156–68.

Cvek M. 1972. Treatment of non-vital permanent incisors with calcium hydroxide. I. Follow-up of periapical repair and apical closure of immature roots. *Odontol Revy* 23(1): 27–44.

Cvek M. 1978. A clinical report on partial pulpotomy and capping with calcium hydroxide in permanent incisors with complicated crown fracture. *J Endod* 4(8): 232–37.

Cvek M, Cleaton-Jones PE, Austin JC, and Andreasen JO. 1982. Pulp reactions to exposure after experimental crown fractures or grinding in adult monkeys. *J Endod* 8(9): 391–97.

Cvek M, Nord CE, and Hollender L. 1976. Antimicrobial effect of root canal debridement in teeth with immature root.

A clinical and microbiologic study. *Odontol Revy* 27(1): 1–10.

d'Aquino R, Graziano A, Sampaolesi M, Laino G, Pirozzi G, De Rosa A, and Papaccio G. 2007. Human postnatal dental pulp cells co-differentiate into osteoblasts and endotheliocytes: A pivotal synergy leading to adult bone tissue formation. *Cell Death Differ* 14(6): 1162–71.

D'Souza R. 2002. Development of the pulpodentin complex. In: *Seltzer and Bender's Dental Pulp.* Carol Stream, IL: Quintessence Publishing Co, Inc.

de Leimburg ML, Angeretti A, Ceruti P, Lendini M, Pasqualini D, and Berutti E. 2004. MTA obturation of pulpless teeth with open apices: Bacterial leakage as detected by polymerase chain reaction assay. *J Endod* 30(12): 883–86.

Derringer KA, Jaggers DC, and Linden RW. 1996. Angiogenesis in human dental pulp following orthodontic tooth movement. *J Dent Res* 75(10): 1761–66.

Duailibi MT, Duailibi SE, Young CS, Bartlett JD, Vacanti JP, and Yelick PC. 2004. Bioengineered teeth from cultured rat tooth bud cells. *J Dent Res* 83(7): 523–28.

El-Meligy OA and Avery DR. 2006. Comparison of apexification with mineral trioxide aggregate and calcium hydroxide. *Pediatr Dent* 28(3): 248–53.

Esteves T, Ramos AL, Pereira CM, and Hidalgo MM. 2007. Orthodontic root resorption of endodontically treated teeth. *J Endod* 33(2): 119–22.

Falster CA, Araujo FB, Straffon LH, and Nor JE. 2002. Indirect pulp treatment: In vivo outcomes of an adhesive resin system vs calcium hydroxide for protection of the dentin-pulp complex. *Pediatr Dent* 24(3): 241–48.

Farsi N, Alamoudi N, Balto K, and Al Mushayt A. 2006. Clinical assessment of mineral trioxide aggregate (MTA) as direct pulp capping in young permanent teeth. *J Clin Pediatr Dent* 31(2): 72–76.

Fava LR. 1999. Apex formation during orthodontic treatment in an adult patient: Report of a case. *Int Endod J* 32(4): 321–27.

Filho PF, Letra A, Carvalhal JC, and Menezes R. 2006. Orthodontically induced inflammatory root resorptions: A case report. *Dent Traumatol* 22(6): 350–53.

Finucane D and Kinirons MJ. 1999. Non-vital immature permanent incisors: Factors that may influence treatment outcome. *Endod Dent Traumatol* 15(6): 273–77.

Fong CD and Davis MJ. 2002. Partial pulpotomy for immature permanent teeth, its present and future. *Pediatr Dent* 24(1): 29–32.

Ford TR, Torabinejad M, Abedi HR, Bakland LK, and Kariyawasam SP. 1996. Using mineral trioxide aggregate as a pulp-capping material. *J Am Dent Assoc* 127(10): 1491–94.

Ford TR, Torabinejad M, McKendry DJ, Hong CU, and Kariyawasam SP. 1995. Use of mineral trioxide aggregate for repair of furcal perforations. *Oral Surg Oral Med Oral Pathol Oral Radiol Endod* 79(6): 756–63.

Fulling HJ and Andreasen JO. 1976. Influence of maturation status and tooth type of permanent teeth upon electrometric and thermal pulp testing. *Scand J Dent Res* 84(5): 286–90.

Fuss Z, Trowbridge H, Bender IB, Rickoff B, and Sorin S. 1986. Assessment of reliability of electrical and thermal pulp testing agents. *J Endod* 12(7): 301–5.

Ghose LJ, Baghdady VS, and Hikmat YM. 1987. Apexification of immature apices of pulpless permanent anterior teeth with calcium hydroxide. *J Endod* 13(6): 285–90.

Goldberg F, Kaplan A, Roitman M, Manfre S, and Picca M. 2002. Reinforcing effect of a resin glass ionomer in the restoration of immature roots in vitro. *Dent Traumatol* 18(2): 70–72.

Goldstein S, Sedaghat-Zandi A, Greenberg M, and Friedman S. 1999. Apexification and apexogenesis. *N Y State Dent J* 65(5): 23–25.

Gronthos S, Brahim J, Li W, Fisher LW, Cherman N, Boyde A, DenBesten P, Robey PG, and Shi S. 2002. Stem cell properties of human dental pulp stem cells. *J Dent Res* 81(8): 531–35.

Gronthos S, Mankani M, Brahim J, Robey PG, and Shi S. 2000. Postnatal human dental pulp stem cells (DPSCs) in vitro and in vivo. *Proc Natl Acad Sci U S A* 97(25): 13625–30.

Hachmeister DR, Schindler WG, Walker WA III, and Thomas DD. 2002. The sealing ability and retention characteristics of mineral trioxide aggregate in a model of apexification. *J Endod* 28(5): 386–90.

Holland R, de Mello W, Nery MJ, Bernabe PF, and de Souza V. 1977. Reaction of human periapical tissue to pulp extirpation and immediate root canal filling with calcium hydroxide. *J Endod* 3(2): 63–67.

Honda MJ, Ohara T, Sumita Y, Ogaeri T, Kagami H, and Ueda M. 2006. Preliminary study of tissue-engineered odontogenesis in the canine jaw. *J Oral Maxillofac Surg* 64(2): 283–89.

Hoshino E, Kurihara-Ando N, Sato I, Uematsu H, Sato M, Kota K, and Iwaku M. 1996. In-vitro antibacterial susceptibility of bacteria taken from infected root dentine to a mixture of ciprofloxacin, metronidazole and minocycline. *Int Endod J* 29(2): 125–30.

Huang GT. 2008. A paradigm shift in endodontic management of immature teeth: Conservation of stem cells for regeneration. *J Dent* 36(6): 379–86.

Huang GT, Sonoyama W, Chen J, and Park SH. 2006. In vitro characterization of human dental pulp cells: Various isolation methods and culturing environments. *Cell Tissue Res* 324(2): 225–36.

Huang GTJ, Sonoyama W, Liu Y, Liu H, Wang S, and Shi S. (2008). The hidden treasure in apical papilla: The potential role in pulp/dentin regeneration and bioroot engineering. *J Endod* 34(6): 645–51.

Iohara K, Nakashima M, Ito M, Ishikawa M, Nakasima A, and Akamine A. 2004. Dentin regeneration by dental pulp stem cell therapy with recombinant human bone morphogenetic protein 2. *J Dent Res* 83(8): 590–95.

Ivanovski S, Gronthos S, Shi S, and Bartold PM. 2006. Stem cells in the periodontal ligament. *Oral Dis* 12(4): 358–63.

Iwaya SI, Ikawa M, and Kubota M. 2001. Revascularization of an immature permanent tooth with apical periodontitis and sinus tract. *Dent Traumatol* 17(4): 185–87.

Javelet J, Torabinejad M, and Bakland LK. 1985. Comparison of two pH levels for the induction of apical barriers in immature teeth of monkeys. *J Endod* 11(9): 375–78.

Jiang J, Zuo J, Chen SH, and Holliday LS. 2003. Calcium hydroxide reduces lipopolysaccharide-stimulated osteoclast formation. *Oral Surg Oral Med Oral Pathol Oral Radiol Endod* 95(3): 348–54.

Kaiser HJ. 1964. *Management of Wide Open Apex Canals with Calcium Hydroxide.* Annual Meeting of the American Association of Endodontists, Washington DC.

Katebzadeh N, Dalton BC, and Trope M. 1998. Strengthening immature teeth during and after apexification. *J Endod* 24(4): 256–59.

Klein H. 1978. Pulp responses to an electric pulp stimulator in the developing permanent anterior dentition. *ASDC J Dent Child* 45(3): 199–202.

Kling M, Cvek M, and Mejare I. 1986. Rate and predictability of pulp revascularization in therapeutically reimplanted permanent incisors. *Endod Dent Traumatol* 2(3): 83–89.

Komine A, Suenaga M, Nakao K, Tsuji T, and Tomooka Y. 2007. Tooth regeneration from newly established cell lines from a molar tooth germ epithelium. *Biochem Biophys Res Commun* 355(3): 758–63.

Kuo T-F, Huang A-T, Chang H-H, Lin F-H, Chen S-T, Chen R-S, Chou C-H, Lin H-C, Chiang H, and Chen M-H. 2007. Regeneration of dentin-pulp complex with cementum and periodontal ligament formation using dental bud cells in gelatin-chondroitin-hyaluronan tri-copolymer scaffold in swine. *J Biomed Mater Res Part A* 86(4): 1062–68.

Kvinnsland I and Heyeraas KJ. 1989. Dentin and osteodentin matrix formation in apicoectomized replanted incisors in cats. *Acta Odontol Scand* 47(1): 41–52.

Lagerstrom L and Kristerson L. 1986. Influence of orthodontic treatment on root development of autotransplanted premolars. *Am J Orthod* 89(2): 146–50.

Laino G, Carinci F, Graziano A, d'Aquino R, Lanza V, De Rosa A, Gombos F, Caruso F, Guida L, Rullo R, Menditti D, and Papaccio G. 2006. In vitro bone production using stem cells derived from human dental pulp. *J Craniofac Surg* 17(3): 511–15.

Laino G, d'Aquino R, Graziano A, Lanza V, Carinci F, Naro F, Pirozzi G, and Papaccio G. 2005. A new population of human adult dental pulp stem cells: A useful source of living autologous fibrous bone tissue (LAB). *J Bone Miner Res* 20(8): 1394–402.

Lee SJ, Monsef M, and Torabinejad M. 1993. Sealing ability of a mineral trioxide aggregate for repair of lateral root perforations. *J Endod* 19(11): 541–44.

Leksell E, Ridell K, Cvek M, and Mejare I. 1996. Pulp exposure after stepwise versus direct complete excavation of deep carious lesions in young posterior permanent teeth. *Endod Dent Traumatol* 12(4): 192–96.

Lieberman J and Trowbridge H. 1983. Apical closure of nonvital permanent incisor teeth where no treatment was performed: Case report. *J Endod* 9(6): 257–60.

Lin L, Shovlin F, Skribner J, and Langeland K. 1984. Pulp biopsies from the teeth associated with periapical radiolucency. *J Endod* 10(9): 436–48.

Mackie IC, Bentley EM, and Worthington HV. 1988. The closure of open apices in non-vital immature incisor teeth. *Br Dent J* 165(5): 169–73.

Marchi JJ, de Araujo FB, Froner AM, Straffon LH, and Nor JE. 2006. Indirect pulp capping in the primary dentition: A 4 year follow-up study. *J Clin Pediatr Dent* 31(2): 68–71.

Maroto M, Barberia E, Planells P, and Vera V. 2003. Treatment of a non-vital immature incisor with mineral trioxide aggregate (MTA). *Dent Traumatol* 19(3): 165–69.

Matusow RJ. 1991a. Acute pulpal-alveolar cellulitis syndrome V. Apical closure of immature teeth by infection control: Case report and a possible microbial-immunologic etiology. Part 1. *Oral Surg Oral Med Oral Pathol* 71(6): 737–42.

Matusow RJ. 1991b. Acute pulpal-alveolar cellulitis syndrome. V. Apical closure of immature teeth by infection control: The importance of an endodontic seal with therapeutic factors. Part 2. *Oral Surg Oral Med Oral Pathol* 72(1): 96–100.

Michanowicz JP and Michanowicz AE. 1967. A conservative approach and procedure to fill an incompletely formed root using calcium hydroxide as an adjunct. *J Dent Child* 34(1): 42–47.

Mitchell DF and Shankwalker GB. 1958. Osteogenic potential of calcium hydroxide and other materials in soft tissue and bone wounds. *J Dent Res* 37(6): 1157–63.

Miura M, Gronthos S, Zhao M, Lu B, Fisher LW, Robey PG, and Shi S. 2003. SHED: Stem cells from human exfoliated deciduous teeth. *Proc Natl Acad Sci U S A* 100(10): 5807–12.

Morsczeck C, Gotz W, Schierholz J, Zeilhofer F, Kuhn U, Mohl C, Sippel C, and Hoffmann KH. 2005. Isolation of precursor cells (PCs) from human dental follicle of wisdom teeth. *Matrix Biol* 24(2): 155–65.

Nakao K, Morita R, Saji Y, Ishida K, Tomita Y, Ogawa M, Saitoh M, Tomooka Y, and Tsuji T. 2007. The development of a bioengineered organ germ method. *Nat Methods* 4(3): 227–30.

Nakashima M and Akamine A. 2005. The application of tissue engineering to regeneration of pulp and dentin in endodontics. *J Endod* 31(10): 711–18.

Nakashima M, Iohara K, Ishikawa M, Ito M, Tomokiyo A, Tanaka T, and Akamine A. 2004. Stimulation of reparative dentin formation by ex vivo gene therapy using dental pulp stem cells electrotransfected with growth/differentiation factor 11 (Gdf11). *Hum Gene Ther* 15(11): 1045–53.

Nelson-Filho P, Leonardo MR, Silva LA, and Assed S. 2002. Radiographic evaluation of the effect of endotoxin (LPS) plus calcium hydroxide on apical and periapical tissues of dogs. *J Endod* 28(10): 694–96.

Nevins A, Finkelstein F, Laporta R, and Borden BG. 1978. Induction of hard tissue into pulpless open-apex teeth using collagen-calcium phosphate gel. *J Endod* 4(3): 76–81.

Nevins A, Wrobel W, Valachovic R, and Finkelstein F. 1977. Hard tissue induction into pulpless open-apex teeth using collagen-calcium phosphate gel. *J Endod* 3(11): 431–33.

Nevins AJ, Finkelstein F, Borden BG, and Laporta R. 1976. Revitalization of pulpless open apex teeth in rhesus monkeys, using collagen-calcium phosphate gel. *J Endod* 2(6): 159–65.

Nör JE. 2006. Tooth regeneration in operative dentistry. *Oper Dent* 31(6): 633–42.

Pace R, Giuliani V, Pini Prato L, Baccetti T, and Pagavino G. 2007. Apical plug technique using mineral trioxide aggregate: Results from a case series. *Int Endod J* 40(6): 478–84.

Pene JR, Nicholls JI, and Harrington GW. 2001. Evaluation of fiber-composite laminate in the restoration of immature, nonvital maxillary central incisors. *J Endod* 27(1): 18–22.

Petersson K, Soderstrom C, Kiani-Anaraki M, and Levy G. 1999. Evaluation of the ability of thermal and electrical tests to register pulp vitality. *Endod Dent Traumatol* 15(3): 127–31.

Rabie G, Trope M, Garcia C, and Tronstad L. 1985. Strengthening and restoration of immature teeth with an acid-etch resin technique. *Endod Dent Traumatol* 1(6): 246–56.

Rafter M. 2005. Apexification: A review. *Dent Traumatol* 21(1): 1–8.

Ritter AL, Ritter AV, Murrah V, Sigurdsson A, and Trope M. 2004. Pulp revascularization of replanted immature dog teeth after treatment with minocycline and doxycycline assessed by laser Doppler flowmetry, radiography, and histology. *Dent Traumatol* 20(2): 75–84.

Robey PG and Bianco P. 2006. The use of adult stem cells in rebuilding the human face. *J Am Dent Assoc* 137(7): 961–72.

Roeykens H, Van Maele G, Martens L, and De Moor R. 2002. A two-probe laser Doppler flowmetry assessment as an exclusive diagnostic device in a long-term follow-up of traumatised teeth: A case report. *Dent Traumatol* 18(2): 86–91.

Rule DC and Winter GB. 1966. Root growth and apical repair subsequent to pulpal necrosis in children. *Br Dent J* 120(12): 586–90.

Rutherford RB. 2001. BMP-7 gene transfer to inflamed ferret dental pulps. *Eur J Oral Sci* 109(6): 422–24.

Rutherford RB and Gu K. 2000. Treatment of inflamed ferret dental pulps with recombinant bone morphogenetic protein-7. *Eur J Oral Sci* 108(3): 202–6.

Saad AY. 1988. Calcium hydroxide and apexogenesis. *Oral Surg Oral Med Oral Pathol* 66(4): 499–501.

Safavi KE, Dowden WE, Introcaso JH, and Langeland K. 1985. A comparison of antimicrobial effects of calcium hydroxide and iodine-potassium iodide. *J Endod* 11(10): 454–56.

Safavi KE and Nichols FC. 1993. Effect of calcium hydroxide on bacterial lipopolysaccharide. *J Endod* 19(2): 76–78.

Safavi KE and Nichols FC. 1994. Alteration of biological properties of bacterial lipopolysaccharide by calcium hydroxide treatment. *J Endod* 20(3): 127–29.

Sato I, Ando-Kurihara N, Kota K, Iwaku M, and Hoshino E. 1996. Sterilization of infected root-canal dentine by topical application of a mixture of ciprofloxacin, metronidazole and minocycline in situ. *Int Endod J* 29(2): 118–24.

Schroder U and Granath LE. 1971. Early reaction of intact human teeth to calcium hydroxide following experimental pulpotomy and its significance to the development of hard tissue barrier. *Odontol Revy* 22(4): 379–95.

Selden HS. 2002. Apexification: An interesting case. *J Endod* 28(1): 44–45.

Seltzer S, Bender IB, and Ziontz M. 1963. The dynamics of pulp inflammation: Correlations between diagnostic data and actual histologic findings in the pulp. *Oral Surg Oral Med Oral Pathol* 16: 969–77.

Seo BM, Miura M, Gronthos S, Bartold PM, Batouli S, Brahim J, Young M, Robey PG, Wang CY, and Shi S. 2004. Investigation of multipotent postnatal stem cells from human periodontal ligament. *Lancet* 364(9429): 149–55.

Shabahang S and Torabinejad M. 2000. Treatment of teeth with open apices using mineral trioxide aggregate. *Pract Periodontics Aesthet Dent* 12(3): 315–20; quiz 322.

Shabahang S, Torabinejad M, Boyne PP, Abedi H, and McMillan P. 1999. A comparative study of root-end induction using osteogenic protein-1, calcium hydroxide, and mineral trioxide aggregate in dogs. *J Endod* 25(1): 1–5.

Sheehy EC and Roberts GJ. 1997. Use of calcium hydroxide for apical barrier formation and healing in non-vital immature permanent teeth: A review. *Br Dent J* 183(7): 241–46.

Shi S and Gronthos S. 2003. Perivascular niche of postnatal mesenchymal stem cells in human bone marrow and dental pulp. *J Bone Miner Res* 18(4): 696–704.

Skoglund A and Tronstad L. 1981. Pulpal changes in replanted and autotransplanted immature teeth of dogs. *J Endod* 7(7): 309–16.

Sonoyama W, Liu Y, Fang D, Yamaza T, Seo BM, Zhang C, Liu H, Gronthos S, Wang CY, Shi S, and Wang S. 2006. Mesenchymal stem cell-mediated functional tooth regeneration in Swine. *PLoS ONE* 1: e79.

Sonoyama W, Liu Y, Yamaza T, Tuan RS, Wang S, Shi S, and Huang GTJ. 2008. Characterization of the apical papilla and its residing stem cells from human immature permanent teeth: A pilot study. *J Endod* 34(2): 166–71.

Stanley HR. 1989. Pulp capping: Conserving the dental pulp—can it be done? Is it worth it? *Oral Surg Oral Med Oral Pathol* 68(5): 628–39.

Steiner JC and Van Hassel HJ. 1971. Experimental root apexification in primates. *Oral Surg Oral Med Oral Pathol* 31(3): 409–15.

Steinig TH, Regan JD, and Gutmann JL. 2003. The use and predictable placement of mineral trioxide aggregate in one-visit apexification cases. *Aust Endod J* 29(1): 34–42.

Thibodeau B and Trope M. 2007. Pulp revascularization of a necrotic infected immature permanent tooth: Case report and review of the literature. *Pediatr Dent* 29(1): 47–50.

Torabinejad M, Hong CU, McDonald F, and Pitt Ford TR. 1995a. Physical and chemical properties of a new root-end filling material. *J Endod* 21(7): 349–53.

Torabinejad M, Hong CU, Pitt Ford TR, and Kettering JD. 1995b. Antibacterial effects of some root end filling materials. *J Endod* 21(8): 403–6.

Torabinejad M, Pitt Ford TR, McKendry DJ, Abedi HR, Miller DA, and Kariyawasam SP. 1997. Histologic assessment of mineral trioxide aggregate as a root-end filling in monkeys. *J Endod* 23(4): 225–28.

Torabinejad M, Watson TF, and Pitt Ford TR. 1993. Sealing ability of a mineral trioxide aggregate when used as a root end filling material. *J Endod* 19(12): 591–95.

Torneck CD. 1982. Effects and clinical significance of trauma to the developing permanent dentition. *Dent Clin North Am* 26(3): 481–504.

Torneck CD, Smith JS, and Grindall P. 1973. Biologic effects of endodontic procedures on developing incisor teeth. IV. Effect of debridement procedures and calcium hydroxide-camphorated parachlorophenol paste in the treatment of experimentally induced pulp and periapical disease. *Oral Surg Oral Med Oral Pathol* 35(4): 541–54.

Trope M, McDougal R, Levin L, May KN, Jr, and Swift EJ, Jr. 2002. Capping the inflamed pulp under different clinical conditions. *J Esthet Restor Dent* 14(6): 349–57.

Tuan RS, Boland G, and Tuli R. 2003. Adult mesenchymal stem cells and cell-based tissue engineering. *Arthritis Res Ther* 5(1): 32–45.

Tziafas D, Smith AJ, and Lesot H. 2000. Designing new treatment strategies in vital pulp therapy. *J Dent* 28(2): 77–92.

Vandevska-Radunovic V. 1999. Neural modulation of inflammatory reactions in dental tissues incident to orthodontic tooth movement. A review of the literature. *Eur J Orthod* 21(3): 231–47.

Vojinovic O and Vojinovic J. 1993. Periodontal cell migration into the apical pulp during the repair process after pulpectomy in immature teeth: An autoradiographic study. *J Oral Rehabil* 20(6): 637–52.

Walia T, Chawla HS, and Gauba K. 2000. Management of wide open apices in non-vital permanent teeth with $Ca(OH)_2$ paste. *J Clin Pediatr Dent* 25(1): 51–56.

Weiland F. 2006. External root resorptions and orthodontic forces: Correlations and clinical consequences. *Prog Orthod* 7(2): 156–63.

Weisleder R and Benitez CR. 2003. Maturogenesis: Is it a new concept? *J Endod* 29(11): 776–78.

Whittle M. 2000. Apexification of an infected untreated immature tooth. *J Endod* 26(4): 245–47.

Windley W III, Teixeira F, Levin L, Sigurdsson A, and Trope M. 2005. Disinfection of immature teeth with a triple antibiotic paste. *J Endod* 31(6): 439–43.

Witherspoon DE and Ham K. 2001. One-visit apexification: Technique for inducing root-end barrier formation in apical closures. *Pract Proced Aesthet Dent* 13(6): 455–60; quiz 462.

Yanpiset K, Vongsavan N, Sigurdsson A, and Trope M. 2001. Efficacy of laser Doppler flowmetry for the diagnosis of revascularization of reimplanted immature dog teeth. *Dent Traumatol* 17(2): 63–70.

Yates JA. 1988. Barrier formation time in non-vital teeth with open apices. *Int Endod J* 21(5): 313–19.

Young CS, Abukawa H, Asrican R, Ravens M, Troulis MJ, Kaban LB, Vacanti JP, and Yelick PC. 2005. Tissue-engineered hybrid tooth and bone. *Tissue Eng* 11(9–10): 1599–610.

Young CS, Terada S, Vacanti JP, Honda M, Bartlett JD, and Yelick PC. 2002. Tissue engineering of complex tooth structures on biodegradable polymer scaffolds. *J Dent Res* 81(10): 695–700.

Zhang W, Walboomers XF, Shi S, Fan M, and Jansen JA. 2006. Multilineage differentiation potential of stem cells derived from human dental pulp after cryopreservation. *Tissue Eng* 12(10): 2813–23.

# Chapter 15

# Prognosis in the Treatment of Teeth with Endodontic Infections

*Shimon Friedman*

## 15.1 Preface: The critical importance of prognosis

When used in the context of health care, the term "prognosis" is defined as the forecast of the course of disease. The disease associated with endodontic infection is apical periodontitis, and the meaning of prognosis is the chance of the infected tooth and surrounding tissues to heal after treatment. The term "prognosis of treatment" is commonly used as a synonym for the expected outcome, mainly the chance for healing. Prognosis is a critically important element in clinical decision making, particularly when multiple options for treatment are available.

The current ethical concepts of health care delivery require that clinical decision making be based on sound evidence and involve the patient. Evidence to support the benefits and risks of available treatment alternatives is shared with the patient, who then selects a specific treatment based on his/her individual values (Pellegrino 1994; Ambrosio and Walkerley 1996; Wertz 1998; Schattner and Tal 2002; Fournier 2005).

To conform to this concept, health care providers must be well versed in the evidence that supports treatments they suggest to patients, and they should be able to project the prognosis for alternative treatments. Clinical decision making in endodontics is no exception to this prevailing concept; therefore, endodontics providers should be well informed about the prognosis of different endodontic treatment modalities.

Information about prognosis of endodontic treatment is available from over 160 studies on nonsurgical and surgical treatment. However, because of methodological and technical variations, this information has been highly inconsistent and answers to the main questions related to the prognosis of endodontic treatment have remained equivocal (Friedman 1998). The methodological and technical variations preclude indiscriminate review of the many available studies. Importantly, the prognosis of teeth with endodontic infections is compromised relative to teeth without infections (Friedman 1998), requiring a specific review of the prognosis of teeth with infections.

It is well established that questions about the prognosis of a disease following state-of-the-art treatment should be addressed in structured reviews that focus on studies selected according to well-defined criteria. Accordingly, the purpose of this chapter is to define the prognosis of nonsurgical and surgical endodontic treatment of teeth with preoperative lesions, which indicates an active infectious process, and to identify prognostic variables, based on selected studies.

## 15.2 Outcome measures and criteria in assessment of endodontic prognosis

Much of the confusion regarding the prognosis of endodontic treatment is caused by inconsistent use of outcome criteria, resulting in highly variable "success" rates among the different studies (Friedman 1998). This inconsistency has affected current studies on initial root canal treatment (Table 15.1), orthograde retreatment (Table 15.2), apical surgery (Table 15.3), and intentional replantation (Table 15.4). The main shortcomings have been the use of ambiguous terms such as "success" and "failure," and lack of calibration in outcome assessment. This section focuses on outcome measures, assessment strategies, and criteria for assessment of prognosis after nonsurgical and surgical endodontic treatment.

### 15.2.1 Clinical outcome measures

Clinical outcome measures have been widely used in endodontic studies to assess the health state of examined teeth. In the majority of studies, patients have been questioned about presence or absence of pain (subjective measure), and teeth were examined clinically for presence or absence of swelling, sinus tract, and tenderness to percussion and palpation (objective measures). Inasmuch as any one of the above can be an expression of persistent endodontic infection, they are not specific signs of apical periodontitis. Therefore, these clinical outcome measures have been coupled with radiographic measures in the majority of endodontic studies. Nevertheless, in many studies on nonsurgical treatment (Ørstavik et al. 1987; Molven and Halse 1988; Murphy et al. 1991; Ørstavik and Hörsted-Bindslev 1993; Smith et al. 1993; Ørstavik 1996; Trope et al. 1999; Heling et al. 2001; Pettiette et al. 2001; Waltimo et al. 2001; Cheung 2002; Huumonen et al. 2003; Peters et al. 2004; Marending et al. 2005) and in at least one apical surgery study (Rapp et al. 1991), clinical outcome measures have not been used and the outcome assessed exclusively by the radiographic appearance. This strategy possibly overestimates the success rate by not noting teeth that could be radiographically normal but symptomatic (Friedman 2002b).

### 15.2.2 Radiographic outcome measures

Assessment of radiographs is subject to bias (Goldman et al. 1972, 1974; Reit and Hollender 1983; Zakariasen et al. 1984; Eckerbom et al. 1986). Calibration and specific observer strategies have been advocated for endodontic studies (Rud et al. 1972; Reit 1987b; Molven et al. 2002a), and they can significantly improve the consistency of assessment. A frequently used strategy is use of the periapical index (PAI) (Ørstavik et al. 1986) for calibration purposes and as reference for assessment of radiographs. Assessed radiographs are compared with five sets of radiographic images and their schematic representations (see Chapter 2). These images are derived from a histologic–radiographic correlation study (Brynolf 1967). They represent a healthy periapex (score 1), minor changes perceived as consistent with a healthy periapex (score 2), and increasing extent and severity of apical periodontitis (scores 3 to 5). Radiographs are assigned a score according to which of the reference

**Table 15.1** Follow-up studies in the past 20 years, reporting specific data on the outcome of initial treatment in teeth with endodontic infection

| Study | Cases observed | Follow-up (years) | Appraisal categories | | | | Outcome (%) | | |
|---|---|---|---|---|---|---|---|---|---|
| | | | Cohort | Exposure | Assessment | Analysis | Healed | Healing | Functional[a] |
| **Byström et al. (1987)** | **79**[b] | **2-5** | y | y | y | n | 85 | 9 | ≥94 |
| Matsumoto et al. (1987) | 85 | 2-3 | n | y | n | n | — | 75 | — |
| **Eriksen et al. (1988)** | **121**[b,c] | **3** | n | y | y | y | 82 | 9 | ≥91 |
| Åkerblom and Hasselgren (1988) | 64[b] | 2-12 | n | y | n | n | 89[d] | — | — |
| Molven and Halse (1988) | 220[b,c] | 10-17 | n | n | y | y | 80 | — | — |
| Shah (1988) | 93 | 0.5-2 | n | n | n | n | — | — | 84 |
| Augsburger and Peters (1990) | 50 | 0.3-6.5 | n | y | n | n | — | 96 | — |
| **Sjögren et al. (1990)** | **204**[b] | **8-10** | y | y | y | y | 86 | — | — |
| Murphy et al. (1991) | 89 | 0.3-2 | n | n | n | n | 46 | 48 | 94 |
| Smith et al. (1993) | 481 | 2-5 | n | n | n | y | — | 81 | — |
| Friedman et al. (1995) | 113 | 0.5-1.5 | y | y | n | n | 69 | 25 | 94 |
| Caliskan and Sen (1996) | 172 | 2-5 | y | y | n | n | 81 | 8 | 89 |
| **Ørstavik (1996)** | **126**[b,c] | **4** | n | y | y | y | 75 | 13 | ≥88 |
| **Sjögren et al. (1997)** | **53** | **≤5** | y | y | y | y | 83 | — | — |
| **Trope et al. (1999)** | **76** | **1** | n | y | y | y | 80[e] | — | — |
| **Weiger et al. (2000)** | **67** | **1-5** | y | y | n | y | 78 | 16 | 94 |
| Waltimo et al. (2001) | — | 1-4 | n | y | y | n | 50 | — | — |
| Chugal et al. (2001) | 177[b] | 4 | n | n | y | y | 63 | — | — |
| Peak et al. (2001) | 280 | ≤1 | n | n | y | n | — | 87 | — |
| Pettiette et al. (2001) | 40 | 1 | n | y | y | n | 60 | — | — |
| Heling et al. (2001) | 319 | 1-12 | n | y | y | n | — | 65 | — |
| **Peters and Wesselink (2002)** | **38** | **1-4.5** | y | y | y | y | 76 | 21 | 97 |
| **Hoskinson et al. (2002)** | **335**[b] | **4-5** | y | y | y | y | 74 | — | — |
| **Friedman et al. (2003)** | **72** | **4-6** | y | n | y | y | 74 | 11 | ≥85 |
| **Huumonen et al. (2003)** | **156** | **1** | y | y | y | n | 76 | 2 | — |
| **Peters et al. (2004)** | **102** | **1-3** | y | y | y | y | 76 | — | — |

*(Continued)*

**Table 15.1**  (*Continued*)

| Study | Cases observed | Follow-up (years) | Appraisal categories | | | | Outcome (%) | | |
|---|---|---|---|---|---|---|---|---|---|
| | | | Cohort | Exposure | Assessment | Analysis | Healed | Healing | Functional[a] |
| **Farzaneh et al. (2004a)** | **144**[c] | **4–6** | y | n | y | y | **79** | **7** | **≥86** |
| **Ørstavik et al. (2004)** | **121**[b,c] | **3** | n | y | y | y | **79** | — | — |
| Caliskan (2004) | 42 | 2–10 | n | y | y | n | 74 | 10 | ≥84 |
| Chu et al. (2005) | 58 | 3–4 | n | y | n | y | 79 | — | 85 |
| **Marquis et al. (2006)** | **209**[c] | **4–6** | y | n | y | y | **80** | **6** | **≥86** |
| Aqrabawi (2006) | 177 | 5 | n | y | y | n | 79 | — | — |
| Suchina et al. (2006) | 22[f] | 0.5–5.5 | n | y | n | n | 55 | 36 | — |
| **Molander et al. (2007)** | **89** | **2** | y | y | y | y | **70** | **20** | **≥90** |
| Conner et al. (2007) | 30 | 1 | n | n | y | n | 50 | 23 | ≥83 |
| Imura et al. (2007) | 694 | 1.5–5 | n | y | n | y | 91 | — | — |
| **Penesis et al. (2008)** | **63** | **1** | y | y | y | y | **68** | **15** | **≥86** |
| de Chevigny et al. (2008b) | 292 | 4–6 | y | n | y | y | 82 | 5 | 96 |

Studies shown in bold have been selected as "best evidence" (adapted from Friedman 2002a).

[a] Asymptomatic, without or with residual radiolucency (≥, data not explicitly reported; rate is sum of healed and healing).
[b] Roots considered as unit of evaluation, rather than teeth.
[c] Includes repeated material.
[d] All canals obliterated to some extent.
[e] Teeth treated in two sessions without intracanal medication excluded.
[f] HIV-positive patients.

y, satisfies criteria of acceptable quality; n, does not satisfy criteria of acceptable quality

**Table 15.2** Follow-up studies in the past 20 years, reporting specific data on the outcome of orthograde retreatment in teeth with endodontic infection

| Study | Examined sample | Follow-up (years) | Appraisal categories | | | | Outcome (%) | | |
|---|---|---|---|---|---|---|---|---|---|
| | | | Cohort | Exposure | Assessment | Analysis | Healed | Healing | Functional[a] |
| Molven and Halse (1988) | 98[b] | 10–17 | n | n | y | y | 71[c] | — | — |
| Allen et al. (1989) | 315[d] | ≥0.5 | n | n | n | n | 73 | 12 | ≥85 |
| **Sjögren et al. (1990)** | **94[b]** | **8–10** | **y** | **y** | **y** | **y** | **62** | **—** | **—** |
| Van Nieuwenhuysen et al. (1994) | 561[d] | ≥0.5 | n | n | n | y | 78 | — | — |
| Friedman et al. (1995) | 86 | 0.5–1.5 | y | y | n | n | 56 | 34 | ≥90 |
| Danin et al. (1996) | 18 | 1 | n | y | n | n | 28 | 28 | ≥56 |
| **Sundqvist et al. (1998)** | **54** | **4** | **n** | **y** | **y** | **y** | **74** | **—** | **—** |
| Piatowska et al. (1997) | 60[d] | — | n | y | n | n | 43 | 42 | ≥85 |
| Abbott (1999) | 432[d] | 0.3–4 | n | n | n | n | 98 | 1 | — |
| **Kvist and Reit (1999)** | **47** | **4** | **y** | **y** | **y** | **n** | **58[e]** | **—** | **—** |
| Chugal et al. (2001) | 85[b,d] | 4 | n | n | y | y | 79 | — | — |
| Hoskinson et al. (2002)[g] | 76[d] | 4–5 | n | y | n | y | 78 | — | — |
| **Farzaneh et al. (2004b)** | **69** | **4–6** | **y** | **n** | **y** | **y** | **81** | **5** | **≥86** |
| Gorni et al. (2004) | 452[d] | 2 | y | n | n | n | 65 | 4 | — |
| Fristad et al. (2004) | 112[b,d,f] | 20–27 | n | n | y | y | 96 | — | — |
| Imura et al. (2007) | 404 | 1.5–5 | n | y | n | y | 81 | — | — |
| Ercan et al. (2007) | 64 | 1 | n | y | n | n | 64 | 14 | ≥78 |
| **de Chevigny et al. (2008a)** | **147** | **4–6** | **y** | **n** | **y** | **y** | **80** | **6** | **93** |

Studies shown in bold have been selected as "best evidence" (adapted from Friedman 2008).
[a] Asymptomatic, without or with residual radiolucency (≥, not reported; rate is sum of healed and healing).
[b] Roots considered as unit of evaluation, rather than teeth.
[c] Cases classified as "uncertain" excluded.
[d] May include unspecified number of teeth without infection.
[e] Approximate figure deducted from graph.
[f] Includes repeated material.
[g] Study selected for initial treatment, but lacking detail on retreatment cases included.
y, satisfies criteria of acceptable quality; n, does not satisfy criteria of acceptable quality.

**Table 15.3** Follow-up studies in the past 20 years, reporting specific data on the outcome of apical surgery in teeth with endodontic infection

| Study | Cases observed | Follow-up (years) | Appraisal categories | | | | Outcome (%) | | |
|---|---|---|---|---|---|---|---|---|---|
| | | | Cohort | Exposure | Assessment | Analysis | Healed | Healing | Functional[a] |
| Forssell et al. (1988) | 358 | 1–4 | n | n | y | n | 69 | 7 | ≥76 |
| Allen et al. (1989) | 695 | ≥0.5 | n | n | n | n | 60 | 27 | ≥87 |
| Amagasa et al. (1989) | 64 | 1–7.5 | y | y | n | n | — | — | 95 |
| Crosher et al. (1989) | 85 | 2 | n | y | n | n | 92 | — | — |
| Dorn and Gartner (1990) | 488 | 0.5 | n | y | n | n | 63 | 18 | ≥81 |
| Grung et al. (1990) | 473 | 1–8 | y | n | y | n | 72 | 12 | ≥84 |
| Friedman et al. (1991a) | 136[b] | 0.5–8 | n | n | y | y | 44 | 23 | ≥67 |
| Lasaridis et al. (1991) | 24 | ≥0.5 | y | y | y | n | 79 | 17 | ≥96 |
| Lustmann et al. (1991)[c] | 124[b] | 0.5–8 | n | n | y | y | 43 | 22 | ≥65 |
| Molven et al. (1991)[d] | 222 | 1–8 | y | n | y | n | 73 | 14 | ≥87 |
| Rapp et al. (1991)[e] | 428 | ≤0.5–≥2 | n | n | n | n | 66 | 29 | ≥95 |
| Rud et al. (1991) | 388 | 0.5–1 | n | y | y | n | 78 | 15 | ≥93 |
| Waikakul and Punwutikorn (1991) | 62 | 0.5–2 | y | y | n | n | 81 | 17 | ≥96 |
| Zetterqvist et al. (1991) | 105 | 1 | y | y | y | n | 61 | 31 | ≥92 |
| Frank et al. (1992) | 104 | ≥10 | n | n | n | n | 58 | — | — |
| Cheung and Lam (1993) | 32 | ≥2 | n | n | y | n | 62 | 22 | ≥84 |
| Pantschev et al. (1994) | 79 | 3 | n | y | y | n | 54 | 21 | ≥75 |
| Jesslen et al. (1995) | 93 | 5 | y | y | y | n | 59 | 28 | ≥87 |
| August (1996) | 39 | 10–23 | n | n | n | n | 74 | 15 | ≥89 |
| Danin et al. (1996) | 19 | 1 | n | y | y | n | 58 | 26 | ≥84 |
| Rud et al. (1996a, b) | 351[b] | 0.5—1.5 | y | y | y | n | 82 | 12 | ≥94 |
| Sumi et al. (1996) | 157 | 0.5–3 | y | y | n | n | — | — | 92 |
| Jansson et al. (1997) | 62 | 0.9–1.3 | y | y | n | n | 31 | 55 | ≥86 |
| Rud et al. (1997)[f] | 551[b] | 0.5–1.5 | n | y | y | n | 79 | 16 | ≥95 |
| Bader and Lejeune (1998) | 254 | 1 | n | y | n | n | — | — | 81 |
| Danin et al. (1999) | 10 | 1 | n | y | y | y | 50 | 50 | 100 |

| Study | n | Evaluation period (yr) | | | | | Rate | | |
|---|---|---|---|---|---|---|---|---|---|
| **Kvist and Reit (1999)** | **45** | **4** | y | y | n | y | **60**[l] | — | — |
| Rubinstein and Kim (1999) | 94 | 1.2 | n | y | n | n | 97 | — | — |
| Testori et al. (1999) | 134 | 1–6 | n | y | n | n | 78 | 9 | ≥87 |
| Von Arx et al. (2001) | 43 | 1 | y | y | n | n | 82 | 14 | ≥96 |
| **Zuolo et al. (2000)** | **102** | **1–4** | **y** | **y** | **y** | **n** | **91** | — | **92** |
| **Rahbaran et al. (2001)** | **129**[g] | **≥4** | **y** | **y** | **y** | **y** | **37** | **33** | **≥80** |
| Rud et al. (2001) | 834[b] | 0.5–12.5 | n | y | y | n | 92 | 1 | ≥93 |
| Von Arx and Kurt (1999) | 25 | 1 | y | y | y | n | 88 | 8 | ≥96 |
| Rubinstein and Kim (2002)[h] | 59 | 5–7 | y | y | y | n | 92 | — | — |
| **Jensen et al. (2002)** | **60**[Rp] | **1** | **y** | **y** | **y** | **y** | **73** | **17** | **≥90** |
| **Chong et al. (2003)** | **108** | **2** | **n** | **y** | **y** | **y** | **90** | **6** | **≥96** |
| Wesson and Gale (2003) | 790 | 5 | y | y | y | n | 57 | 5 | ≥62 |
| Maddalone and Gagliani (2003) | 120 | 0.3–3 | n | y | y | n | 93 | 3 | ≥96 |
| Schwartz-Arad et al. (2003) | 262 | 0.5–0.9 | n | y | y | y | 44 | 21 | ≥65 |
| **Wang et al. (2004)** | **94** | **4–8** | **y** | **y** | **y** | **y** | **74** | — | **91** |
| **Gagliani et al. (2005)** | **231**[b] | **5** | **y** | **y** | **n** | **y** | **78** | **10** | **89** |
| Lindeboom et al. (2005) | 100 | 1 | y | y | n | y | 89 | 10 | 99 |
| **Tsesis et al. (2006)** | **45**[i] | **0.5–1.5** | **y** | **y** | **n** | **n** | **91** | **4** | **≥95** |
| Taschieri et al. (2006)[j] | 71 | 1 | y | y | n | n | 93 | — | — |
| Von Arx et al. (2007) | 191 | 1 | y | y | n | y | 84 | 10 | ≥94 |
| **Yazdi et al. (2007)**[k] | **60**[Rp] | **6.5–9** | **y** | **y** | **y** | **y** | **78** | **17** | **≥90** |

Studies shown in bold have been selected as "best evidence" (adapted from Friedman 2005).

[a] Asymptomatic, without or with residual radiolucency (≥, not reported; rate is sum of healed and healing).

[b] Roots considered as unit of evaluation, rather than teeth.

[c] Same sample as in Friedman et al. (1991a, b).

[d] Same sample as in Grung et al. (1990).

[e] Same sample as in Allen et al. (1989).

[f] Same sample as in Rud et al. (1991, 1996a, b).

[g] Includes only treatments performed in the endodontic clinic.

[h] Same sample as in Rubinstein and Kim (1999).

[i] Cases treated by "modern technique."

[j] Randomized controlled trial with inadequate analysis.

[k] Same sample as Jensen et al. (2002).

[l] Approximate figure deducted from graph.

[Rp] Includes only teeth treated with Retroplast.

y, satisfies criteria of acceptable quality; n, does not satisfy criteria of acceptable quality.

287

**Table 15.4**    Follow-up studies in the past 20 years, reporting specific data on the outcome of intentional replantation in teeth with endodontic infection

| Study | Cases observed | Follow-up (years) | Treatment outcome (%) | | |
|---|---|---|---|---|---|
| | | | Success | Root resorption | Persistent infection |
| Koenig et al. (1988) | 177 | 0.5–4 | 82 | 5 | 11 |
| Kahnberg (1988) | 58 | 2–7 | 71 | 0 | 29 |
| Keller (1990) | 34 | 3 | 91 | 0 | 9 |
| Bender and Rossman (1993) | 31 | 0–22 | 81 | 6 | 10 |

images they match best. Such "blinded" assessment reduces the bias and improves the sensitivity of the assessment, compared with the common assessment of success/failure (Ørstavik et al. 1986, 2004). The PAI has been used mainly in studies on nonsurgical treatment (Ørstavik and Hörsted-Bindslev 1993; Ørstavik 1996; Trope et al. 1999; Waltimo et al. 2001; Friedman et al. 2003; Huumonen et al. 2003; Farzaneh et al. 2004a, b; Ørstavik et al. 2004; Peters et al. 2004; Marending et al. 2005; Waltimo et al. 2005; Aqrabawi 2006; Marquis et al. 2006; de Chevigny et al. 2008a, b; Penesis et al. 2008), but it was used also in one apical surgery study (Wang et al. 2004) to minimize bias and to facilitate comparisons with nonsurgical treatment studies from the same group (Friedman et al. 2003; Farzaneh et al. 2004a, b; Marquis et al. 2006; de Chevigny et al. 2008a, b). Although the PAI was originally designed to observe changes in mean scores as the main outcome, it can be used to dichotomize outcomes as health (scores 1 and 2) and disease (scores 3, 4, and 5) (Trope et al. 1999; Waltimo et al. 2001; Boucher et al. 2002; Dugas et al. 2003; Friedman et al. 2003; Huumonen et al. 2003; Farzaneh et al. 2004a, b; Peters et al. 2004; Waltimo et al. 2005; Kirkevang et al. 2006; Marquis et al. 2006; de Chevigny et al. 2008a, b; Penesis et al. 2008).

### 15.2.3 Outcome criteria

Criteria and terminology used for outcome assessment in endodontic studies have varied, mainly in the use of "strict" and "lenient" classifications of success. Whereas in the majority of the studies "success" or "complete healing" is defined as complete radiographic and clinical normalcy, in many current studies, success is defined primarily as clinical normalcy that may be accompanied by a residual radiolucency, that is either decreased in size (Matsumoto et al. 1987; Smith et al. 1993; Piatowska et al. 1997; Peak et al.

2001; Pettiette et al. 2001) or unchanged (Shah 1988; Amagasa et al. 1989; Sumi et al. 1996). The difference in success between these two sets of criteria can be approximately 15% (Friedman et al. 1995; Wang et al. 2004). Adding to the confusion, outcome categories of "uncertain," "doubtful," "questionable," and "improved" have been used inconsistently to imply uncertainty of the outcome in some of the current studies (Chugal et al. 2001; Hoskinson et al. 2002), improved outcomes in other studies on nonsurgical treatment (Åkerblom and Hasselgren 1988; Molven and Halse 1988; Friedman et al. 1995; Caliskan and Sen 1996; Weiger et al. 2000; Peters and Wesselink 2002) and apical surgery (Allen et al. 1989; Lasaridis et al. 1991; Pantschev et al. 1994; von Arx and Kurt 1999; Zuolo et al. 2000 [up to 4 years]; Rahbaran et al. 2001; von Arx et al. 2001; Cheung 2002; Schwartz-Arad et al. 2003; Wesson and Gale 2003; Lindeboom et al. 2005; von Arx et al. 2007), and nonimproved outcomes in many other studies on apical surgery (Forssell et al. 1988; Grung et al. 1990; Friedman et al. 1991a; Halse et al. 1991; Molven et al. 1991, 1996; Rud et al. 1991b, 1996a, b, 1997, 2001; Danin et al. 1996, 1999; Rud and Rud 1998; Testori et al. 1999; Zuolo et al. 2000 [over 4 years]; Maddalone and Gagliani 2003). These inconsistencies result for the major part from the use of the ambiguous and value-laden terms "success" and "failure" (Ørstavik 1996). Therefore, these terms should be replaced with neutral expressions to facilitate communication with patients. The terms used should preferably relate to the specific goals of treatment.

In teeth with endodontic infection, the primary aim of treatment is to heal the tissues supporting the tooth affected by apical periodontitis (Ørstavik and Pitt Ford 1998). Thus, the primary outcome of treatment should be related to healing (Rud et al. 1972; Byström et al. 1987; Ørstavik 1996; Friedman 2002a, 2005; Friedman and Mor 2004). The term "healed" is used for complete clinical and radiographic

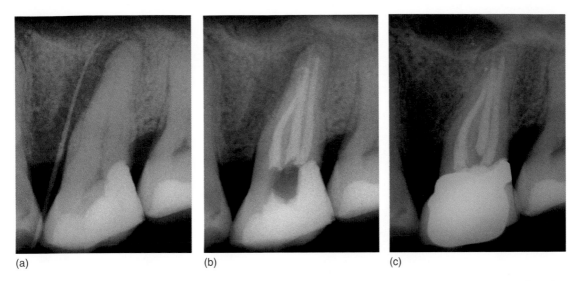

(a)                                    (b)                                    (c)

**Fig. 15.1**   Primary infection healed after initial treatment. (a) Maxillary second molar with apical periodontitis extending along the mesial root surface and associated sinus tract (traced with a gutta-percha cone). (b) Completed treatment. (c) At 8 years, radiographic and clinical normalcy suggest that the tooth has healed. (a, reprinted with permission from Friedman 2002b.)

normalcy (no signs, symptoms, residual radiolucency) (Figs 15.1–15.3). This category includes the typical appearance of a scar after apical surgery (Andreasen and Rud 1972a, b; Rud et al. 1972; Molven et al. 1996) (Fig. 15.4). The term "healing" is used for decreased radiolucency and clinical normalcy after a follow-up period shorter than 4 years (Fig. 15.3). The term "nonhealed" or "persistent disease" is used for persistent radiolucency regardless of clinical presentation (Fig. 15.5) or persistent symptoms. For individual patients who may be satisfied with just elimination of symptoms, particularly when clinical conditions suggest a poor prognosis for healing, the aim of treatment may be moderated. In such cases, the secondary outcome should be related to the retention of the tooth in a symptom-free state, and the term "functional retention" can be used for clinical normalcy albeit with persistent radiolucency.

## 15.3 Levels of evidence in assessment of endodontic prognosis

Reports on prognosis are frequently inconsistent in methodology and in the level of evidence they provide (Sackett et al. 1991). Consequently, structured analysis of the literature is necessary to differentiate clinical studies according to the level of evidence and to gather valid evidence from selected studies.

### 15.3.1 Study designs

Design categories of clinical studies are defined by the Cochrane Collaboration (http://www.cochrane.org) as follows:

- *Clinical trial:* "A clinical trial involves administering a treatment to test it. It is an experiment. Clinical trial is an umbrella term for a variety of health care trials...Types include uncontrolled trials, controlled clinical trials (CCT), community trials, and randomized controlled trials (RCT). A randomized controlled trial is always prospective."
- *Cohort study:* "A 'cohort' is a group of people clearly identified; a cohort study follows that group over time, and reports on what happens to them. A cohort study ... can be prospective or retrospective."
- *Case-control study:* "Compares people with a disease or condition ('cases') to another group of people from the same population who do not have that disease or condition ('controls'). A case-control study can identify risks and trends, and suggest some possible causes for disease, or for particular outcomes. A case-control study is retrospective."
- *Cross-sectional study:* "Also called a prevalence study. It is an observational study. It is like taking a snapshot of a group of people at one point in time

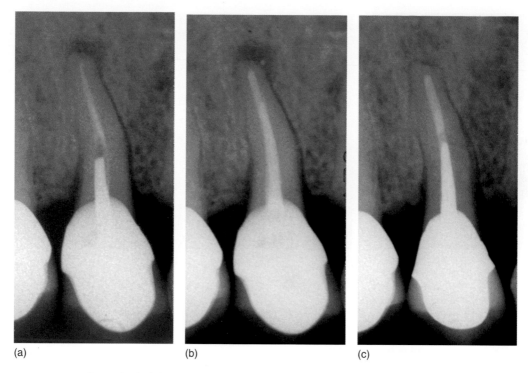

**(a)**          **(b)**          **(c)**

**Fig. 15.2**    Persistent infection healed after orthograde retreatment. (a) Maxillary first premolar with persistent apical periodontitis, restored with a cast post and crown. (b) Completed retreatment with the original crown recemented in place. (c) At 4 years, radiographic and clinical normalcy suggest that the tooth has healed. (Reprinted with permission from Friedman 2002a.)

and seeing the prevalence of diseases or actions in that population."

• *Case series:* "A case study is a report of a single experience. A case series is a description of a number of 'cases'."

The different designs suit different aims, such as assessment of effectiveness, prognosis, or risks associated with interventions. Therefore, reviews geared to answer specific questions should focus on studies with matched design (Fletcher et al. 1996). While RCT is the suitable design for comparing benefits of different interventions, the suitable design for assessment of prognosis is a cohort study (Green and Byar 1984; Fletcher et al. 1996).

### 15.3.2 Methodological rigor

A primary concern in clinical studies is different forms of bias. Data may be distorted so that differences are demonstrated between groups that may not really exist, while existing differences may not be shown

(Fletcher et al. 1996). Bias can occur during assembly of the study cohort, when the characteristics of groups differ in variables that may influence the outcome or in capacity to heal (Fletcher et al. 1996). Bias can also occur during assessment of the outcome (Fletcher et al. 1996), particularly if assessment is done by the providers of treatment, who may be biased toward favorable outcome (Goldman et al. 1972). A structured checklist can be used to identify bias in studies on prognosis (Department of Clinical Epidemiology and Biostatistics 1981; Sackett et al. 1991; Laupacis et al. 1994; Fletcher et al. 1996; Sutherland 2001):

• Was the study cohort defined, assembled at the inception of the study, described in detail, entered at a similar point in the course of the disease?
• Was the referral pattern described?
• Were baseline features measured reproducibly?
• Was the follow-up achieved in at least 80% of the inception cohort, the follow-up period described, and long enough for the outcome to occur?

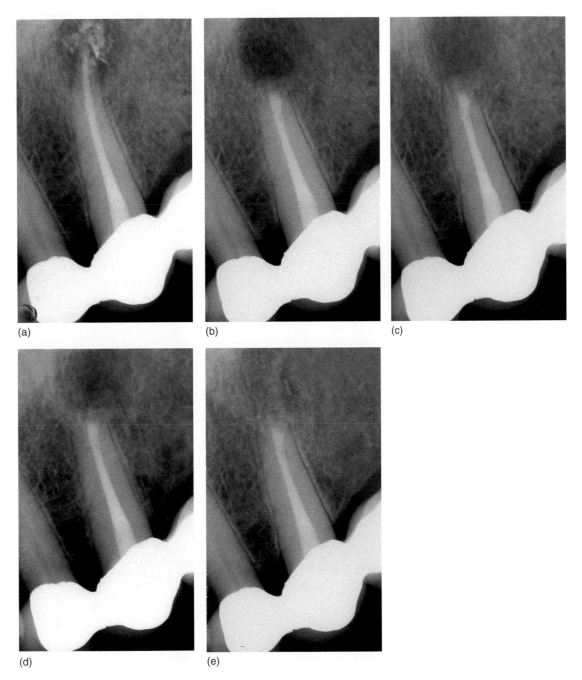

(a)  (b)  (c)

(d)  (e)

**Fig. 15.3** Persistent infection healed after apical surgery. (a) Maxillary canine with a large excess of sealer and persistent apical periodontitis. (b) Completed surgery, including root-end filling with MTA. (c) At 3 months, some bone deposition is suggested, but the lesion is not reduced. (d) At 6 months, the tooth is symptom free and the lesion appears to be healing. (e) At 1 year and 8 months, radiographic and clinical normalcy suggest that the tooth has healed. (Reprinted with permission from Friedman 2005.)

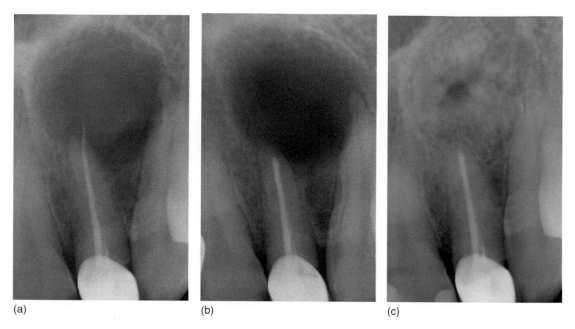

(a)          (b)          (c)

**Fig. 15.4**   Persistent infection healed by scar formation (incomplete healing) after apical surgery. (a) Maxillary lateral incisor with a root filling extruded beyond the root end and persistent apical periodontitis. (b) Completed surgery, including root-end filling with Super-EBA. (c) At 1 year, radiographic and clinical normalcy suggest that the tooth has healed with a small scar formed several millimeters from the root end. (Courtesy of Dr Richard Rubinstein, Farmington Hills, MI. Reprinted with permission from Friedman 2005.)

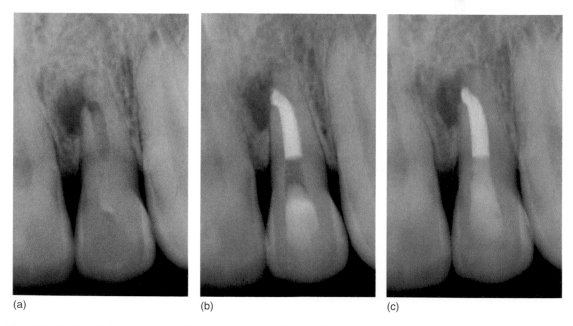

(a)          (b)          (c)

**Fig. 15.5**   Persistent infection after initial treatment. (a) Maxillary lateral incisor with primary apical periodontitis. (b) Completed treatment. (c) At 1 year, unchanged radiolucency suggests persistence of the infection. (Reprinted with permission from Friedman 2002b.)

- Were the criteria used for outcome assessment described, objective, clinically important and reproducibly measured?
- Was the outcome assessment blind?
- Was adjustment for extraneous prognostic factors carried out?

The checklist criteria can be grouped into four general categories, used as the basis for appraisal of the endodontic studies below.

### 15.3.2.1 Cohort, at inception and end point of the study

The inception cohort should be clearly characterized for variables that can potentially influence the outcome, the pattern of referral of the treated subjects, type of cases treated, and case selection criteria used (Fletcher et al. 1996). Case selection is likely to determine the results (Ingle et al. 1994) because subjects are included or excluded according to perceived prognosis.

At the end point of the study, failure to examine the majority of treated subjects may skew and invalidate the results (Strindberg 1956; Fletcher et al. 1996); therefore, at least 80% of the treated subjects should be examined (Department of Clinical Epidemiology and Biostatistics 1981; Laupacis et al. 1994; Sutherland 2001). Those who are not examined should be explicitly accounted as "dropouts" who do not present for follow-up at their own volition (their absence may be related to the outcome of interest) or "discontinuers" who are excluded from the study for accountable reasons, for example, death or relocation (their absence is not related to the outcome of interest).

The examined sample is a determinant of the study's validity (Fletcher et al. 1996) and of its statistical power when associations of the outcome with different variables are analyzed. Small differences in outcome require large samples to achieve significance (Fletcher et al. 1996).

### 15.3.2.2 Exposure (treatment, intervention)

Treatment providers should be characterized, as their expertise may determine the results (Ingle et al. 1994). Treatment procedures performed should be explicitly described to avoid the need for interpretations. Studies may be excluded if the treatment procedures are considered irrelevant or unacceptable.

### 15.3.2.3 Outcome assessment

To minimize bias (Fletcher et al. 1996), objective outcome measures should be used consistently in a blinded manner; therefore, examiners should be independent and calibrated with established reliability. The follow-up period should be long enough to capture the outcome of interest. Specifically for endodontic studies, the conclusion of the dynamic healing processes must be captured in the majority of the study sample (Fig. 15.6).

### 15.3.2.4 Data reporting and analysis

Data pertaining to the study cohort, intervention, outcome assessment, and analysis should be reported in detail to allow identification of potential bias and assessment of validity. Statistical analyses should be designed to minimize potential bias, and take into account extraneous factors and their potential confounding effects. Preferably, multivariate analyses should be used to account for all the variables.

### 15.3.3 The "best evidence" for the prognosis of endodontic treatment

Based on the research design and methodological rigor, clinical studies can be ranked by descending hierarchy of evidence as follows (Oxman 1994): (1) Rigorous randomized controlled trials (RCTs) and their systematic reviews. (2) Rigorous cohort studies and their systematic reviews; compromised RCTs. (3) Rigorous case-control studies and their systematic reviews. (4) Compromised cohort or case-control studies; cross-sectional studies; case series. (5) Expert opinions; case reports; narrative literature reviews. Evidence-based practice is "...the conscientious, explicit and judicious use of current best evidence in making decisions about the care of individual patients" (Sackett et al. 1991). Accordingly, evidence-based endodontic care requires the best evidence available for endodontic prognosis to be identified.

In a series of review articles and a textbook chapter, Friedman reviewed the clinical studies in endodontics to identify the best evidence for endodontic prognosis (Friedman 2002b, 2005, 2008). These reviews have been based on the premise that the methodological rigor of clinical studies is a crucial consideration (Barton 2000), so much so that rigorous cohort studies can outweigh compromised RCTs and structured

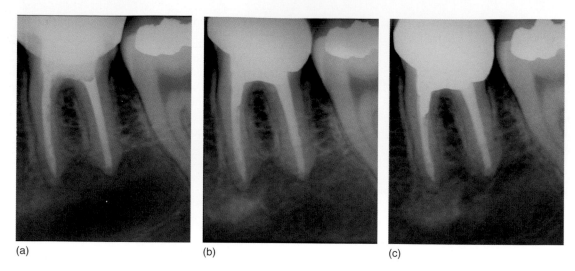

(a)                                  (b)                                  (c)

**Fig. 15.6** Healing dynamics after initial treatment. (a) Immediate postoperative radiograph of mandibular first molar with extensive primary infection, included in a clinical study (Friedman et al. 1995). (b) At 18 months (termination of the study), residual small radiolucency at the mesial root tip suggests that healing is incomplete. (c) At 3.5 years, the area is completely healed. Extension of the study to 4 years would have captured the completion of healing. (Reprinted with permission from Friedman 2002b.)

reviews of rigorous cohort studies can yield consistent conclusions with those of systematic reviews of RCTs (Benson and Hartz 2000; Concato et al. 2000). Therefore, cohort studies have been included in the review process, and those that comply with three of the four criteria—cohort, intervention, assessment, and analysis/reporting—have been selected as methodologically adequate. Excluding studies whose samples have been repeated, 14 relatively current studies are selected (Table 15.1) on initial root canal treatment in which distinct groups of teeth with primary apical periodontitis are identified (Byström et al. 1987; Sjögren et al. 1990, 1997; Ørstavik 1996; Trope et al. 1999; Weiger et al. 2000; Hoskinson et al. 2002; Peters and Wesselink 2002; Huumonen et al. 2003; Ørstavik et al. 2004; Peters et al. 2004; Marending et al. 2005; Penesis et al. 2008; de Chevigny et al. 2008b). In addition, 4 current studies on orthograde retreatment (Table 15.2) are selected ( Sjögren et al. 1990; Sundqvist et al. 1998; Kvist and Reit 1999; de Chevigny et al. 2008a) and 10 current studies (Table 15.3) on apical surgery (Kvist and Reit 1999; Zuolo et al. 2000; Rahbaran et al. 2001; Chong et al. 2003; Wang et al. 2004; Gagliani et al. 2005; Lindeboom et al. 2005; Taschieri et al. 2006; von Arx et al. 2007; Yazdi et al. 2007) of teeth with persistent apical periodontitis. None of the current

studies on intentional replantation (Table 15.4) meet the selection criteria; thus, the level of evidence to support the prognosis of this specific treatment modality is the lowest. The selected studies represent the "best evidence" for the prognosis of treatment in teeth with endodontic infections. They also serve as reference for identifying significant prognostic variables.

## 15.4 The prognosis of primary apical periodontitis after initial treatment

Treatment providers are expected to advise patients of the prognosis of treatment and to maximize the prognosis by using treatment methods based on solid evidence. Therefore, the prognosis is reported along with variables that may affect it. Although the methodology among the 14 current studies reviewed in this section (Byström et al. 1987; Sjögren et al. 1990, 1997; Ørstavik 1996; Trope et al. 1999; Weiger et al. 2000; Hoskinson et al. 2002; Peters and Wesselink 2002; Huumonen et al. 2003; Ørstavik et al. 2004; Peters et al. 2004; Marending et al. 2005; de Chevigny et al. 2008b; Penesis et al. 2008) is rather uniform, they still differ considerably in case selection, study materials and, consequently, in results.

### 15.4.1 Potential for healing

The proportion of completely healed teeth after initial treatment ranges from 74% (Hoskinson et al. 2002) to 86% (Sjögren et al. 1990). Considering the uniform outcome assessment among the selected studies, the results may have varied because of differences in tooth types, definition of the tooth or root as the evaluated unit (Friedman 2002b), case selection (Ingle et al. 1994), and restoration.

In addition to the healed teeth, progressive or incomplete healing has been reported in 2% (Huumonen et al. 2003) to 21% (Peters and Wesselink 2002) of the teeth. Typically, the proportion of incomplete healing is inversely proportional to the follow-up period, because the completion of the healing process cannot be captured after a short follow-up (Friedman 2002b). Thus, the potential of teeth with apical periodontitis to heal after initial treatment is 74–86%, while additional 10–20% may demonstrate incomplete healing. It has also been suggested that regardless of healing (complete or incomplete), 85–97% of the teeth may remain symptom free and functional (Byström et al. 1987; Eriksen et al. 1988; Ørstavik 1996; Weiger et al. 2000; Peters and Wesselink 2002; de Chevigny et al. 2008b).

### 15.4.2 Time course of healing

The healing process of primary apical periodontitis, indicative of an endodontic infection, is initiated within the first year after treatment (Reit 1987a; Kvist and Reit 1999); however, its completion often requires a longer time. Therefore, of all the teeth that heal eventually, only about 50% appear completely healed by 1 year (Adenubi and Rule 1976), and a small percentage appear healed only after 4–5 years (Byström et al. 1987; Sjögren et al. 1990; Ørstavik 1996; Kvist and Reit 1999), or even longer (Molven et al. 2002b). As long as 4 years after treatment, about 13% of the teeth may still show incomplete healing (Ørstavik 1996), compared to 5% at 6 years (de Chevigny et al. 2008b). Apparently, about 6% of teeth that were not healed 10–17 years after treatment were completely healed a decade later (Molven et al. 2002b). Infrequently, very extensive lesions can heal without total resolution of the radiolucency, when fibrous tissue occupies the periapical space (apical scar) (Penick 1961; Bhaskar 1966; Byström et al. 1987; Nair et al. 1999; Selden 1999) (Fig. 15.7).

Reversal of the healing process is uncommon (Ørstavik 1996; Kvist and Reit 1999), suggesting that extended follow-up of teeth that demonstrate signs of healing after 1 year may be unnecessary (Ørstavik 1996). Nevertheless, root-filled teeth remain constantly at risk of recurrent infection in the long-term. For example, over 1% of teeth, observed to be healed 10–17 years after treatment, reverted to disease a decade later (Molven et al. 2002b). To address this long-term risk, periodic follow-up of root-filled teeth is advocated.

### 15.4.3 Prognostic variables

The prognostic variables are divided into those that have been identified as significant predictors of outcome in multivariate analyses, those that appear to be nonsignificant, and those that are equivocal and require further study.

#### 15.4.3.1 Significant outcome predictors

*Systemic health.* One selected study (Marending et al. 2005) suggests that a compromised nonspecific immune system impairs the outcome of treatment. Similarly, a poorer outcome is reported in diabetic patients in a nonselected study (Fouad and Burleson 2003). Even though further evidence is required to confirm these associations, the prognosis appears to be poorer in immune-compromised patients. The compromised prognosis, however, does not preclude treatment, as healing is still possible in these patients. Also, it is possible that healing in the immune-compromised patients requires a longer time than the 2-year observation in these studies.

*Number of roots.* The majority of the selected current studies have shown no association between tooth location and the outcome of treatment (Weiger et al. 2000; Hoskinson et al. 2002; Peters and Wesselink 2002; de Chevigny et al. 2008b). However, because the risk of persistent disease in multi-rooted teeth is multiplied by the number of roots, the outcome in multi-rooted teeth is poorer than in single-rooted teeth (de Chevigny et al. 2008b). Thus, the prognosis is somewhat compromised in multi-rooted teeth compared with single-rooted ones.

#### 15.4.3.2 Nonsignificant variables

*Age, gender.* These variables have not been significantly associated with treatment outcome (Sjögren

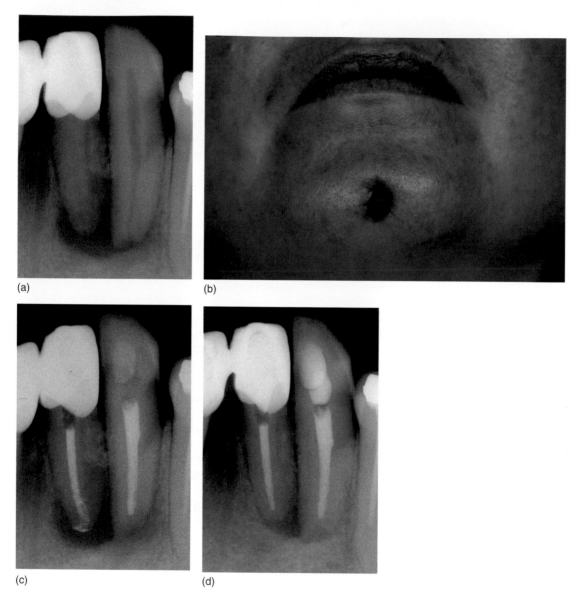

(a)    (b)

(c)    (d)

**Fig. 15.7**    Primary infection healed by scar formation after initial treatment. (a and b) Mandibular lateral incisor and canine with primary apical periodontitis associated with an orofacial fistula. (c) Completed treatment. (d and e) At 2 years, the fistula has healed with minimal scarring of the skin. The residual radiolucency may suggest persistence of infection. (f) Clinical view after reflection of a full-thickness flap reveals a thick fibrous bundle connecting the periapical lesion and the soft tissues over the chin. Histological examination of the dissected bundle confirmed it to be fibrous (scar) tissue. (g) At 6 months after surgery, further decreased radiolucency and better defined periodontal ligament space suggest healing in progress. (Reprinted with permission from Friedman 2002b.)

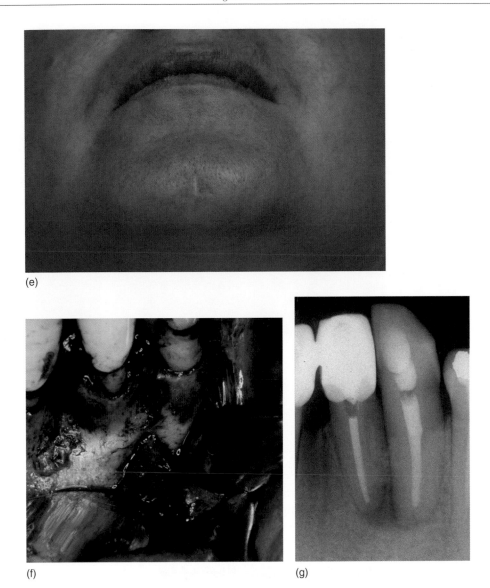

(e)

(f)                                    (g)

**Fig. 15.7** (*Continued*)

et al. 1990; Hoskinson et al. 2002; Peters and Wesselink 2002; de Chevigny et al. 2008b), and they do not appear to influence the prognosis.

*Symptoms.* Symptoms have not been significantly associated with treatment outcome (Byström et al. 1987; Sjögren et al. 1990; Weiger et al. 2000; Ørstavik et al. 2004; de Chevigny et al. 2008b), and their presence or absence does not appear to influence the prognosis.

*Periodontal support.* Except for one study (Ørstavik et al. 2004), the periodontal status has not been significantly associated with treatment outcome (Sjögren et al. 1990; de Chevigny et al. 2008b), and it does not appear to influence the prognosis. Nevertheless, if present at the time of treatment, advanced periodontal disease has a poor prognosis and should be expected to progress over time (Fig. 15.8).

*Flare-up during treatment.* Flare-ups have not been significantly associated with treatment outcome (Byström et al. 1987; Sjögren et al. 1990; Peters and

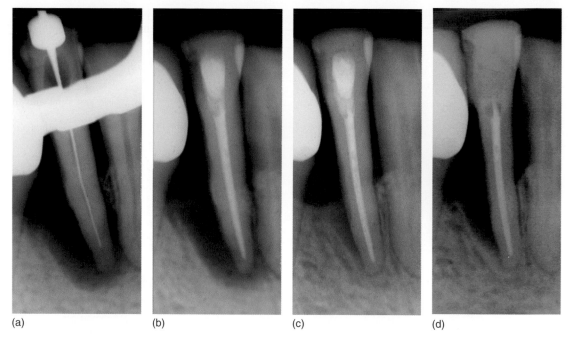

(a)        (b)        (c)        (d)

**Fig. 15.8**  Effect of advanced periodontal disease on the prognosis. (a) Mandibular lateral incisor with primary apical periodontitis and advanced marginal periodontitis, resulting in extensive bone loss. (b) Completed treatment. (c) At 8 months, clinical normalcy and drastically decreased radiolucency suggest incomplete healing (tooth is still restored with a temporary filling). (d) At 3 years, recurrent bone loss because of advancing periodontal disease. (Reprinted with permission from Friedman 2002b.)

Wesselink 2002; de Chevigny et al. 2008b), and they do not appear to influence the prognosis.

*Restoration.* The type of restoration (temporary, definitive, filling, cast) has not been significantly associated with treatment outcome (Sjögren et al. 1997; de Chevigny et al. 2008b), except in one selected study (Sjögren et al. 1990), where healing occurred less frequently in teeth restored with crowns and those serving as bridge abutments, than in teeth restored with fillings. Posts have not been associated with the outcome either (Sjögren et al. 1990; de Chevigny et al. 2008b). Thus, the type of the intraradicular, intracoronal, and extracoronal restoration does not appear to influence the prognosis.

### 15.4.3.3 Equivocal variables

*Radiolucency size.* Better outcome in teeth with small lesions ($\leq$5 mm diameter) than larger lesions has been reported in three selected current studies (Weiger et al. 2000; Hoskinson et al. 2002; Ørstavik et al. 2004), while comparable outcomes have been reported for small and large lesions in five other studies (Byström et al. 1987; Sjögren et al. 1990, 1997; Peters and Wesselink 2002; de Chevigny et al. 2008b). Thus, the influence of the size of the radiolucency on the prognosis is equivocal.

*Length of root filling.* Association of the outcome with root filling length has been reported in two studies (Sjögren et al. 1990; Ørstavik et al. 2004). Specifically, a poorer outcome was reported when root fillings were either extruded (Sjögren et al. 1990) or excessively short (2 mm or shorter than the root end) (Sjögren et al. 1990). In contrast, comparable outcomes for adequate and inadequate root-filling lengths have been reported in eight studies (Byström et al. 1987; Weiger et al. 2000; Hoskinson et al. 2002; Peters and Wesselink 2002; Huumonen et al. 2003; Peters et al. 2004; Marending et al. 2005; de Chevigny et al. 2008b). Thus, the influence of root-filling length on the prognosis is equivocal.

*Apical enlargement.* Although it has been suggested that extensive apical enlargement may enhance disinfection in the apical portion of the root canal (Ørstavik et al. 1991; Yared and Dagher 1994; Card et al. 2002), extensive apical enlargement has not been significantly associated with treatment outcome (Hoskinson et al. 2002; Peters and Wesselink 2002; Ørstavik et al. 2004). Considering the inability to assess the extent of apical enlargement without knowing the initial canal dimensions in the different studies, there are insufficient data available to assess the influence of apical enlargement on the prognosis, and no clinical evidence to support extensive apical enlargement.

*Bacterial culture before root filling.* Bacteriologic root canal samples showing no growth before root filling were associated with better treatment outcome in a study where the methodology did not address the anaerobic endodontic microflora (Engström et al. 1964), but also in a later study that used advanced anaerobic bacteriologic techniques (Sjögren et al. 1997). Better outcome has also been reported in teeth with no bacterial growth at the beginning of the second treatment session after medication with calcium hydroxide compared with teeth with growth (Waltimo et al. 2005). These findings, however, are apparently disputed by another study (Peters and Wesselink 2002). The contradiction may have resulted from the limited ability of the sampling techniques to culture root canal bacteria (Paquette et al. 2007). Thus, the data available on the influence of growth or no growth in root canal samples on the prognosis are inconclusive.

*Number of treatment sessions.* Intracanal medication applied between treatment sessions has been suggested to improve root canal disinfection and the outcome of treatment (Byström and Sundqvist 1981, 1983, 1985; Byström et al. 1985; Molander et al. 1990; Ørstavik et al. 1991; Yared and Dagher 1994; Shuping et al. 2000). Studies have reported nonsignificant differences in healing after treatment in one or two sessions, in the range of 10% in favor of either method (Trope et al. 1999; Weiger et al. 2000; Peters and Wesselink 2002; de Chevigny et al. 2008b); however, the samples in all these studies are too small to show significance of the observed differences. A systematic review of these studies concluded that "the biological benefit of multi-session treatment has not been supported by clinical evidence" (Sathorn et al. 2005). However, there may be risks associated with multiple-session treatment; teeth treated in one or two sessions may survive longer than teeth treated in three sessions or more (Cheung 2002), and teeth treated in multiple sessions may have an increased risk of infection with *E. faecalis* and persistent apical periodontitis (Siren et al. 1997). Taken together, the data available on the influence of one or two treatment sessions on the prognosis are inconclusive, but prolongation of treatment beyond two sessions should be avoided.

*Materials and techniques.* There is only limited information available on the association of specific treatment regimens and treatment outcome. There is no significant association between the outcome and the use of hand or engine-driven instruments (Marending et al. 2005), different engine-driven instrument systems (Peters et al. 2004), and the degree of taper (Hoskinson et al. 2002). Comparable outcomes have been reported for different sealer types (Eriksen et al. 1988; Waltimo et al. 2001; Huumonen et al. 2003), as well as for vertical compaction of the root filling when compared with hybrid compaction (Hoskinson et al. 2002) and lateral compaction (Peters et al. 2004). The Toronto Study series (de Chevigny et al. 2008b) have reported a significantly better outcome (10% more healed) in teeth treated with flared canal preparation and vertically compacted warm gutta-percha (Schilder technique) than in teeth treated with step-back instrumentation and lateral compaction of gutta-percha; however, the authors emphasize the requirement to validate this finding in a randomized controlled trial. Thus, the instrumentation technique, the type of sealer, and the root-filling technique do not appear to influence the prognosis, while the apparent superiority of the Schilder technique requires validation.

*Complications.* Perforation, file breakage, and massive extrusion of filling materials have all impaired healing (Sjögren et al. 1990; de Chevigny et al. 2008b), except in one study (Ørstavik et al. 2004). Endodontists currently may successfully manage complications associated with perforation (Main et al. 2004; Ghoddusi et al. 2007) and broken instruments (Spili et al. 2005). Thus, the negative influence of mid-treatment complications on the prognosis may be mitigated by current management strategies. Nevertheless, complications can impair healing; therefore, they should be avoided.

## 15.5 The prognosis of persistent apical periodontitis after orthograde retreatment

Similarly to the studies on initial treatment, considerable differences exist in case selection and composition of study materials among the four selected current studies on retreatment (Sjögren et al. 1990; Sundqvist et al. 1998; Kvist and Reit 1999; de Chevigny et al.

2008a), and their results vary even more than those of initial treatment.

### 15.5.1 Potential for healing

The proportion of completely healed teeth after orthograde retreatment ranges from 56% (Kvist and Reit 1999) to 84% (in teeth without perforation) (de Chevigny et al. 2008a). Falling below the range reported in the other studies, the reported 56% (approximate; the exact rate was not provided) in one study (Kvist and Reit 1999) and 62% in another study (Sjögren et al. 1990) appear to be outliers. The variability of the results may be attributed to the same factors as those suggested above for initial treatment. In addition to the healed teeth, progressive or incomplete healing has been reported in 5% of the teeth (Byström et al. 1987). Thus, the potential of teeth with persistent apical periodontitis to heal after orthograde retreatment is 74–84% (excluding the outliers), while additional 5 may demonstrate incomplete healing. It has also been suggested that even if complete healing does not occur, 93% of the teeth may remain symptom free and functional (de Chevigny et al. 2008a).

### 15.5.2 Time course of healing

Same as healing of primary apical periodontitis after initial treatment, healing of persistent apical periodontitis requires considerable time after orthograde retreatment. Apparently, as many as 50% of teeth that were not healed 10–17 years after retreatment were completely healed a decade later (Fristad et al. 2004). The late healing was mainly characteristic of teeth with surplus root-filling material (Fristad et al. 2004).

### 15.5.3 Prognostic variables

The similarities between initial treatment and orthograde retreatment justify the consideration of the same variables as outcome predictors. It is noteworthy that the evidence base for retreatment is limited to only few selected studies with relatively small samples, that preclude conclusive assessment of the influence of many factors on the prognosis. Also, specifically for retreatment, characteristics of the previous root canal treatment history have to be considered, including the previous root filling, a perforation that may be present in a minority of retreated teeth, and the time elapsed since initial treatment.

#### 15.5.3.1 Significant outcome predictors

*Apparent quality of the previous root filling.* A significantly better outcome (36% difference) in teeth with inadequate length or density of the previous root filling, compared with teeth with apparently adequate root fillings, has been reported (de Chevigny et al. 2008a). Thus, the prognosis appears to be better in teeth with an apparent poor quality of the previous root filling.

*Previous perforation.* A significantly poorer outcome (50% difference) was reported in teeth where previous perforations were repaired with resin-modified glass ionomer cement (Farzaneh et al. 2004b). In contrast, 93–100% of teeth where perforations were repaired with MTA healed (Main et al. 2004; Ghoddusi et al. 2007; de Chevigny et al. 2008a) (Fig. 15.9). Thus, the prognosis may be impaired by a previous perforation, depending on the material used for repair.

*Number of treatment sessions.* Unlike initial treatment (see above), a significantly better outcome (23% difference) in teeth retreated in one session, compared with two sessions, has been reported (de Chevigny et al. 2008a). This evidence, however, requires validation from a randomized controlled trial.

#### 15.5.3.2 Nonsignificant variables

*Age, gender.* As in initial treatment (see above), these variables are unlikely to influence the prognosis.

*Number of roots.* Unlike initial treatment (see above), the number of roots did not influence the prognosis in the most recent report (de Chevigny et al. 2008a).

*Symptoms.* As in initial treatment (see above), presence or absence of symptoms is unlikely to influence the prognosis.

*Periodontal support.* The periodontal status has not been significantly associated with treatment outcome (de Chevigny et al. 2008a), and it does not appear to influence the prognosis. Nevertheless, advanced periodontal disease has a poor prognosis and should be expected to progress over time when present at the time of treatment.

*Elapsed time after previous treatment.* The time elapsed after initial treatment has not been significantly associated with treatment outcome (de Chevigny et al. 2008a), and it does not appear to influence the prognosis.

*Radiolucency size.* The size of the lesion has not been significantly associated with treatment outcome

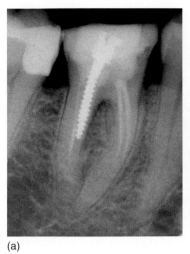 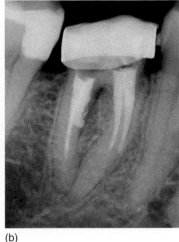 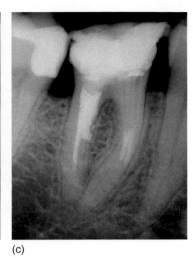

(a)                                     (b)                                     (c)

**Fig. 15.9**   Infection associated with perforation healed after orthograde retreatment. (a) Mandibular molar with a distal root perforation and associated bone loss. (b) Completed retreatment and perforation seal with MTA. (c) At 1.5 years, radiographic and clinical normalcy suggest that the tooth has healed. Regrettably, the tooth is not properly restored.

(de Chevigny et al. 2008a), and it does not appear to influence the prognosis.

*Materials and techniques.* A nonsignificant difference in outcome is reported in one study (de Chevigny et al. 2008a), between teeth treated with flared canal preparation and vertically compacted warm gutta-percha (Schilder technique) and teeth treated with step-back instrumentation and lateral compaction of gutta-percha. Thus, the treatment technique does not appear to influence the prognosis.

*Length of root filling.* The new filling length has not been significantly associated with treatment outcome (de Chevigny et al. 2008a), and it does not appear to influence the prognosis.

*Flare-up during treatment.* As in initial treatment (see above), flare-ups are unlikely to influence the prognosis.

*Complications.* As in initial treatment (see above), mid-treatment complications should be avoided even if their negative influence on the prognosis has not been demonstrated.

*Restoration.* The type of the intraradicular, intracoronal, and extracoronal restoration does not appear to influence the prognosis (de Chevigny et al. 2008a).

#### 15.5.3.3 Equivocal variables

*Bacterial culture before root filling.* Better outcome is reported in teeth where root canal samples, obtained

with advanced anaerobic bacteriologic techniques, showed no growth before root filling, when compared with teeth with growth in the samples (Sundqvist et al. 1998). Nevertheless, considering the limitations of the sampling technique, there are insufficient data available to assess the influence of growth and no growth in root canal samples on the prognosis.

## 15.6 The prognosis of persistent apical periodontitis after apical surgery

The selected studies on apical surgery (Kvist and Reit 1999; Zuolo et al. 2000; Rahbaran et al. 2001; Chong et al. 2003; Wang et al. 2004; Gagliani et al. 2005; Lindeboom et al. 2005; Taschieri et al. 2006; von Arx et al. 2007; Yazdi et al. 2007) differ considerably in their cohorts, interventions, and observation times. For example, in two studies all or most teeth received previous orthograde retreatment ( Zuolo et al. 2000; Chong et al. 2003), in other two studies 40% of the teeth were previously retreated (Rahbaran et al. 2001; Wang et al. 2004), and the remaining studies (Kvist and Reit 1999; Jensen et al. 2002; Gagliani et al. 2005; Lindeboom et al. 2005; von Arx et al. 2007) did not mention the previous treatment history of their cohorts. Also, the proportions of teeth in the study cohort that received previous apical surgery comprised

44% (Rahbaran et al. 2001), 33% (Gagliani et al. 2005), close to 10% (Wang et al. 2004; von Arx et al. 2007), none (Jensen et al. 2002), or an unspecified percentage (Kvist and Reit 1999; Chong et al. 2003; Lindeboom et al. 2005). As highlighted below, the outcome of apical surgery is expected to be better when it is preceded by orthograde retreatment, and for first-time surgery compared with repeat surgery. In addition, the 1-year observation in three of the studies (Lindeboom et al. 2005; Taschieri et al. 2006; von Arx et al. 2007) is too short to demonstrate reversal of healing, reported in over 5% of teeth that were initially considered as healed (Halse et al. 1991; Frank et al. 1992; Jesslen et al. 1995; Kvist and Reit 1999; Rubinstein and Kim 2002; Wesson and Gale 2003; Yazdi et al. 2007). These and other differences among the selected studies result in the considerable inconsistency in the reported outcomes.

### 15.6.1 Potential for healing

The proportion of completely healed teeth after apical surgery ranges from 37% (Rahbaran et al. 2001) to 91% (Zuolo et al. 2000). Considering the uniform outcome assessment among the selected studies, the results may have varied because of differences in proportions of previously retreated teeth, in observation periods and in treatment procedures. The reported rates above 90% (Zuolo et al. 2000; Chong et al. 2003) that may have been affected by previous orthograde retreatment, and those below 60% (Rahbaran et al. 2001) that fall considerably below the range reported in the remaining studies, appear to be outliers. Also, the high rates of healed teeth in the studies with only 1-year observation (Lindeboom et al. 2005; Taschieri et al. 2006; von Arx et al. 2007) have to be viewed with caution, because regression may occur in some of the healed teeth in the longer term.

In addition to the healed teeth, progressive healing has been reported in 6% (Chong et al. 2003) to 33% (Rahbaran et al. 2001) of the teeth. Thus, the potential of teeth to heal after apical surgery is 74–78% (excluding the outliers), while additional 5–30% may demonstrate incomplete healing. It has also been suggested that regardless of healing (complete or incomplete), over 90% of the teeth may remain symptom free and functional (Zuolo et al. 2000; Chong et al. 2003; Wang et al. 2004; Gagliani et al. 2005; Lindeboom et al. 2005; von Arx et al. 2007; Yazdi et al. 2007).

### 15.6.2 Time course of healing

Healing progresses rapidly within the first year after surgical treatment (Halse et al. 1991; Kvist and Reit 1999). Of all the teeth that heal eventually, 35–60% appear completely healed by 1 year (Grung et al. 1990; Halse et al. 1991; Molven et al. 1996; Maddalone and Gagliani 2003; Wesson and Gale 2003), while approximately 85% appear healed by 3 years (Grung et al. 1990). Healing by a fibrous scar is quite common after apical surgery (Rud et al. 1972; Molven et al. 1996) (Fig. 15.4), particularly when both the buccal and lingual bone plates are perforated at the conclusion of the surgical procedure (Molven et al. 1991). Postsurgery apical scars appear to remain stable over time (Molven et al. 1996); therefore, they are considered a surrogate for a healed site (Rud et al. 1972; Molven et al. 1996).

Although outcomes observed at 1 year have been suggested to remain stable upto 3–5 years (Halse et al. 1991; Rubinstein and Kim 2002), some 5–40% of healed teeth may revert to disease in the long-term (Halse et al. 1991; Frank et al. 1992; Jesslen et al. 1995; Kvist and Reit 1999; Rubinstein and Kim 2002; Wesson and Gale 2003; Yazdi et al. 2007) (Fig. 15.10). To address the risk of recurrent disease, it is advisable to periodically examine the treated teeth in the long-term.

### 15.6.3 Prognostic variables

The majority of the selected studies on apical surgery have attempted to associate observed outcomes with different variables. For the major part, these variables are different from those that have been examined in relation to nonsurgical treatment.

#### 15.6.3.1 Significant outcome predictors

*Radiolucency size.* Better outcome in teeth with small lesions (≤5 mm diameter) than larger lesions is reported in two studies (Wang et al. 2004; von Arx et al. 2007), while comparable outcomes are reported for small and large lesions in another study (Rahbaran et al. 2001). Small crypts are surgically enlarged forming an excisional wound in the surrounding bone (Harrison and Jurosky 1992), while large lesions are merely curetted without formation of such wound (Wang et al. 2004). Healing by scar tissue frequently occurs in very large lesions (≤10 mm diameter) (Molven et al. 1991). Thus, the prognosis appears to be poorer in teeth with lesions exceeding 5 mm compared with smaller lesions.

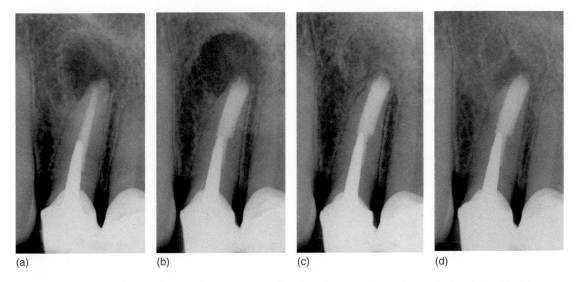

(a)          (b)          (c)          (d)

**Fig. 15.10**   Recurrent infection after apical surgery. (a) Maxillary lateral incisor with persistent apical periodontitis. (b) Completed surgery, including a retrograde root filling with sealer and injectable gutta-percha. (c) At 6 months, radiographic and clinical normalcy suggest that the tooth has healed. (d) At 2.5 years, renewed radiolucency suggests recurrent infection. (Reprinted with permission from Friedman 2005.)

*Supporting bone loss.* Several nonselected studies (Skoglund and Persson 1985; Forssell et al. 1988; Wesson and Gale 2003) have reported poor outcomes in teeth with considerable bone loss, either vertical or marginal, which can compromise periodontal reattachment. Thus, the prognosis appears to be compromised by considerable attachment loss of the treated tooth.

*Level of apical resection.* Better outcome is reported after resection at the mid-root level than at a more apical level in an older, nonselected study (Altonen and Mattila 1976). A more coronal resection (3 mm from the apex) may avoid exposing canal ramifications that can allow intracanal bacteria to sustain disease after surgery (Carr and Bentkover 1998), while also facilitating preparation of the root-end cavity and filling. Thus, a more coronal resection appears to improve the prognosis.

*Presence or absence of a root-end filling.* A root-end filling is placed to establish an effective barrier against interaction of intracanal bacteria with the periapical tissues (Friedman 1991). Better outcomes have been reported with root-end fillings than without (Friedman et al. 1991a; Rapp et al. 1991). Thus, placement of a root-end filling improves the prognosis, particularly when persistent root canal infection is presumed.

*Root-end filling material.* Studies based on an in vivo model developed to simulate clinical conditions (Friedman et al. 1991b), have suggested that IRM (Andreasen and Pitt Ford 1994; Pitt Ford et al. 1994), Super-EBA (Pitt Ford et al. 1995; Trope et al. 1996), MTA (Torabinejad et al. 1995), and Diaket (Witherspoon and Gutmann 2000) are the best root-end filling materials. However, animal studies do not rank as evidence to supporting the clinical effectiveness of these materials. According to recent randomized controlled trials (Chong et al. 2003; Lindeboom et al. 2005; Yazdi et al. 2007) and a selected nonrandomized trial (von Arx et al. 2007), Retroplast is significantly better than glass-ionomer cement for apical capping (Yazdi et al. 2007), and is comparable to root-end filling with MTA (von Arx et al. 2007). When root-end fillings are compared, MTA appears to be comparable to IRM (Chong et al. 2003; Lindeboom et al. 2005) and superior to Super-EBA (von Arx et al. 2007). Thus, the choice of root-end filling material can influence the prognosis. Retroplast used for apical capping and MTA and IRM used for intracanal root-end filling appear to offer a better prognosis than alternative materials.

*Concurrent surgical and orthograde treatment.* When apical surgery and initial treatment are performed concurrently "infection is eliminated and reinfection is prevented" (Molven et al. 1991). Therefore, the outcome is usually better when compared to apical surgery alone (Friedman 1991; Hepworth and Friedman 1997). However, because currently apical surgery is performed alone as an alternative to orthograde retreatment, the better prognosis after concurrent surgical and orthograde treatment is merely of academic interest. Concurrent non-surgical and surgical intervention is indicated in selected situations where orthograde retreatment cannot manage the complexity of the infectious process and combination with surgery does address the problem more comprehensively.

*Retrograde root canal retreatment.* Retrograde retreatment is an attempt to instrument, irrigate, and fill the root canal as far coronally as can be reached

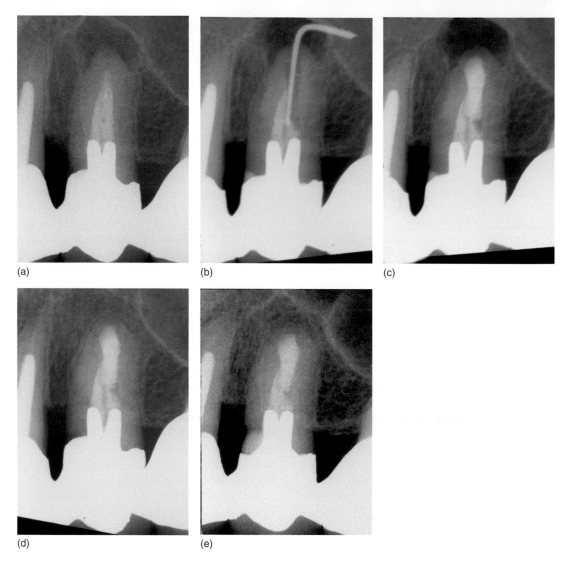

(a)                    (b)                    (c)

(d)                    (e)

**Fig. 15.11**   Persistent infection healed after retrograde retreatment. (a) Maxillary second premolar with persistent apical periodontitis. (b) Retrograde retreatment is carried out with ultrasonic files. (c) Completed surgery, including root filling with sealer and injectable gutta-percha. (d and e) At 1 and 7 years, respectively, radiographic and clinical normalcy suggest that the tooth has healed. (Reprinted with permission from Friedman 2005.) See also Fig. 15.10.

from the apical end (Fig. 15.11) (Nygaard-Östby 1971; Storms 1978; Serota and Krakow 1983; Reit and Hirsch 1986; Flath and Hicks 1987; Amagasa et al. 1989; Goldberg et al. 1990). Complete healing has been reported in 71–100% of the teeth (Reit and Hirsch 1986; Amagasa et al. 1989; Goldberg et al. 1991; Wang et al. 2004). The deeper barrier placed using this approach, between intracanal bacteria and the periapical tissue, offers an advantage over the standard root-end filling. Nevertheless, continued bacterial ingress into the canal, under the restoration and along the post, may still result in recurrence of disease (Fig. 15.10). Thus, retrograde retreatment improves the prognosis, even though it may not entirely prevent recurrence of disease.

### 15.6.3.2 Nonsignificant variables

*Age, gender.* These variables have not been significantly associated with treatment outcome (Zuolo et al. 2000; Rahbaran et al. 2001; Wang et al. 2004; von Arx et al. 2007), and they do do not appear to influence the prognosis.

*Tooth location.* Tooth location has not been significantly associated with treatment outcome (Zuolo et al. 2000; Rahbaran et al. 2001; Wang et al. 2004; Gagliani et al. 2005; von Arx et al. 2007); however, healing by scar tissue—consistent with complete healing— frequently occurs in maxillary lateral incisors (Molven et al. 1991). Apparently, convenience of access and the specific root anatomy influence the prognosis to a greater extent than the location of the tooth (Friedman 1998).

*Material and density of previous root filling.* The type of the previous root-filling material and the filling density are not associated with the outcome (Rahbaran et al. 2001; Wang et al. 2004), and they do not appear to influence the prognosis.

*Hemostasis.* Good hemostasis is critically important for the quality of the root-end filling (Carr and Bentkover 1998) and bonding of an apical cap (Jensen et al. 2002). Nevertheless, use of hemostatic agents is not significantly associated with treatment outcome (Wang et al. 2004), and they do not appear to influence the prognosis.

*Laser irradiation.* Although laser irradiation of the resected root surface and surrounding crypt has been suggested to render the dentin surface nonpermeable (Stabholz et al. 1992a, b; Arens et al. 1993; Paghdiwala 1993; Wong et al. 1994) and to improve disinfection and hemostasis (Melcer 1986; Miserendino 1988;

Komori et al. 1997), it has not been significantly associated with treatment outcome both in animal studies (Takeda 1989; Friedman et al. 1991b) and in a non-selected clinical trial (Bader and Lejeune 1998). Thus, laser irradiation applied in different steps of the surgical procedure does not appear to influence the prognosis.

*Barriers and bone-grafting substances.* Guided regeneration barriers (Pecora et al. 1995; Pecora et al. 1997b) and various bone-grafting substances (Sikri et al. 1986; Pecora et al. 1997a, 2001; Murashima et al. 2002; Yoshikawa et al. 2002) have been advocated for use in apical surgery; however, case series reports (Saad and Abdellatief 1991; Grimes 1994; Pecora et al. 1995; Rankow and Krasner 1996; Tobon et al. 2002) do not provide evidence to support these procedures. Thus, barriers and bone-grafting substances do not appear to influence the prognosis. When such are applied, care must be taken to avoid infection.

*Complications.* Different complications that may occur during apical surgery have not been significantly associated with the outcome, including perforation of the opposing bone plate (Grung et al. 1990; Molven et al. 1991; Wang et al. 2004), perforation into the sinus (Rud and Rud 1998), and paresthesia, which has been reported in 20% of patients after apical surgery in mandibular molars (Wesson and Gale 2003). The majority of patients who experienced paresthesia regained normal sensation within 3 months after surgery, in 1% the sensory deficit lingered for 2 years, and in 1% it lingered beyond 2 years (Wesson and Gale 2003). Thus, procedural complications do not appear to influence the prognosis. Sensory deficit, although not related to healing, must be considered as a risk when mandibular molars are treated.

*Antibiotics.* A course of systemic antibiotics before and after treatment has not been significantly associated with the outcome (Rahbaran et al. 2001; Wang et al. 2004; von Arx et al. 2007), and it does not appear to influence the prognosis.

### 15.6.3.3 Equivocal variables

*Symptoms.* Poorer outcome in symptomatic teeth is reported in one study (von Arx et al. 2007), while comparable outcomes have been reported for symptomatic and symptom-free teeth in two other studies (Rahbaran et al. 2001; Wang et al. 2004). Thus, the influence of presence or absence of symptoms on the prognosis is equivocal.

*Restoration.* Poorer outcome in teeth with a defective restoration or with a post is reported in one study (Rahbaran et al. 2001), but not supported by two other studies (Wang et al. 2004; von Arx et al. 2007). Thus, the influence of the type of the restoration on the prognosis is equivocal.

*Length of the previous root filling.* A significantly better outcome in teeth where the previous root-filling length is inadequate ($\geq$2 mm short of the root end or extruded) is reported in one study (Wang et al. 2004), but not supported by two other studies (Rahbaran et al. 2001; von Arx et al. 2007). Thus, the influence of the length of the filling on the prognosis is equivocal.

*Second-time surgery.* Poorer outcome of second-time surgery than first-time surgery has been reported in one selected study (Gagliani et al. 2005), and a systematic review of several nonselected studies (Peterson and Gutmann 2001), but not supported by three other selected studies (Rahbaran et al. 2001; Wang et al. 2004; von Arx et al. 2007). Modified case

selection criteria and techniques have been suggested to improve the outcome of second-time surgery in the latter studies (Wang et al. 2004) (Fig. 15.12). Thus, the potentially poorer prognosis of second-time surgery may be mitigated by use of an improved strategy.

*Root-end cavity design.* The classical root-end cavity drilled with a small round bur has been replaced by two current techniques: (1) After capping of the cut root surface with Retroplast (Rud et al. 1991a, b; Andreasen et al. 1993) (Fig. 15.13), 78% of the teeth have healed (Yazdi et al. 2007) (outlier studies excluded). (2) After cavity preparation with ultrasonic tips (Carr 1992; Carr and Bentkover 1998) (Fig. 15.14), 74–78% of the teeth have healed (Wang et al. 2004; Gagliani et al. 2005) (outlier studies excluded). Because comparison of outcomes with reports on root-end cavities drilled with burs is precluded, there is insufficient evidence to suggest that both modifications of the classical root-end cavity improve the prognosis; however, there is a sound clinical rationale for using

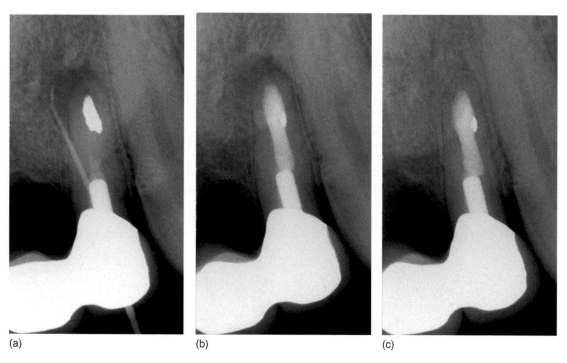

(a)                              (b)                              (c)

**Fig. 15.12**   Persistent infection healed after repeated (second-time) surgery. (a) Maxillary lateral incisor with persistent apical periodontitis after previous surgery, and a gutta-percha cone tracing the sinus tract. (b) Completed repeat surgery, comprising retrograde retreatment and filling with MTA. (c) At 3 years, radiographic and clinical normalcy suggest that the tooth has healed. (Reprinted with permission from Friedman 2002a.)

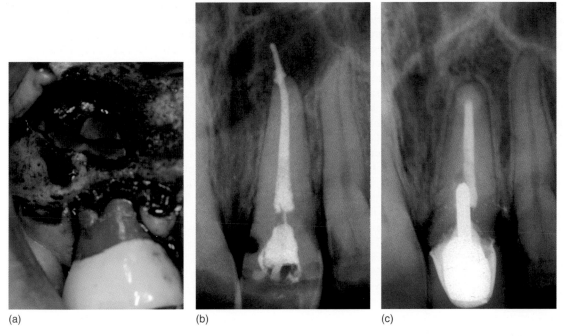

**Fig. 15.13**   Root-end management with bonded Retroplast. (a) Clinical view of maxillary first molar with Retroplast "caps" bonded to the three roots. (b) Maxillary central incisor with persistent apical periodontitis and extruded root filling. (c) At 9 years after surgery and bonding Retroplast at the root end, radiographic and clinical normalcy suggest that the tooth has healed. (Courtesy of Dr Vibe Rud, Copenhagen, Denmark. Reprinted with permission from Friedman 2005.)

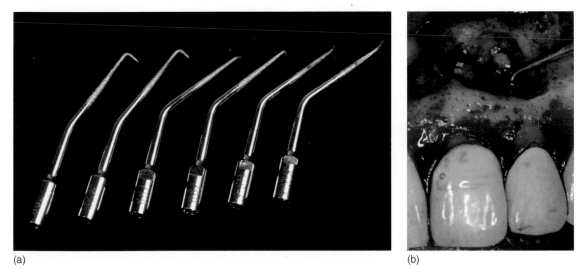

**Fig. 15.14**   Root-end cavity preparation with ultrasonic tips. (a) Assortment of ultrasonic tips for root-end cavity preparation. (b) Clinical view of root-end cavity preparation with an ultrasonic tip. (Reprinted with permission from Friedman 2005.)

both approaches. Both approaches also facilitate the procedure and improve its consistency relative to the outdated drilling of the root-end cavity with burs.

*Depth of root-end filling.* Better outcome after adequate placement and depth of the root-end filling has been suggested in one study (Rahbaran et al. 2001), but not supported in another study (Wang et al. 2004). Thus, the influence of the root-end filling depth, ranging from 1 to 4 mm, on the prognosis is equivocal.

*Magnification and illumination.* Loupes (Lindeboom et al. 2005), operating microscopes (Carr 1992; Rubinstein 1997; Mines et al. 1999), and endoscopes (von Arx et al. 2003, 2007; von Arx 2005; Taschieri et al. 2006) enhance vision, illumination, operator convenience, and identification of root and canal anatomy, and they improve control of all aspects of the surgical procedure (Carr 1992; Pecora and Andreana 1993; Kim 1997; Carr and Bentkover 1998; von Arx et al. 2003; Kim and Baek 2004; Velvart and Peters 2005; von Arx 2005; Kim and Kratchman 2006). One nonselected study (Rubinstein and Kim 1999) implies that the outcome can be improved by use of the operating microscope and Super-EBA cement as the root-end filling. In the selected studies, loupes (Wang et al. 2004; Gagliani et al. 2005; Lindeboom et al. 2005), microscope, or endoscope were used (Chong et al. 2003; Taschieri et al. 2006; von Arx et al. 2007). Comparable outcomes were reported for use of microscope or endoscope (Taschieri et al. 2006), but these magnification devices were not compared with loupes or no magnification at an adequate level of evidence. Thus, there are insufficient data available to assess the influence of magnification and illumination on the prognosis. Nevertheless, there is a sound rationale for their use to improve the quality of the surgical procedure.

*Operator's skill.* The individual operator's skill has been significantly associated with treatment outcome in a nonselected study (Lustmann et al. 1991), but reference to operator skill in the selected studies is scarce. Thus, even though it appears that the operator's skill may influence the prognosis, there are insufficient data available to assess this variable.

*Results of biopsy.* Theoretically, the biopsy results defining the pathological diagnosis of the lesion— granuloma, abscess or cyst—might be used as indicators of the potential for healing. A better outcome was reported when biopsy indicated "granuloma" than "cyst" in one study (Jensen et al. 2002), but this find-ing was not supported in two other studies ( Zuolo et al. 2000; Wang et al. 2004). Thus, the influence of the biopsy results on the prognosis is equivocal, mainly because routine biopsies (comprising few non-serial sections) seldom reflect the true nature of the pathological lesion.

## 15.7 The prognosis of persistent apical periodontitis after intentional replantation

In accordance with current concepts, intentional replantation may be used as an alternative to extraction, when both retreatment and apical surgery are not feasible in situ (Guy and Goerig 1984; Dumsha and Gutmann 1985) (Fig. 15.15). The expected goal is survival of the replanted tooth, considered as success (Grossman 1982) even if pathological processes persist. Healing of the attachment apparatus without root resorption depends on survival of the periodontal ligament and cementum along the root surface (Andreasen 1985; Andreasen et al. 1995), and prevention of infection (Tronstad 1988; Trope and Friedman 1992). Reattachment without resorption is affected by the trauma associated with extraction, extraoral manipulation and replantation, the extraoral time and storage medium, and the type and duration of splinting (Hammarstrom et al. 1986; Oikarinen 1993).

The relatively current studies on the prognosis of intentional replantation (Table 15.4) are consistent with the lowest level of evidence. According to these studies, the success/survival rate of intentional replantation ranges from 71% (Kahnberg 1988) to 91% (Keller 1990), but the assessment criteria have been vague. Importantly, the clinical protocols of a few studies listed in Table 15.4 did not meet the currently recognized requirements for successful replantation of teeth. For example, on occasion roots were only sealed apically without a root filling, predisposing them to inflammatory root resorption (Koenig et al. 1988; Keller 1990). The available studies do not allow assessment of the realistic outcome of intentional replantation adhering to current guidelines. In the absence of suitable clinical studies, clues about the prognosis may be obtained from animal studies. These studies suggest that the incidence of progressive root resorption may not exceed 4%, when treatment conditions are optimal (Hammarstrom et al. 1986). Indeed, two of the clinical studies on intentional replantation

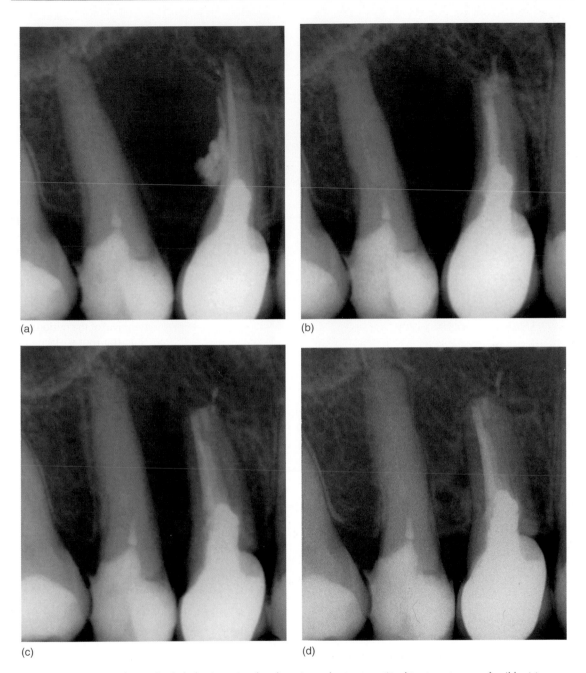

**Fig. 15.15** Persistent infection healed after intentional replantation, where conventional treatment was unfeasible. (a) Maxillary first premolar with extensive persistent apical periodontitis, associated with palatal root perforation at the distal aspect, untreated palatal canal, buccal post perforation and total loss of the buccal bone plate. (b) Completed intentional replantation that included repair of two perforations and two root-end fillings, all with MTA. (c) At 3 months, the roots of both teeth have become realigned, and the radiolucency considerably decreased. (d) At 1 year and 4 months, radiographic and clinical normalcy suggest that the tooth has healed with two small scars remaining.

have reported root resorption in only 5% and 6% of the replanted teeth (Koenig et al. 1988; Bender and Rossman 1993), because of the optimal extraoral time and conditions. Persistent infection ranges from 9% (Keller 1990) to 29% (Kahnberg 1988), because the infected root canal can be effectively sealed with a root-end filling that is easily placed. Indeed, many case reports (Feldman et al. 1971; Rosenberg et al. 1980; Solomon and Abelson 1981; Stroner and Laskin 1981; Kaufman 1982; Lubin 1982; Guy and Goerig 1984; Nosonowitz and Stanley 1984; Ross 1985; Dryden 1986; Lindeberg et al. 1986; Lu 1986; Madison 1986; Messkoub 1991) have suggested that the incidence of resorption and persistent infection is low. These reports have demonstrated the predictable healing potential after intentional replantation performed in well-controlled conditions, albeit at the lowest level of evidence.

### 15.7.1 Dynamics of resorption

Resorption after intentional replantation is usually discernible within 1 year (Emmertsen and Andreasen 1966). This is definitely true regarding inflammatory resorption. Replacement resorption may be first observed radiographically several years after replantation (Andreasen et al. 1995). However, it may be indicated clinically much earlier than radiographically, by a specific pitch upon percussion.

## 15.8 Etiology of persistent apical periodontitis after endodontic treatment

Persistent apical periodontitis after endodontic treatment may occur in response to nonmicrobial factors, including foreign materials and true cysts (Nair et al. 1990b, 1993; Nair 2003a, b); however, this is not a common occurrence (Sjögren et al. 1990). For the major part, persistent apical periodontitis is sustained by persistent or recurrent infection (Siqueira 2001; Friedman 2002a). The sites colonized by bacteria and the pathways of bacteria–host interactions may differ after nonsurgical and surgical treatment. Furthermore, the reason for nonhealing or delayed healing seen in some cases may be related to the presence of specific virulent microorganisms that exert significant irritation to periapical tissues and elicit a pronounced host response.

### 15.8.1 Persistent infection after nonsurgical treatment

The disease process is sustained by microorganisms, mainly bacteria, which colonize the root canal system or extraradicular sites.

*Root canal system.* This is the most frequently colonized site in teeth with persistent infection (Borssen and Sundqvist 1981; Fukushima et al. 1990; Nair et al. 1990a, 1999; Baumgartner and Falkler 1991; Lin et al. 1991, 1992; Molander et al. 1998; Sundqvist et al. 1998; Peciuliene et al. 2000; Cheung and Ho 2001; Hancock et al. 2001; Peciuliene et al. 2001; Rolph et al. 2001; Siqueira 2001; Haapasalo et al. 2003; Pinheiro et al. 2003a, b; Sundqvist and Figdor 2003; Nair 2004; Fouad et al. 2005; Kaufman et al. 2005; Zerella et al. 2005; Gomes et al. 2006). The bacteria have either survived the root canal treatment procedures (Sjögren et al. 1997), or invaded the filled canal space after treatment, possibly through a coronal pathway (Friedman et al. 2000).

*Extraradicular sites.* Specific bacteria, particularly *Actinomyces israelii* and *Propionibacterium propionicum*, can colonize in the periapical tissues (Sundqvist and Reuterving 1980; Nair and Shroeder 1984; Happonen et al. 1985; Happonen 1986; Haapasalo et al. 1987; O'Grady and Reade 1988; Sjögren et al. 1988; Figures and Douglas 1991; Sakellariou 1996; Kalfas et al. 2001; Siqueira 2003; Sundqvist and Figdor 2003; Figdor 2004), after penetrating the host tissues during a long-term infection of the root canal system or when inoculated periapically during treatment. Other bacterial species have also been implicated in extraradicular infection (Gatti et al. 2000; Sunde et al. 2000a, b, 2002, 2003; Tronstad and Sunde 2003). They can colonize the cementum on the root surface (Pitt Ford 1982; Nair 1987) and around the apical foramina (Tronstad et al. 1987, 1990a, b; Siqueira and Lopes 2001; Leonardo et al. 2002; Noiri et al. 2002; Tronstad and Sunde 2003), or in dentin debris inadvertently extruded during treatment (Yusuf 1982). Whether these bacteria can exclusively sustain infection is unclear; however, current knowledge suggests that the predominant cause of persistent apical periodontitis is root canal infection, while exclusive extraradicular infection comprises a small percentage of cases (Friedman 2002a).

Even after orthograde retreatment, bacteria can still survive in the root canal system and sustain persistent apical periodontitis (Sundqvist et al. 1998;

Peciuliene et al. 2001; Zerella et al. 2005). Nevertheless, there is an increased probability that disease is sustained either by foreign body reaction and true cysts (Nair et al. 1990a, 1993), or by extraradicular infection (Sundqvist and Reuterving 1980; Pitt Ford 1982; Yusuf 1982; Nair and Shroeder 1984; Happonen et al. 1985; Happonen 1986; Haapasalo et al. 1987; Nair 1987; Tronstad et al. 1987, 1990a, b; O'Grady and Reade 1988; Sjögren et al. 1988; Figures and Douglas 1991; Sakellariou 1996; Gatti et al. 2000; Sunde et al. 2000a, b, 2002, 2003; Kalfas et al. 2001; Siqueira and Lopes 2001; Siqueira 2003; Tronstad and Sunde 2003; Figdor 2004). As apical surgery can effectively eliminate these etiologic factors, it is the treatment of choice for persistent apical periodontitis after orthograde retreatment.

### 15.8.2 Persistent infection after apical surgery

Persistence of apical periodontitis after surgery usually suggests that root canal bacteria are not effectively enclosed within the canal space by the root-end filling (Friedman 1991). Placement of a root-end filling is a challenging procedure, and several pathways may remain that allow continued interaction of root canal bacteria with the host tissues, resulting in persistent or recurrent infection.

*Margins of the root-end filling.* Compromised placement, adaptation to the canal walls, and sealing ability of the filling material—all can compromise the seal of the root-end filling (Friedman 1991).

*Accessory canals or isthmuses.* Accessory foramina and, in specific teeth, isthmuses are frequently present in the apical portion of root canals. A root-end filling may not seal these pathways, particularly when it is placed without magnification and illumination aids (Hsu and Kim 1997; Carr and Bentkover 1998).

*Exposed dentinal tubules.* Apical resection is typically performed with a bevel, exposing dentinal tubules at the cut surface (Gilheany et al. 1994). A shallow root-end filling does not internally seal all the tubules. The exposed tubules provide a pathway for root canal bacteria to interact with the tissues (Vertucci and Beatty 1986; Tidmarsh and Arrowsmith 1989; Gilheany et al. 1994).

*Vertical root crack or fracture.* Oral bacteria can colonize the crack or fracture line and sustain the infection even if the root canal is effectively sealed by a root-end filling. Presence of a root crack or fracture should be ruled out before further treatment.

Because persistence of disease is likely to be sustained by persistent root canal infection (Friedman 1991), orthograde retreatment is the treatment of choice for persistent apical periodontitis after apical surgery. When retreatment is not feasible, apical surgery should be repeated with an emphasis on retrograde retreatment, or effective sealing of the infected root canal.

## 15.9 Concluding remarks

The projected prognosis of treatment is a key consideration in selection of cases for endodontic treatment. Well-informed clinicians can and should project a specific prognosis for every tooth considered for treatment. As highlighted in this chapter, the chance of teeth with primary and persistent apical periodontitis to heal after appropriate endodontic treatment ranges from good to very good, depending on specific outcome predictors. Furthermore, the chance for functional retention of the tooth in the long-term is excellent. Therefore, whenever patients consider it feasible and acceptable, nonsurgical or surgical endodontic treatment should be attempted before considering tooth extraction and replacement. This is certainly the most conservative, and frequently, the most economical manner in which to treat endodontic infections. Given that the lack of healing is primarily mediated by a persistence of the infectious process, more research is critical in defining microbial virulence factors that mediate the disease, as well as the microbial loads necessary to sustain clinically detectable disease. In addition, a better definition of the acceptable follow-up periods should be determined to help the practitioner make treatment decisions for teeth that do not respond to treatment with complete radiographic and clinical healing.

## 15.10 References

Adenubi JO and Rule DC. 1976. Success rate for root fillings in young patients. A retrospective analysis of treated cases. *Br Dent J* 141(8): 237–41.

Åkerblom A and Hasselgren G. 1988. The prognosis for endodontic treatment of obliterated root canals. *J Endod* 14(11): 565–67.

Allen RK, Newton CW, and Brown CE, Jr. 1989. A statistical analysis of surgical and nonsurgical endodontic retreatment cases. *J Endod* 15(6): 261–66.

Altonen M and Mattila K. 1976. Follow-up study of apicoectomized molars. *Int J Oral Surg* 5(1): 33–40.

Amagasa T, Nagase M, Sato T, and Shioda S. 1989. Apicoectomy with retrograde gutta-percha root filling. *Oral Surg Oral Med Oral Pathol* 68(3): 339–42.

Ambrosio E and Walkerley S. 1996. Broadening the ethical focus: A community perspective on patient autonomy. *Hum Health Care Int* 12(2): E10.

Andreasen JO 1985. External root resorption: Its implication in dental traumatology, paedodontics, periodontics, orthodontics and endodontics. *Int Endod J* 18(2): 109–18.

Andreasen JO, Borum MK, Jacobsen HL, and Andreasen FM. 1995. Replantation of 400 avulsed permanent incisors. 4. Factors related to periodontal ligament healing. *Endod Dent Traumatol* 11(2): 76–89.

Andreasen JO, Munksgaard L, and Rud J. 1993. Periodontal tissue regeneration including cementogenesis adjacent to dentin-bonded retrograde composite fillings in humans. *J Endod* 19: 151–53.

Andreasen JO and Pitt Ford TR. 1994. A radiographic study of the effect of various retrograde fillings on periapical healing after replantation. *Endod Dent Traumatol* 10(6): 276–81.

Andreasen JO and Rud J. 1972a. Correlation between histology and radiography in the assessment of healing after endodontic surgery. *Int J Oral Surg* 1(3): 161–73.

Andreasen JO and Rud J. 1972b. Modes of healing histologically after endodontic surgery in 70 cases. *Int J Oral Surg* 1(3): 148–60.

Aqrabawi JA. 2006. Outcome of endodontic treatment of teeth filled using lateral condensation versus vertical compaction (Schilder's technique). *J Contemp Dent Pract* 7(1): 17–24.

Arens DL, Levy GC, and Rizoiu IM. 1993. A comparison of dentin permeability after bur and laser apicoectomies. *Compendium* 14(10): 1290, 1292, 1294 passim; quiz 1298.

Bader G and Lejeune S. 1998. Prospective study of two retrograde endodontic apical preparations with and without the use of $CO_2$ laser. *Endod Dent Traumatol* 14: 75–78.

Barton S. 2000. Editorial. Which clinical studies provide the best evidence? *Br Med J* 321: 255–56.

Baumgartner JC and Falkler WA, Jr. 1991. Bacteria in the apical 5 mm of infected root canals. *J Endod* 17(8): 380–83.

Bender IB and Rossman LE. 1993. Intentional replantation of endodontically treated teeth. *Oral Surg Oral Med Oral Pathol* 76(5): 623–30.

Benson K and Hartz AJ. 2000. A comparison of observational studies and randomized, controlled trials. *N Engl J Med* 342(25): 1878–86.

Bhaskar SN. 1966. Oral surgery—oral pathology conference No. 17, Walter Reed Army Medical Center. Periapical lesions—types, incidence, and clinical features. *Oral Surg Oral Med Oral Pathol* 21(5): 657–71.

Borssen E and Sundqvist G. 1981. Actinomyces of infected dental root canals. *Oral Surg Oral Med Oral Pathol* 51(6): 643–48.

Boucher Y, Matossian L, Rilliard F, and Machtou P. 2002. Radiographic evaluation of the prevalence and technical quality of root canal treatment in a French subpopulation. *Int Endod J* 35(3): 229–38.

Brynolf L. 1967. Histological and roentgenological study of periapical region of human upper incisors. *Odontol Revy* 18(Suppl. 11): 1–176.

Byström A, Claesson R, and Sundqvist G. 1985. The antibacterial effect of camphorated paramonochlorophenol, camphorated phenol and calcium hydroxide in the treatment of infected root canals. *Endod Dent Traumatol* 1(5): 170–75.

Byström A, Happonen RP, Sjögren U, and Sundqvist G. 1987. Healing of periapical lesions of pulpless teeth after endodontic treatment with controlled asepsis. *Endod Dent Traumatol* 3(2): 58–63.

Byström A and Sundqvist G. 1981. Bacteriologic evaluation of the efficacy of mechanical root canal instrumentation in endodontic therapy. *Scand J Dent Res* 89(4): 321–28.

Byström A and Sundqvist G. 1983. Bacteriologic evaluation of the effect of 0.5 percent sodium hypochlorite in endodontic therapy. *Oral Surg Oral Med Oral Pathol* 55(3): 307–12.

Byström A and Sundqvist G. 1985. The antibacterial action of sodium hypochlorite and EDTA in 60 cases of endodontic therapy. *Int Endod J* 18(1): 35–40.

Caliskan MK and Sen BH. 1996. Endodontic treatment of teeth with apical periodontitis using calcium hydroxide: A long-term study. *Endod Dent Traumatol* 12(5): 215–21.

Card SJ, Sigurdsson A, Ørstavik D, and Trope M. 2002. The effectiveness of increased apical enlargement in reducing intracanal bacteria. *J Endod* 28(11): 779–83.

Carr GB. 1992. Microscopes in endodontics. *J Calif Dent Assoc* 20(11): 55–61.

Carr GB and Bentkover SK. 1998. Surgical Endodontics. In: Cohen S and Burns RC (eds), *Pathways of the Pulp*. St Louis, MO: Mosby, 608–56.

Cheung GS. 2002. Survival of first-time nonsurgical root canal treatment performed in a dental teaching hospital. *Oral Surg Oral Med Oral Pathol Oral Radiol Endod* 93(5): 596–604.

Cheung GS and Ho MW. 2001. Microbial flora of root canal-treated teeth associated with asymptomatic periapical radiolucent lesions. *Oral Microbiol Immunol* 16(6): 332–37.

Chong BS, Pitt Ford TR, and Hudson MB. 2003. A prospective clinical study of mineral trioxide aggregate and IRM when used as root-end filling materials in endodontic surgery. *Int Endod J* 36(8): 520–26.

Chugal NM, Clive JM, and Spangberg LS. 2001. A prognostic model for assessment of the outcome of endodontic treatment: Effect of biologic and diagnostic variables. *Oral Surg Oral Med Oral Pathol Oral Radiol Endod* 91(3): 342–52.

Concato J, Shah N, and Horwitz RI. 2000. Randomized, controlled trials, observational studies, and the hierarchy of research designs. *N Engl J Med* 342(25): 1887–92.

Conner DA, Caplan DJ, Teixeira FB, and Trope M. 2007. Clinical outcome of teeth treated endodontically with a nonstandardized protocol and root filled with Resilon. *J Endod* 33 (11): 1290–92.

Danin J, Linder LE, Lundqvist G, Ohlsson L LO, Ramskold, and Stromberg T. 1999. Outcomes of periradicular surgery in cases with apical pathosis and untreated canals. *Oral Surg Oral Med Oral Pathol Oral Radiol Endod* 87(2): 227–32.

Danin J, Stromberg T, Forsgren H LE, Linder, and Ramskold LO. 1996. Clinical management of nonhealing periradicular pathosis. Surgery versus endodontic retreatment. *Oral Surg Oral Med Oral Pathol Oral Radiol Endod* 82(2): 213–17.

de Chevigny C, Dao TT, Basrani BR, Marquuis V, Farzaneh M, Abitbol S, and Friedman S. 2008a. Treatment outcome in endodontics: The Toronto Study—Phases 3 and 4: Orthograde retreatment. *J Endod* 34(2): 131–37.

de Chevigny C, Dao TT, Basrani BR, Marquuis V, Farzaneh M, Abitbol S, and Friedman S. 2008b. Treatment outcome in endodontics: The Toronto Study—Phase 4: Initial treatment. *J Endod* 34(3): 258–63.

Department of Clinical Epidemiology and Biostatistics, McMaster University Health Science Centre. 1981. How to read clinical journals: III. To learn the clinical course and prognosis of disease. *Can Med Assoc J* 124: 869–72.

Dryden JA. 1986. Ten-year follow-up of intentionally replanted mandibular second molar. *J Endod* 12(6): 265–67.

Dugas NN, Lawrence HP, Teplitsky PE, Pharoah MJ, and Friedman S. 2003. Periapical health and treatment quality assessment of root-filled teeth in two Canadian populations. *Int Endod J* 36(3): 181–92.

Dumsha TC and Gutmann JL. 1985. Clinical guidelines for intentional replantation. *Compend Contin Educ Dent* 6(8): 604–8.

Eckerbom M, Andersson JE, and Magnusson T. 1986. Interobserver variation in radiographic examination of endodontic variables. *Endod Dent Traumatol* 2(6): 243–46.

Emmertsen E and Andreasen JO. 1966. Replantation of extracted molars. A radiographic and histological study. *Acta Odontologica Scandinavica* 24(3): 327–46.

Engström B, Hard AF, Segerstad L, Ramstrom G, and Frostell G. 1964. Correlation of positive cultures with the prognosis for root canal treatment. *Odontol Revy* 15: 257–70.

Ercan E., Dalli M, Dülgergil T, and Yaman F. 2007. Effect of intracanal medication with calcium hydroxide and 1% chlorhexidine in endodontic retreatment cases with periapical lesions: An *in vivo* study. *J Formos Med Assoc* 106(3): 217–24.

Eriksen HM, Ørstavik D, and Kerekes K. 1988. Healing of apical periodontitis after endodontic treatment using three different root canal sealers. *Endod Dent Traumatol* 4(3): 114–17.

Farzaneh M, Abitbol S, and Friedman S. 2004b. Treatment outcome in endodontics: The Toronto study. Phases I and II: Orthograde retreatment. *J Endod* 30(9): 627–33.

Farzaneh M, Abitbol S, Lawrence HP, and Friedman S. 2004a. Treatment outcome in endodontics-the Toronto Study. Phase II: Initial treatment. *J Endod* 30(5): 302–9.

Feldman G, Solomon C, Notaro P, and Moskowitz E. 1971. Intentional replantation of a molar tooth. *N Y J Dent* 41(10): 352–53.

Figdor D. 2004. Microbial aetiology of endodontic treatment failure and pathogenic properties of selected species. *Aust Endod J* 30(1): 11–14.

Figures KH and Douglas CW. 1991. Actinomycosis associated with a root-treated tooth: Report of a case. *Int Endod J* 24(6): 326–29.

Flath RK and Hicks ML. 1987. Retrograde instrumentation and obturation with new devices. *J Endod* 13(11): 546–49.

Fletcher RH, Fletcher SW, and Wagner EH. 1996. *Clinical epidemiology: The essentials*. Baltimore: Williams & Wilkins.

Forssell H, Tammisalo T, and Forssell K. 1988. A follow-up study of apicectomized teeth. *Proc Fin Dent Soc* 84(2): 85–93.

Fouad A and Burleson J. 2003. The effect of diabetes mellitus on endodontic treatment outcome: Data from an electronic patient record. *J Am Dent Assoc* 134: 43–51.

Fouad AF, Zerella J, Barry J, and Spangberg LS. 2005. Molecular detection of Enterococcus species in root canals of therapy-resistant endodontic infections. *Oral Surg Oral Med Oral Pathol Oral Radiol Endod* 99(1): 112–18.

Fournier V. 2005. The balance between beneficence and respect for patient autonomy in clinical medical ethics in France. *Camb Q Healthc Ethics* 14(3): 281–86.

Frank AL, Glick DH, Patterson SS, and Weine FS. 1992. Long-term evaluation of surgically placed amalgam fillings. *J Endod* 18(8): 391–98.

Friedman S. 1991. Retrograde approaches in endodontic therapy. *Endod Dent Traumatol* 7(3): 97–107.

Friedman S. 1998. Treatment outcome and prognosis of endodontic therapy. In: Ørstavik D and Pitt Ford TR (eds), *Essential Endodontology: Prevention and treatment of apical periodontitis*. Oxford: Blackwell Science.

Friedman S. 2002a. Considerations and concepts of case selection in the management of post-treatment endodontic disease (treatment failure). *Endod Topics* 1: 54–78.

Friedman S. 2002b. Prognosis of initial endodontic therapy. *Endod Topics* 2: 59–88.

Friedman S. 2005. The prognosis and expected outcome of apical surgery. *Endod Topics* 11: 219–62.

Friedman S. 2008. Endodontic treatment outcome: The potential for healing and retained function. In: Ingle IJ, Bakland LK, and Baumgartner JC (eds), *Ingle's Endodontics*, 2nd edn. Hamilton: BC Decker Inc.

Friedman S, Abitbol S, and Lawrence HP. 2003. Treatment outcome in endodontics: The Toronto Study. Phase 1: Initial treatment. *J Endod* 29(12): 787–93.

Friedman S, Komorowski R, Maillet W, Klimaite R, Nguyen HQ, and Torneck CD. 2000. In vivo resistance of coronally induced bacterial ingress by an experimental glass ionomer cement root canal sealer. *J Endod* 26(1): 1–5.

Friedman S, Löst C, Zarrabian M, and Trope M. 1995. Evaluation of success and failure after endodontic therapy using a glass ionomer cement sealer. *J Endod* 21(7): 384–90.

Friedman S, Lustmann J, and Shaharabany V. 1991a. Treatment results of apical surgery in premolar and molar teeth. *J Endod* 17(1): 30–33.

Friedman S and Mor C. 2004. The success of endodontic therapy—healing and functionality. *Calif Dent Assoc J* 32: 493–503.

Friedman S, Rotstein I, and Mahamid A. 1991b. In vivo efficacy of various retrofills and of $CO_2$ laser in apical surgery. *Endod Dent Traumatol* 7(1): 19–25.

Fristad I, Molven, O, and Halse A. 2004. Nonsurgically retreated root-filled teeth—radiographic findings after 20–27 years. *Int Endod J* 37: 12–18.

Fukushima H, Yamamoto K, Hirohata K, Sagawa H, Leung KP, and Walker CB. 1990. Localization and identification of root canal bacteria in clinically asymptomatic periapical pathosis. *J Endod* 16(11): 534–38.

Gagliani MM, Gorni FGM, and Strohmenger L. 2005. Periapcial resurgery versus periapical surgery: A 5-year longitudinal comparison. *Int Endod J* 38: 320–27.

Gatti JJ, Dobeck JM, Smith C, White RR, Socransky SS, and Skobe Z. 2000. Bacteria of asymptomatic periradicular endodontic lesions identified by DNA-DNA hybridization. *Endod Dent Traumatol* 16(5): 197–204.

Ghoddusi J., Sanaan A, and Shahrami F. 2007. Clinical and radiographic evaluation of root perforation repair using MTA. *New York State Dent J* 73(3): 46–49.

Gilheany PA, Figdor D, and Tyas MJ. 1994. Apical dentin permeability and microleakage associated with root end resection and retrograde filling. *J End* 20(1): 22–26.

Goldberg F, Torres MD, and Bottero C. 1990. Thermoplasticized gutta-percha in endodontic surgical procedures. *Endod Dent Traumatol* 6(3): 109–13.

Goldberg F, Torres MD, Bottero C, and Alvarez AF. 1991. The use of thermoplasticized gutta-percha as a retrograde filling material. *Endodontology* 3(2): 1–6.

Goldman M, Pearson AH, and Darzenta N. 1972. Endodontic success—who's reading the radiograph? *Oral Surg Oral Med Oral Pathol* 33(3): 432–37.

Goldman M, Pearson AH, and Darzenta N. 1974. Reliability of radiographic interpretations. *Oral Surg Oral Med Oral Pathol* 38(2): 287–93.

Gomes BP, Jacinto RC, Pinheiro ET, Sousa EL, Zaia AA, Ferraz CC, and Souza FJ-Filho. 2006. Molecular analysis of *Filifactor alocis*, *Tannerella forsythia*, and *Treponema denticola* associated with primary endodontic infections and failed endodontic treatment. *J Endod* 32(10): 937–40.

Green SB and Byar DP. 1984. Using observational data from registries to compare treatments: The fallacy of omnimetrics. *Stat Med* 3(4): 361–73.

Grimes, W E. 1994. A use of freeze-dried bone in endodontics. *J Endod* 20(7): 355–56.

Grossman, I L. 1982. Intentional replantation of teeth: A clinical evaluation. *J Am Dent Assoc* 104(5): 633–39.

Grung B, Molven O, and Halse A. 1990. Periapical surgery in a Norwegian county hospital: Follow-up findings of 477 teeth. *J End* 16(9): 411–17.

Guy SC and Goerig AC. 1984. Intentional replantation: Technique and rationale. *Quintessence Int* 15(6): 595–603.

Haapasalo M, Ranta K, and Ranta H. 1987. Mixed anaerobic periapical infection with sinus tract. *Endod Dent Traumatol* 3(2): 83–85.

Haapasalo M, Udnaes T, and Endal U. 2003. Persistent, recurrent, and acquired infection of the root canal system post-treatment. *Endodontic Topics* 6: 29–56.

Halse A, Molven O, and Grung B. 1991. Follow-up after periapical surgery: The value of the one-year control. *Endod Dent Traumatol* 7(6): 246–50.

Hammarstrom L, Pierce A, Blomlof L, Feiglin B, and Lindskog S. 1986. Tooth avulsion and replantation—a review. *Endod Dent Traumatol* 2(1): 1–8.

Hancock HH III, Sigurdsson A, Trope M, and Moiseiwitsch J. 2001. Bacteria isolated after unsuccessful endodontic treatment in a North American population. *Oral Surg Oral Med Oral Pathol Oral Radiol Endod* 91(5): 579–86.

Happonen, P R. 1986. Periapical actinomycosis: A follow-up study of 16 surgically treated cases. *Endod Dent Traumatol* 2(5): 205–9.

Happonen RP, Soderling E, Viander M, Linko L-Kettunen, and Pelliniemi LJ. 1985. Immunocytochemical demonstration of *Actinomyces* species and *Arachnia propionica* in periapical infections. *J Oral Pathol* 14(5): 405–13.

Harrison J and Jurosky K. 1992. Wound healing in the tissues of the periodontium following periradicular surgery. 3. The excisional wound. *J Endod* 18: 76–81.

Heling I, Bialla S-S, Turetzky A, Horwitz J, and Sela J. 2001. The outcome of teeth with periapical periodontitis treated with nonsurgical endodontic treatment: A computerized morphometric study. *Quintessence Int* 32(5): 397–400.

Hepworth MJ and Friedman S. 1997. Treatment outcome of surgical and non-surgical management of endodontic failures. *J Can Dent Assoc* 63(5): 364–71.

Hoskinson SE, Ng YL, Hoskinson AE, Moles DR, and Gulabivala K. 2002. A retrospective comparison of outcome of root canal treatment using two different protocols. *Oral Surg Oral Med Oral Pathol Oral Radiol Endod* 93(6): 705–15.

Hsu YY and Kim S. 1997. The resected root surface. The issue of canal isthmuses. *Dent Clin North Am* 41(3): 529–40.

Huumonen S, Lenander-Lumikari, M, Sigurdsson, A, and Ørstavik D. 2003. Healing of apical periodontitis after endodontic treatment: A comparison between a silicone-based and a zinc oxide-eugenol-based sealer. *Int Endod J* 36: 296–301.

Imura N., Pinheiro ET, Gomes BPFA, Zaia AA, Ferraz CCR, and Souza-Filho FJ. 2007. The outcome of endodontic treatment: A retrospective study of 2000 cases performed by a specialist. *J Endod* 33(11): 1278–82.

Ingle JI, Beveridge EE, Glick DH, and Weichman JA. 1994. Modern endodontic therapy. In: Ingle *JI* and Taintor JF (eds), *Endodontics*. Baltimore: Williams & Wilkins, 27–53.

Jensen SS, Nattestad A, Egdo P, Sewerin I, Munksgaard EC, and Schou S. 2002. A prospective, randomized, comparative clinical study of resin composite and glass ionomer cement for retrograde root filling. *Clin Oral Investig* 6(4): 236–43.

Jesslen P, Zetterqvist L, and Heimdahl A. 1995. Long-term results of amalgam versus glass ionomer cement as apical sealant after apicectomy. *Oral Surg Oral Med Oral Pathol Oral Radiol Endod* 79(1): 101–3.

Kahnberg KE. 1988. Surgical extrusion of root-fractured teeth—a follow-up study of two surgical methods. *Endod Dent Traumatol* 4(2): 85–89.

Kalfas S, Figdor D, and Sundqvist G. 2001. A new bacterial species associated with failed endodontic treatment: Identification and description of *Actinomyces radicidentis*. *Oral Surg Oral Med Oral Pathol Oral Radiol Endod* 92(2): 208–14.

Kaufman AY. 1982. Intentional replantation of a maxillary molar. A 4-year follow-up. *Oral Surg Oral Med Oral Pathol* 54(6): 686–88.

Kaufman B, Spangberg L, Barry J, and Fouad AF. 2005. *Enterococcus* spp. in endodontically treated teeth with and without periradicular lesions. *J Endod* 31(12): 851–56.

Keller U. 1990. A new method of tooth replantation and autotransplantation: Aluminum oxide ceramic for extraoral retrograde root filling. *Oral Surg Oral Med Oral Pathol* 70(3): 341–44.

Kim S. 1997. Principles of endodontic microsurgery. *Dent Clin North Am* 41(3): 481–97.

Kim S and Baek S. 2004. The microscope and endodontics. *Dent Clin North Am* 48(1): 11–18.

Kim S and Kratchman S. 2006. Modern endodontic surgery concepts and practice: A review. *J Endod* 32(7): 601–23.

Kirkevang L-L, Vaeth M, Hörsted-Bindslev P, and Wenzel A. 2006. Longitudinal study of periapical and endodontic status in a Danish population. *Int Endod J* 39: 100–107.

Koenig KH, Nguyen NT, and Barkhordar RA. 1988. Intentional replantation: A report of 192 cases. *Gen Dent* 36(4): 327–31.

Komori T, Yokoyama K, Takato T, and Matsumoto K. 1997. Clinical application of the erbium: YAG laser for apicoectomy. *J Endod* 23(12): 748–50.

Kvist T and Reit C. 1999. Results of endodontic retreatment: A randomized clinical study comparing surgical and non-surgical procedures. *J Endod* 25(12): 814–17.

Lasaridis N, Zouloumis L, and Antoniadis K. 1991. Bony lid approach for apicoectomy of mandibular molars. *Aust Dent J* 36(5): 366–68.

Laupacis A, Wells G, Richardson WS, and Tugwell P. 1994. Users' guides to the medical literature. V. How to use an article about prognosis. Evidence-Based Medicine Working Group. *JAMA* 272(3): 234–37.

Leonardo MR, Rossi MA, Silva LA, Ito IY, and Bonifacio KC. 2002. EM evaluation of bacterial biofilm and microorganisms on the apical external root surface of human teeth. *J Endod* 28(12): 815–18.

Lin LM, Pascon EA, Skribner J, Gangler P, and Langeland K. 1991. Clinical, radiographic, and histologic study of endodontic treatment failures. *Oral Surg Oral Med Oral Pathol* 71(5): 603–11.

Lin LM, Skribner JE, and Gaengler P. 1992. Factors associated with endodontic treatment failures. *J Endod* 18(12): 625–27.

Lindeberg RW, Girardi AF, and Troxell JB. 1986. Intentional replantation: Management in contraindicated situations. *Compend Contin Educ Dent* 7(4): 248.

Lindeboom JA, Frenken JW, Kroon FH, and van HP den Akker. 2005. A comparative prospective randomized clinical study of MTA and IRM as root-end filling materials in single-rooted teeth in endodontic surgery. *Oral Surg Oral Med Oral Pathol Oral Radiol Endod* 100(4): 495–500.

Lu, P D. 1986. Intentional replantation of periodontally involved and endodontically mistreated tooth. *Oral Surg Oral Med Oral Pathol* 61(5): 508–13.

Lubin H. 1982. Intentional reimplantation: Report of case. *J Am Den Assoc* 104(6): 858–59.

Lustmann J, Friedman S, and Shaharabany V. 1991. Relation of pre- and intraoperative factors to prognosis of posterior apical surgery. *J Endod* 17(5): 239–41.

Maddalone M and Gagliani M. 2003. Periapical endodontic surgery: A 3-year follow-up study. *Int Endod J* 36(3): 193–98.

Madison S. 1986. Intentional replantation. *Oral Sur Oral Med Oral Pathol* 62(6): 707–9.

Main C, Mirzayan N, Shabahang S, and Torabinejad M. 2004. Repair of root perforations using mineral trioxide aggregate: A long-term study. *J Endod* 30(2): 80–83.

Marending M, Peters OA, and Zehnder M. 2005. Factors affecting the outcome of orthograde root canal therapy in a general dentistry hospital practice. *Oral Surg Oral Med Oral Pathol Oral Radiol Endod* 99: 119–24.

Marquis V, Dao TT, Farzaneh M, Abitbol S, and Friedman S. 2006. Treatment outcome in endodontics: The Toronto Study. Phase III: Initial treatment. *J Endod* 32: 299–306.

Matsumoto T, Nagai T, Ida K, Ito M, Kawai Y, Horiba N, Sato R, and Nakamura H. 1987. Factors affecting successful prognosis of root canal treatment. *J Endod* 13(5): 239–42.

Melcer J. 1986. Latest treatment in dentistry by means of the $CO_2$ laser beam. *Lasers Surg Med* 6(4): 396–98.

Messkoub M. 1991. Intentional replantation: A successful alternative for hopeless teeth. *Oral Surg Oral Med Oral Pathol* 71(6): 743–47.

Mines P, Loushine RJ, West LA, Liewehr FR, and Zadinsky JR. 1999. Use of the microscope in endodontics: A report based on a questionnaire. *J Endod* 25(11): 755–58.

Miserendino LJ. 1988. The laser apicoectomy: Endodontic application of the $CO_2$ laser for periapical surgery. *Oral Surg Oral Med Oral Pathol* 66(5): 615–19.

Molander A, Reit C, and Dahlèn G. 1990. Microbiological evaluation of clindamycin as a root canal dressing in teeth with apical periodontitis. *Int Endod J* 23(2): 113–18.

Molander A, Reit C, Dahlèn G, and Kvist T. 1998. Microbiological status of root-filled teeth with apical periodontitis. *Int Endod J* 31(1): 1–7.

Molander A., Warfvinge J, Reit C, and Kvist T. 2007. Clinical and radiographic evaluation of one- and two-visit endodontic treatment of asymptomatic necrotic teeth with apical periodontitis: A randomized clinical trial. *J Endod* 33 (10): 1145–49.

Molven O and Halse A. 1988. Success rates for gutta-percha and Kloroperka N-0 root fillings made by undergraduate students: Radiographic findings after 10–17 years. *Int Endod J* 21(4): 243–50.

Molven O, Halse A, and Fristad I. 2002a. Long-term reliability and observer comparisons in the radiographic diagnosis of periapical disease. *Int Endod J* 35(2): 142–47.

Molven O, Halse A, Fristad I, and MacDonald-Jankowksi D. 2002b. Periapical changes following root-canal treatment observed 20–27 years postoperatively. *Int Endod J* 35: 784–90.

Molven O, Halse A, and Grung B. 1991. Surgical management of endodontic failures: Indications and treatment results. *Int Dent J* 41(1): 33–42.

Molven O, Halse A, and Grung B. 1996. Incomplete healing (scar tissue) after periapical surgery—radiographic findings 8 to 12 years after treatment. *J End* 22(5): 264–68.

Murashima Y, Yoshikawa G, Wadachi R, Sawada N, and Suda H. 2002. Calcium sulfate as a bone substitute for various osseous defects in conjunction with apicectomy. *Int Endod J* 35: 768–74.

Murphy WK, Kaugars GE, Collett WK, and Dodds RN. 1991. Healing of periapical radiolucencies after nonsurgical endodontic therapy. *Oral Surg Oral Med Oral Pathol* 71(5): 620–24.

Nair P. 2003a. Non-microbial etiology: Foreign body reaction maintaining post-treatment apical periodontitis. *Endod Topics* 6: 114–34.

Nair P. 2003b. Non-microbial etiology: Periapical cysts sustain post-treatment apical periododontitis. *Endod Topics* 6: 96–113.

Nair P. 2004. Pathogenesis of apical periodontitis and the causes of endodontic failures. *Crit Rev Oral Biol Med* 15: 348–81.

Nair N P. 1987. Light and electron microscopic studies of root canal flora and periapical lesions. *J Endod* 13: 29–39.

Nair RN and Shroeder JH. 1984. Periapical actinomycosis. *J Endod* 10: 567–70.

Nair RN, Sjögren U, Figdor D, and Sundqvist G. 1999. Persistent periapical radiolucencies of root-filled human teeth, failed endodontic treatments, and periapical scars. *Oral Surg Oral Med Oral Pathol Oral Radiol Endod* 87(5): 617–27.

Nair RN, Sjögren U, Krey G, Kahnberg KE, and Sundqvist G. 1990a. Intraradicular bacteria and fungi in root-filled, asymptomatic human teeth with therapy-resistant periapical lesions: A long-term light and electron microscopic follow-up study. *J Endod* 16(12): 580–88.

Nair RN, Sjögren U, Krey G, and Sundqvist G. 1990b. Therapy-resistant foreign body giant cell granuloma at the periapex of a root-filled human tooth. *J Endod* 16(12): 589–95.

Nair RN, Sjögren U, Schumacher E, and Sundqvist G. 1993. Radicular cyst affecting a root-filled human tooth: A long-term post-treatment follow-up. *Int Endod J* 26(4): 225–33.

Noiri Y, Ehara A, Kawahara T, Takemura N, and Ebisu S. 2002. Participation of bacterial biofilms in refractory and chronic periapical periodontitis. *J Endod* 28(10): 679–83.

Nosonowitz DM and Stanley HR. 1984. Intentional replantation to prevent predictable endodontic failures. *Oral Surg Oral Med Oral Pathol* 57(4): 423–32.

Nygaard-Östby B. 1971. Introduction to endodontics. Oslo, Norway: Universitetsforlaget.

O'Grady JF and Reade PC. 1988. Periapical actinomycosis involving *Actinomyces israelii*. *J Endod* 14(3): 147–49.

Oikarinen K. 1993. Dental tissues involved in exarticulation, root resorption and factors influencing prognosis in relation to replanted teeth. A review. *Proc Fin Dent Soc* 89(1–2): 29–44.

Ørstavik D. 1996. Time-course and risk analyses of the development and healing of chronic apical periodontitis in man. *Int Endod J* 29(3): 150–55.

Ørstavik D and Hörsted-Bindslev P. 1993. A comparison of endodontic treatment results at two dental schools. *Int Endod J* 26(6): 348–54.

Ørstavik D, Kerekes K, and Eriksen HM. 1986. The periapical index: A scoring system for radiographic assessment of apical periodontitis. *Endod Dent Traumatol* 2(1): 20–34.

Ørstavik D, Kerekes K, and Eriksen HM. 1987. Clinical performance of three endodontic sealers. *Endod Dent Traumatol* 3(4): 178–86.

Ørstavik D, Kerekes K, and Molven O. 1991. Effects of extensive apical reaming and calcium hydroxide dressing on bacterial infection during treatment of apical periodontitis: A pilot study. *Int Endod J* 24(1): 1–7.

Ørstavik D and Pitt Ford TR. 1998. Apical periodontitis: Microbial infection and host responses. In: Ørstavik D and Pitt Ford TR (eds), *Essential endodontology: Prevention of treatment of apical periodontitis*. Oxford, Blackwell Science.

Ørstavik D and Pitt Ford TR, Eds. 1998. *Essential Endodontology: Prevention and Treatment of Apical Periodontitis*, 2nd edn. Oxford: Blackwell Science.

Ørstavik D, Qvist, V, and Stoltze K. 2004. A multivariate analysis of the outcome of endodontic treatment. *Eur J Oral Sci* 112: 224–30.

Oxman, DA. 1994. Checklists for review articles. *BMJ* 309(6955): 648–51.

Paghdiwala, FA. 1993. Root resection of endodontically treated teeth by erbium: YAG laser radiation. *J Endod* 19(2): 91–94.

Pantschev A, Carlsson AP, and Andersson L. 1994. Retrograde root filling with EBA cement or amalgam. A comparative clinical study. *Oral Surg Oral Med Oral Pathol* 78(1): 101–4.

Paquette L, Legner, M, Fillery, ED, and Friedman S. 2007. Antibacterial efficacy of chlorhexidine gluconate intracanal medication in vivo. *J Endod* 33 (7): 788–95.

Peak JD, Hayes SJ, Bryant ST, and Dummer PM. 2001. The outcome of root canal treatment. A retrospective study within the armed forces (Royal Air Force). *Br Dent J* 190(3): 140–44.

Peciuliene V, Balciuniene I, Eriksen HM, and Haapasalo M. 2000. Isolation of *Enterococcus faecalis* in previously root-filled canals in a Lithuanian population. *J Endod* 26(10): 593–95.

Peciuliene V, Reynaud AH, Balciuniene I, and Haapasalo M. 2001. Isolation of yeasts and enteric bacteria in root-filled teeth with chronic apical periodontitis. *Int Endod J* 34(6): 429–34.

Pecora G and Andreana S. 1993. Use of dental operating microscope in endodontic surgery. *Oral Surg Oral Med Oral Pathol* 75(6): 751–58.

Pecora G, Andreana S, Margarone JE III, Covani U, and Sottosanti JS. 1997a. Bone regeneration with a calcium sulfate barrier. *Oral Surg Oral Med Oral Pathol Oral Radiol Endod* 84(4): 424–29.

Pecora G, Baek SH, Rethnam S, and Kim S. 1997b. Barrier membrane techniques in endodontic microsurgery. *Dent Clin North Am* 41(3): 585–602.

Pecora G, De D Leonardis, Ibrahim N, Bovi M, and Cornelini R. 2001. The use of calcium sulphate in the surgical treatment of a "through and through" periradicular lesion. *Int Endod J* 34(3): 189–97.

Pecora G, Kim S, Celletti R, and Davarpanah M. 1995. The guided tissue regeneration principle in endodontic surgery: One-year postoperative results of large periapical lesions. *Int Endod J* 28(1): 41–46.

Pellegrino, DE. 1994. Patient autonomy and the physician's ethics. *Ann R Coll Physicians Surg Can* 27(3): 171–73.

Penesis VA, Fitzgerlad PI, Fayad MI, Wenckus CS, Begole EA, and Johnson BR. 2008. Outcome of one-visit and two-visit endodontic treatment of necrotic teeth with apical periodontitis: A randomized controlled trial with one-year evaluation. *J Endod* 34(3): 252–57.

Penick, CE. 1961. Periapical repair by dense fibrous connective tissue following conservative endodontic therapy. *Oral Surg Oral Med Oral Pathol* 76: 239–42.

Peters LB and Wesselink PR. 2002. Periapical healing of endodontically treated teeth in one and two visits obturated in the presence or absence of detectable microorganisms. *Int Endod J* 76 (8): 660–67.

Peters O, Barbakow F, and Peters CI. 2004. An analysis of endodontic treatment with three nickel-titanium rotary root canal preparation techniques. *Int Endod J* 76: 849–59.

Peterson J and Gutmann JL. 2001. The outcome of endodontic resurgery: A systematic review. *Int Endod J* 34(3): 169–75.

Pettiette MT, Delano EO, and Trope M. 2001. Evaluation of success rate of endodontic treatment performed by students with stainless-steel K-files and nickel-titanium hand files. *J Endod* 27(2): 124–27.

Piatowska D, Pawlicka H, Taskiewicz J, Boltacz E-R, and Brauman S-F. 1997. Evaluation of endodontic retreatment. *Czas Stomat* L: 451–58.

Pinheiro ET, Gomes BP, Ferraz CC, Sousa EL, Teixeira FB, and Souza FJ-F. 2003a. Microorganisms from canals of root-filled teeth with periapical lesions. *Int Endod J* 36(1): 1–11.

Pinheiro ET, Gomes BP, Ferraz CC, Teixeira FB, Zaia AA, and Souza FJ-F. 2003b. Evaluation of root canal microorganisms isolated from teeth with endodontic failure and their antimicrobial susceptibility. *Oral Microbiol Immunol* 18(2): 100–103.

Pitt Ford TR. 1982. The effects on the periapical tissues of bacterial contamination of the filled root canal. *Int Endod J* 15(1): 16–22.

Pitt Ford TR, Andreasen JO, Dorn SO, and Kariyawasam SP. 1994. Effect of IRM root end fillings on healing after replantation. *J Endod* 20(8): 381–85.

Pitt Ford TR, Andreasen JO, Dorn SO, and Kariyawasam SP. 1995. Effect of super-EBA as a root end filling on healing after replantation. *J Endod* 21(1): 13–15.

Rahbaran S, Gilthorpe MS, Harrison SD, and Gulabivala K. 2001. Comparison of clinical outcome of periapical surgery in endodontic and oral surgery units of a teaching dental hospital: A retrospective study. *Oral Surg Oral Med Oral Pathol Oral Radiol Endod* 91(6): 700–709.

Rankow HJ and Krasner PR. 1996. Endodontic applications of guided tissue regeneration in endodontic surgery. *Oral Health* 86(12): 33–35, 37–40, 43.

Rapp EL, Brown CE, Jr, and Newton CW. 1991. An analysis of success and failure of apicoectomies. *J Endod* 17(10): 508–12.

Reit C. 1987a. Decision strategies in endodontics: On the design of a recall program. *Endod Dent Traumatol* 3(5): 233–39.

Reit C. 1987b. The influence of observer calibration on radiographic periapical diagnosis. *Int Endod J* 20(2): 75–81.

Reit C and Hirsch J. 1986. Surgical endodontic retreatment. *Int Endod J* 19(3): 107–12.

Reit C and Hollender L. 1983. Radiographic evaluation of endodontic therapy and the influence of observer variation. *Scand J Dent Res* 91(3): 205–12.

Rolph HJ, Lennon A, Riggio MP, Saunders WP, MacKenzie D, Coldero L, and Bagg J. 2001. Molecular identification of microorganisms from endodontic infections. *J Clin Microbiol* 39(9): 3282–89.

Rosenberg ES, Rossman LE, and Sandler AB. 1980. Intentional replantation: A case report. *J Endod* 6(6): 610–13.

Ross WJ. 1985. Intentional replantation: An alternative. *Compend Contin Educ Dent* 6(10): 734.

Rubinstein R. 1997. The anatomy of the surgical operating microscope and operating positions. *Dent Clin North Am* 41(3): 391–413.

Rubinstein RA and Kim S. 1999. Short-term observation of the results of endodontic surgery with the use of a surgical operation microscope and Super-EBA as root-end filling material. *J Endod* 25: 43–48.

Rubinstein RA and Kim S. 2002. Long-term follow-up of cases considered healed one year after apical microsurgery. *J Endod* 28(5): 378–83.

Rud J, Andreasen JO, and Jensen JE. 1972. Radiographic criteria for the assessment of healing after endodontic surgery. *Int J Oral Surg* 1(4): 195–214.

Rud J, Munksgaard EC, Andreasen JO, and Rud V. 1991b. Retrograde root filling with composite and a dentin-bonding agent. 2. *Endod Dent Traumatol* 7(3): 126–31.

Rud J, Munksgaard EC, Andreasen JO, Rud V, and Asmussen E. 1991a. Retrograde root filling with composite and a dentin-bonding agent. 1. *Endod Dent Traumatol* 7(3): 118–25.

Rud J and Rud V. 1998. Surgical endodontics of upper molars: Relation to the maxillary sinus and operation in acute state of infection. *J Endod* 24(4): 260–61.

Rud J, Rud V, and Munksgaard EC. 1996a. Long-term evaluation of retrograde root filling with dentin-bonded resin composite. *J Endod* 22(2): 90–93.

Rud J, Rud V, and Munksgaard EC. 1996b. Retrograde root filling with dentin-bonded modified resin composite. *J Endod* 22(9): 477–80.

Rud J, Rud V, and Munksgaard EC. 1997. Effect of root canal contents on healing of teeth with dentin-bonded resin composite retrograde seal. *J Endod* 23(8): 535–41.

Rud J, Rud V, and Munksgaard EC. 2001. Periapical healing of mandibular molars after root-end sealing with dentine-bonded composite. *Int Endod J* 34(4): 285–92.

Saad AY, and Abdellatief E-SM 1991. Healing assessment of osseous defects of periapical lesions associated with failed endodontically treated teeth with use of freeze-dried bone allograft. *Oral Surg Oral Med Oral Pathol* 71: 612–17.

Sackett DL, Haynes RB, Guyatt GH, and Tugwell P. 1991. *Clinical Epidemiology: A Basic Science of Clinical Medicine.* Boston: Little, Brown.

Sakellariou, LP. 1996. Periapical actinomycosis: Report of a case and review of the literature. *Endod Dent Traumatol* 12(3): 151–54.

Sathorn C, Parashos P, and Messer HH. 2005. Effectiveness of single-versus multiple-visit endodontic treatment of teeth with apical periodontitis: A systematic review and meta-analysis. *Int Endod J* 38(6): 347–55.

Schattner A and Tal M. 2002. Truth telling and patient autonomy: The patient's point of view. *Am J Med* 113(1): 66–69.

Schwartz-Arad D, Yaorm N, Lustig JP, and Kaffe I. 2003. A retrospective radiographic study of root-end surgery with amalgam and intermediate restorative material. *Oral Surg Oral Med Oral Pathol Oral Radiol Endod* 96: 472–77.

Selden HS. 1999. Periradicular scars: A sometime diagnostic conundrum. *J Endod* 25(12): 829–30.

Serota KS and Krakow AA. 1983. Retrograde instrumentation and obturation of the root canal space. *J Endod* 9(10): 448–51.

Shah N. 1988. Nonsurgical management of periapical lesions: A prospective study. *Oral Surg Oral Med Oral Pathol* 66(3): 365–71.

Shuping GB, Ørstavik D, Sigurdsson A, and Trope M. 2000. Reduction of intracanal bacteria using nickel-titanium rotary instrumentation and various medications. *J Endod* 26(12): 751–55.

Sikri K, Dua SS, and Kapur R. 1986. Use of tricalcium phosphate ceramic in apicoectomized teeth and in their periapical areas—clinical and radiological evaluation. *J Ind Dent Assoc* 58: 441–47.

Siqueira J. 2003. Periapical actinomycosis and infection with *Propionibacterium Propionicum*. *Endod Topics* 6: 78–95.

Siqueira JF and Lopes HP. 2001. Bacteria on the apical root surfaces of untreated teeth with periradicular lesions: A scanning electron microscopy study. *Int Endod J* 34(3): 216–20.

Siqueira JF. 2001. Aetiology of root canal treatment failure: Why well-treated teeth can fail. *Int Endod J* 34(1): 1–10.

Siren EK, Haapasalo MP, Ranta K, Salmi P, and Kerosuo EN. 1997. Microbiological findings and clinical treatment procedures in endodontic cases selected for microbiological investigation. *Int Endod J* 30(2): 91–95.

Sjögren U, Figdor D, Persson S, and Sundqvist G. 1997. Influence of infection at the time of root filling on the outcome of endodontic treatment of teeth with apical periodontitis. *Int Endod J* 30(5): 297–306.

Sjögren U, Hagglund B, Sundqvist G, and Wing K. 1990. Factors affecting the long-term results of endodontic treatment. *J Endod* 16(10): 498–504.

Sjögren U, Happonen RP, Kahnberg KE, and Sundqvist G. 1988. Survival of Arachnia propionica in periapical tissue. *Int Endod J* 21(4): 277–82.

Skoglund A and Persson G. 1985. A follow-up study of apicoectomized teeth with total loss of the buccal bone plate. *Oral Surg Oral Med Oral Pathol* 59(1): 78–81.

Smith CS, Setchell DJ, and Harty FJ. 1993. Factors influencing the success of conventional root canal therapy—a five-year retrospective study. *Int Endod J* 26(6): 321–33.

Solomon CS and Abelson J. 1981. Intentional replantation: Report of case. *J Endod* 7(7): 317–19.

Spili P, Parashos P, and Messer HH. 2005. The impact of instrument fracture on outcome of endodontic treatment. *J Endod* 31(12): 845–50.

Stabholz A, Khayat A, Ravanshad SH, McCarthy DW, Neev J, and Torabinejad M. 1992a. Effects of Nd:YAG laser on apical seal of teeth after apicoectomy and retrofill. *J Endod* 18(8): 371–75.

Stabholz A, Khayat A, Weeks DA, Neev J, and Torabinejad M. 1992b. Scanning electron microscopic study of the apical dentine surfaces lased with ND:YAG laser following apicectomy and retrofill. *Int Endod J* 25(6): 288–91.

Storms JL. 1978. Root canal therapy via the apical foramen—radical or conservative? *Oral Health* 68(10): 60–65.

Strindberg LZ. 1956. The dependence of the results of pulp therapy on certain factors. An analytic study based on radiographic and clinical follow-up examination. *Acta Odontol Scand* 14(Suppl.): 2–101.

Stroner WF and Laskin DM. 1981. Replantation of a mandibular molar: Report of case. *J Am Dent Assoc* 103(5): 730–31.

Suchina JA, Levine D, Flaitz CM, Nichols CM, and Hicks MJ. 2006. Restrosepctive clinical and radiologic evaluation of nonsurgical endodontic treatment in human immunodeficiency virus (HIV) infection. *J Contemp Dent Practice* 7(1): 1–11.

Sumi Y, Hattori H, Hayashi K, and Ueda M. 1996. Ultrasonic root-end preparation: Clinical and radiographic evaluation of results. *J Oral Maxillofac. Surg* 54(5): 590–93.

Sunde PT, Olsen I, Debelian GJ, and Tronstad L. 2002. Microbiota of periapical lesions refractory to endodontic therapy. *J Endod* 28(4): 304–10.

Sunde PT, Olsen I, Gobel UB, Theegarten D, Winter S, Debelian GJ, Tronstad L, and Moter A. 2003. Fluorescence in situ hybridization (FISH) for direct visualization of bacteria in periapical lesions of asymptomatic root-filled teeth. *Microbiology* 149(Pt 5): 1095–1102.

Sunde PT, Olsen I, Lind PO, and Tronstad L. 2000a. Extraradicular infection: A methodological study. *Endod Dent Traumatol* 16(2): 84–90.

Sunde PT, Tronstad L, Eribe ER, Lind PO, and Olsen I. 2000b. Assessment of periradicular microbiota by DNA-DNA hybridization. *Endod Dent Traumatol* 16(5): 191–96.

Sundqvist G, Figdor D, Persson S, and Sjögren U. 1998. Microbiologic analysis of teeth with failed endodontic treatment and the outcome of conservative re-treatment. *Oral Surg Oral Med Oral Pathol Oral Radiol Endod* 85(1): 86–93.

Sundqvist G and Figdor D. 2003. Life as an endodontic pathogen. Ecological differences between the untreated and the root-filled root canals. *Endodontic Topics* 6: 3–28.

Sundqvist G and Reuterving CO. 1980. Isolation of *Actinomyces israelii* from periapical lesion. *J Endod* 6(6): 602–6.

Sutherland SE. 2001. Evidence-based dentistry: Part VI. Critical appraisal of the dental literature: Papers about diagnosis, etiology and prognosis. *J Can Dent Assoc* 67(10): 582–85.

Takeda A. 1989. An experimental study upon the application of Nd:YAG laser to surgical endodontics. *Jpn J Conser Dent* 32(2): 541–553.

Taschieri S, Del Fabbro M, Testori T, Francetti L, and Weinstein R. 2006. Endodontic surgery using 2 different magnification devices: Preliminary results of a randomized controlled study. *J Oral Maxillofac Surg* 64(2): 235–42.

Testori T, Capelli M, Milani S, and Weinstein RL. 1999. Success and failure in periradicular surgery: A longitudinal retrospective analysis. *Oral Surg Oral Med Oral Pathol Oral Radiol Endod* 87(4): 493–98.

Tidmarsh BG and Arrowsmith MG. 1989. Dentinal tubules at the root ends of apicected teeth: A scanning electron microscopic study. *Int Endod J* 22(4): 184–89.

Tobon SI, Arismendi JA, Marin ML, Mesa AL, and Valencia JA. 2002. Comparison between a conventional technique and two bone regeneration techniques in periradicular surgery. *Int Endod J* 35(7): 635–41.

Torabinejad M, Hong CU, Lee SJ, Monsef M, and Pitt TR Ford. 1995. Investigation of mineral trioxide aggregate for root-end filling in dogs. *J Endod* 21(12): 603–8.

Tronstad L. 1988. Root resorption—etiology, terminology and clinical manifestations. *Endod Dent Traumatol* 4(6): 241–52.

Tronstad L, Barnett F, and Cervone F. 1990a. Periapical bacterial plaque in teeth refractory to endodontic treatment. *Endod Dent Traumatol* 6(2): 73–77.

Tronstad L, Barnett F, Riso K, and Slots J. 1987. Extraradicular endodontic infections. *Endod Dent Traumatol* 3(2): 86–90.

Tronstad L, Kreshtool D, and Barnett F. 1990b. Microbiological monitoring and results of treatment of extraradicular endodontic infection. *Endod Dent Traumatol* 6(3): 129–36.

Tronstad L and Sunde PT. 2003. The evolving new understanding of endodontic infections. *Endod Topics* 6: 57–77.

Trope M, Delano EO, and Ørstavik D. 1999. Endodontic treatment of teeth with apical periodontitis: Single vs. multivisit treatment. *J Endod* 25(5): 345–50.

Trope M and Friedman S. 1992. Periodontal healing of replanted dog teeth stored in Viaspan, milk and Hank's balanced salt solution. *Endod Dent Traumatol* 8(5): 183–88.

Trope M, Löst C, Schmitz HJ, and Friedman S. 1996. Healing of apical periodontitis in dogs after apicoectomy and retrofilling with various filling materials. *Oral Surg Oral Med Oral Pathol Oral Radiol Endod* 81(2): 221–28.

Tsesis I., Rosen E, and Schwartz-Arad D. 2006. Retrospective evaluation of surgical endodontic treatment: Traditional versus modern technique. *J Endod* 32(5): 412–16.

Velvart P and Peters CI. 2005. Soft tissue management in endodontic surgery. *J Endod* 31(1): 4–16.

Vertucci FJ and Beatty RG. 1986. Apical leakage associated with retrofilling techniques: A dye study. *J Endod* 12(8): 331–36.

von Arx T. 2005. Frequency and type of canal isthmuses in first molars detected by endoscopic inspection during periradicular surgery. *Int Endod J* 38(3): 160–68.

von Arx T, Gerber C, and Hardt N. 2001. Periradicular surgery of molars: A prospective clinical study with a one-year follow-up. *Int Endod J* 34(7): 520–25.

von Arx T, Jensen SS, and Hanni S. 2007. Clinical and radiographic assessment of various predictors for healing outcome 1 year after periapical surgery. *J Endod* 33(2): 123–28.

von Arx T and Kurt B. 1999. Root-end cavity preparation after apicoectomy using a new type of sonic and diamond-surfaced retrotip: A 1-year follow-up study. *J Oral Maxillofac Surg* 57(6): 656–61.

von Arx T, Montagne D, Zwinggi C, and Lussi A. 2003. Diagnostic accuracy of endoscopy in periradicular surgery—a comparison with scanning electron microscopy. *Int Endod J* 36(10): 691–99.

Waltimo T, Boiesen, J, Eriksen HM, and Ørstavik D. 2001. Clinical performance of 3 endodontic sealers. *Oral Surg Oral Med Oral Pathol Oral Radiol Endod* 92: 89–92.

Waltimo T, Trope, M, Haapasalo, M, and Ørstavik D. 2005. Clinical efficacy of treatment procedures in Endodontic infection control and one year follow-up of periapical healing. *J Endod* 31: 863–66.

Wang N, Knight K, Dao TT, and Friedman S. 2004. Treatment outcome in endodontics—the Toronto Study. Phases I and II: Apical surgery. *J Endod* 30(11): 751–61.

Weiger R, Rosendahl R, and Löst C. 2000. Influence of calcium hydroxide intracanal dressings on the prognosis of teeth with endodontically induced periapical lesions. *Int Endod J* 33(3): 219–26.

Wertz DC. 1998. Patient and professional views on autonomy: A survey in the United States and Canada. *Health Law Rev* 7(3): 9–10.

Wesson CM and Gale TM. 2003. Molar apicectomy with amalgam root-end filling: Results of a prospective study in two district general hospitals. *Br Dent J* 195(12): 707–14; discussion 698.

Witherspoon DE and Gutmann JL. 2000. Analysis of the healing response to gutta-percha and Diaket when used as root-end filling materials in periradicular surgery. *Int Endod J* 33(1): 37–45.

Wong SW, Rosenberg PA, Boylan RJ, and Schulman A. 1994. A comparison of the apical seal achieved using retrograde amalgam fillings and the Nd:YAG laser. *J Endod* 20: 595–97.

Yared GM and Dagher FE. 1994. Influence of apical enlargement on bacterial infection during treatment of apical periodontitis. *J Endod* 20(11): 535–37.

Yazdi PM, Schou S, Jensen SS, Stoltze K, Kenrad B, and Sewerin I. 2007. Dentine-bonded resin composite (Retroplast) for root-end filling: A prospective clinical and radiographic study with a mean follow-up period of 8 years. *Int Endod J* 40(7): 1–11.

Yoshikawa G, Murashima Y, Wadachi R, *Sawada N, and Suda H*. 2002. Guided bone regeneration (GBR) using membranes and calcium sulphate after apicectomy: A comparative histomorphometrical study. *Int End J* 35: 255–64.

Yusuf H. 1982. The significance of the presence of foreign material periapically as a cause of failure of root treatment. *Oral Surg Oral Med Oral Pathol* 54(5): 566–74.

Zakariasen KL, Scott DA, and Jensen JR. 1984. Endodontic recall radiographs: How reliable is our interpretation of endodontic success or failure and what factors affect our reliability? *Oral Surg Oral Med Oral Pathol* 57(3): 343–47.

Zerella JA, Fouad AF, and Spangberg LS. 2005. Effectiveness of a calcium hydroxide and chlorhexidine digluconate mixture as disinfectant during retreatment of failed endodontic cases. *Oral Surg Oral Med Oral Pathol Oral Radiol Endod* 100(6): 756–61.

Zuolo ML, Ferreira MO, and Gutmann JL. 2000. Prognosis in periradicular surgery: A clinical prospective study. *Int End J* 33(2): 91–98.

# Chapter 16
# Endodontic Infections and Systemic Disease

*Ashraf F. Fouad*

## 16.1 Introduction

In the past decade, there has been an increasing awareness and recognition of the remarkable interaction between oral and systemic disease. Various systemic diseases have been found to not only have oral manifestations, but also influence the presentation and healing of oral diseases. Likewise, a number of oral diseases, particularly periodontal disease, have been linked to the pathogenesis of some systemic diseases. Some studies have even suggested that periodontal therapy may contribute to the overall improvement of the systemic condition of the patient.

Endodontic pathosis is the result of the interplay of infectious agents and host response in the dental pulp and periapical tissues. Recent evidence suggests that certain systemic conditions may play an important role in modulating this interaction. Likewise, the root canal system may act as a pathway for, and/or reservoir of, certain unique microbial communities to contribute to, or cause systemic diseases. This chapter outlines the available information on endodontic–systemic interrelationship and provides some hypotheses for future exploration.

In the first section of this chapter, a brief presentation of several nonendodontic painful conditions or jawbone radiolucencies will be presented. The clinician is frequently confronted with these conditions, and has to make a decision on whether or not they involve pathosis of the pulp and periapical tissues. In this instance, the endodontic and the systemic or nonendodontic entities are not linked, but may be similar in their presentation such that they present a diagnostic dilemma. Clearly, the recognition of these entities is essential to allow the clinician to perform adequate diagnosis and to manage the patient effectively.

## 16.2 Systemic pain syndromes than mimic endodontic pathosis

In this section, the most common orofacial pain entities are reviewed. While some of these may have an inflammatory or infectious origin, the tooth pulp is usually not involved in the pathogenesis of these conditions.

### 16.2.1 Myofascial pain

Myofascial pain is a type of chronic orofacial pain that is associated with inflammation of muscles in the head and neck area, primarily the muscles of mastication. The accumulation of inflammatory mediators such as cytokines, eicosanoids, and neuropeptides in certain areas of these muscles create a state of chronic pain that may mimic endodontic pain. Careful palpation of the muscles in the region and examination of alteration of function reveals the source of pain.

### 16.2.2 Sinus mucosal pain

The maxillary sinus lies in close proximity to maxillary posterior teeth. Sinus mucosa is frequently inflamed as a result of infections, allergies, or other forms of irritation. The inflamed sinus is frequently filled with serous exudates. This results in feeling of fullness, pain with sudden head movement, and headaches particularly early in the morning. Imaging of the sinuses by CT-scans, a panoramic radiograph or a Waters' projection reveals the unilateral or bilateral opacity of the maxillary sinus.

### 16.2.3 Neurovascular pain

This type of pain involves hemodynamic changes in vasculature within hard unyielding structures, such as the skull, and causes a variety of headaches that are frequently accompanied by referred pain to the dental tissues. Migraines are a form of neurovascular pain, and so are cluster headache and tension headache.

### 16.2.4 Neuropathic pain

Neuropathic pain involves a pathological change within the neural elements supplying a particular tissue, particularly sensory neurons. Neuralgia, neuroma, neuropathy, and neuritis are forms of neuropathic pain.

### 16.2.5 Angina pectoris

Anginal pain is a form of chest pain that arises from ischemia to the cardiac muscles. The pain is ill-defined, associated with exercise, and is frequently attributed to a gastric reflux or indigestion by the patient. It is often referred to the left shoulder, arm, neck, and face (Kreiner and Okeson 1999). Anginal pain may also be manifested as pain in the left mandible (Batchelder et al. 1987) and this may be the first presentation of the disease. In a recent multicenter trial of 186 patients with cardiac ischemia, it was found that 11 of them had pain exclusively in the craniofacial area and 60 had craniofacial pain concomitant with other types of pain. Craniofacial pain was most commonly in the throat, left and right mandible, left temporomandibular region, and teeth (Kreiner et al. 2007).

There are several other pain syndromes, whose signs and symptoms may mimic endodontic pain, although to a lesser degree that those described before. These syndromes include fibromyalgia, primary and metastatic malignancies, arteritis syndromes, and viral infections such as herpes zoster (see below).

## 16.3 Jawbone radiolucencies than mimic endodontic pathosis

There are many other (usually nonpainful) jawbone radiolucencies that are similar to endodontic infections in their presentation (Table 16.1). The infections listed in Table 16.1 are likely to be sequelae of endodontic infections such as actinomycosis, while the remaining lesions are not likely to be of endodontic origin. Traditionally, the presence of an intact lamina dura around the root in a periapical radiograph has been a major sign that the radiolucency is nonendodontic in origin. However, the importance of the lamina dura in defining an endodontic lesion has recently been questioned (Ricucci et al. 2006a). Common examples of nonendodontic lesions include the periapical cemental dysplasia (Wilcox and Walton 1989), odontogenic keratocyst (Garlock et al. 1998), central giant cell granuloma (Dahlkemper et al. 2000), and metastatic carcinoma (Nevins et al. 1988). While pulp testing is the simplest and most effective tool to determine if a particular radiolucency is of endodontic origin, occasionally these lesions present in conjunction with teeth that have already been endodontically

**Table 16.1**  Unusual periapical diagnoses.

| Category | Type | Number of cases reported |
|---|---|---|
| Cysts | Odontogenic keratocyst | 22 |
| | Nasopalatine duct cyst | 4 |
| | Lateral periodontal cyst | 4 |
| | Residual cyst | 3 |
| | Globulomaxillary cyst (which is no longer a valid diagnosis) | 1 |
| Infections | Actinomycosis | 15 |
| | Histoplasmosis | 1 |
| | Aspergillosis | 3 |
| Benign aggressive lesions | Central giant cell granuloma | 24 |
| | Central ossifying fibroma | 1 |
| | Myxomas | 2 |
| | Central odontogenic fibroma | 1 |
| | Pindborg tumor | 1 |
| | Osteoblastomas | 2 |
| | Langerhans cell disease | 3 |
| Benign fibro-osseous lesions | Periapical cemental dysplasia | 30 |
| | Other | 2 |
| Granulomatous inflammation | Foreign body | 40 |
| | Pulse granuloma | 22 |
| Malignant lesions | Carcinoma, including adenocarcinoma and metastatic lesions | 10 |
| | Sarcoma | 4 |
| | Lymphoma | 7 |
| | Multiple myeloma | 2 |
| | Leukemia | 1 |

Data from Peters and Lau (2003) with permission.

treated (possibly due to missed diagnosis) (Nevins et al. 1988; Wilcox and Walton 1989). Surveys of biopsy analyses following endodontic surgery, in large populations, show that the incidence of nonendodontic pathosis ranges from 1 to 6.5% (Spatafore et al. 1990; Nobuhara and del Rio 1993; Kuc et al. 2000). In addition to these lesions, systemic metabolic diseases such as primary or secondary hyperparathyroidism (Loushine et al. 2003) or osteoporosis may lead to erroneous diagnosis of endodontic pathosis (Fig. 16.1).

### 16.3.1 Malignant neoplasms

The astute clinician should, therefore, obtain a detailed history, perform all the necessary tests, obtain the necessary consultations, formulate a differential diagnosis list, and send surgically excised tissues for histopathological examination in order to reach an accurate diagnosis.

## 16.4 Systemic diseases or conditions that may influence the pathogenesis or healing of endodontic pathosis

### 16.4.1 Diabetes mellitus

Diabetes represents a group of diseases characterized by increased serum glucose due to decreased production or action of insulin. Type 1 diabetes (formerly called insulin-dependent diabetes) is the result of destruction of the pancreatic islet cells due to autoimmune, genetic, or environmental causes, and represents 5–10% of all diabetics. Type 2 diabetes, which is the most prevalent form of diabetes, is associated with increased age, obesity, lack of exercise, and race/ethnicity. In the United States, African American, Hispanic, and Native American populations have about twice the prevalence of type 2 diabetes as Caucasians. The most recent statistical estimates (see http://diabetes.niddk.nih.gov/dm/pubs/statistics/index.htm# 7) reveal that about 7% of the U.S.

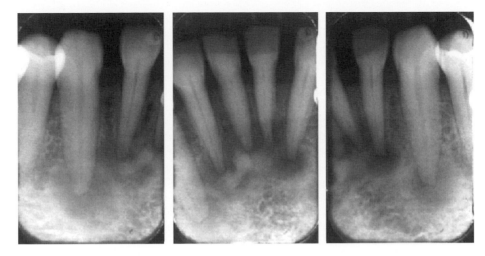

**Fig. 16.1**   Periapical radiographs of multiple teeth with periapical radiolucencies that mimic endodontic pathosis. All these teeth responded normally to pulp testing. This patient had hyperparathyroidism secondary to kidney failure.

population or 20.8 million individuals are diabetic. This figure consists of 14.6 million individuals who are diagnosed diabetics, and the balance (about one-third) is undiagnosed. Diabetes is relatively rare among individuals less than 20 years of age; however, it affects 9.6% (1 in 10) of people over 20 years of age and 20.9% (1 in 5) of people over 60 years of age.

Numerous studies have shown that diabetes mellitus is associated with increased periodontal disease at all age groups and after controlling for many confounding factors (for recent review see Mealey and Rose 2008). Some studies have shown improvement in glycemic control of patients after periodontal treatment (Grossi et al. 1997; Iwamoto et al. 2001). However, a more recent meta-analysis of interventional studies on this topic concluded that although there appears to be some improvement of the glycated hemoglobin (HbA1c) measure of patients after periodontal treatment, particularly type 2 diabetics, the differences are not statistically significant (Janket et al. 2005).

Interest in the relationship of endodontic pathosis and its treatment to diabetes mellitus has a long history. An older paper based on subjective findings concluded that for well-controlled diabetics there does not appear to be a risk involved with respect to postoperative healing (Bender et al. 1960). Another study among endodontic patients showed a reduction in postoperative healing of about 26% after 30 months follow-up with increase of about 20 mg/dL in postprandial glycemia. This shows the possible relationship of the healing to the glycemic measure, regardless of the diagnosis of diabetes. More recently, a number of cross-sectional studies have been reported on the prevalence of endodontic pathosis in diabetic and nondiabetic patients. Diabetic patients seemed to have a disproportionately higher percentage of acute odontogenic infections according to one study (Ueta et al. 1993). Compared to nondiabetics, periapical radiolucencies were more prevalent in relation to at least one tooth (Segura-Egea et al. 2005) or endodontically treated teeth of men (Britto et al. 2003) in type 2 diabetic patients. Likewise, long-duration insulin-dependent diabetes was associated with a higher percentage of nonhealing lesions in relation to endodontically treated teeth (Falk et al. 1989).

Clearly, cross-sectional studies provide a limited perspective on endodontic healing because of the long duration necessary for healing of periapical lesions (see Chapter 15). Therefore, prospective cohort studies with adequate follow-up periods and a high recall rate are optimal to describe the relationship of diabetes and endodontic pathosis.

One such study was made possible by the availability of an electronic patient record for endodontics in a dental school patient population that kept information on all patients from 1995 till 2001(Fouad and Burleson 2003). At that time the type 1 and type 2 classifications were not in use, so the patients were classified according to whether they were on insulin.

A total of 5,210 endodontic cases were completed for nondiabetic patients, 70 cases for insulin-dependent (IDDM), and 214 cases from non-insulin-dependent diabetics (NIDDM). There was a strong trend for increased periapical pain in the IDDM group ($p = 0.058$), which may be related to the neuropathy that is reported as one of the sequelae of diabetes. Despite the presence of an active follow-up program in that institution in which all patients were contacted and offered the opportunity to return for follow-up, about one-third of all patients did return for follow-up at any time after treatment, and only 544 cases (about 10%) were available for follow-up 2 years or longer after treatment. This is clearly a limitation of this study; however, given the large number there is merit to considering this data from an epidemiological perspective. As had been determined in numerous other endodontic outcome studies and most relevant to this textbook, the presence of infection in the form of a preoperative lesion was significantly associated with reduced outcomes. However, cases with infections that had 2 years or more follow-up were only 189, including 17 from diabetic patients. When the entire patient population that reported for the 2 years or longer period was considered, there were no differences between the diabetic and nondiabetic patients. However, when only cases with infections were considered (and both IDDM and NIDDM had to be pooled due to the low numbers), diabetics had a significantly higher risk of lack of complete healing compared to nondiabetics. This was true even after controlling for gender, age, presence of a permanent restoration at the time of follow-up, primary treatment versus retreatment, provider category (dental student vs resident), time to last follow-up, and the presence of periodontal disease on the tooth that was treated endodontically. Interestingly, this latter factor by itself showed a significant increase in diabetics, consistent with the data that were described earlier.

These data are consistent with earlier animal research data that have shown that compared to nondiabetics, diabetic animal models develop larger periapical lesions (Kohsaka et al. 1996), and are more prone to have increased morbidity and mortality in response to endodontic infections (Fouad et al. 2002).

There are many potential mechanisms that can explain these findings about the diabetic host. It has long been known that diabetics have increased levels of a glycated form of tissue proteins called advanced glycated end products (AGE) (Yan et al. 2004; Janket et al. 2008). These molecules interact with receptors

(called RAGE) to produce a number of local and systemic inflammatory mediators, such as IL-1β, IL-6, and TNF-α. The chronic nature of the release of these cytokines creates a host environment that is susceptible to increased bone resorption and tissue damage. Some studies have also shown that in the diabetic host, immune cells such as monocytes develop a form that is hyperexcitable in that they respond to external irritation by releasing a large amount of inflammatory mediators, which would produce the same chronic inflammation described before (Salvi et al. 2000). Finally, it is hypothesized that the diabetic host may favor more bacteria, particularly virulent species, which may compromise the host responses more so than the nondiabetic host. In endodontic infections, *Eubacterium infirmum*, an anaerobic Gram-positive organism, was found to be more prevalent in diabetic patients (Fouad et al. 2003).

### 16.4.2 Smoking

There is no doubt that smoking as a general social habit is on the decline in western countries. Major tobacco manufacturers have been successfully prosecuted in recent years on the premise that significant health information was withheld or misrepresented to consumers. There is clear evidence of the association and contribution of smoking with cardiovascular disease (CVD), neoplasia, chronic bronchopulmonary disease, and periodontal disease. The association between smoking and periodontal disease has been shown in numerous well-designed studies in recent years. In this regard, a study assessing the extent of periodontal disease or evaluating the effectiveness of periodontal therapy would be lacking significant information, if it did not control for smoking.

It is only recently that the association of endodontic pathosis and its treatment with smoking has been explored. In a cross-sectional Swedish study of 247 individuals, a significant association was found between smoking and the presence of apical periodontitis or root canal treatment (Bergstrom et al. 2004). However, this association disappeared after controlling for the age of the patient. More recently, a study was reported from Spain in which apical periodontitis in at least one tooth was found in 74% of smokers and in 41% of nonsmokers ($P < 0.01$) (Segura-Egea et al. 2008). Among smokers 5% of the teeth had apical periodontitis, whereas in nonsmokers 3% of teeth were affected ($P = 0.008$).

The results of a larger longitudinal study provided even more compelling data (Krall et al. 2006). In this study, known as the longitudinal Veterans Administration (VA) dental study, 811 men were recruited from 1968 to 1973 and continue to be followed till the present time. The age-adjusted incidence of root canal treatment—which would be a surrogate for pulp pathosis—was greater in current cigarette smokers relative to never-smokers and in men who stopped smoking less than 9 years compared to more than 9 years before. Incident root canal treatment was also more in men who smoked for 12 years than in those who smoked for 5–12 years, which was more than in those who smoked for less than 4 years. These findings clearly show a dose–effect relationship and bode well for individuals who stop smoking.

Finally, smoking increased postoperative pain following endodontic surgery in a prospective Swiss cohort of 102 patients (Garcia et al. 2007). In this study, the number of cigarettes smoked was not significant, but patients who had also poor oral hygiene had significant postoperative pain and swelling. Therefore, taken together, it appears that smoking may have an influence on the pathogenesis of pulpal disease, the prevalence of apical periodontitis, and postoperative discomfort. It remains to be determined in longitudinal studies if smoking affects long-term endodontic treatment outcomes.

### 16.4.3 Systemic viral infections

#### 16.4.3.1 HIV/AIDS

Human immunodeficiency virus infects CD4 cells, resulting in significant deficiency in specific immunity. It is generally accepted that patients whose CD4 count decrease below 200/mm$^3$ have more severe clinical manifestations of the disease, and frequently present with a number of comorbid bacterial and fungal infections. When the HIV virus was first identified and the mechanism of its action first described, there was a concern that patients with endodontic infections would have significant perioperative morbidity because they are immunocompromised. Earlier case reports of endodontic postoperative symptoms and flare-ups in HIV-infected individuals raised some concerns about this issue (Hillman 1986; Gerner et al. 1988). It became clear shortly thereafter that the virus can easily be detected in pulpal (Glick et al. 1989) and periapical (Elkins et al. 1994) tissues, as would be expected, and therefore these tissues are likely to be compromised as well.

A series of animal studies examined the influence of specific immune response on the pathogenesis of periapical lesions. These studies are relevant to mention here, although they did not specifically study HIV infection, because they illustrate the role of innate and specific immunity on the pathogenesis of endodontic lesions. Pulp necrosis and the development of periapical lesions were shown to occur at comparable rates in normal and SCID (severe combined immune deficiency) mice, when pulp exposures were left open to the oral cavity (Fouad 1997). SCID mice lack all types of T and B cells, and therefore, have no form of specific immunity. It was also shown that RAG-2 SCID mice that are exposed to a large bacterial load of virulent endodontic pathogens do develop disseminating oral infections, although the lesion size was not larger than those in normal animals (Teles et al. 1997). To further isolate the specific reason for the immune deficiency, a later study revealed that T-cell-deficient animals did not demonstrate any morbidity in response to virulent endodontic pathogens. However, B-cell-deficient animals did have the significant morbidity identified in the earlier study (Hou et al. 2000). This clearly shows that T-cell deficiency, as is the case in HIV infection, does not seem to compromise the ability of the host to mount an effective immune response to endodontic infections.

Consistent with these results, a large clinical study of over 330 patients reporting for dental treatment was showed that infection with HIV together with a CD4 count of less than 200/mm$^3$ was not a risk factor in causing more extensive or aggressive endodontic infections (Glick et al. 1994). Another study was reported in which endodontic treatment of 57 HIV-positive patients (and 17 patients who did not report HIV infection) was endodontically treated. In the 1–3 months postoperative period, there were no differences in the complications reported between the two groups (Cooper 1993). Finally, an endodontic treatment outcome paper was more recently reported, in which 33 HIV-positive patients were compared to an equivalent number of controls using the periapical index at 1 year postoperatively. The results revealed no differences between the groups (Quesnell et al. 2005).

#### 16.4.3.2 Herpes zoster

Herpes zoster is caused by varicella-zoster virus, which causes chicken pox in childhood, then remains dormant with periodic exacerbations in the form of herpes zoster in adults. The viral infection presents as vesiculobullous lesions, which typically follow the innervation pathway of a nerve such as a somatic nerve or a branch of the trigeminal nerve. The herpes lesions are self-limiting, and the infection subsides within 1–2 weeks. Occasionally, the viral infection is followed by neuropathic pain along the pathway of the affected nerve, causing postherpetic neuralgia that may be protracted in type and extent.

There have been a few case reports in the endodontic literature, in which well-documented herpetic lesions were followed by pulpal pathosis in teeth that are in the path of affection by the viral infection. A case was reported in which a 23-year-old Asian man, who had herpes zoster infection 5 months earlier, developed pulpitis in teeth # 9 and 10. These teeth were endodontically treated, and the patient was followed up. Later, he presented with pulp necrosis of teeth # 11–13 (Fig. 16.2) (Goon and Jacobsen 1988). It was surmised from this report that the viral infection had initiated pulpal pathosis in several teeth that were in

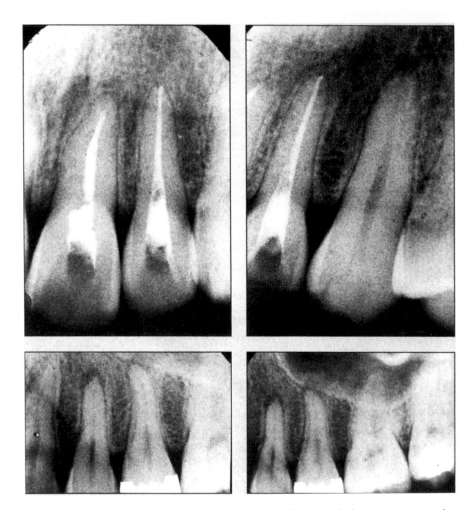

**Fig. 16.2**   Periapical radiograph taken 5 months after the herpes zoster infection. Endodontic treatment can be seen in maxillary left lateral and central incisors. Periapical radiolucent lesions can also be seen at the apices of left canine, first premolar and second premolar. Except for the endodontic access openings, all teeth were intact with negligible carious involvement or minimal restorations. (Reprinted with permission from Goon and Jacobsen 1988. Copyright © 1988 American Dental Association. All rights reserved.)

the path of the affected nerve. However, what is more common for patients with herpes zoster infection or postherpetic neuralgia is to present with symptoms of pulpal pain, but do not have any documented pulp disease. In these cases accurate diagnosis and management on the underlying condition, rather than endodontic treatment, remain the appropriate course of action (Sigurdsson and Jacoway 1995).

### 16.4.3.3 Other viral infections

As was seen in the discussion presented in Chapter 8, other viruses such as cytomegalovirus (CMV) and Epstein–Barr virus, but not herpes simplex virus, may be involved in the pathogenesis of symptomatic periapical lesions, and lesions that are larger in size. CMV was recently shown to be equally prevalent in both periapical cysts as well as odontogenic keratocysts (Andric et al. 2007). The relationship of viral infections and endodontic pathosis is still poorly understood. For example, it is not known if viral infections (herpes zoster or other viruses) can initiate endodontic pathosis, in the absence of bacterial infection. As noted in Chapter 8, the exact nature of viral involvement in the pathogenesis of periapical lesions that follow caries or traumatic injuries is also poorly understood.

### *16.4.4 Sickle cell anemia*

Sickle cell anemia is a congenital disease inherited through an autosomal recessive transmission. It is a homozygous disease in which at least both parents are carriers. Carrier or heterozygous state is expressed as sickle cell trait and is much milder in its clinical presentation. Sickle cell anemia is characterized by sickle shape of the red cells due to substitution of a single amino acid—valine for glutamic acid—at the sixth residue of the β-chain of hemoglobin (Little et al. 2008). Interestingly, sickle cell anemia tends to be prevalent in communities and countries where malaria (which results from infection by the parasite *Plasmodium falciparum*) is endemic, such as Central African countries. It has been shown that this type of anemia confers protection against malaria by reducing adhesion of parasite-infected erythrocytes to endothelial cells of microvasculature of the brain and other organs, which causes the life-threatening complications of malaria (Cholera et al. 2008).

Two case series have been reported in which sickle cell anemia seemed to be involved in the pathogenesis of pulpal and periapical disease. In the first series (Andrews et al. 1983) 22 patients with sickle cell anemia were examined. Five of 22 (23%) had jawbone radiolucencies. Three of 22 (14%) had pulp necrosis with periradicular lesions in 8 teeth. Five of these eight lesions were in noncarious teeth that had no other known etiology for pulpal pathosis, indicating a possible role of sickle cell defect in initiating spontaneous endodontic infections (Fig. 16.3). The second series, which was more recently reported, had a case-control design and included 36 sickle cell cases and an equivalent control group (Demirbas Kaya et al. 2004). In this study, it was shown that sickle cell anemia can be involved in initiating pulpal pathosis, endodontic pain, as well as jawbone radiolucencies (Table 16.2.)

Primary malignant neoplasms may arise in a number of oral tissues, including keratinized and nonkeratinized oral mucosa, mandibular and maxillary bone, and salivary glands. Locally invasive neoplasms such as ameloblastoma, which arises from odontogenic tissues, like other jawbone radiolucencies, may present in close proximity to teeth, thus mimicking the presentation of endodontic infections. In this context, it is of interest that primary neoplasms have rarely if ever been reported in the dental pulp. It is not known why this may be the case.

As mentioned before, reports have been published of metastatic neoplasms in the maxilla or mandible, causing radiolucent lesions, occasionally with invasive resorption of the root, pain, parasthesia, including numbness. While the tooth root and pulp may be affected in these situations, clearly, proper diagnosis and appropriate corrective therapy, which may or may not include endodontic therapy, must be performed in consultation with the treating physician or surgeon. Examples of these neoplasms include those that arise in the lung, prostate, breast, pancreas, or colon. Primary neoplasms of the immunologic cells that manifest as bone radiolucencies include multiple myeloma, eosinophilic granuloma, Letterer Siwe's disease, and Gaucher's disease.

### *16.4.5 Other systemic disease or abnormalities*

Most clinicians appreciate that if a patient's immune system is compromised, this patient may be more susceptible to perioperative pain and flare-ups, disseminated infections, and/or reduced postoperative

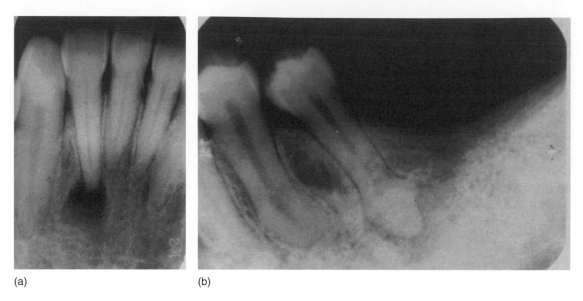

(a)                                         (b)

**Fig. 16.3**   Examples of teeth in which no apparent etiology other than sickle cell anemia was reported for the development of pulp necrosis and periapical lesions. (Reprinted with permission from Andrews et al. (1983).)

healing. One study showed that patients with systemic allergies were more likely to have endodontic flare-ups than normal patients (Torabinejad et al. 1988). In another study, patients undergoing bone marrow transplantation and having asymptomatic periapical lesions in relation to endodontically treated teeth were divided into two groups: in one group retreatment, root-end resection or extraction was performed while the control group had no treatment done before the bone marrow transplant. The dependent variables were the numbers of days in the hospital with neutropenia (<500/mL) or elevated temperature (>100°F). The results showed no significant differences between the two groups (Peters et al. 1993).

With respect to long-term outcomes, in one study 66 patients, who had normal or immunocompromised medical status, were followed for a minimum of 30 months (Marending et al. 2005). The medical conditions of the patients included insulin- and non-insulin-dependent diabetes, kidney insufficiency, breast cancer and concomitant chemotherapeutic drugs, gout, steroids, ulcerative colitis, gastroesophageal reflux disease, and rheumatoid arthritis. The study showed that endodontic treatment outcome was significantly dependant on whether they were in the normal or immunocompromised group. Conversely, another study examined the effects of osteoradionecrosis on endodontic treatment outcomes (Lilly et al. 1998). In this study, endodontic

**Table 16.2**   The effects of sickle cell anemia (SCA) on orofacial pain, bone radiolucency, and pulpal and periapical lesions

|  | SCA ($N = 36$) | Normal ($N = 36$) |
|---|---|---|
| Number of patients with orofacial pain (not during SCA crises) | 30 (83%) | 5 (14%) |
| Number of patients with "step ladder" bone pattern | 10 (28%) | 0 |
| Number of teeth examined | 827 | 1,084 |
| Number of noncarious, unrestored teeth with pulp necrosis | 51 (6%) | 0 |

Data summarized from Demirbas Kaya et al. (2004).

treatment of 22 patients was performed on teeth, which were in the direct path of irradiation with a mean of 5,000 cGy, in the preceding 6 months. The mean postoperative evaluation period was 19 months. The study showed a success rate of 91%, which was consistent with that of other endodontic outcome studies of noncompromised patients. This particular finding is important given the recent recognition of cases of osteonecrosis of the jaw following oral surgical procedures in patients on significant doses of bisphosphonate treatment. These patients typically receive intravenous bisphosphonates for the treatment of multiple myeloma, Paget's disease, metastatic breast cancer, or osteoporosis (Badros et al. 2006; Gutta and Louis 2007). It would be of interest to determine if endodontic treatment outcome may be compromised in these patients.

## 16.5 Can endodontic infections contribute to the pathogenesis of systemic disease?

### 16.5.1 Severe spreading endodontic infections

As was described in Chapter 10, acute endodontic infections can spread to regional lymph nodes, fascial spaces, and distant organs, particularly when left untreated for long durations. In developing countries, and regions in industrialized countries where there is limited access to dental care, these types of infection are seen with some prevalence and contribute to numerous hospital emergency room visits. The literature contains several reports of spreading endodontic infections that caused mediastinitis (Garatea-Crelgo and Gay-Escoda 1991; Bonapart et al. 1995), fatal necrotizing fasciitis (Stoykewych et al. 1992), and brain abscess (Corson et al. 2001). In the United States there was a case in 2007 of a 12-year-old child in the Washington, DC, area, who had a dental abscess, and whose treatment was delayed for 2 years because of access to care issues. He then developed a brain abscess and was treated for 6 weeks in a hospital emergency room, but eventually died as a result of complications of this condition (http://www.washingtonpost.com/wp-dyn/content/ article/ 2007/02/27/AR2007022702116.html). This incident has created a national outcry about access to care issues, and led to the introduction in the U.S. Congress of the Deamonte's Bill. If passed into law, this would amend the Public Health Service Act to

expand and improve the provision of pediatric dental services to medically underserved populations. Therefore, it is clear that endodontic infections can result in spreading, potentially life-threatening, infections.

Common acute endodontic infections, which are seen frequently in dental and endodontic practices, may result in a number of signs and symptoms of systemic inflammation, albeit with less morbidity. These include fever, malaise, regional lymphadenopathy, anorexia, somnolence, trismus, hypotension and hepatic synthesis of a number of proteins such as complement and coagulation proteins (Kumar et al. 2003). Like other types of acute infections, these infections may also result in leukocytosis and increased erythrocyte sedimentation rate. Systemic inflammation results in an increase in serum cytokines such as TNF-$\alpha$, IL-1, and IL-6. IL-1 and TNF-$\alpha$ are responsible for temperature elevation due to their action on the thermoregulatory center of the hypothalamus through local prostaglandin production (thus the effectiveness of NSAIDs to reduce fever). IL-6 causes an increase in serum acute-phase proteins. Acute-phase proteins, such as C-reactive protein (CRP), serum amyloid A (SAA), and others (see below), are produced in the liver and mediate many of the systemic effects of inflammation described before. CRP has been implicated in the development of atherosclerotic plaque by inducing adhesion molecule expression in human endothelial cells (Pasceri et al. 2000), activating vascular smooth muscle cells causing induction of monocyte chemoattractant peptide (Hattori et al. 2003), and the uptake of low-density lipoprotein by macrophages (Zwaka et al. 2001).

In endodontic infections, CRP was investigated in a number of studies in the past two decades. In an older study, the serum concentration of acute-phase proteins, CRP, $\alpha$2-macroglobulin (AMG), $\alpha$1-antitripsin (AAT), haptoglobin (HPT), complement component C3, and ceruloplasmin (CER) were measured following endodontic root and surgery of cases with periapical lesions, associated with previous endodontic treatment. There was a slight increase in CRP, AAT and CER in these chronic granuloma patients preoperatively. The levels of AMG and AAT fell significantly as early as 7 days after endodontic surgery. All investigated acute-phase proteins decreased significantly 3 months after treatment (Marton et al. 1988). In another study, CRP increased in the pulp of teeth with pulpitis, but not in serum (Proctor et al. 1991) indicating that symptomatic pulpal inflammation is a localized phenomenon and does not produce a systemic reaction. In addition, systemic CRP and SAA did not increase in a dog model of chronic apical periodontitis (Buttke et al. 2005). However, it was more recently shown that the number of patients with acute apical abscess, who have a significant increase in CRP, decreases significantly within 1 week of treatment (Ren and Malmstrom 2007). Interestingly, in this study the number of patients with periodontal abscess or acute osteitis, who also had elevated CRP, did not change significantly after 1 week. Taken together, these results suggest that acute-phase proteins, but mainly CRP, increase systemically in acute periapical infections (but not acute pulpitis) and possibly in chronic mildly systematic lesions. They decrease gradually within 1–9 weeks following treatment.

### 16.5.2 Bacteremia as a result of endodontic pathosis and/or treatment

It has been known for a long time that instrumentation of the root canal in a case with endodontic infection, or periapical surgery can lead to bacteremia (Table 16.3). It appears that extrusion of bacteria occurs whether or not the instruments in the necrotic pulp cases are maintained within the canal or extend past the working length (Debelian et al. 1995). One study showed that the instrumentation itself appears to be causative of an increase in the incidence of bacteremia perioperatively, compared to preoperatively (Savarrio et al. 2005). It is of interest in this study that instrumentation of cases with vital pulp resulted in no bacteremia and that the incidence of bacteremia was not different in cases with primary or persistent endodontic infections. Preoperative bacteremia in cases with chronic infections was found in only 7% of all cases (Savarrio et al. 2005). While this incidence is much less than that which occurs during instrumentation, the duration of lesion induction, until it is discovered and treated, can take a long time measured in months or years, compared to the brief period that it takes during instrumentation. Furthermore, it is known that 10–20% of cases with preoperative lesions will not respond to nonsurgical treatment (as was discussed in Chapter 15), and about 50% or more of persistent lesions may have bacteria in the periapical lesions (as was discussed in Chapter 6). Therefore, the amount of bacteremia that is caused by chronic endodontic infections preoperatively, and to some degree postoperatively if they do not respond to treatment, remains largely unknown.

The interest in bacteremia is related to the effects that endodontic pathogens can create in distant systemic sites. While it is evident as described before that acute periapical infections can and do result in transmission of bacteria in distant locations, the situation is less clear with chronic infections that are of longer duration, and may result in chronic bacteremia. The effect may also be cumulative if the patient has multiple teeth with endodontic infections. Chronic bacteremia may contribute to diseases such as infective endocarditis, in susceptible patients (see Chapter 12), or atherosclerosis (see below). A number of oral and endodontic pathogens have been implicated in the etiology of infective endocarditis (Lockhart and Durack 1999) (Table 16.4). Furthermore, virulence genes that are critical in the pathogenesis of endocarditis, such as those for fibrinogen-binding protein and fibronectin-binding protein, have been identified in endodontic bacteria (Bate et al. 2000). However, the impact of bacteremia due to dental procedures or any one oral source on the pathogenesis of chronic

**Table 16.3**  Bacteremia associated with endodontic treatment.

| Treatment | Incidence | Study |
|---|---|---|
| Nonsurgical treatment | 25% | Bender et al. (1960) |
| Nonsurgical treatment | 17% | Baumgartner et al. (1976) |
| Nonsurgical treatment | 20% | Heimdahl et al. (1990) |
| Nonsurgical treatment (intracanal) | 31% | Debelian et al. (1995) |
| Nonsurgical treatment (overinstrumentation) | 54% | Debelian et al. (1995) |
| Nonsurgical treatment | 23–30%[a] | Savarrio et al. (2005) |
| Flap reflection | 83% | Baumgartner et al. (1977) |
| Periradicular curettage | 33% | Baumgartner et al. (1977) |

[a] 30% of patients had positive blood culture, but in 23% the same organism was identified in the endodontic and blood cultures.

**Table 16.4** Number of studies in which a particular oral pathogen has been detected in bacteremia specimens

| | Single tooth extraction | Multiple extractions (or not specified) | Brushing or flossing | Periodontal scaling | Endodontic treatment |
|---|---|---|---|---|---|
| Gram-positive cocci | | | | | |
| Viridans streptococci | | | | 1 | |
| S. oralis group | 2 | 1 | 1 | 2 | |
| S. sanguis (S. sanguis I) | 4 | 2 | 1 | | 1[a] |
| S. oralis (S. mitior, S. sanguis II) | 2 | 1 | | 1 | 1[a] |
| S. milleri group | | | | | |
| S. intermedius | 5 | 1 | 1 | 1 | |
| S. mutans | 3 | 1 | | | |
| S. salivarius group | | | | | |
| S. salivarius | 3 | 1 | 1 | 1 | 3[b] |
| Parvimonas micros, Peptostreptococcus asaccharolyticus, P. evolutus, P. anaerobius | 5 | 2 | 1 | 1 | |
| Staphylococcus aureus | 1 | | 1 | 1 | |
| Staphylococcus epidermidis | 1 | 1 | 1 | 1 | |
| Gram-negative bacilli/rods | | | | 3 | |
| Fusobacterium nucleatum, Fusobacterium fusiforme | 5 | 3 | 1 | | |
| Aggregatibacter actinomycetemcomitans | 1 | | | | |
| Prevotella melaninogenica | 5 | 2 | 1 | 1 | |
| Bacteroides (Tannerella forsythensis, Eikenella[c] corrodens, Prevotella ruminicola, Prevotella oris, Parabacteroides[c] distasonis, Mitsuokella[c] multiacidus) | 1 | 1 | 1 | 1 | 1[a] |
| Porphyromonas gingivalis | | 1 | | | |

Data from Lockhart and Durack (1999) with permission.

[a] Heimdahl et al. (1990).

[b] Bender and Pressman (1956), Baumgartner et al. (1976), Beighton et al. (1994).

[c] Taxonomic update is to the best of this author's knowledge.

systemic diseases is difficult to ascertain. To illustrate this difficulty, a recent study showed that tooth extraction was associated with a cumulative incidence of bacteremia (measured in several perioperative specimens) of 60%, which was reduced to 33% if the patient was on amoxicillin, given according to the most recent American Heart Association prophylaxis guidelines (Lockhart et al. 2008). However, the cumulative incidence of bacteremia following toothbrushing in this study was 23%, raising the concern that this daily routine may be more hazardous for the susceptible patients in the long term than a single tooth extraction.

### 16.5.3 Endodontic pathosis and cardiovascular disease

Recent studies have shown with some degree of certainty that CVD is a result of the interaction of genetic predisposition and environmental risk factors. The pathogenesis of atherosclerosis, which is the underlying disease that leads to CVD mortality, stroke, and myocardial infarction, involves the formation of an atheromatous plaque on the intimal walls of major vessels. The inflammatory etiology of this lesion may start with the invasion of vascular wall with pathogenic strains of bacteria, thus inducing an inflammatory response. An early lesion is composed of fatty streaks or plaques, and it eventually matures into established lesions that include in their composition lymphocytes, macrophages, and bacteria. In an in vitro experiment, it was shown that *Porphyromonas gingivalis* strain 381 (an invasive periodontal and endodontic pathogen) and *Porphyromonas endodontalis* ATCC 35406 (an endodontic pathogen), but not other *P. endodontalis* strains or *Prevotella intermedia* and *Prevotella nigrescens* strains, are capable of invading human coronary artery endothelial cells (Fig. 16.4) (Dorn et al. 2002).

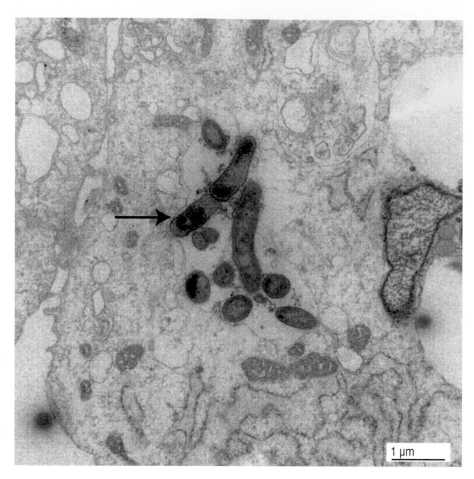

**Fig. 16.4**  Transmission electron micrograph of internalized *P. endodontalis* (arrow) in human coronary artery endothelial cell following a 90 min infection. (Reprinted with permission from Dorn et al. 2002.)

As was noted in the section entitled "Severe Spreading Endodontic Infections", acute endodontic infections, endodontic treatment procedures and presumably, endodontic infections that go through phases of exacerbation and remission without appropriate treatment, may raise systemic inflammation, and acute-phase proteins.

The release of bacteria from the necrotic pulp space into the periapex and systemically in chronic endodontic infections is less clear. As was noted in Chapter 6, a significant proportion of persistent endodontic infections appear to have bacteria within the periapical lesions (Sunde et al. 2003). In primary endodontic infections, the physical presence of bacteria in the periapical lesion has been shown in a number of older

studies but not with high consistency, presumably because of the limitations of the older techniques used on human and animal specimens (Nair 1987; Fouad et al. 1992; Ricucci et al. 2006b). More recently, molecular techniques have shown more consistently the presence of bacteria periapically in models of primary chronic periapical lesions (Russo et al. 2008) (Fig. 16.5). Clearly, once the bacteria escape to the periapical region, there is no barrier to prevent slow systemic dissemination. This could occur from biofilms located within the root canal, on the root apex (Ricucci et al. 2005), or within the lesion itself such as with periapical actinomycosis (see before, and also Chapter 6).

In recent years, a number of investigations have shown the presence of oral bacterial pathogens in

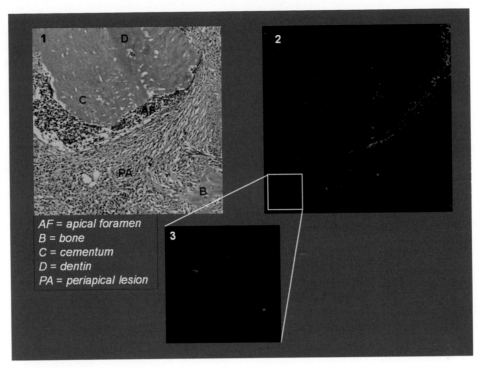

**Fig. 16.5**  Periapical lesion of the distal root of a mouse molar stained with universal bacterial PNA-FISH probe positive fluorescence (red) for bacteria in PA tissues (1) H&E stain of lesion (×200); (2) FISH stain of the same region at the same magnification (×200); (3) magnification of (2) (×1,000) (Russ. et al. 2008).

atheromatous plaques removed from patients who underwent endarterectomies for the management of carotid artery and other major vessel thrombosis. In one study *Tannerella forsythia* was prevalent in 30%, *P. gingivalis* in 26%, and *Prevotella intermedia* in 14% of the endarterectomy specimens from 50 patients (Haraszthy et al. 2000). Interestingly, *Chlamydia pneumoniae*, which is one of the key organisms that are thought to be involved in the pathogenesis of atherosclerosis, was present in only 18% of these patients in this study. In another report, 23% of the same type of specimens in 26 patients had *Treponema denticola*. This organism was not found in any of the nondiseased aorta specimens that were used as controls (Okuda et al. 2001). More recently, atheromatous specimens from two groups of patients were investigated: a young group that died of CVD who was sampled postmortem ($N = 20$), and an elderly patient group following endarterectomy ($N = 9$) (Kozarov et al. 2006). Quantitative real-time PCR was used to identify the following organisms: *Aggregatibacter acti-*

*nomycetemcomitans, P. gingivalis, P. intermedia, S. mutans, T. denticola, Eikenella corrodens*, and the nonoral *Staphylococcus aureus, Staphylococcus epidermidis*, and *C. pneumoniae*. The prevalence of these organisms in both groups ranged from 40 to 80%, with *P. gingivalis, P. intermedia, A. actinimycetemcomitans, C. pneumoniae*, and *S. epidermidis* being the most prevalent in this order. These studies have not documented the periodontal or endodontic status of these patients; however, most of these organisms have been consistently identified in endodontic infections, as stated in the previous chapters.

It is also noteworthy that the oral cavity, and indeed the pulp space of teeth with endodontic infections, may act as a reservoir for some bacteria, which are not common to the oral cavity, but are capable of causing significant systemic disease. These bacteria could then escape into systemic organ systems either directly or through a hematogenous route. For example, *Streptococcus pneumoniae* and *C. pneumoniae* are organisms involved in the initiation of pneumonia, bronchitis, and sinusitis among other respiratory diseases, and have been identified in the oral cavity (Mantyla et al. 2004; Suzuki et al. 2006). However, a recent analysis of 40 endodontic infections failed to identify either of these organisms from endodontic specimens (Nandakumar et al. 2008).

### 16.5.3.1 Models of cardiovascular disease

The apolipoprotein E-deficient (ApoE-/-) mouse is one of the best models available to study the pathogenesis of atherosclerosis. It was created by disrupting the *apo*E gene. ApoE, a glycoprotein, is a structural component of very low-density lipoproteins (VLDLs), which are synthesized by the liver, and a constituent of a subclass of high-density lipoproteins (HDLs) involved in cholesterol transport activity among cells. One of the most important roles of ApoE is to mediate high-affinity binding of chylomicrons and VLDL particles that contain ApoE to the low-density lipoprotein (LDL) receptor. This allows for the specific uptake of these particles by the liver, which is necessary for transport preventing the accumulation in plasma of cholesterol-rich remnants. The homozygous inactivation of the *apo*E gene results in animals that are devoid of ApoE in their sera. The mice appear to develop normally; however, they exhibit five times the normal serum plasma cholesterol and spontaneous atherosclerotic lesions over a period of about 10–20 weeks of age, necessitating the need for long observation periods. This is similar to a disease in people who have a variant form of the *apo*E gene that is defective in binding to the LDL receptor and are at risk for early development of atherosclerosis, and increased plasma triglyceride and cholesterol levels. Using this model, it has been shown that *P. gingivalis*, whether inoculated intravenously (Li et al. 2002) or through an oral gavage (Lalla et al. 2003), can lead to an increase in the size of and speed with which an atheromatous plaque is formed. More recently, this was shown to be fundamentally related to the fimbrae of *P. gingivalis*, which is an adhesion molecule that mediates invasion and inflammatory responses of this organism (Chou et al. 2005). Interestingly, immunization with *P. gingivalis* prevented the local inflammatory responses in the vascular intima of this animal model, indicating that upregulation of localized innate immune responses mediated by macrophages, rather than systemic inflammation contribute to the pathogen-related atherosclero-

sis (Miyamoto et al. 2006). Taken together, these findings suggest that an oral virulent pathogen may be involved in the pathogenesis of atherosclerosis. To corroborate this hypothesis, it was recently demonstrated in a clinical study that diabetic patients, who had moderate-to-severe periodontitis, showed significant reduction in CRP, and in the monocyte/macrophage capacity to secrete proinflammatory cytokines, following periodontal therapy (Lalla et al. 2007).

While these studies show the potential for the involvement of periodontal/endodontic pathogens in the pathogenesis of CVD, it is really important to show actual link between the oral and the systemic diseases through epidemiological studies. In the last decade there have been a number of studies that provided compelling evidence for the association between the presence and severity of periodontal disease and CVD. A recent meta-analysis on the subject showed statistically significant overall associations in 15 prospective, cross-sectional, or case-control studies that were included and analyzed (Bahekar et al. 2007). It has been proposed that periodontal disease translates to the exposure of a large area of sulcular surface to pathogenic organisms, estimated to be about the area of the palm of a hand (Page 1998). The available surface area in endodontic lesions is likely much smaller than that, unless the patient has multiple teeth involved, and has large periapical lesions. However, there is a high prevalence of periapical lesions in one or more teeth or root-filled teeth, as reported in cross-sectional studies. This has been estimated to range from 22 to 72% of research subjects/patients (Frisk and Hakeberg 2005). Therefore, there may be some contributions to systemic disease that this bacterial load exerts on patients, which has hitherto not been sufficiently studied.

Several epidemiological studies on the association between endodontic pathosis and CVD have been reported. In a study of 1,056 older women in Sweden, recruited in 1992–1993, endodontic variables were examined in relation to the incidence of coronary heart disease (CHD) (angina pectoris and myocardial infarction) (Frisk et al. 2003). The independent variables included were: number of root-filled teeth, number of teeth with periapical radiolucencies, tooth loss, age, marital status, smoking, alcohol habits, body mass index, waist/hip ratio, serum cholesterol and triglyceride concentrations, hypertension, and diabetes. A multivariate logistic regression analysis did not prove the endodontic variables to be predictive of CHD but showed that age and tooth loss (of more than 16 teeth) were significantly associated with CHD. A bivariate logistic regression analysis showed a positive significant association between subjects with two root canal fillings and CHD, but for teeth with periapical radiolucencies, the bivariate analysis did not show an association with CHD. The effects of endodontic treatment (as a surrogate variable for pulpal disease) on CHD were evaluated in a large longitudinal study of older male health care professionals (Joshipura et al. 2006). In this study, the subjects were recruited from 1986 to 2000, and were surveyed about the history and timing of root canal therapy (RCT), as well as the presence of caries. Medical records were also evaluated, including cause of death for subjects who died during this period. Data for 34,683 participants were analyzed. Multivariate analysis showed that men with RCT had significantly (albeit small) CVD risk, were older, and were slightly more likely to be current smokers. There was a stronger association between history of RCT and CVD when men with two or more RCTs were included than any RCT, indicating a dose–response effect. There was also a stronger association between history of RCT and CVD when men less than 56 years old at baseline or current smokers (vs past smokers or had never smoked) were analyzed separately (Joshipura et al. 2006). Interestingly, the strongest associations with CVD in this study were seen among dentists who were current smokers or had two or more RCTs.

Another epidemiological study sought to link the incidence of CVD more directly to endodontic infections (referred to as lesions of endodontic origin) (Caplan et al. 2006). In this study, 708 men (mean age 47.4) who were a part of the VA Dental Longitudinal Study in the Boston area, and who were not VA patients themselves, were followed for 24 years. Thirty-five percent of the patients had periapical lesions, and 23% were eventually diagnosed with CVD after being recruited in the study. A multivariate regression analysis showed a significant association between lesion years and CHD, but only in men equal to or less than 40 years of age. Taken together these findings suggest that endodontic infections may pose a significant risk of CVD, at least among men of younger age groups. Finally, it is noteworthy that a patient history (but not a family history) of CVD seemed to correlate with the presence of pulp stones in one pilot study (Edds et al. 2005). More studies with control of some of the confounding factors and in different patient populations need to be performed to determine if these findings are consistent and could be of diagnostic or prognostic value.

Despite an interest that spans over a century in the relationship of endodontic pathosis and systemic disease, this area of knowledge is still in its infancy. More epidemiological data, which are based on large patient and nonpatient populations, are clearly needed to determine if specific associations can be established. Furthermore, better models need to be developed in order to identify the mechanisms of disease, the pathogenicity of the involved endodontic microorganisms, and the contributions of the host.

## 16.6 References

Andrews CH, England MC, Jr, and Kemp WB. 1983. Sickle cell anemia: An etiological factor in pulpal necrosis. *J Endod* 9(6): 249–52.

Andric M, Milasin J, Jovanovic T, and Todorovic L. 2007. Human cytomegalovirus is present in odontogenic cysts. *Oral Microbiol Immunol* 22(5): 347–51.

Badros A, Weikel D, Salama A, Goloubeva O, Schneider A, Rapoport A, Fenton R, Gahres N, Sausville E, Ord R, and Meiller T. 2006. Osteonecrosis of the jaw in multiple myeloma patients: Clinical features and risk factors. *J Clin Oncol* 24(6): 945–52.

Bahekar AA, Singh S, Saha S, Molnar J, and Arora R. 2007. The prevalence and incidence of coronary heart disease is significantly increased in periodontitis: A meta-analysis. *Am Heart J* 154(5): 830–37.

Batchelder BJ, Krutchkoff DJ, and Amara J. 1987. Mandibular pain as the initial and sole clinical manifestation of coronary insufficiency: Report of case. *J Am Dent Assoc* 115(5): 710–12.

Bate AL, Ma JK, and Pitt Ford TR. 2000. Detection of bacterial virulence genes associated with infective endocarditis in infected root canals. *Int Endod J* 33(3): 194–203.

Baumgartner JC, Heggers JP, and Harrison JW. 1976. The incidence of bacteremias related to endodontic procedures. I. Nonsurgical endodontics. *J Endod* 2(5): 135–40.

Baumgartner JC, Heggers JP, and Harrison JW. 1977. Incidence of bacteremias related to endodontic procedures. II. Surgical endodontics. *J Endod* 3(10): 399–402.

Beighton D, Carr AD, and Oppenheim BA. 1994. Identification of viridans streptococci associated with bacteraemia in neutropenic cancer patients. *J Med Microbiol* 40(3): 202–4.

Bender IB and Pressman RS. 1956. Antibiotic treatment of the gingival sulcus in prevention of postextraction bacteremia. *J Oral Surg (Chic)* 14(1): 20–28.

Bender IB, Seltzer S, and Yermish M. 1960. Incidence of bacteremia in endodontic manipulation. *Oral Surg Oral Med Oral Pathol* 13: 353–56.

Bergstrom J, Babcan J, and Eliasson S. 2004. Tobacco smoking and dental periapical condition. *Eur J Oral Sci* 112(2): 115–20.

Bonapart IE, Stevens HP, Kerver AJ, and Rietveld AP. 1995. Rare complications of an odontogenic abscess:

Mediastinitis, thoracic empyema and cardiac tamponade. *J Oral Maxillofac Surg* 53(5): 610–13.

Britto LR, Katz J, Guelmann M, and Heft M. 2003. Periradicular radiographic assessment in diabetic and control individuals. *Oral Surg Oral Med Oral Pathol Oral Radiol Endod* 96(4): 449–52.

Buttke TM, Shipper G, Delano EO, and Trope M. 2005. C-reactive protein and serum amyloid A in a canine model of chronic apical periodontitis. *J Endod* 31(10): 728–32.

Caplan DJ, Chasen JB, Krall EA, Cai J, Kang S, Garcia RI, Offenbacher S, and Beck JD. 2006. Lesions of endodontic origin and risk of coronary heart disease. *J Dent Res* 85(11): 996–1000.

Cholera R, Brittain NJ, Gillrie MR, Lopera-Mesa TM, Diakite SA, Arie T, Krause MA, Guindo A, Tubman A, Fujioka H, Diallo DA, Doumbo OK, Ho M, Wellems TE, and Fairhurst RM. 2008. Impaired cytoadherence of *Plasmodium falciparum*-infected erythrocytes containing sickle hemoglobin. *Proc Natl Acad Sci USA* 105(3): 991–96.

Chou HH, Yumoto H, Davey M, Takahashi Y, Miyamoto T, Gibson FC, III, and Genco CA. 2005. *Porphyromonas gingivalis* fimbria-dependent activation of inflammatory genes in human aortic endothelial cells. *Infect Immun* 73(9): 5367–78.

Cooper H. 1993. Root canal treatment on patients with HIV infection. *Int Endod J* 26(6): 369–71.

Corson MA, Postlethwaite KP, and Seymour RA. 2001. Are dental infections a cause of brain abscess? Case report and review of the literature. *Oral Dis* 7(1): 61–65.

Dahlkemper P, Wolcott JF, Pringle GA, and Hicks ML. 2000. Periapical central giant cell granuloma: A potential endodontic misdiagnosis. *Oral Surg Oral Med Oral Pathol Oral Radiol Endod* 90(6): 739–45.

Debelian GJ, Olsen I, and Tronstad L. 1995. Bacteremia in conjunction with endodontic therapy. *Endod Dent Traumatol* 11(3): 142–49.

Demirbas Kaya A, Aktener BO, and Unsal C. 2004. Pulpal necrosis with sickle cell anaemia. *Int Endod J* 37(9): 602–6.

Dorn BR, Harris LJ, Wujick CT, Vertucci FJ, and Progulske-Fox A. 2002. Invasion of vascular cells in vitro by *Porphyromonas endodontalis*. *Int Endod J* 35(4): 366–71.

Edds AC, Walden JE, Scheetz JP, Goldsmith LJ, Drisko CL, and Eleazer PD. 2005. Pilot study of correlation of pulp stones with cardiovascular disease. *J Endod* 31(7): 504–6.

Elkins DA, Torabinejad M, Schmidt RE, Rossi JJ, and Kettering JD. 1994. Polymerase chain reaction detection of human immunodeficiency virus DNA in human periradicular lesions. *J Endod* 20(8): 386–88.

Falk H, Hugoson A, and Thorstensson H. 1989. Number of teeth, prevalence of caries and periapical lesions in insulin-dependent diabetics. *Scand J Dent Res* 97: 198–206.

Fouad AF. 1997. IL-1 alpha and TNF-alpha expression in early periapical lesions of normal and immunodeficient mice. *J Dent Res* 76(9): 1548–54.

Fouad AF, Barry J, Russo J, Radolf J, and Zhu Q. 2002. Periapical lesion progression with controlled microbial inoculation in a type I diabetic mouse model. *J Endod* 28(1): 8–16.

Fouad AF and Burleson J. 2003. The effect of diabetes mellitus on endodontic treatment outcome: Data from an electronic patient record. *J Am Dent Assoc* 134(1): 43–51; quiz 117–18.

Fouad AF, Kum KY, Clawson ML, Barry J, Abenoja C, Zhu Q, Caimano M, and Radolf JD. 2003. Molecular characterization of the presence of *Eubacterium* spp. and *Streptococcus* spp. in endodontic infections. *Oral Microbiol Immunol* 18: 249–55.

Fouad AF, Walton RE, and Rittman BR. 1992. Induced periapical lesions in ferret canines: Histologic and radiographic evaluation. *Endod Dent Traumatol* 8(2): 56–62.

Frisk F and Hakeberg M. 2005. A 24-year follow-up of root filled teeth and periapical health amongst middle aged and elderly women in Goteborg, Sweden. *Int Endod J* 38(4): 246–54.

Frisk F, Hakeberg M, Ahlqwist M, and Bengtsson C. 2003. Endodontic variables and coronary heart disease. *Acta Odontol Scand* 61(5): 257–62.

Garatea-Crelgo J and Gay-Escoda C. 1991. Mediastinitis from odontogenic infection. Report of three cases and review of the literature. *Int J Oral Maxillofac Surg* 20(2): 65–68.

Garcia B, Penarrocha M, Marti E, Gay-Escodad C, and von Arx T. 2007. Pain and swelling after periapical surgery related to oral hygiene and smoking. *Oral Surg Oral Med Oral Pathol Oral Radiol Endod* 104(2): 271–76.

Garlock JA, Pringle GA, and Hicks ML. 1998. The odontogenic keratocyst: A potential endodontic misdiagnosis. *Oral Surg Oral Med Oral Pathol Oral Radiol Endod* 85(4): 452–56.

Gerner NW, Hurlen B, Dobloug J, and Brandtzaeg P. 1988. Endodontic treatment and immunopathology of periapical granuloma in an AIDS patient. *Endod Dent Traumatol* 4(3): 127–31.

Glick M, Abel SN, Muzyka BC, and DeLorenzo M. 1994. Dental complications after treating patients with AIDS. *J Am Dent Assoc* 125(3): 296–301.

Glick M, Trope M, and Pliskin ME. 1989. Detection of HIV in the dental pulp of a patient with AIDS. *J Am Dent Assoc* 119(5): 649–50.

Goon WW and Jacobsen PL. 1988. Prodromal odontalgia and multiple devitalized teeth caused by a herpes zoster infection of the trigeminal nerve: Report of case. *J Am Dent Assoc* 116(4): 500–504.

Grossi SG, Skrepcinski FB, DeCaro T, Robertson DC, Ho AW, Dunford RG, and Genco RJ. 1997. Treatment of periodontal disease in diabetics reduces glycated hemoglobin. *J Periodontol* 68(8): 713–19.

Gutta R and Louis PJ. 2007. Bisphosphonates and osteonecrosis of the jaws: Science and rationale. *Oral Surg Oral Med Oral Pathol Oral Radiol Endod* 104(2): 186–93.

Haraszthy VI, Zambon JJ, Trevisan M, Zeid M, and Genco RJ. 2000. Identification of periodontal pathogens in atheromatous plaques. *J Periodontol* 71(10): 1554–60.

Hattori Y, Matsumura M, and Kasai K. 2003. Vascular smooth muscle cell activation by C-reactive protein. *Cardiovasc Res* 58(1): 186–95.

Heimdahl A, Hall G, Hedberg M, Sandberg H, Soder PO, Tuner K, and Nord CE. 1990. Detection and quantitation by lysis-filtration of bacteremia after different oral surgical procedures. *J Clin Microbiol* 28(10): 2205–9.

Hillman D. 1986. Combination treatment for a patient with AIDS-related complex and a chronic periapical lesion. *J Conn State Dent Assoc* 60(3): 165–70.

Hou L, Sasakj H, and Stashenko P. 2000. B-cell deficiency predisposes mice to disseminating anaerobic infections: Protection by passive antibody transfer. *Infect Immun* 68(10): 5645–51.

Iwamoto Y, Nishimura F, Nakagawa M, Sugimoto H, Shikata K, Makino H, Fukuda T, Tsuji T, Iwamoto M, and Murayama Y. 2001. The effect of antimicrobial periodontal treatment on circulating tumor necrosis factor-alpha and glycated hemoglobin level in patients with type 2 diabetes. *J Periodontol* 72(6): 774–78.

Janket SJ, Jones JA, Meurman JH, Baird AE, and Van Dyke TE. 2008. Oral infection, hyperglycemia, and endothelial dysfunction. *Oral Surg Oral Med Oral Pathol Oral Radiol Endod* 105(2): 173–79.

Janket SJ, Wightman A, Baird AE, Van Dyke TE, and Jones JA. 2005. Does periodontal treatment improve glycemic control in diabetic patients? A meta-analysis of intervention studies. *J Dent Res* 84(12): 1154–59.

Joshipura KJ, Pitiphat W, Hung HC, Willett WC, Colditz GA, and Douglass CW. 2006. Pulpal inflammation and incidence of coronary heart disease. *J Endod* 32(2): 99–103.

Kohsaka T, Kumazawa M, Yamasaki M, and Nakamura H. 1996. Periapical lesions in rats with streptozotocin-induced diabetes. *J Endod* 22(8): 418–21.

Kozarov E, Sweier D, Shelburne C, Progulske-Fox A, and Lopatin D. 2006. Detection of bacterial DNA in atheromatous plaques by quantitative PCR. *Microbes Infect* 8(3): 687–93.

Krall EA, Abreu Sosa C, Garcia C, Nunn ME, Caplan DJ, and Garcia RI. 2006. Cigarette smoking increases the risk of root canal treatment. *J Dent Res* 85(4): 313–17.

Kreiner M and Okeson JP. 1999. Toothache of cardiac origin. *J Orofac Pain* 13(3): 201–7.

Kreiner M, Okeson JP, Michelis V, Lujambio M, and Isberg A. 2007. Craniofacial pain as the sole symptom of cardiac ischemia: A prospective multicenter study. *J Am Dent Assoc* 138(1): 74–79.

Kuc I, Peters E, and Pan J. 2000. Comparison of clinical and histologic diagnoses in periapical lesions. *Oral Surg Oral Med Oral Pathol Oral Radiol Endod* 89(3): 333–37.

Kumar V, Cotran RS, and Robbins SL. 2003. *Robbins Basic Pathology*. Philadelphia, Saunders: Elsevier Science.

Lalla E, Kaplan S, Yang J, Roth GA, Papapanou PN, and Greenberg S. 2007. Effects of periodontal therapy on serum C-reactive protein, sE-selectin, and tumor necrosis factor-alpha secretion by peripheral blood-derived macrophages in diabetes. A pilot study. *J Periodontal Res* 42(3): 274–82.

Lalla E, Lamster IB, Hofmann MA, Bucciarelli L, Jerud AP, Tucker S, Lu Y, Papapanou PN, and Schmidt AM. 2003. Oral infection with a periodontal pathogen accelerates early atherosclerosis in apolipoprotein E-null mice. *Arterioscler Thromb Vasc Biol* 23(8): 1405–11.

Li L, Messas E, Batista EL, Jr, Levine RA, and Amar S. 2002. *Porphyromonas gingivalis* infection accelerates the progression of atherosclerosis in a heterozygous apolipoprotein E-deficient murine model. *Circulation* 105(7): 861–67.

Lilly JP, Cox D, Arcuri M, and Krell KV. 1998. An evaluation of root canal treatment in patients who have received irradiation to the mandible and maxilla. *Oral Surg Oral Med Oral Pathol Oral Radiol Endod* 86(2): 224–26.

Little JW, Falace DA, Miller CS, and Rhodus NL. 2008. *Dental Management of the Medically Compromised Patient.* St Louis, MO: Mosby Inc., Elsivier Inc.

Lockhart PB, Brennan MT, Sasser HC, Fox PC, Paster BJ, and Bahrani-Mougeot FK. 2008. Bacteremia associated with toothbrushing and dental extraction. *Circulation* 117(24): 3118–25.

Lockhart PB and Durack DT. 1999. Oral microflora as a cause of endocarditis and other distant site infections. *Infect Dis Clin North Am* 13(4): 833–50, vi.

Loushine RJ, Weller RN, Kimbrough WF, and Liewehr FR. 2003. Secondary hyperparathyroidism: A case report. *J Endod* 29(4): 272–74.

Mantyla P, Stenman M, Paldanius M, Saikku P, Sorsa T, and Meurman JH. 2004. *Chlamydia pneumoniae* together with collagenase-2 (MMP-8) in periodontal lesions. *Oral Dis* 10(1): 32–35.

Marending M, Peters OA, and Zehnder M. 2005. Factors affecting the outcome of orthograde root canal therapy in a general dentistry hospital practice. *Oral Surg Oral Med Oral Pathol Oral Radiol Endod* 99(1): 119–24.

Marton I, Kiss C, Balla G, Szabo T, and Karmazsin L. 1988. Acute phase proteins in patients with chronic periapical granuloma before and after surgical treatment. *Oral Microbiol Immunol* 3(2): 95–96.

Mealey BL and Rose LF. 2008. Diabetes mellitus and inflammatory periodontal diseases. *Curr Opin Endocrinol Diabetes Obes* 15(2): 135–41.

Miyamoto T, Yumoto H, Takahashi Y, Davey M, Gibson FC, III, and Genco CA. 2006. Pathogen-accelerated atherosclerosis occurs early after exposure and can be prevented via immunization. *Infect Immun* 74(2): 1376–80.

Nair P. 1987. Light and electron microscopic studies of root canal flora and periapical lesions. *J Endod* 13: 29–39.

Nandakumar R, Whiting J, and Fouad AF. 2008. Identification of selected respiratory pathogens in endodontic infections. *Oral Surg Oral Med Oral Pathol Oral Radiol Endod* 106(1): 969–75.

Nevins A, Ruden S, Pruden P, and Kerpel S. 1988. Metastatic carcinoma of the mandible mimicking periapical lesion of endodontic origin. *Endod Dent Traumatol* 4(5): 238–39.

Nobuhara WK and del Rio CE. 1993. Incidence of periradicular pathoses in endodontic treatment failures. *J Endod* 19(6): 315–18.

Okuda K, Ishihara K, Nakagawa T, Hirayama A, and Inayama Y. 2001. Detection of *Treponema denticola* in atherosclerotic lesions. *J Clin Microbiol* 39(3): 1114–7.

Page RC. 1998. The pathobiology of periodontal diseases may affect systemic diseases: Inversion of a paradigm. *Ann Periodontol* 3(1): 108–20.

Pasceri V, Willerson JT, and Yeh ET. 2000. Direct proinflammatory effect of C-reactive protein on human endothelial cells. *Circulation* 102(18): 2165–8.

Peters E and Lau M. 2003. Histopathologic examination to confirm diagnosis of periapical lesions: A review. *J Can Dent Assoc* 69(9): 598–600.

Peters E, Monopoli M, Woo SB, and Sonis S. 1993. Assessment of the need for treatment of postendodontic asymptomatic periapical radiolucencies in bone marrow transplant recipients. *Oral Surg Oral Med Oral Pathol* 76(1): 45–48.

Proctor ME, Turner DW, Kaminski EJ, Osetek EM, and Heuer MA. 1991. Determination and relationship of C-reactive protein in human dental pulps and in serum. *J Endod* 17(6): 265–70.

Quesnell BT, Alves M, Hawkinson RW, Jr, Johnson BR, Wenckus CS, and BeGole EA. 2005. The effect of human immunodeficiency virus on endodontic treatment outcome. *J Endod* 31(9): 633–36.

Ren YF and Malmstrom HS. 2007. Rapid quantitative determination of C-reactive protein at chair side in dental emergency patients. *Oral Surg Oral Med Oral Pathol Oral Radiol Endod* 104(1): 49–55.

Ricucci D, Martorano M, Bate AL, and Pascon EA. 2005. Calculus-like deposit on the apical external root surface of teeth with post-treatment apical periodontitis: Report of two cases. *Int Endod J* 38(4): 262–71.

Ricucci D, Mannocci F, and Ford TR. 2006a. A study of periapical lesions correlating the presence of a radiopaque lamina with histological findings. *Oral Surg Oral Med Oral Pathol Oral Radiol Endod* 101(3): 389–94.

Ricucci D, Pascon EA, Ford TR, and Langeland K. 2006b. Epithelium and bacteria in periapical lesions. *Oral Surg Oral Med Oral Pathol Oral Radiol Endod* 101(2): 239–49.

Russo JA, Mirchandani R, Nandakumar R, and Fouad AF. 2008. Molecular identification of *E. faecalis* and other bacteria in periapical lesions. *J Endod* 34(3): 353.

Salvi GE, Lawrence HP, Offenbacher S, and Beck JD. 2000. Influence of risk factors on the pathogenesis of periodontitis. *Periodontol* 14: 173–201.

Savarrio L, Mackenzie D, Riggio M, Saunders WP, and Bagg J. 2005. Detection of bacteraemias during non-surgicalroot canal treatment. *J Dent* 33(4): 293–303.

Segura-Egea JJ, Jimenez-Pinzon A, Rios-Santos JV, Velasco-Ortega E, Cisneros-Cabello R, and Poyato-Ferrera M. 2005. High prevalence of apical periodontitis amongst type 2 diabetic patients. *Int Endod J* 38(8): 564–69.

Segura-Egea JJ, Jimenez-Pinzon A, Rios-Santos JV, Velasco-Ortega E, Cisneros-Cabello R, and Poyato-Ferrera MM. 2008. High prevalence of apical periodontitis amongst smokers in a sample of Spanish adults. *Int Endod J* 41(4): 310–16.

Sigurdsson A and Jacoway JR. 1995. Herpes zoster infection presenting as an acute pulpitis. *Oral Surg Oral Med Oral Pathol Oral Radiol Endod* 80(1): 92–95.

Spatafore CM, Griffin JA, Jr, Keyes GG, Wearden S, and Skidmore AE. 1990. Periapical biopsy report: An analysis of over a 10-year period. *J Endod* 16(5): 239–41.

Stoykewych AA, Beecroft WA, and Cogan AG. 1992. Fatal necrotizing fasciitis of dental origin. *J Can Dent Assoc* 58(1): 59–62.

Sunde PT, Olsen I, Gobel UB, Theegarten D, Winter S, Debelian GJ, Tronstad L, and Moter A. 2003. Fluorescence in situ hybridization (FISH) for direct visualization of bacteria in periapical lesions of asymptomatic root-filled teeth. *Microbiology* 149(Pt 5): 1095–1102.

Suzuki N, Yuyama M, Maeda S, Ogawa H, Mashiko K, and Kiyoura Y. 2006. Genotypic identification of presumptive *Streptococcus pneumoniae* by PCR using four genes highly specific for *S. pneumoniae*. *J Med Microbiol* 55(Pt 6): 709–14.

Teles R, Wang CY, and Stashenko P. 1997. Increased susceptibility of RAG-2 SCID mice to dissemination of endodontic infections. *Infect Immun* 65(9): 3781–87.

Torabinejad M, Kettering JD, McGraw JC, Cummings RR, Dwyer TG, and Tobias TS. 1988. Factors associated with endodontic interappointment emergencies of teeth with necrotic pulps. *J Endod* 14(5): 261–66.

Ueta E, Osaki T, Yoneda K, and Yamamoto T. 1993. Prevalence of diabetes mellitus in odontogenic infections and oral candidiasis: An analysis of neutrophil suppression. *J Oral Pathol Med* 22(4): 168–74.

Wilcox LR and Walton RE. 1989. Case of mistaken identity: Periapical cemental dysplasia in an endodontically treated tooth. *Endod Dent Traumatol* 5(6): 298–301.

Yan SF, Ramasamy R, Bucciarelli LG, Wendt T, Lee LK, Hudson BI, Stern DM, Lalla E, DU Yan S, Rong LL, Naka Y, and Schmidt AM. 2004. RAGE and its ligands: A lasting memory in diabetic complications? *Diab Vasc Dis Res* 1(1): 10–20.

Zwaka TP, Hombach V, and Torzewski J. 2001. C-reactive protein-mediated low density lipoprotein uptake by macrophages: Implications for atherosclerosis. *Circulation* 103(9): 1194–7.

# Glossary

**Abscess.** A pus-containing lesion encapsulated by a fibrotic capsule and an inflammatory reaction.

**Antibody.** A glycoprotein synthesized in a plasma cell, which is derived from a B cell that has interacted with a specific antigen. The antibody molecule can bind specifically to this antigen.

**Antigen.** A molecule that (a) triggers syntheses of antibody and/or a T-cell response and (b) binds specifically to an antibody or a lymphocyte receptor.

**Apical periodontitis.** Used for the inflammatory lesion in the apical area of the tooth that can be seen on radiographs.

**Apoptosis.** Programmed cell death. Cell suicide. A process controlled by a cell that results in the death of the cell.

**Assembly.** The stage (in virology) in the virus replication cycle, when components come together to form virions.

**Bacterial artificial chromosome (BAC).** A cloning vector able to carry very large inserts of DNA up to 150 kb.

**Bacterial competence.** The ability of a bacterium to take up DNA from the environment.

**Bactericidal.** Bactericidal is the destruction of bacteria.

**Bacteriostatic.** Bacteriostatic is the inhibition of the growth of bacteria without their destruction.

**Biofilm.** Adherent microbial community that forms at a solid/liquid interface.

**B-lymphocyte.** Cell with surface receptors that can recognize a specific antigen. Antigen binding can trigger a B cell to develop into an antibody-secreting plasma cell.

**Blastospore.** An asexual fungal spore produced by budding.

**Capsid.** The protein coat that encloses the nucleic acid of a virus.

**Capsomere.** A discrete component of a capsid, constructed from several identical protein molecules.

**Chemokine.** A cytokine that stimulates the migration and activation of cells in the animal body, especially cells involved in inflammation.

**Chlamydospore.** The thick-walled, big resting spore of several kinds of fungi.

**Codon.** Three base units of DNA that code for amino acids.

**Commensal.** Microorganism living on or in a host but causing the host no harm.

**Complement.** A series of proteins that are activated when the body is infected. Activation has a number of antiviral effects, including lysis of infected cells and enhancement of phagocytosis.

**Conjugation.** Conjugation is the transfer of DNA from one bacterium to another through sex pili.

**Cytokine.** Protein that is secreted from a cell and has a specific effect on other cells, including cells of the immune system.

**Disinfection.** A procedure to achieve a state without any pathogenic microorganisms.

**DNA dependent RNA polymerase.** An enzyme that synthesizes RNA from a DNA template.

**Domain.** Discrete portion of a protein or a nucleic acid with its own structure and function.

**Envelope.** A lipid bilayer and associated protein, forming the outer component of an enveloped virion.

**Fluorescent in situ hybridization (FISH).** A histologic technique in which specific DNA probes hybridize to complimentary DNA in tissues and are seen by visualizing a fluorescent marker using a confocal microscope.

**Fosmid.** A type of plasmid that can carry large inserts of foreign DNA.

**Genome.** The DNA or RNA that encodes the genes of an organism or virus.

**Genotype.** The complete set of genes of an organism or virus.

**Genus.** Defined for a microorganism in the first part of the Latin name, for example, *Streptococcus*, which cover all streptococcal species group.

**Hapten.** Hapten is a specific nonprotein substance that does not itself elicit antibody formation, but does elicit the immune response when coupled with a carrier protein.

**Horizontal gene transfer.** DNA with the potential to transfer between bacteria by transformation, transduction, or conjugation, including plasmids, transposons, insertion sequences, bacteriophages genomes, and chromosomal DNA.

**Host.** A cell or an organism in which a virus or a plasmid can replicate.

**Hypha (plural hyphae).** Septate or aseptate filament of a fungus. Hyphae are the main mode of vegetative growth and are collectively called a mycelium.

**Icosahedral symmetry.** A type of symmetry present in viruses where the capsid is constructed from protein molecules arranged to form 20 triangular faces.

**Immunoglobulin.** The glycoprotein that functions as an antibody.

**Latent infection.** Infection of a cell where the replication cycle is not completed, but the virus genome is maintained in the cell.

**Lipopolysaccharide.** Molecule consisting of lipid and polysaccharide, found in the cell walls of gram-negative bacteria.

**Metagenomics.** The study of all of the constituent genomes of a community as a single entity.

**Major histocompatibility complex (MHC).** A region of the vertebrate genome that encodes major histocompatibility proteins. MHC class I and class II molecules are cell surface proteins that play important roles in immune responses.

**Microbiome.** All of the constituent members of a microbial community, considered as a single entity.

**Minimal bactericidal concentration (MBC).** MBC is the antimicrobial concentration that results in 99.9% reduction of CFU/mL compared with the original inoculum.

**Minimal inhibitory concentration (MIC).** MIC is the lowest antimicrobial concentration that completely inhibits visible bacterial growth.

**Mutation.** An alteration in the sequence of DNA or RNA. Mutations may occur.

**Nuclear envelope.** The structure composed of two membranes that separate the nucleus from the cytoplasm in a eukaryotic cell.

**Open reading frame (ORF).** A fragment of DNA that encodes a protein, delimited by start and stop codons.

**Operon.** A group of genes that are able to produce messenger RNA from one or more structural genes within the operon.

**Osteomyelitis.** Infection of the bone and bone marrow.

**Pathogen.** A microorganism that produces disease in another organism.

**Peptidoglycan.** Polymer making up part of the cell wall in bacteria.

**Periapical osteitis.** An older term that was used synonymous to apical periodontitis; however, it means strictly that the inflammation is within the bone tissue and that cannot be evaluated on radiographs.

**Phylotype.** A term that indicates species or taxon (plural taxa) particularly for bacteria that have not been cultivated in vitro, and for which the phenotype has not been studied.

**Plasmid.** A self-replicating, extrachromosomal DNA molecule that is generally circular though can be linear and can replicate autonomously. Plasmids are common in prokaryotes and rare in eukaryotes. Artificial plasmids can be created to clone DNA sequence. They can carry genes for antibiotic resistance and virulence factors.

**Polymerase.** An enzyme that synthesizes RNA from a DNA template.

**Polymerase chain reaction (PCR).** An in vitro technique for amplifying specific DNA sequences.

**Primer.** A molecule (often short strand RNA or DNA) that provides template for formation of another (usually longer) molecule such as in a PCR reaction.

**Probe.** A short single strand DNA sequence attached to a marker and designed to hybridize to complimentary DNA.

**Pseudohypha (plural pseudohyphae).** They are not true septate hyphae. They are most often found in yeasts as the result of a sort of incomplete budding where the cells remain attached after division.

**Quorum sensing.** The ability of a bacterial community to determine the total number of organisms in the community.

**Reverse transcriptase PCR (RT-PCR).** An in vitro technique for converting data in RNA to complementary DNA (cDNA), using a reverse transcriptase. The DNA is then amplified by a PCR.

**Reverse transcription.** Syntheses of RNA from a DNA.

**Real-time PCR.** Quantitative PCR in which every PCR cycle is calibrated to known concentrations of DNA to calculate the number of amplified copies.

**Resistance transfer factor (R factor).** R factor is a plasmid carrying bacterial chromosomal DNA that is passed on to daughter cells during replication.

**Resuscitation-promoting factors (Rpf).** Proteins produced by bacteria that stimulate the growth of other bacteria and can revive dormant bacteria.

**Retrovirus.** A member of the family Retroviridae, so named because these viruses carry out reverse transcription.

**Ribonuclease RNase.** An enzyme that hydrolyses RNA. (RNase H is a ribonuclease that specifically digests the RNA in an RNA-DNA duplex.)

**Sinus tract.** A spontaneous drainage of an abscess, sometimes designated fistula.

**Species.** Defined for a microorganism in the last part of the Latin name, for example, *S. intermedius*, where S. stands for *Streptococcus* and is usually abbreviated once it has been given in full in the text. Undefined species are given as spp. (plural) or sp. (singular), for example, *Streptococcus* spp. for several undefined streptococcal species.

**Sterilization.** A procedure used for obtaining a state with absence of any living (viable) microorganism. Microbiologically, this includes the very resistant bacterial spores. It will in practice also include virus even if its absence cannot be controlled. In this book, sterilization is used for absence of culturable microorganisms.

**Strain.** Strain is used for all isolates of a particular microorganism obtained from a sample. Different strains may belong to the same species.

**Superinfection.** Superinfection is a new infection complicating the course of antimicrobial therapy of an existing infection, due to invasion by bacteria or fungi resistant to the drug(s) in use.

**T cell (T lymphocyte).** A cell with surface receptors that can recognize a specific antigen. Antigen binding can trigger a T cell to perform one of several roles, including helper T cell or cytotoxic T cell.

**Tegument.** A layer of protein and RNA between the capsid and the envelope of a herpes virus particle.

**Transcriptase.** An enzyme that carries out transcription.

**Transcription.** Synthesis of RNA from complementary DNA or an RNA template.

**Transduction.** Transfer of genetic material (and its phenotypic expression) from one cell to another by viral infection or the injection of genetic material by a virus (bacteriophage) into a bacterium.

**Transformation.** Transformation is the uptake of extracellular DNA and incorporation of the DNA into the recipient's chromosome where there is a similar base sequence.

**Translation.** Synthesis of protein from the genetic information in mRNA.

**Transposition.** Transposition is the insertion of DNA transposons into a chromosome without the need for homology between the donor and the recipient cells.

**Transposon.** A type of mobile genetic element, a section of DNA that can move within the genome of an organism, or between organisms. A transposable element encoding a function such as antibiotic resistance.

**Virion.** Virus particle.

**Virulence.** A measure of the severity of disease or pathogenicity that a microorganism is capable of causing.

**Virulence factors.** Factors enabling a microorganism to establish itself on or within a host, thereby contributing to the disease process.

**Yeast.** Any of various unicellular fungi.

# Index